Lecture Notes on Data Engineering and Communications Technologies 204

Series Editor

Fatos Xhafa, *Technical University of Catalonia, Barcelona, Spain*

The aim of the book series is to present cutting edge engineering approaches to data technologies and communications. It will publish latest advances on the engineering task of building and deploying distributed, scalable and reliable data infrastructures and communication systems.

The series will have a prominent applied focus on data technologies and communications with aim to promote the bridging from fundamental research on data science and networking to data engineering and communications that lead to industry products, business knowledge and standardisation.

Indexed by SCOPUS, INSPEC, EI Compendex.

All books published in the series are submitted for consideration in Web of Science.

Leonard Barolli
Editor

Advanced Information Networking and Applications

Proceedings of the 38th International Conference on Advanced Information Networking and Applications (AINA-2024), Volume 6

Editor
Leonard Barolli
Department of Information and Communication
Engineering
Fukuoka Institute of Technology
Fukuoka, Japan

ISSN 2367-4512 ISSN 2367-4520 (electronic)
Lecture Notes on Data Engineering and Communications Technologies
ISBN 978-3-031-57941-7 ISBN 978-3-031-57942-4 (eBook)
https://doi.org/10.1007/978-3-031-57942-4

This Springer imprint is published by the registered company Springer Nature Switzerland AG
The registered company address is: Gewerbestrasse 11, 6330 Cham, Switzerland

Welcome Message from AINA-2024 Organizers

Welcome to the 38th International Conference on Advanced Information Networking and Applications (AINA-2024). On behalf of AINA-2024 Organizing Committee, we would like to express to all participants our cordial welcome and high respect.

AINA is an International Forum, where scientists and researchers from academia and industry working in various scientific and technical areas of networking and distributed computing systems can demonstrate new ideas and solutions in distributed computing systems. AINA is a very open society and is always welcoming international volunteers from any country and any area in the world.

AINA International Conference is a forum for sharing ideas and research work in the emerging areas of information networking and their applications. The area of advanced networking has grown very rapidly and the applications have experienced an explosive growth, especially in the area of pervasive and mobile applications, wireless sensor and ad-hoc networks, vehicular networks, multimedia computing, social networking, semantic collaborative systems, as well as IoT, big data, cloud computing, artificial intelligence, and machine learning. This advanced networking revolution is transforming the way people live, work, and interact with each other and is impacting the way business, education, entertainment, and health care are operating. The papers included in the proceedings cover theory, design and application of computer networks, distributed computing, and information systems.

Each year AINA receives a lot of paper submissions from all around the world. It has maintained high-quality accepted papers and is aspiring to be one of the main international conferences on the information networking in the world.

We are very proud and honored to have two distinguished keynote talks by Prof. Fatos Xhafa, Technical University of Catalonia, Spain, and Dr. Juggapong Natwichai, Chiang Mai University, Thailand, who will present their recent work and will give new insights and ideas to the conference participants.

An international conference of this size requires the support and help of many people. A lot of people have helped and worked hard to produce a successful AINA-2024 technical program and conference proceedings. First, we would like to thank all authors for submitting their papers. We are indebted to Program Track Co-chairs, Program Committee Members and Reviewers, who carried out the most difficult work of carefully evaluating the submitted papers.

We would like to thank AINA-2024 General Co-chairs, PC Co-chairs, Workshops Organizers for their great efforts to make AINA-2024 a very successful event. We have special thanks to the Finance Chair and Web Administrator Co-chairs.

We do hope that you will enjoy the conference proceedings and readings.

AINA-2024 Organizing Committee

Honorary Chair

Makoto Takizawa	Hosei University, Japan

General Co-chairs

Minoru Uehara	Toyo University, Japan
Euripides G. M. Petrakis	Technical University of Crete (TUC), Greece
Isaac Woungang	Toronto Metropolitan University, Canada

Program Committee Co-chairs

Tomoya Enokido	Rissho University, Japan
Mario A. R. Dantas	Federal University of Juiz de Fora, Brazil
Leonardo Mostarda	University of Perugia, Italy

International Journals Special Issues Co-chairs

Fatos Xhafa	Technical University of Catalonia, Spain
David Taniar	Monash University, Australia
Farookh Hussain	University of Technology Sydney, Australia

Award Co-chairs

Arjan Durresi	Indiana University Purdue University in Indianapolis (IUPUI), USA
Fang-Yie Leu	Tunghai University, Taiwan
Marek Ogiela	AGH University of Science and Technology, Poland
Kin Fun Li	University of Victoria, Canada

Publicity Co-chairs

Markus Aleksy	ABB Corporate Research Center, Germany
Flora Amato	University of Naples "Federico II", Italy
Lidia Ogiela	AGH University of Science and Technology, Poland
Hsing-Chung Chen	Asia University, Taiwan

International Liaison Co-chairs

Wenny Rahayu	La Trobe University, Australia
Nadeem Javaid	COMSATS University Islamabad, Pakistan
Beniamino Di Martino	University of Campania "Luigi Vanvitelli", Italy

Local Arrangement Co-chairs

Keita Matsuo	Fukuoka Institute of Technology, Japan
Tomoyuki Ishida	Fukuoka Institute of Technology, Japan

Finance Chair

Makoto Ikeda	Fukuoka Institute of Technology, Japan

Web Co-chairs

Phudit Ampririt	Fukuoka Institute of Technology, Japan
Ermioni Qafzezi	Fukuoka Institute of Technology, Japan
Shunya Higashi	Fukuoka Institute of Technology, Japan

Steering Committee Chair

Leonard Barolli	Fukuoka Institute of Technology, Japan

Tracks Co-chairs and Program Committee Members

1. Network Architectures, Protocols and Algorithms

Track Co-chairs

Spyropoulos Thrasyvoulos	Technical University of Crete (TUC), Greece
Shigetomo Kimura	University of Tsukuba, Japan
Darshika Perera	University of Colorado at Colorado Springs, USA

TPC Members

Thomas Dreibholz	Simula Metropolitan Center for Digital Engineering, Norway
Angelos Antonopoulos	Nearby Computing SL, Spain
Hatim Chergui	i2CAT Foundation, Spain
Bhed Bahadur Bista	Iwate Prefectural University, Japan
Chotipat Pornavalai	King Mongkut's Institute of Technology Ladkrabang, Thailand
Kenichi Matsui	NTT Network Innovation Center, Japan
Sho Tsugawa	University of Tsukuba, Japan
Satoshi Ohzahata	University of Electro-Communications, Japan
Haytham El Miligi	Thompson Rivers University, Canada
Watheq El-Kharashi	Ain Shams University, Egypt
Ehsan Atoofian	Lakehead University, Canada
Fayez Gebali	University of Victoria, Canada
Kin Fun Li	University of Victoria, Canada
Luis Blanco	CTTC, Spain

2. Next Generation Mobile and Wireless Networks

Track Co-chairs

Purav Shah	School of Science and Technology, Middlesex University, UK
Enver Ever	Middle East Technical University, Northern Cyprus
Evjola Spaho	Polytechnic University of Tirana, Albania

TPC Members

Burak Kizilkaya	Glasgow University, UK
Muhammad Toaha	Middle East Technical University, Turkey
Ramona Trestian	Middlesex University, UK
Andrea Marotta	University of L'Aquila, Italy
Adnan Yazici	Nazarbayev University, Kazakhstan
Orhan Gemikonakli	Final International University, Cyprus
Hrishikesh Venkataraman	Indian Institute of Information Technology, Sri City, India
Zhengjia Xu	Cranfield University, UK
Mohsen Hejazi	University of Kashan, Iran
Sabyasachi Mukhopadhyay	IIT Kharagpur, India
Ali Khoshkholghi	Middlesex University, UK
Admir Barolli	Aleksander Moisiu University of Durres, Albania
Makoto Ikeda	Fukuoka Institute of Technology, Japan
Yi Liu	Oita National College of Technology, Japan
Testuya Oda	Okayama University of Science, Japan
Ermioni Qafzezi	Fukuoka Institute of Technology, Japan

3. Multimedia Networking and Applications

Track Co-chairs

Markus Aleksy	ABB Corporate Research Center, Germany
Francesco Orciuoli	University of Salerno, Italy
Tomoyuki Ishida	Fukuoka Institute of Technology, Japan

TPC Members

Hadil Abukwaik	ABB Corporate Research Center, Germany
Thomas Preuss	Brandenburg University of Applied Sciences, Germany
Peter M. Rost	Karlsruhe Institute of Technology (KIT), Germany
Lukasz Wisniewski	inIT, Germany
Angelo Gaeta	University of Salerno, Italy
Angela Peduto	University of Salerno, Italy
Antonella Pascuzzo	University of Salerno, Italy
Roberto Abbruzzese	University of Salerno, Italy
Tetsuro Ogi	Keio University, Japan

Yasuo Ebara	Osaka Electro-Communication University, Japan
Hideo Miyachi	Tokyo City University, Japan
Kaoru Sugita	Fukuoka Institute of Technology, Japan

4. Pervasive and Ubiquitous Computing

Track Co-chairs

Vamsi Paruchuri	University of Central Arkansas, USA
Hsing-Chung Chen	Asia University, Taiwan
Shinji Sakamoto	Kanazawa Institute of Technology, Japan

TPC Members

Sriram Chellappan	University of South Florida, USA
Yu Sun	University of Central Arkansas, USA
Qiang Duan	Penn State University, USA
Han-Chieh Wei	Dallas Baptist University, USA
Ahmad Alsharif	University of Alabama, USA
Vijayasarathi Balasubramanian	Microsoft, USA
Shyi-Shiun Kuo	Nan Kai University of Technology, Taiwan
Karisma Trinanda Putra	Universitas Muhammadiyah Yogyakarta, Indonesia
Cahya Damarjati	Universitas Muhammadiyah Yogyakarta, Indonesia
Agung Mulyo Widodo	Universitas Esa Unggul Jakarta, Indonesia
Bambang Irawan	Universitas Esa Unggul Jakarta, Indonesia
Eko Prasetyo	Universitas Muhammadiyah Yogyakarta, Indonesia
Sunardi S. T.	Universitas Muhammadiyah Yogyakarta, Indonesia
Andika Wisnujati	Universitas Muhammadiyah Yogyakarta, Indonesia
Makoto Ikeda	Fukuoka Institute of Technology, Japan
Tetsuya Oda	Okayama University of Science, Japan
Evjola Spaho	Polytechnic University of Tirana, Albania
Tetsuya Shigeyasu	Hiroshima Prefectural University, Japan
Keita Matsuo	Fukuoka Institute of Technology, Japan
Admir Barolli	Aleksander Moisiu University of Durres, Albania

5. Web-Based Systems and Content Distribution

Track Co-chairs

Chrisa Tsinaraki	Technical University of Crete (TUC), Greece
Yusuke Gotoh	Okayama University, Japan
Santi Caballe	Open University of Catalonia, Spain

TPC Members

Nikos Bikakis	Hellenic Mediterranean University, Greece
Ioannis Stavrakantonakis	Ververica GmbH, Germany
Sven Schade	European Commission, Joint Research Center, Italy
Christos Papatheodorou	National and Kapodistrian University of Athens, Greece
Sarantos Kapidakis	University of West Attica, Greece
Manato Fujimoto	Osaka Metropolitan University, Japan
Kiki Adhinugraha	La Trobe University, Australia
Tomoki Yoshihisa	Shiga University, Japan
Jordi Conesa	Open University of Catalonia, Spain
Thanasis Daradoumis	Open University of Catalonia, Spain
Nicola Capuano	University of Basilicata, Italy
Victor Ströele	Federal University of Juiz de Fora, Brazil

6. Distributed Ledger Technologies and Distributed-Parallel Computing

Track Co-chairs

Alfredo Navarra	University of Perugia, Italy
Naohiro Hayashibara	Kyoto Sangyo University, Japan

TPC Members

Serafino Cicerone	University of L'Aquila, Italy
Ralf Klasing	LaBRI Bordeaux, France
Giuseppe Prencipe	University of Pisa, Italy
Roberto Tonelli	University of Cagliari, Italy
Farhan Ullah	Northwestern Polytechnical University, China

Leonardo Mostarda	University of Perugia, Italy
Qiong Huang	South China Agricultural University, China
Tomoya Enokido	Rissho University, Japan
Minoru Uehara	Toyo University, Japan
Lucian Prodan	Polytechnic University of Timisoara, Romania
Md. Abdur Razzaque	University of Dhaka, Bangladesh

7. Data Mining, Big Data Analytics and Social Networks

Track Co-chairs

Pavel Krömer	Technical University of Ostrava, Czech Republic
Alex Thomo	University of Victoria, Canada
Eric Pardede	La Trobe University, Australia

TPC Members

Sebastián Basterrech	Technical University of Denmark, Denmark
Tibebe Beshah	University of Addis Ababa, Ethiopia
Nashwa El-Bendary	Arab Academy for Science, Egypt
Petr Musilek	University of Alberta, Canada
Varun Ojha	Newcastle University, UK
Alvaro Parres	ITESO, Mexico
Nizar Rokbani	ISSAT-University of Sousse, Tunisia
Farshid Hajati	Victoria University, Australia
Ji Zhang	University of Southern Queensland, Australia
Salimur Choudhury	Lakehead University, Canada
Carson Leung	University of Manitoba, Canada
Syed Mahbub	La Trobe University, Australia
Osama Mahdi	Melbourne Institute of Technology, Australia
Choiru Zain	La Trobe University, Australia
Rajalakshmi Rajasekaran	La Trobe University, Australia
Nawfal Ali	Monash University, Australia

8. Internet of Things and Cyber-Physical Systems

Track Co-chairs

Tomoki Yoshihisa	Shiga University, Japan
Winston Seah	Victoria University of Wellington, New Zealand
Luciana Pereira Oliveira	Instituto Federal da Paraiba (IFPB), Brazil

TPC Members

Akihiro Fujimoto	Wakayama University, Japan
Akimitsu Kanzaki	Shimane University, Japan
Kazuya Tsukamoto	Kyushu Institute of Technology, Japan
Lei Shu	Nanjing Agricultural University, China
Naoyuki Morimoto	Mie University, Japan
Teruhiro Mizumoto	Chiba Institute of Technology, Japan
Tomoya Kawakami	Fukui University, Japan
Adrian Pekar	Budapest University of Technology and Economics, Hungary
Alvin Valera	Victoria University of Wellington, New Zealand
Chidchanok Choksuchat	Prince of Songkla University, Thailand
Jyoti Sahni	Victoria University of Wellington, New Zealand
Murugaraj Odiathevar	Sungkyunkwan University, South Korea
Normalia Samian	Universiti Putra Malaysia, Malaysia
Qing Gu	University of Science and Technology Beijing, China
Tao Zheng	Beijing Jiaotong University, China
Wenbin Pei	Dalian University of Technology, China
William Liu	Unitec, New Zealand
Wuyungerile Li	Inner Mongolia University, China
Peng Huang	Sichuan Agricultural University, PR China
Ruan Delgado Gomes	Instituto Federal da Paraiba (IFPB), Brazil
Glauco Estacio Goncalves	Universidade Federal do Pará (UFPA), Brazil
Eduardo Luzeiro Feitosa	Universidade Federal do Amazonas (UFAM), Brazil
Paulo Ribeiro Lins Júnior	Instituto Federal da Paraiba (IFPB), Brazil

9. Intelligent Computing and Machine Learning

Track Co-chairs

Takahiro Uchiya	Nagoya Institute of Technology, Japan
Flavius Frasincar	Erasmus University Rotterdam, The Netherlands
Miltos Alamaniotis	University of Texas at San Antonio, USA

TPC Members

Kazuto Sasai	Ibaraki University, Japan
Shigeru Fujita	Chiba Institute of Technology, Japan
Yuki Kaeri	Mejiro University, Japan
Jolanta Mizera-Pietraszko	Military University of Land Forces, Poland
Ashwin Ittoo	University of Liège, Belgium
Marco Brambilla	Politecnico di Milano, Italy
Alfredo Cuzzocrea	University of Calabria, Italy
Le Minh Nguyen	JAIST, Japan
Akiko Aizawa	National Institute of Informatics, Japan
Natthawut Kertkeidkachorn	JAIST, Japan
Georgios Karagiannis	Durham University, UK
Leonidas Akritidis	International Hellenic University, Greece
Athanasios Fevgas	University of Thessaly, Greece
Yota Tsompanopoulou	University of Thessaly, Greece
Yuvaraj Munian	Texas A&M-San Antonio, USA

10. Cloud and Services Computing

Track Co-chairs

Salvatore Venticinque	University of Campania "Luigi Vanvitelli", Italy
Shigenari Nakamura	Tokyo Denki University, Japan
Sajal Mukhopadhyay	National Institute of Technology, Durgapur, India

TPC Members

Giancarlo Fortino	University of Calabria, Italy
Massimiliano Rak	University of Campania "Luigi Vanvitelli", Italy
Jason J. Jung	Chung-Ang University, Korea

Dimosthenis Kyriazis	University of Piraeus, Greece
Geir Horn	University of Oslo, Norway
Dario Branco	University of Campania "Luigi Vanvitelli", Italy
Dilawaer Duolikun	Cognizant Technology Solutions, Hungary
Naohiro Hayashibara	Kyoto Sangyo University, Japan
Tomoya Enokido	Rissho University, Japan
Sujoy Saha	NIT Durgapur, India
Animesh Dutta	NIT Durgapur, India
Pramod Mane	IIM Rohtak, India
Nanda Dulal Jana	NIT Durgapur, India
Banhi Sanyal	NIT Kurukshetra, India

11. Security, Privacy and Trust Computing

Track Co-chairs

Ioannidis Sotirios	Technical University of Crete (TUC), Greece
Michail Alexiou	Georgia Institute of Technology, USA
Hiroaki Kikuchi	Meiji University, Japan

TPC Members

George Vasiliadis	Hellenic Mediterranean University, Greece
Antreas Dionysiou	University of Cyprus, Cyprus
Apostolos Fouranaris	Athena Research Center, Greece
Panagiotis Ilia	Technical University of Crete, Greece
George Portokalidis	IMDEA, Spain
Nikolaos Gkorgkolis	University of Crete, Greece
Zeezoo Ryu	Georgia Institute of Technology, USA
Muhammad Faraz Karim	Georgia Institute of Technology, USA
Yunjie Deng	Georgia Institute of Technology, USA
Anna Raymaker	Georgia Institute of Technology, USA
Takamichi Saito	Meiji University, Japan
Kazumasa Omote	University of Tsukuba, Japan
Masakatsu Nishigaki	Shizuoka University, Japan
Mamoru Mimura	National Defense Academy of Japan, Japan
Chun-I Fan	National Sun Yat-sen University, Taiwan
Aida Ben Chehida Douss	National School of Engineers of Tunis, ENIT Tunis, Tunisia
Davinder Kaur	IUPUI, USA

12. Software-Defined Networking and Network Virtualization

Track Co-chairs

Flavio de Oliveira Silva	Federal University of Uberlândia, Brazil
Ashutosh Bhatia	Birla Institute of Technology and Science, Pilani, India

TPC Members

Rui Luís Andrade Aguiar	Universidade de Aveiro (UA), Portugal
Ivan Vidal	Universidad Carlos III de Madrid, Spain
Eduardo Coelho Cerqueira	Federal University of Pará (UFPA), Brazil
Christos Tranoris	University of Patras (UoP), Greece
Juliano Araújo Wickboldt	Federal University of Rio Grande do Sul (UFRGS), Brazil
Haribabu K.	BITS Pilani, India
Virendra Shekhavat	BITS Pilani, India
Makoto Ikeda	Fukuoka Institute of Technology, Japan
Farookh Hussain	University of Technology Sydney, Australia
Keita Matsuo	Fukuoka Institute of Technology, Japan

AINA-2024 Reviewers

Admir Barolli
Aida ben Chehida Douss
Akimitsu Kanzaki
Alba Amato
Alberto Postiglione
Alex Thomo
Alfredo Navarra
Amani Shatnawi
Anas AlSobeh
Andrea Marotta
Angela Peduto
Anne Kayem
Antreas Dionysiou
Arjan Durresi
Ashutosh Bhatia
Beniamino Di Martino
Bhed Bista
Burak Kizilkaya
Carson Leung
Chidchanok Choksuchat
Christos Tranoris
Chung-Ming Huang
Dario Branco
David Taniar
Elinda Mece
Enver Ever
Eric Pardede
Euripides Petrakis
Evjola Spaho
Fabrizio Messina
Feilong Tang
Flavio Silva
Francesco Orciuoli
George Portokalidis

Giancarlo Fortino
Giorgos Vasiliadis
Glauco Gonçalves
Hatim Chergui
Hiroaki Kikuchi
Hiroki Sakaji
Hiroshi Maeda
Hiroyuki Fujioka
Hyunhee Park
Isaac Woungang
Jana Nowaková
Jolanta Mizera-Pietraszko
Junichi Honda
Jyoti Sahni
Kazunori Uchida
Keita Matsuo
Kenichi Matsui
Kiki Adhinugraha
Kin Fun Li
Kiyotaka Fujisaki
Leonard Barolli
Leonardo Mostarda
Leonidas Akritidis
Lidia Ogiela
Lisandro Granville
Lucian Prodan
Luciana Oliveira
Mahmoud Elkhodr
Makoto Ikeda
Mamoru Mimura
Manato Fujimoto
Marco Antonio To
Marek Ogiela
Masaki Kohana
Minoru Uehara
Muhammad Karim
Muhammad Toaha Raza Khan
Murugaraj Odiathevar
Nadeem Javaid
Naohiro Hayashibara
Nobuo Funabiki
Nour El Madhoun
Omar Darwish
Panagiotis Ilia
Petr Musilek
Philip Moore
Purav Shah
R. Madhusudhan
Raffaele Guarasci
Ralf Klasing
Roberto Tonelli
Ronald Petrlic
Sabyasachi Mukhopadhyay
Sajal Mukhopadhyay
Salvatore D'Angelo
Salvatore D'Angelo
Salvatore D'Angelo
Salvatore Venticinque
Santi Caballé
Satoshi Ohzahata
Serafino Cicerone
Shigenari Nakamura
Shinji Sakamoto
Sho Tsugawa
Sriram Chellappan
Stephane Maag
Takayuki Kushida
Tetsuya Oda
Thomas Dreibholz
Tomoki Yoshihisa
Tomoya Enokido
Tomoya Kawakami
Tomoyuki Ishida
Vamsi Paruchuri
Victor Ströele
Vikram Singh
Wei Lu
Wenny Rahayu
Winston Seah
Yong Zheng
Yoshitaka Shibata
Yusuke Gotoh
Yuvaraj Munian
Zeezoo Ryu
Zhengjia Xu

AINA-2024 Keynote Talks

Agile Edge: Harnessing the Power of the Intelligent Edge by Agile Optimization

Fatos Xhafa

Technical University of Barcelona, Barcelona, Spain

Abstract. The digital cloud ecosystem comprises various degrees of computing granularity from large cloud servers and data centers to IoT devices, leading to the cloud-to-thing continuum computing paradigm. In this context, the intelligent edge aims at placing intelligence to the end devices, at the edges of the Internet. The premise is that collective intelligence from the IoT data deluge can be achieved and used at the edges of the Internet, offloading the computation burden from the cloud systems and leveraging real-time intelligence. This, however, comes with the challenges of processing and analyzing the IoT data streams in real time. In this talk, we will address how agile optimization can be useful for harnessing the power of the intelligent edge. Agile optimization is a powerful and promising solution, which differently from traditional optimization methods, is able to find optimized and scalable solutions under real-time requirements. We will bring real-life problems and case studies from Smart City Open Data Repositories to illustrate the approach. Finally, we will discuss the research challenges and emerging vision on the agile intelligent edge.

Challenges in Entity Matching in AI Era

Juggapong Natwichai

Chiang Mai University, Chiang Mai, Thailand

Abstract. Entity matching (EM) is to identify and link entities originating from various sources that correspond to identical real-world entities, thereby constituting a foundational component within the realm of data integration. For example, in order to counter-fraud detection, the datasets from sellers, financial services providers, or even IT infrastructure service providers might be in need for data integration, and hence, the EM is highly important here. This matching process is also recognized for its pivotal role in data augmenting to improve the precision and dependability of subsequent tasks within the domain of data analytics. Traditionally, the EM procedure composes of two integral phases, namely blocking and matching. The blocking phase associates with the generation of candidate pairs and could affect the size and complexity of the data. Meanwhile, the matching phase will need to trade-off between the accuracy and the efficiency. In this talk, the challenges of both components are thoroughly explored, particularly with the aid of AI techniques. In addition, the preliminary experiment results to explore some important factors which affect the performance will be presented.

Contents

Survival Strategies for IT Companies During Crisis: A Case Study of Russia

Mohammad Khalil(✉) and Manuel Mazzara

Innopolis University, Faculty of Computer Science and Engineering, Innopolis, Russia
m.khalil@innopolis.university

Abstract. This thesis addresses the decision-making and survival strategies of IT companies in Russia amid socio-political conflicts. It investigates why IT firms choose to stay or leave Russia and the survival tactics they employ. Employing mixed methods, including qualitative interviews and quantitative analysis, two key experiments form the core of this research. The first experiment uncovers reasons driving IT migration, including economic sanctions, financial issues, regulatory complexities, and reputational risks. The second experiment assesses strategies used by IT companies that remain, involving business model innovation, remote work, risk management, and Asian market expansion. This study offers insights into IT industry resilience in conflict zones, with implications for operational planning in Russia and similar contexts, contributing to a broader understanding of adaptability to socio-political conflicts.

1 Introduction

The software development industry is an essential component of the digital economy, driving innovation and contributing significantly to economic growth. The industry has become a vital aspect of modern society, with an ever-increasing demand for software products and services across different sectors. However, despite the industry's potential for growth and stability, it is not immune to the impacts of global crises [1].

The COVID-19 pandemic is a prime example of a global crisis that has had significant disruptions on the software industry. Shifting to remote work suddenly has forced software companies adapting and implementing new technologies to enable remote work, leading to increased costs and technical challenges [2].

In Russia, the software industry has also been impacted by various global crises and sanctions. These include economic recessions, political conflicts, and sanctions imposed by western countries. These crises have affected the market landscape, leading to changes in demand and supply chain disruptions. In response, IT companies have been forced to employ new survival strategies to stay afloat [3].

L. Barolli (Ed.): AINA 2024, LNDECT 204, pp. 1–10, 2024.
https://doi.org/10.1007/978-3-031-57942-4_1

Outline: The research commences with an overview of global sanctions' impact on IT firms, incorporating insights from prior cases like Pandamec. The methodology outlines the qualitative and quantitative approaches and leads to two key experiments: analyzing factors driving company exits from Russia and investigating strategies adopted by those staying. Implementation discusses challenges and outcomes, followed by a focused analysis of business, management, and IT aspects. The conclusion summarizes findings, contributions, and suggests avenues for future research.

Problem Statement: In this paper, we investigate the impact of global crises on the software development industry, with a focus on the case of Russia. The software industry is a vital component of the digital economy, contributing significantly to economic growth and innovation. However, like other sectors, the software industry is not immune to the effects of global crises. Sothat, we aim to address two research questions:

RQ1: What are the reasons behind IT companies' decisions to either leave or stay in Russia during this conflict?
RQ2: What survival strategies are companies in Russia employing, especially in terms of business, management, and technical field?

To find answers to the mentioned questions, we adopt a mixed-methods approach that involves surveys and interviews with decision-makers in the software industry. Our study provides insights into the impacts of global crises on the software industry, particularly in the case of Russia, and offers recommendations for software businesses to develop effective survival strategies during times of uncertainty.

2 Literature Review

Research has shown that when the world faces a global crisis, the financial landscape is severely affected, with global markets experiencing turbulence and widespread uncertainty [4]. One of the primary reasons for this is the decline in consumer and institutional purchasing due to a sudden and sharp drop in asset values, leading to a lack of financial liquidity in banks and some other financial organisation [5]. The stock market, as a result, becomes increasingly unstable, with significant fluctuations in share prices and a heightened sense of apprehension among investors and market participants [6].

2.1 Pandemic 2020 as a Crisis (Russia)

In 2020, Russia saw a decline in the number of new startups entering the market, with less than 10,000 startups joining, representing a 21% decrease compared to 2015. This downturn can largely be attributed to the impact of Covid-19 pandemic on the economic disruptions [7]. However, it is worth noting that specific industries experienced growth during this period, with the number of pharmaceutical startups doubling and the creation of medical devices increasing by 30% [8]. This trend suggests a shift in entrepreneurial focus towards sectors that became more critical in response to the pandemic of COVID-19.

The majority of these startups primarily employ business-to-business (B2B) models and offer information technology services [9]. Geographically, most of these startups are concentrated in large cities, for example Moscow and St. Petersburg, reflecting the greater availability of resources and support in these metropolitan areas and some other areas [7].

According to the Total Early-Stage Entrepreneurial Activity (TEA) index, Russia experienced a 0.8% decrease in the percentage of people creating or maintaining projects during this period [8]. The decline can be attributed to several factors, including the closure of international markets, reduced household incomes, and the widespread implementation of Covid-19 quarantine measures [10].

2.2 Global Sanctions Impact on IT Field

Global sanctions significantly impact IT companies, affecting their operations and global competitiveness. For example, Iran faced extensive economic sanctions from the United States and the European Union, severely limiting technology access, funding, and market reach [11]. To cope, Iranian IT companies developed indigenous technologies to reduce foreign reliance [12]. Russian IT companies encountered challenges due to sanctions imposed by the USA and the EU after Crimea's annexation and the conflict in Ukraine. Sanctions restricted capital and technology access, compelling Russian IT firms to focus on domestic markets, expand into less-affected sectors (e.g., healthcare, agriculture, and public administration), and form partnerships with non-sanctioning countries like China and India [13–15].

In conclusion, the literature review has provided valuable insights into the impact of global crises on the software business and the survival strategies of Russian software companies. Throughout the literature, we have observed how global crises can significantly affect various industries, including the software sector. The case studies and examples presented have highlighted the adaptability and resilience of the software industry, especially in the face of challenges brought on by pandemics and economic sanctions.

3 Methodology

In this section, there will be an overview of the methodology employed in our study, which consists of two main components: (1) identifying the common reasons why companies have left Russia, and (2) determining the survival strategies implemented by companies that have remained in the Russian market. We aim to provide a rigorous and clear framework that supports the validity and reliability of our findings and recommendations.

Chosen Approaches: In this study, we employed a mixed-methods research approach, combining both qualitative and quantitative methods to provide a comprehensive understanding of the survival strategies of software companies in Russia amidst recent challenges.

- Qualitative research: Represented by interviews with key representatives from various software companies, unveiling unique insights into their survival strategies during the recent changes in the business environment.
- Quantitative research: Represented by surveys offering a more objective perspective and statistics related to the number of companies that have left Russia or established new businesses, as well as changes in the number of clients for these companies. We also analyzed trends and patterns in the data to provide a more comprehensive understanding of the software industry's response to the latest challenges in Russia.

Data Collection and Analysis: Structured interviews were held with 10 several companies' representatives (CEO, Business manager, account manager, etc.). The topics of discussion were mainly about the business part, focusing on the following:

- Alternative strategies within the local and international companies.
- The changes in the number, type, and behavior of clients.
- The marketing and reputation.

Surveys were provided for the technical team (developers, tech lead, CTO, etc.) about some practical clarification, focusing on the following:

- The changes in the management methodologies of work.
- The most affected stage in the software life-cycle.
- Replacing the banned technical tools and applications.

The data were collected, analyzed, and sing non-parametric inferential statistics.

4 Implementation

In the initial phase, this study conducted an experiment to uncover the reasons for the departure of diverse IT companies from Russia, including small businesses, startups, game development firms, and branches of international IT corporations. Subsequently, the second experiment focused on elucidating survival strategies employed by companies operating within Russia, with responses collected from a varied group differing in size, sector, and business model.

4.1 Analysing Factors Influencing Companies' Decisions to Leave

Small Companies and Startups: Primary among these was the desire to become international companies, with a strong focus on the European Union (EU) and North American markets. Economic sanctions from these regions created high-risk conditions for pursuing this goal within Russia, as noted by Vyacheslav, Founder & CEO of Edge Vision: "To build an international company targeting the EU and North America, you have to be outside of Russia; otherwise, sanctions pose significant risks and slow down business processes." Additionally, these companies faced challenges in securing investment due to Russian legislation and economic conditions. Migration led to a decline in the professional community within Russia, resulting in a shortage of quality industry conferences, further straining small businesses and startups. Social factors, such as the potential mobilization of male employees for war, posed personal and operational threats to these companies.

Game Development Companies: Game development companies, in addition to the previously mentioned challenges, contended with industry-specific obstacles. Anton Skudarnov, a former CEO of a game development firm, highlighted how government programs meant to "support" game development came with burdensome regulations. These regulations, coupled with restrictions on dealing with "unfriendly countries," obstructed working with non-government funds. Furthermore, government-funded projects primarily catered to the CIS region, a small part of the global market. Reputational risks also loomed large for these companies, as players sometimes reacted negatively to games originating from Russia. Additionally, payment hindrances for players in Russia, especially on mobile and Steam platforms, added to their difficulties.

International IT Companies: The international IT sector, intricately connected to global dynamics, grappled with significant challenges in the Russian business environment. Numerous international IT companies that had established branches in Russia eventually opted to cease their operations due to these multifaceted difficulties. Foremost among these challenges were financial hardships, exacerbated by economic sanctions that hindered international transactions and the company's daily functions.

Faced with these obstacles, many international companies chose to consolidate their operations in more favorable regions such as the United States and the European Union, known for their stable economic environments and predictable regulatory frameworks.

4.2 Adapting Strategies of IT Companies Remaining in Russia

Impact on Business Operations. Responses varied notably across different types of businesses. Large companies, both governmental and private, reported significant changes in demand. Albert, the CBDO of Friflex, mentioned that while SaaS providers saw no change, PaaS providers experienced a significant shift.

Interestingly, some businesses saw a surge in demand as international companies left Russia. Nikita Gurov, CEO of Mirai Vision, pointed out that their B2G and B2B businesses had grown. On the other hand, B2C businesses did not meet the expected results. The cost of services remained unchanged for 90%

of the companies interviewed, and there were no changes in employee salaries, creating an inevitable tension in the context of the current inflation.

Strategic Changes and Supply Chain Effects. In terms of strategic changes, companies that were trying to establish offices in the EU and USA are now attempting to work in GCC or CIS. International companies like Soramitsu reported no impact. Some saw this as a chance to start businesses in Africa, the Middle East, and CIS, like Friflex.

There was a clear dependency on China for hardware supply, as reported by Nikita Gurov. Some companies mentioned a lack of Macbooks and 50% of startups funded by external sources noticed significant challenges, but it was still acceptable. Ksenia, head of the robotics lab at Innopolis University, noted that product delivery time and prices had increased and that many international opportunities had closed.

Impact on Management and Employee Policies. Companies also adapted their management strategies and employee policies in response to the conflict. One prominent trend was the shift towards remote work. A majority of companies transitioned to a hybrid model, with 30% increased of employees choosing to work remotely. Employee turnover remained consistent, suggesting acceptance of the remote work arrangement.

Surprisingly, despite the economic challenges, there were no noticed changes in employee salaries. Recruitment strategies were also impacted, with companies relying on employee loyalty and a bit increasing salaries to retain staff. In terms of work methodology and future plans, companies are reducing product development and seeking projects with quicker returns due to the significant business scarcity.

Impact on IT Tools and Practices. Small companies reported no changes in the IT tools needed for their work processes, but a few large companies had to change some database tools due to some attacks. PaaS companies reported that tools had become more expensive, leading to attempts to import from China, and this was confirmed by Nikita. In terms of cybersecurity, little change was reported except for a few cases like Suramitsu, where some employees moved to different time zones, which matters for security emergency issues. Regarding cloud usage, 70% of the small companies started using Yandex Cloud, while big companies like MTS started developing their own data centers. Companies with international offices continued using their current cloud providers.

General recommendations: The participants advised aspiring entrepreneurs to start in Singapore as a preferable alternative or GCC. They also recommended not spending too much time watching the news, being brave, and looking for opportunities with an international scale.

5 Results

This part presents the results obtained from the two experiments in the study:

Result from the First Experiment. The first experiment involved studying various types of IT companies, including small businesses, startups, game development firms, and branches of international IT companies in Russia, that decided to leave the country:

Table 1. Reasons for IT Companies Leaving Russia

Type of Company	Reasons for Leaving Russia
Small Businesses and Startups	• Aspiration to expand internationally • Economic sanctions • Lack of investment • Shrinkage of professional community • Threat of male employees' mobilization for war
Game Development Companies	• Regulatory complexities • Financial difficulties • reputational risks • impediments in receiving payments
Branches of International IT Companies	• Regulatory complexities • Financial difficulties • reputational risks • Consolidating operations in the United States and European Union

Result from the Second Experiment. The second experiment involved interviewing companies still operating in Russia to understand their survival strategies across three aspects: business, management, and IT.

Table 2. Changes and Strategies in Business Aspect

Business Aspect	Change/Strategy
Demand for Products/Services	• Big companies experienced significant changes • Small-medium companies reported mixed results
Strategic Changes	• Varying strategies depending on company size and international presence
Hardware Supply Chain	• Dependence on China • Lack of certain products • Increased delivery time and prices

Table 3. Changes and Strategies in Management Aspect

Management Aspect	Change/Strategy
Working Modes	• Adoption of remote or hybrid work models • Some employees leaving due to offline work requirements
Salary Adjustments	• No significant changes in salary despite the ongoing inflation
Recruitment Strategies	• Dependence on employee loyalty • Limited salary increases
Future Work Methodologies	• Decrease in product-focused work • Emphasis on faster money-generating projects

Table 4. Changes and Strategies in IT Aspect

IT Aspect	Change/Strategy
IT Tools	• Small companies reported no change • Big companies reported changes in some database tools
Cybersecurity Measures	• No significant changes • Except for timezone adjustments for remote employees
Cloud Usage	• Small companies started using Yandex cloud • big companies developed their own data centers
Development Frameworks	• Continued use of pre-existing frameworks

6 Discussion and Analysis

The results of this study provide significant insights into the current landscape of IT companies in Russia. For the first experiment, the departure of IT companies from Russia is not a singular event, but a complex phenomenon driven by a confluence of factors such as economic sanctions, regulatory complexities, reputational risks, and social factors. The desire for international expansion, the lack of investment, shrinking professional community, and the social instability due to the threat of mobilization for war are particularly notable. For the second experiment, the adaptive strategies adopted by companies in Russia vary widely. This variation can be attributed to factors such as the size of the company, their market (B2G, B2B, B2C), and whether they provide SaaS or PaaS. The observation that larger companies and those serving the government have experienced significant changes while smaller companies and SaaS providers have not, suggests that the impact of the conflict is uneven across the sector.

Comparison with Previous Research: Our findings corroborate previous research that has highlighted the detrimental impact of economic sanctions and political instability on businesses. However, our study further enriches this understanding by illustrating the specific ways in which these factors affect IT companies in Russia. The findings also align with research on business resilience, demonstrating how companies adapt to adverse circumstances by shifting their markets, developing new strategies, or changing their operating models.

Discrepancies and Unexpected Findings: One discrepancy between our findings and previous research is the resilience of the IT sector despite significant challenges. While previous studies suggest a sharp decline in business activities under similar conditions, our study found that many companies have adapted and continue to operate. These results underscore the dynamic nature of the IT sector in Russia and its ability to resist the changes in the business environment.

Limitations of the Research: While this study provides valuable insights, it is not without limitations. The sample size for both experiments was relatively small, which may limit the generalizability of the findings. Additionally, due to the volatile nature of the situation in Russia, the findings of this study reflect a specific point in time and may not capture future developments.

Implications for Future Research and Applications: This study findings have significant implications for future research and applications. They underscore the need for ongoing research to monitor the evolving situation in Russia and its impact on the IT sector. In terms of applications, the findings can inform strategies for IT companies operating in conflict zones or under economic sanctions. Moreover, the study also highlights the potential for IT companies to adapt and thrive in challenging business environments, which can be a valuable lesson for businesses in other sectors as well.

7 Conclusion

This thesis extensively explored the exodus of IT companies from Russia, providing insights into the forces that guided their migration decisions. It thoroughly addressed the research questions, revealing a variety of factors that drove companies away, including economic sanctions, investment shortages, regulatory intricacies, and reputational risks, affecting businesses across the spectrum. The research's significance lay in its examination of a phenomenon with far-reaching implications for the Russian economy and the global IT industry. It underscored the critical role of a supportive business environment in nurturing the growth of IT companies, a sector pivotal for global economic development. While limited by data availability and a focus on departing companies, future research could include the experiences of those who chose to stay. This study underscored the complexity of the issue and its significance for policy-making in similarly challenged regions.

References

1. Ma, D., Zhu, Q.: Innovation in emerging economies: research on the digital economy driving high-quality green development. J. Bus. Res. **145**, 801–813 (2022)
2. Svabova, L., Kramarova, K., Chabadova, D.:: Impact of the COVID-19 pandemic on the business environment in Slovakia. Economies **10**(10), 244 (2022)
3. Gurvich, E., Prilepskiy, I.: The impact of financial sanctions on the Russian economy. Russian J. Econ. **1**, 359–385 (2015). https://doi.org/10.1016/j.ruje.2016.02.002

4. Reinhart, C.M., Rogoff, K.S.: The aftermath of financial crises. Am. Econ. Rev. **99**(2), 466–472 (2009). http://www.jstor.org/stable/25592442
5. Claessens, S., Kose, A.: Financial crises: explanations, types, and implications. In: IMF Working Papers, vol. 13 (2013). https://doi.org/10.2139/ssrn.2295201
6. Aliber, R., Kindleberger, C.: Manias, panics, and crashes: a history of financial crises. In: Manias, Panics, and Crashes: A History of Financial Crises, Seventh Edition (2015). https://doi.org/10.1007/978-1-137-52574-1
7. Zemtsov, S., Chepurenko, A., Mikhaylov, A.: Pandemic challenges for the technological startups in the Russian regions. Foresight STI Governance **15**, 61–77 (2021). https://doi.org/10.17323/2500-2597.2021.4.61.77
8. Kudrin, ., Sinelnikov-Murylev, S., Radygin, A. (eds.): Russian Economy in 2020. Trends and Outlooks. (Issue 42) In English, vol. 42(1) (2021). https://EconPapers.repec.org/RePEc:gai:gbooks:re42-2021-en
9. Gurkov, I.: Innovative actions and innovation (in)capabilities of Russian industrial companies: a further extension of observations. Post-Communist Eco. **23**, 507–516 (2011). https://doi.org/10.1080/14631377.2011.622572
10. Coibion, O., Gorodnichenko, Y., Weber, M.: The Cost of the Covid-19 Crisis: Lockdowns, Macroeconomic Expectations, and Consumer Spending. National Bureau of Economic Research, Working Paper Series, vol. 27141 (2020). http://www.nber.org/papers/w27141https://doi.org/10.3386/w27141
11. Sandle, L.: Book Review: Seyed Hossein Mousavian with Shahir Shahidsaless, Iran and the United States: an insider's view on the failed past and the road to peace. Polit. Stud. Rev. **14**(4), 583–584 (2016). https://doi.org/10.1177/1478929916660502
12. Tarikhi, P.: Sanctions and the scientific community of Iran. Sociology Islam **8**, 225–264 (2020). https://doi.org/10.1163/22131418-00802005
13. Connolly, R.: The empire strikes back: economic statecraft and the securitisation of political economy in Russia. Europe-Asia Stud. **68**, 1–24 (2016). https://doi.org/10.1080/09668136.2016.1156056
14. Gurvich, E., Prilepskiy, I.: The impact of financial sanctions on the Russian economy. Russian J. Econ. **1**(4), 359–385 (2015). https://doi.org/10.1016/j.ruje.2016.02.002
15. Vinokurov, E.: Eurasian economic union: current state and preliminary results. Russian J. Econ. **3**, 54–70 (2017). https://doi.org/10.1016/j.ruje.2017.02.004

Comparative Analysis of Ontologies for Archival Representation

Alba Amato(✉) and Giuseppe Cirillo

University of Campania "Luigi Vanvitelli", Department of Political Science, Caserta, Italy
{alba.amato,giuseppe.cirillo}@unicampania.it

Abstract. Ontologies are valuable tools for structuring and organizing knowledge, improving communication and data integration, enhancing information retrieval and search, and supporting various applications in different fields. They play a crucial role in promoting shared understanding and knowledge management in complex and information-rich domains. An examination of the initiatives conducted over the past decade in the field of cultural heritage has clearly highlighted the lack of an established framework in the conceptual modeling of information resources, despite the numerous ontologies created as a function of the many projects for the publication of linked open data. As a result, it is far from easy to know exhaustively all the ontologies available in relation to one's field of interest and to obtain in an easy and systematic way a reliable assessment about their representative capacity and their degree of semantic interoperability.

1 Introduction

Archiving historical materials helps to preserve the cultural heritage of a society allowing researchers, scholars, and the general public to understand the historical background of various topics and to understand how societies have evolved over time. This can lead to a more informed perspective on contemporary issues [2]. Ontologies are valuable tools for structuring and organizing knowledge, improving communication and data integration, enhancing information retrieval and search, and supporting various applications in different fields. They play a crucial role in promoting shared understanding and knowledge management in complex and information-rich domains [3]. The number of metadata standards in the cultural heritage sector is overwhelming, and their inter-relationships further complicate the situation. This visual map of the metadata landscape [14] is intended to assist planners with the selection and implementation of metadata standards. The archival discipline has produced numerous standards and corresponding markup languages for the description of archives that help archivists, librarians, and information professionals to create consistent and interoperable descriptions of archival materials [7]. An archival description standard can be defined as the set of norms and conventions adopted by a community of experts

L. Barolli (Ed.): AINA 2024, LNDECT 204, pp. 11–19, 2024.
https://doi.org/10.1007/978-3-031-57942-4_2

with the aim of standardizing the behavior of those who create representations and those who seek representations. They contribute to the preservation, accessibility, and discoverability of valuable historical records and cultural heritage materials. Properly encoding archival information in these standards ensures that it can be shared, searched, and accessed effectively by researchers and the public. In this paper we propose a comparative analysis of the most recent ontologies and vocabularies in the archiving field with the aim to provide a practical source of information to consult in order to make an informed and conscious choice about the already modelled and reusable pieces of knowledge provided by other ontologies. The paper is organised as follows: Sect. 2 describes the state of the art of the ontology evaluation, in Sect. 3 are explained the ontologies we analysed and a description of the main features. In Sect. 4, future work directions end the paper.

2 Related Works

Several ontologies have been proposed in literature for the archiving domain. According to Clover[1] the repositories for the ontologies in cultural domain, there are 9 ontologies dealing with archiving domain and more than 80 dealing with cultural heritage. Among them, we selected the more diffused ontologies in order to make an useful comparison.

This visual map of the metadata landscape, shown in Fig. 1 is intended to assist planners [14] with the selection and implementation of metadata standards is shown in 1 where we selected the section dealing with archives. Ontology comparison is an hard task as it is necessary to evaluate several elements [12]. Since ontologies are considered as reference models, one must insure their evaluation in the view of two important perspectives [10] quality and correctness. These two perspectives address several criteria [16] that are explained in [8,9].

According to [12], it is necessary to evaluate the following criteria:

- Accuracy is a criterion that states if the definitions, descriptions of classes, properties, and individuals in an ontology are correct.
- Completeness measures if the domain of interest is appropriately covered in this ontology.
- Conciseness is the criteria that states if the ontology includes irrelevant elements with regards to the domain to be covered.
- Adaptability measures how far the ontology anticipates its uses. An ontology should offer the conceptual foundation for a range of anticipated tasks.
- Clarity measures how effectively the ontology communicates the intended meaning of the defined terms. Definitions should be objective and independent of the context.
- Computational efficiency measures the ability of the used tools to work with the ontology, in particular the speed that reasoners need to fulfil the required tasks.

[1] http://arco.istc.cnr.it:8081/ontologies.

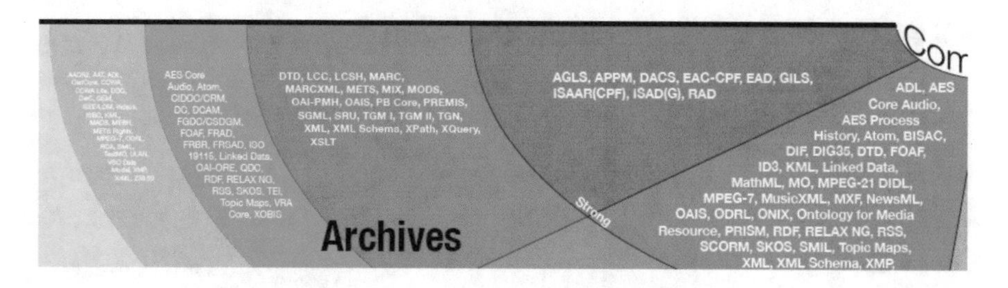

Fig. 1. http://jennriley.com/metadatamap/

- Consistency describes that the ontology does not include or allow for any contradictions.

In order to evaluate how the selected ontologies can fit the requirements of archival representation, we do not consider computational efficency and consistency.

3 Ontology

3.1 CIDOC CRM

The CIDOC [1] Conceptual Reference Model (CRM)[2] is an ontology developed by the International Committee for Documentation of the International Council of Museums (ICOM) that provides definitions and a formal structure for describing implicit and explicit concepts and relationships used in cultural heritage documentation. Thus, it is a formal ontology intended to facilitate the integration and exchange of heterogeneous information in the field of Cultural Heritage. As such, it constitutes the de facto standard model for data description. CIDOC CRM is an official ISO standard (21127: 2006). It was created to enable information exchange and integration between heterogeneous databases in the field of cultural heritage. The ontology is focused on the relationships between the physical cultural object and the relationships between it and the events that involved it, as well as with the entities that interacted with it (people, organizations). As shown in Fig. 2 Format: RDF/XML Syntax : RDFS Ontology type: upper level ontology Main classes: E1 CRM Entity, E77 Persistent Item, E5 Event, E2 Temporal Entity, E52 Time Span, E53 Place, E54 Dimension, E71 Man made thing, E72 Legal Object, E28 Conceptual Object. The CIDOC ontology consists of 84 classes, 275 object properties and 12 datatype properties and does not materializes in RDFS any equivalence or semantic alignment with other ontologies either domain nor foundational. CIDOC-CRM was obtained from more than 10 years' work by the CIDOC Documentation Standards Working Group. As of December 2006, the conceptual model became an ISO208 standard, and is intended to become the reference for the semantic representation of information related to

[2] https://cidoc-crm.org/.

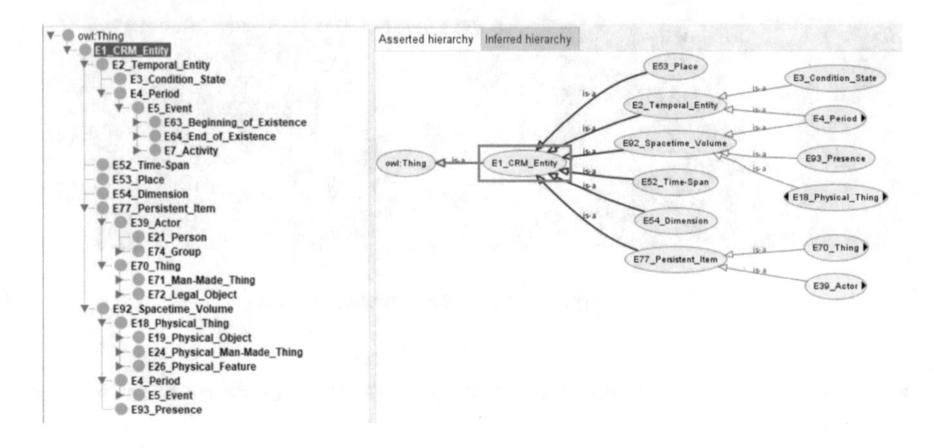

Fig. 2. CIDOC

cultural heritage. In 2014 the ontology was technically revised by the relevant ISO/TC 46, "Information and Documentation," subcommittee SC 4, "Technical Interoperability," in collaboration with the International Council of Museums - Documentation Committee (ICOM CIDOC). CIDOC-CRM defines in the terms of a formal ontology the implicit semantics present in the schemas and structures of databases and documents used in the cultural field, with particular reference to museum documentation. The scope is defined as "the exchange and integration of heterogeneous scientific documentation regarding museum collections."

3.2 EDM

The Europeana data model (EDM)[3] is the ontological key evolution of the Europeana Semantic Elements (ESE), the data model with which in the Europeana portal is managed the import of data related to descriptions of cultural heritage objects [5]. It makes direct reuse of the Dublin Core, DC Terms, OAI ORE, CIDOC ontologies and defines additional classes and specific properties declaring any equivalences with other ontologies. As shown in Fig. 3 Format: RDF/XML Syntax : RDFS Ontology type : domain ontology Main classes: Europeana Object, Event, Information Resource, PhysicalThing, ProvidedCHO, Place, Agent. It consists of 41 classes and 51 object properties and 12 datatype properties, makes direct reuse of the ontologies Dublin Core, DC Terms, OAI ORE and materializes in OWL the equivalences and conceptual relationships with CIDOC, Dolce Lite, Abc, DCTems.

[3] https://pro.europeana.eu/page/edm-documentation.

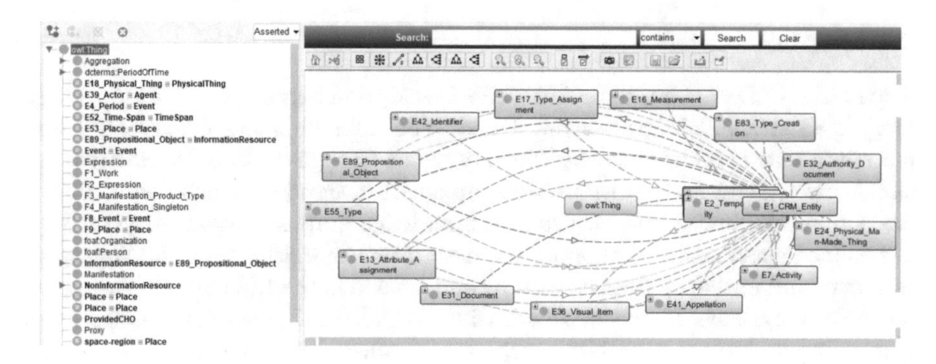

Fig. 3. EDM

3.3 RiC-O

RiC-O (Records in Contexts Ontology)[4] is an OWL ontology for describing archival record resources. As the second part of Records in Contexts standard, it is a formal representation of Records in Contexts Conceptual Model (RiC-CM). RiC-O provides a generic vocabulary and formal rules for creating RDF datasets [11] (or generating them from existing archival metadata) that describe in a consistent way any kind of archival record resource.

As shown in Fig. 4 consists of 125 classes, 423 object properties and 56 datatype properties. Alignments with several ontologies like CIDOC-CRM, Europeana Data Model (EDM) are in progress. Format: RDF / XML Author: International Council on Archives Expert Group on Archival Description (ICA EGAD) Ontology type: domain ontology Main classes: Record Resource (with subclasses RecordSet, Record, RecordPart), Agent, Role, Relation, Event, Function, Process, Place

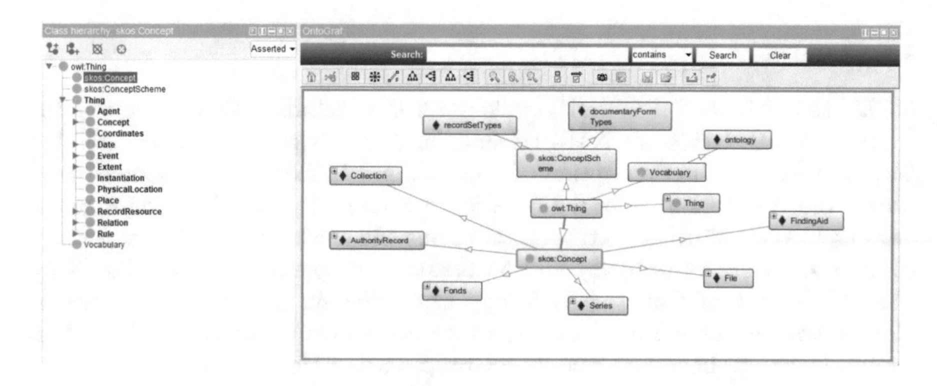

Fig. 4. RiC-O

[4] https://www.ica.org/standards/RiC.html.

3.4 MADS

The MADS/RDF[5] (Metadata Authority Description Schema in RDF) [6] vocabulary is a data model that was developed to describe bibliographic concepts and metadata and is maintained by the Library of Congress. Its authoritative version is given as an XML schema based on an XML mindset which means that it has significant limitations for use in a knowledge graphs context. MADS/RDF is designed specifically to support the description of cultural and bibliographic resources and authority data as used by and needed in the LIS community and its technology systems. Data described using MADS/RDF, therefore, assists with identifying and annotating bibliographic and cultural resources.

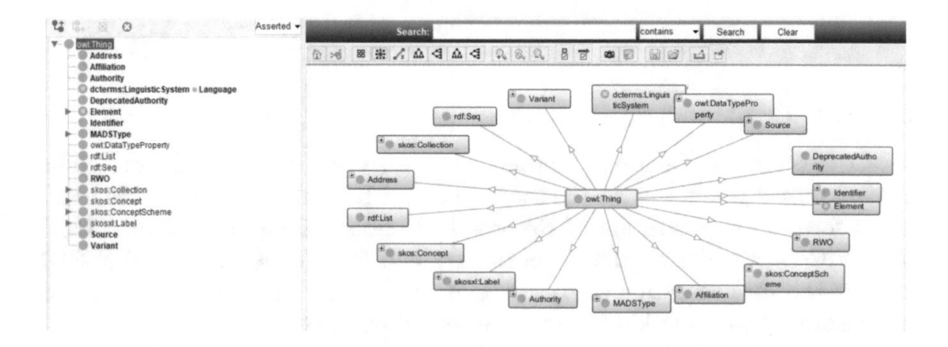

Fig. 5. MADS

As shown in Fig. 5 consists of 92 classes, 62 object property e 27 datatype property. Format: XML/RDF Author: Library of Congress Ontology type: domain ontology Main classes: MADS Collection, Authority, Real World Object, Variant, DeprecatedAuthority, Name, MADS Type.

3.5 MODS

MODS RDF[6] is an OWL/RDF ontology for the Metadata Object Description Schema (MODS) [13], an XML schema for a bibliographic element set that may be used for the description of cultural and bibliographic resources used within the library and information science community, which includes museums, archives, and other cultural institutions. As shown in Fig. 6 it consists of 25 classes, 43 object property and 55 datatype property. Format: RDF XML Author: Library of Congress Ontology type: domain ontology Main classes: ModsResource, AdminMetadata, Cartographics, ClassificationGroup, IdentifierGroup, Location, LocationCopy, NoteGroup, Part, RoleRelationship.

[5] https://www.loc.gov/standards/mods/modsrdf-primer.html.
[6] https://www.loc.gov/standards/mods/modsrdf/.

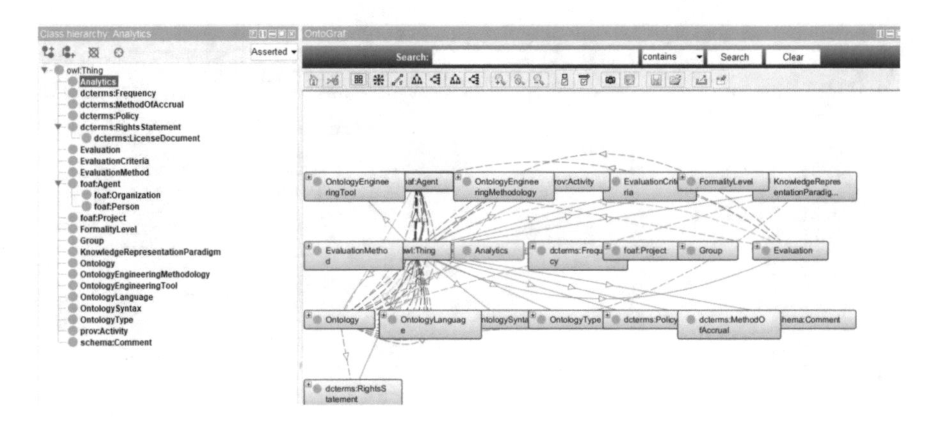

Fig. 6. MODS

3.6 SAN

In the context of the National Archival System (SAN) Linked Open Data development and publication project[7], is a conceptual model of the SAN formalized in an ontology expressed in an OWL language fully corresponding to the information structure conveyed by the XML schemas proposed to the member systems as exchange paths for the contribution to the Catalog of Archival Resources - CAT of their own data related to archival complexes, producer subjects, research tools and preserving subjects [4]; the definition of a series of extensions to the Basic Ontology for the integration of entities, information elements and reationships between objects originally not provided in the CAT SAN tracks, but retrievable from some source systems for a more complete description of SAN LOD resources; the creation of thesauri, repertories, tools formalized according to Semantic web standards that can be a support to the control of the description of archival heritage and at the same time constitute general information frames capable of integrating data from various sources contextualizing them on the basis of time (history and institutions) and space (territory, historical place names); the production and publication of SAN data in LOD format, made available mainly through two channels: a dedicated SPARQL endpoint; a download area organized into specific datasets made in consideration of the main object classes defined by the model and their conceptual organization.

As shown in Fig. 7 consists of 40 classes, 34 object properties and 50 datatype properties and materializes equivalences and semantic relationships with SRO, ORG, EAC-CPF, Geonames, SKOS Acronym: SAN Format: RDF XML

According to our analysis, the most complete ontology is CIDOC-CRM. In the cultural heritage sector, the CIDOC ontology could be taken as foundational with respect to the domain, and the conceptual mapping towards its classes and properties could perform an action comparable to the semantic alignment of classes and properties of more specific ontologies [15]. EDM, SAN and RiC-O

[7] http://www.san.beniculturali.it/web/san/dati-san-lod.

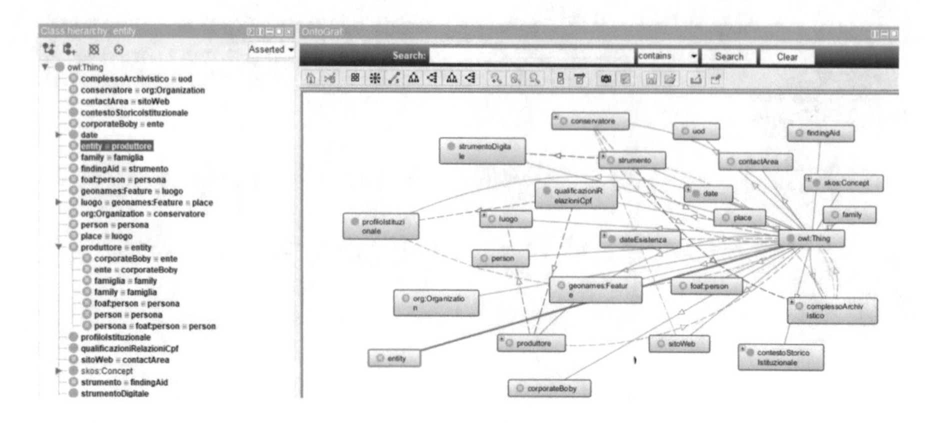

Fig. 7. SAN

	Accuracy	Completeness	Adaptability	Clarity	Computational efficiency	Consistency
Cidoc CRM	XXX	XXX	XXX	XXX	N/A	N/A
EDM	XX	XX	XX	XX	N/A	N/A
RiC-O	XX	X	XXX	XX	N/A	N/A
MADS	XX	X	XXX	X	N/A	N/A
MODS	X	X	X	XX	N/A	N/A
SAN	XX	XXX	XX	X	N/A	N/A

are also good in terms on Completeness and Adaptability while MODS, MADS and designed for use with controlled values for names.

4 Conclusions

The existence of numerous ontologies dedicated to cultural heritage indicates the wide interest in this filed. Nevertheless it is not easy to understand their strengths and weakness as well as salient differences. In this paper we have presented a preliminary attempt to classify ontologies related to the domain of cultural heritage according to a model present in the literature. As future work we plan to extend this study by taking into consideration other ontologies present in the literature.

Acknowledgements. The work described in this paper has been supported by the research project RASTA: Realtá Aumentata e Story-Telling Automatizzato per la valorizzazione di Beni Culturali ed Itinerari; Italian MUR PON Proj. ARS01 00540.

References

1. Aalberg, T., et al.: Conceptual reference model
2. Amato, A.: Towards the interoperability of metadata for cultural heritage. In: Barolli, L. (ed.) Complex, Intelligent and Software Intensive Systems, pp. 309–317. Springer Nature Switzerland, Cham (2023). https://doi.org/10.1007/978-3-031-35734-3_31
3. Amato, A., Branco, D., Venticinque, S., Renda, G., Mataluna, S.: Metadata and semantic annotation of digital heritage assets: a case study. In: 2023 IEEE International Conference on Cyber Security and Resilience (CSR), pp. 516–522 (2023). https://doi.org/10.1109/CSR57506.2023.10224935
4. Amato, A., Venticinque, S.: Multiobjective optimization for brokering of multicloud service composition. ACM Trans. Internet Technol. **16**(2) (2016). https://doi.org/10.1145/2870634
5. Doerr, M., Gradmann, S., Hennicke, S., Isaac, A., Meghini, C., Van de Sompel, H.: The europeana data model (edm). In: World Library and Information Congress: 76th IFLA General Conference and Assembly, vol. 10, p. 15 (2010)
6. Feliciati, P.: Archives in a graph. the records in contexts ontology within the framework of standards and practices of archival description. JLIS.it **12**(1), 92–101 (2021). https://doi.org/10.4403/jlis.it-12675. https://jlis.fupress.net/index.php/jlis/article/view/20
7. Formenton, D., Gracioso, L.: Metadata standards in web archiving: technological resources for ensuring the digital preservation of archived websites. RDBCI Revista Digital de Biblioteconomia e Ciência da Informaç ão **20**, 1–29 (2022). https://doi.org/10.20396/rdbci.v20i00.8666263/27830
8. Gangemi, A., Catenacci, C., Ciaramita, M., Lehmann, J.: A theoretical framework for ontology evaluation and validation. In: SWAP, vol. 166, p. 16 (2005)
9. Gruber, T.R.: Toward principles for the design of ontologies used for knowledge sharing? Int. J. Hum Comput Stud. **43**(5–6), 907–928 (1995)
10. Hlomani, H., Stacey, D.A.: Approaches , methods , metrics , measures , and subjectivity in ontology evaluation: a survey (2014). https://api.semanticscholar.org/CorpusID:51371006
11. Llanes-Padrón, D., Pastor-Sánchez, J.A.: Records in contexts: the road of archives to semantic interoperability. Program **51** (2017). https://doi.org/10.1108/PROG-03-2017-0021
12. Raad, J., Cruz, C.: A survey on ontology evaluation methods. In: Proceedings of the International Conference on Knowledge Engineering and Ontology Development, part of the 7th International Joint Conference on Knowledge Discovery, Knowledge Engineering and Knowledge Management, Lisbonne, Portugal (2015). https://doi.org/10.5220/0005591001790186, https://hal.science/hal-01274199
13. Rayan, R., Shimizu, C., Sieverding, H., Hitzler, P.: A modular ontology for mods – metadata object description schema (2023)
14. Riley, J.: Seeing Standards: A Visualization of the Metadata Universe (2018). https://doi.org/10.5683/SP2/UOHPVH
15. Veninata, C.: Linked open data e ontologie per la descrizione del patrimonio culturale: criteri per la progettazione di un registro ragionato. Universitá la Sapienza
16. Vrandečić, D.: Ontology Evaluation, pp. 293–313. Springer, Berlind (2009).

Augmented Reality for Cyberphisical Exploration of Archeological Sites

Luigi Alberico[1], Dario Branco[1], Antonio Coppa[2], Salvatore D'Angelo[1], Stefania Quilici Gigli[2], Giuseppina Renda[2], and Salvatore Venticinque[1](✉)

[1] Università della Campania "Luigi Vanvitelli", Aversa, Italy
luigi.alberico3@studenti.unicampania.it,
{dario.branco,salvatore.dangelo,salvatore.venticinque}@unicampania.it
[2] Università della Campania "Luigi Vanvitelli", Caserta, Italy
{antonio.coppa,stefania.gigli,giuseppina.renda}@unicampania.it

Abstract. The interaction with archaeological heritage by smart devices allows for a more immersive experience and can improve understanding by users, increasing interest and enjoyment. However, the comprehension of delivered multimedia content can be affected by its complexity, which makes difficult the exploration and retrieval of relevant information. In this paper, we propose a cyber-physical exploration of the archaeological site that is based on the collaborative interaction of users and software agents with 3D models, s in an augmented or virtual environment.

1 Introduction

The scenarios and possibilities for enjoying cultural heritage are many and very different, therefore it needs to handle the integrated utilization of heterogeneous, complementary, or alternative technological solutions. Visitors can enjoy museums and archaeological sites, supported by dedicated or personal devices, which enrich the live experiences with additional content and services, exploiting advanced technologies such as virtual, augmented, and hybrid reality [3]. Technology is even more relevant for the delivery of virtual exhibitions and digital archives, which are available for those users who connect remotely from a personal computer at home or using a public facility. Despite the increasing availability of multimedia content, data providers and aggregators, intelligent services, and innovative technologies which provide an effective and immersive experience in real scenarios, are still missing. At least they do not work with heterogeneous off-the-shelf devices. For example, the rendering of a 3D model can be affected by the limited resources of the user's device. Advanced services could require specific peripherals or connection modes. Dedicated applications are not compliant with the user's device or cannot be installed on a public computer. On the other hand, some experiences cannot be enjoyed by users without minimal support because of a lack of infrastructure, missing human guides, difficult comprehension, overload of unreliable and not focused information. This

L. Barolli (Ed.): AINA 2024, LNDECT 204, pp. 20–28, 2024.
https://doi.org/10.1007/978-3-031-57942-4_3

paper focuses on the design and development of technologies that allow users to augment the live experience in archaeological sites by interacting with virtual content through images, 3D models, and textual information. The overall visitor experience is enhanced by delivering information about cultural artifacts, offering images, and presenting 3D models related to cultural content through innovative technologies. Using their smartphones or headsets, users can embark on virtual journeys, navigating the archaeological site from their hands. We present two alternative scenarios where, according to the available devices, the user is still supported in augmented interaction with the reality that surrounds him. A web application is available to address the interoperability with a large class of devices, while a more immersive experience is proposed to exploit smart headsets or high-end smartphones.

2 Related Work

The research context explored in this article is closely related to the applications of information technology in the cultural heritage sector [2], focusing specifically on the potential of collaborative augmented reality (AR) in the context of cyber-physical exploration of archaeological sites [11]. To date, although augmented reality is gaining more and more credence and interest in various fields such as medicine, education, logistics, etc. in the field of cultural heritage few researchers and institutions are investing and investigating these technologies. An element in favor of the development of augmented reality in the field of cultural heritage is the growing number of 3D models available online thanks to institutions that are committed to publishing their models on public repositories: among the most important institutions we find the 3D Virtual Museum with its 95 3D models but also the British Museum that offers the possibility of downloading models of its 3D reproductions of physical works and 3D printing them. Another important element is the change in fruition systems, since with today's technology it is possible to experience augmented reality and collaborative augmented reality through the use of a simple smartphone making this technology much more accessible [6]. This is the context for the RASTA (Augmented Reality and Automated Storytelling for Cultural Heritage and Tourist Routes) research project. The goal of the project is to create a framework that succeeds in integrating the potential of augmented and virtual reality with advanced intelligent storytelling technologies that manage to stimulate user curiosity by recommending content consonant with their interests through recommendation mechanisms and algorithms [1].

3 The Domus of Norba

The focus of this case study is the Norba archaeological park situated in the Lazio region, within the province of Latina. Nestled amid the scenic Lepini mountains, Norba stands as a testament to the ancient Roman civilization, positioned approximately 60 km southeast of Rome. Abandoned during the Middle Ages due to barbarian incursions, the once vibrant city succumbed to the gradual encroachment of vegetation and debris. However, owing to dedicated political and scientific endeavors aimed at research and conservation, the Norba Archaeological Park has emerged in recent decades as a tangible outcome of these commitments. Private dwellings lined the side streets along *Norba's* main axis. We will focus on the *Domus IV*, which has reached us in its final form, having been separated from its adjacent house, *Domus V*, to make it independent.

As it is shown in Fig. 1, *Domus IV* exhibits the canonical organization of Roman houses with an atrium: a spacious central area, the atrium, with a central pool (*impluvium*), symmetrically arranged rooms on either side, followed by two larger rooms (*alae*), and three rooms at the back, with the central one, open to the atrium, serving as the reception and representation area for the homeowner (*tablinum*).

A renovation of the room on the left-hand side of the *tablinum* was prompted by the social need to provide the house with a room for dining, and reclining on beds. The discovery of two bronze bed feet in a corner of the expanded room, which is shown in Fig. 2, confirms that the renovation in this house was driven by the need to create a *triclinium*: a social custom that spread in the Roman world following wars in the East, propagating from Rome to Italian cities from the mid-2nd century BCE [9].

Fig. 1. View from above of the *domus* along crossing street II, with a highlight of the triclinium of *domus* IV (red arrow). (Color figure online)

Notably, the room shows a more elaborate decorative scheme compared to others in the house. Its floor is made of cement with fragments of terracotta interspersed with small pieces of white and gray limestone and dark gray stone, bound by mortar [5]. A hole in the floor is believed to have housed a drain cover: a feature often found in *triclinia*, designed to channel water used for

Fig. 2. The finding of the bronze bed feet.

cleaning after meals. The wall decoration consists of a tall base topped with panels divided by vertical red and light-colored bands, featuring decorated panels with perspective cubes, reminiscent of numerous decorations from the so-called "first style," dating back to the 2nd century BCE [7] (Fig. 3). A limestone base found almost at the center of the back wall of the room may have served as support for a piece of furniture placed between the beds.

Fig. 3. The room of the triclinium in the domus IV.

The triclinium, furniture, and bed are significant testimony to the spread of new social customs even in more secluded areas like Norba, and among individuals of not high social status, such as the owner of *Domus IV* in Norba.

4 Use Case Scenario

The RASTA project involves both the visit to the archaeological park in virtual and assisted mode with the aid of a smartphone with active GPS. The visit itineraries are in any case conceived as a series of geolocalized points of interest (POI) enriched with multimedia content. Based on the preferences expressed by the user, the system dynamically proposes new content and personalized routes. In case the user shows disinterest or wants to explore something else, the system may adapt promptly to his requests. This flexibility allows the user to explore points of interest at their pleasure while maintaining an overall guide that directs them along the recommended main route. For assisted use of the archaeological sites of the project, three-dimensional models and virtual reconstructions are being prepared which improve the understanding of the archaeological remains. Mapping and 3D scans were carried out on monuments and archaeological remains of the Norba site. Let's suppose that the user is looking at the room of the triclinium in the Domus IV.

The recommendation system, according to the kind of user's device suggests the interaction with 3D models through their smartphone or to wear the headset for a more immersive experience.

4.1 Augmented Creation of Virtual Reconstruction

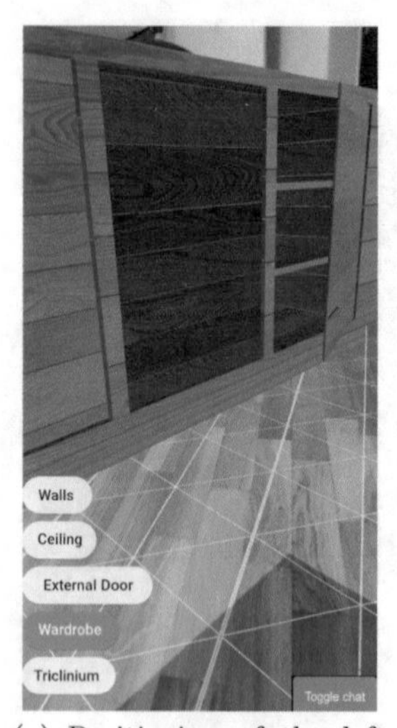

(a) Positioning of the left wall

(b) Complete view

Fig. 4. Room reconstruction in augmented reality

Let's suppose that the user wears smart glasses or has a virtual/augmented reality viewer for his smartphone. In Fig. 4(a) he can use the toggle button of a mobile app to drag and anchor the 3D reconstruction of missing elements, such as walls, floor, ceiling, furniture, etc. In Fig. 4(b) the room is completely rebuilt, and the user can move and explore the 3D objects or the 3D environment. The mobile application has been developed with the Flutter multi-platform framework[1], and executes on an Android Phone/Tablet. It's based on the ARCore libraries and uses glb/gltf files for the representation of 3D objects.

4.2 Web Based Interaction

In Fig. 5, a web-based viewer[2] supports the visualization of both 2D images and 3D models made available through IIIF APIs [10]. In particular, users are allowed to enjoy the 3D scan of the discovered bed feet, despite the original artifacts being preserved and displayed in the Norba Archaeological Museum. Users can seamlessly interact with the content, employing features such as zooming, rotation, and panning for 2D images, while also exploring three-dimensional models with the same fluidity [4]. This functionality not only enhances user engagement but also serves as a crucial component in the broader user interface.

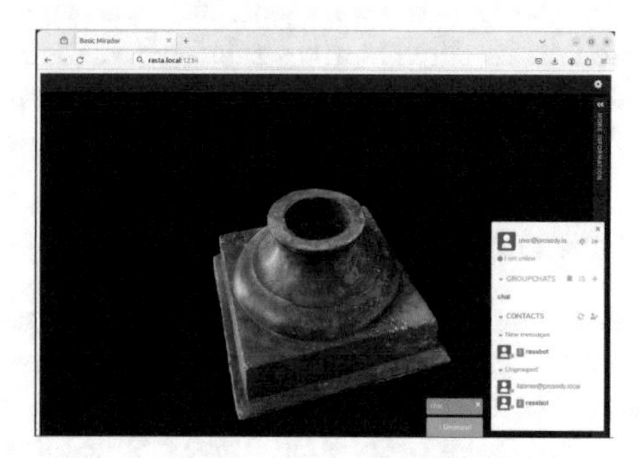

Fig. 5. 3D visualization of the bronze bed feet

5 Assisted Multimedia Interaction

Both the mobile application and the viewer are designed to accommodate virtual assistants, implemented by intelligent software agents. The system incorporates a chat feature, allowing users to log in with credentials and engage in a collaborative and interactive chat environment. To make agents aware of the context any modifications made by the user to an image or a 3D model, such as zoom and rotation, are automatically tracked. This ensures that user interactions remain persistently recorded, allowing for seamless retrieval and reference during the conversation and in future sessions. Moreover, virtual assistants can perform

[1] https://flutter.dev/.
[2] https://universalviewer.io/.

movements, zooming, and rotations on both images and 3D models within the viewer, for presentation purposes when they answer the user or they proactively recommend focusing on specific details.

In the presented use case, two conversational agents participate in the conversation, performing different roles. A virtual guide uses an NLU model which has been trained to identify well-defined intentions whose activation allows to guide the user along the itinerary and to provide answers to well-defined questions. The second agent reacts to all other requests as a fallback behavior for all the intentions that are not understood. To accomplish this task it exploits a pre-trained Large Language Model (LLM), however, this choice introduces a critical issue in the use of this kind of services which are characterized by none or limited explainability [8]. In fact, in the context of the delivery of cultural content, the correctness of the generated responses and the references to the source of information cannot be neglected. We addressed this problem by integrating two distinct approaches. In the case that the response comes from the first agent, the correctness of the information is guaranteed directly by the expert who has contributed to building the knowledge base. Nevertheless building the knowledge base is an expensive task and requires technical skills. Regarding the responses generated by the pre-trained agent, the challenge was to carry out a second training process on certified sources related to cultural sites. This allows us to perform a check when the response has been generated, comparing the result with the certified documents, to evaluate the accuracy of the answer.

5.1 Implementation Details

The multi-agent chatbot has been developed through the use of different technologies. First of all, every message is read by a first software agent. It performs an intent evaluation on the user's message. The intent evaluation is carried out using Spacy, a natural language processing library that allows for the comparison of the similarity between two sentences. Such conversational agents has been developed by the RASA framework[3] and integrates an original recommendation system to guide the user through the most convenient cultural itinerary, which has been modeled as a *String of Pearls*. Each Pearl contains more cultural contents related to one PoI along the visit itinerary. The second agent uses a Large Language Model, which has higher capabilities of natural language understanding. GPT 3.5 turbo implementation has been integrated, because of its efficiency and fast generation capabilities at a cheaper cost. To train GPT 3.5 on the topic of interest, the LangChain framework has been used. LangChain[4] connects a language model to sources of context (prompt instructions, a few shot examples, pdf, CSV, doc file, etc.). LangChain allows for creation of a chain between the original GPT 3.5 model and the certified text document, giving specialized knowledge to the language model. The GPT 3.5 agents gives evidence to the user about the relevance of his response respect to these certified sources. The

[3] https://rasa.com/.

[4] https://python.langchain.com/.

embeddings of each response generated by GPT 3.5 and the embeddings of all the documents inserted into Langchain for the second training are calculated. The cosine similarity of the response with each document is computed as a similarity metric. Given $\vec{a}$ the vector containing the embeddings of the GPT response, and $\vec{b}_i$ the vector containing the embeddings of each document b_i, the metric is calculated computing the dot product of the two vectors $\vec{a}$ and $\vec{b}_i$ and dividing by the product of the Euclidean norm of the two vectors.

$$\text{Cosine Similarity}_i = \frac{\vec{\mathbf{a}} \cdot \vec{\mathbf{b_i}}}{\|\vec{\mathbf{a}}\| \cdot \|\vec{\mathbf{b_i}}\|}$$

If the cosine similarity is greater than 0.8, a reference to the document is recommended to the user as a source of the provided information.

In Fig. 6 the system chat is shown. The conversation among many human users is supported, but the software agents are not able to handle this kind of social interaction yet. Both agents, the RASA bot and the GPT 3.5 bot, are shown in the multichat. After the response has been related to a certified source, the link to the source document is sent in a private chat to the user. The user can download the document and eventually focus the conversation on that topic with the GPT 3.5 agent in the private chat.

Fig. 6. GPT-agent's response

6 Conclusion

The augmented fruition of cultural heritage through innovative technologies and smart applications can enhance the user's experience. The collaborative interaction with multimedia content through virtual assistants can leverage the delivery of cultural contents and can increase the degree of interactivity. A challenging issue is the interoperability with a wide variety of devices, ensuring a good level of immersion and interactivity, concerning the technological characteristics and available computational and network resources. We demonstrated how the integrated utilization of heterogeneous solutions can be exploited in a real scenario.

Acknowledgments. The work has been supported by ARS01_00540 - RASTA project, funded by the Italian Ministry of Research PNR 2015–2020.

Thanks to Alessandro Ciancio for the 3D virtual model of the *domus* IV of *Norba*.

References

1. Amato, A., Branco, D., Venticinque, S., Renda, G., Mataluna, S.: Metadata and semantic annotation of digital heritage assets: a case study. In: 2023 IEEE International Conference on Cyber Security and Resilience (CSR), pp. 516–522 (2023)
2. Ambrisi, A., et al.: Intelligent agents for diffused cyber-physical museums. In: Camacho, D., Rosaci, D., Sarné, G.M.L., Versaci, M. (eds.) Intelligent Distributed Computing XIV, IDC 2021. Studies in Computational Intelligence, vol. 1026, pp. 285–295. Springer, Cham (2022). https://doi.org/10.1007/978-3-030-96627-0_26. Cited by: 1
3. Bekele, M.K., Pierdicca, R., Frontoni, E., Malinverni, E.S., Gain, J.: A survey of augmented, virtual, and mixed reality for cultural heritage. J. Comput. Cult. Herit. **11**(2), 1–36 (2018)
4. Branco, D., Aversa, R., Venticinque, S.: A tool for creation of virtual exhibits presented as IIIF collections by intelligent agents. In: Barolli, L. (ed.) Advanced Information Networking and Applications, pp. 241–250. Springer, Cham (2023). https://doi.org/10.1007/978-3-031-28694-0_22
5. Carfora, P., Ferrante, S., Quilici Gigli, S.: Pavimenti in cementizio dagli scavi di norba. In: Angelelli, C. (ed.) Atti del XVI Colloquio dell'AISCOM, pp. 397–407 (2011)
6. Chen, C.A., Lai, H.I.: Application of augmented reality in museums - factors influencing the learning motivation and effectiveness. Sci. Prog. **104**, 003685042110,590 (2021)
7. Fogagnolo, S.: Resti di decorazione pittorica a cubi prospettici dalla domus iv. Atlante Tematico di Topografia Antica **23**, 280–282 (2013)
8. Preece, A.: Asking 'why' in AI: explainability of intelligent systems-perspectives and challenges. Intell. Syst. Account. Finance Manag. **25**(2), 63–72 (2018)
9. Quilici Gigli, S.: Triclini e letti bronzei: testimonianze da norba sulla recezione culturale e materiale in ambito locale. Orizzonti **XVI**, 21–30 (2014)
10. Snydman, S., Sanderson, R., Cramer, T.: The international image interoperability framework (IIIF): a community & technology approach for web-based images. In: Archiving Conference, vol. 1, pp. 16–21. Society for Imaging Science and Technology (2015)
11. Wang, C., Zhu, Y.: A survey of museum applied research based on mobile augmented reality. Comput. Intell. Neurosci. **2022**, 1–22 (2022). https://doi.org/10.1155/2022/2926241

Research on Archive Digital Editing and Studying Mode

Jingyi Zeng(✉)

Nankai University Business School, 94 Weijin Road, Nankai District, Tianjin 300071, People's Republic of China
zengjingyi@nankai.edu.cn

Abstract. The digital era presents new challenges for academic development in archive editing and studying. A proposed archive digital editing and studying mode aims to achieve specific goals in the department, operating with theoretical guidance and maximizing the potential of technological empowerment and narrative intervention. The mode revolves around four business procedures and emphasizes the integration of technology and narrative to better support archive editing and studying work.

Keywords: Digital Editing and Studying · Digital Humanities · Archival Storytelling

1 Introduction

In recent years, the field of archive editing and studying has witnessed new phenomena, primarily characterized by the empowerment of archive editing and studying through the application of digital technologies and a shift in narrative focus towards the revitalization of archival content organization. This has introduced new topics for research, some of which require further reflection and exploration. From the perspective of academic development, archive editing and studying face new problems and challenges in the digital age. Traditional archive editing and studying theories struggle to provide effective guidance amid evolving practical work, necessitating adaptive theoretical development for adequate support.

The driving forces behind the development of archive editing and studying theory in the digital age can be attributed to two main aspects. Firstly, as archive editing and studying work is grounded in the archives of archival institutions, changes in the objects of archive editing and studying in the digital age (shifting from predominantly paper-based to predominantly digital) necessitate corresponding changes in the means and presentation of archive editing and studying outcomes. Secondly, recent innovations in archive editing and studying theory have been discussed, with the introduction of new concepts and technologies infusing vitality and dynamism into archive editing and studying. However, a systematic theoretical framework has yet to be established, requiring precise positioning within the context of the digital civilization to further research. The

L. Barolli (Ed.): AINA 2024, LNDECT 204, pp. 29–36, 2024.
https://doi.org/10.1007/978-3-031-57942-4_4

ecological landscape of archive editing and studying has undergone significant changes due to the influence of technological empowerment and narrative intervention, making the field more dynamic. However, challenges are evident, as the effectiveness of both technological and narrative applications in optimizing archive editing and studying remains limited. As digital technology continues to evolve, narrative orientation deepens, and archival practices persist, there is a need for a scientific exploration of the development of archive editing and studying theory in the digital age. The development of archive editing and research theory is eagerly anticipated.

This study is conducted against this backdrop to improve the level of archive editing and studying work. It addresses how to effectively integrate technology, narrative, editing, and studying under the dual forces of technological empowerment and narrative intervention. The goal is to explore the digital transformation and innovative development of archive editing and studying, construct an archive digital editing and studying mode, and enrich archive editing and studying theory and applied research for the reference and guidance of practical departments.

2 Literature Review

2.1 Technological Empowerment in Archive Editing and Studying

Digital technology empowerment is the most direct factor driving the digital development of archive editing and studying. As research on digital technology in the field of archive editing and studying continues to deepen, discussions related to the concept of 'archive digital editing and studying ' have emerged. Summarizing the digital development forms of archive editing and studying from various perspectives is a consequence of this exploration.

The first aspect involves discussions on the forms and development of archive editing and studying under technological influence. In the context of the information age, Xiaji H and Yan L (2016) propose that the new era of archive editing and studying is characterized by variations in increased collaboration among subjects, expansion, socialization of topic selection, diversification of material selection, digitization of methods, semi-automation of platforms, enrichment of outcome types, and diversification of dissemination [1]. In the context of the data age, Siqi D and Ying L (2020) propose archive editing and studying exhibit new characteristics such as the emergence of short series editing and studying outcomes, widespread recognition of electronic editing and studying methods, and increasing integration of editing and studying work with the public [2].

The second aspect involves discussions on the technical support system for archive editing and studying. Zhijie W (2019) suggests the active application of new technologies to construct an integrated archive editing and studying platform [3]. Xiaoke F (2020) introduces digital humanities to various levels of archive editing and studying. The third aspect involves discussions on the application of technological empowerment in archive editing and studying [4]. Jingyi Z (2021) focuses on photo archive editing and studying, proposing approaches that include utilization-oriented photo archive content analysis and storage, relation-based discovery and aggregation of photo archive resources, narrative-oriented clues acquisition and construction for photo archive editing and studying, and organization and presentation strategies for display-oriented photo archive editing and studying outcomes [5]. Jiming H (2022) proposes that the integration of new media and archive editing and studying has given rise to a new editing and studying mode called "micro-editing and studying"[6].

2.2 Narrative Intervention in Archive Editing and Studying

Narrative intervention primarily impacts the organization of content and the presentation of outcomes in archive editing and studying at the methodological level. Currently, there is only one direct exploration of the application of narrative in archive editing and studying. Jingy Z (2021) proposes that at the research level, it is essential to analyze the context, identify narrative and memory cues, and formulate a narrative mainline (framework) [5]. This narrative mainline serves as the basis for the storytelling approach in editing and studying photo archive research outcomes. To comprehensively and precisely understand the current directions and key points of narrative application in the archival domain, the scope of archive editing and studying has been broadened in this review. It is observed that narrative in archival development and utilization tends to manifest in two main research tendencies.

Firstly, there is an increasing emphasis on incorporating narrative techniques during the content development stage. Linxing Z and Yunping C (2021) propose the application of narrative in various avenues of archival cultural communication, such as archival theme exhibitions, archival cultural lectures, archival promotional videos, and archival content dissemination [7]. Battad Z and Si M (2016; 2019) introduce the use of storytelling to position and utilize archival data [8][9]. They developed a system capable of automatically generating narratives from a thematic relational information network. Their research involves the exploration of technologies that intertwine multiple narrative storylines, using this system to achieve new knowledge discovery based on knowledge graph technology.

Secondly, during the outcome presentation stage, the focus shifts to achieving goals through visual storytelling. Yuxue X (2019) applies visual storytelling methods to digital archival resource services, proposing a model for visual storytelling services for digital archival resources [10]. Davidson A. and Reid P. (2018) suggest constructing an image website that can house mobile image archives related to the Scottish town of Fraserburgh [11].

In summary, both the research on technological empowerment in archive editing and studying and the research on narrative intervention in archive editing and studying has made some achievements, providing a certain foundation for this study. However, there are still issues that need further exploration. The impacts of technological empowerment and narrative intervention on archive editing and studying lack a scientific generalization, and the pathways of their roles in archive editing and studying are not yet clear. Additionally, the overall logic, theory, and methods of archive digital editing and studying are still awaiting systematic development. These gaps provide opportunities for further research in this paper. To address these challenges, this paper proposes a archive digital editing and studying mode.

3 Research Design

This paper adopts theoretical construction and case study methods.

The application of theoretical construction has two main aspects in this paper: firstly, it constructs the theoretical framework of the archive digital editing and studying mode from the perspective of the dual forces of technological empowerment and narrative intervention. This involves building the theoretical foundation, core, elements, and model of the archive digital editing and studying mode. Secondly, the paper utilizes theoretical construction to develop the business logic, technical architecture, and specific content of application exploration for the archive digital editing and studying mode.

The case study method is employed in two instances: first, in the section where the mode is proposed, an analysis of the explorations in existing digital editing and studying concepts is conducted; second, in the section on the application exploration of the mode, multiple typical cases are analyzed and summarized, facilitating closed-loop research from theory to practice.

The main goal of the construction of the mode of archive digital editing and studying is to face the digital transformation of archives work, effectively exert the dual forces of technology and narrative, realizing archive digital editing and studying, to give full play to the supporting role of archive editing and studying work, which is focused on the three main points, including the transformation, promotion, and service. Two basic principles should be adhered to in the construction of archive digital editing and studying mode. First, the archive digital editing and studying mode is not a castle in the air from the traditional editing and studying mode but an extension based on integrity. Second, the archive digital editing and studying mode is not only the supplement and extension of traditional archive editing and studying in digital space but also forms a new inside logic through the integration of development. Under the goal-oriented and logical approach, this paper adopts the idea of "from theory to practice, from method to application" to construct the archive digital editing and studying mode.

4 Results and Findings

4.1 Theoretical Construction

This paper begins by discussing the theoretical construction of the archive digital editing and studying mode in three aspects: theoretical base, theoretical core, and theoretical elements. It proposes a theoretical model for the archive digital editing and studying mode (see Fig. 1). The theoretical base introduces the concept of ternary space, emphasizing the expansion of digital space for archive editing and studying. The theoretical core focuses on technology empowerment and narrative intervention, using the theory of five states and four modernizations to examine the archive digital editing and studying mode. The theoretical elements redefine the work content of archive editing and studying, providing a foundation for the shift from traditional compilation to digital narrative.

The paper constructs a theoretical model at three levels: "business-technology- application." It includes the business logic construction based on human-computer interaction, the reconstruction of archive editing and studying infrastructure for digital narrative, and the re-engineering of digital application scenarios driven by value. These parts form a comprehensive theoretical framework for the archive digital editing and studying mode, clarifying its practical application orientation.

Secondly, the paper delves into the detailed discussion of "business logic," "technical framework," and "application exploration" of archive digital editing and studying mode. It aims to clarify the "what" and "how" of archive digital editing and studying mode from the perspectives of epistemology and methodology.

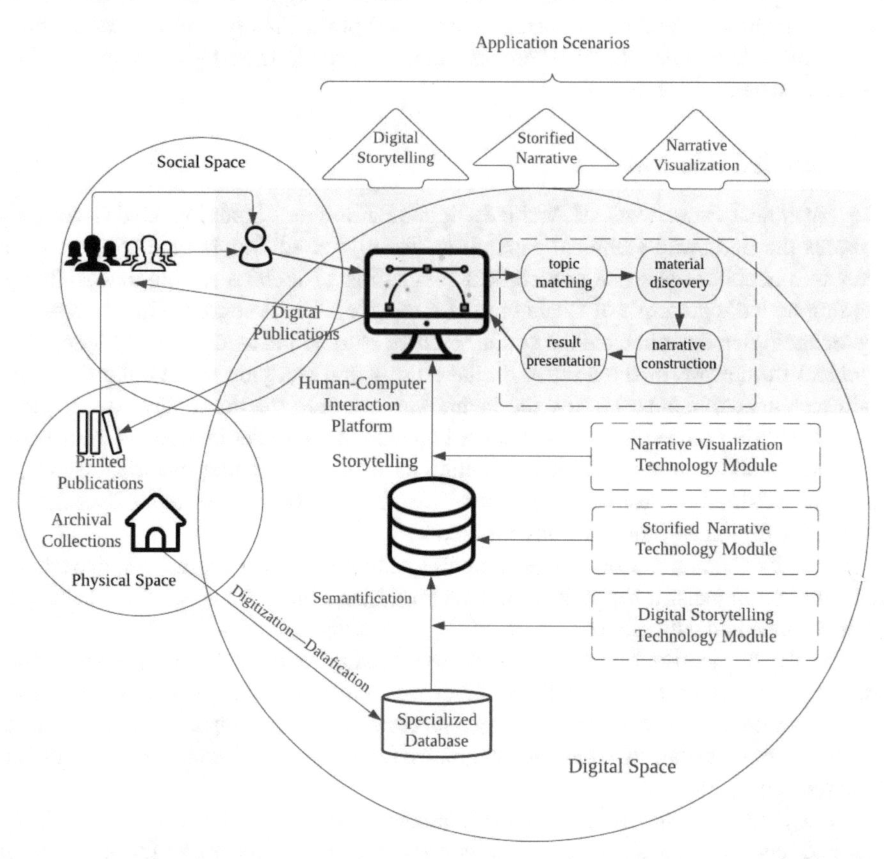

Fig. 1. This model is based on the interaction of ternary space, with a focus on the functional expansion of archive editing and studying in the digital space. The functional expansion of archive digital editing and studying includes three parts: business logic, technical framework, and application exploration.

4.2 Business Logic

The "business logic of archive digital editing and studying mode" is grounded in the fundamental steps of traditional archive editing and studying, namely "topic selection-material selection-processing and editing-publishing and distribution." It emphasizes four key aspects: "from demand to business, from carrier to content, from discrete to sequence, from selection to combination."

Following the chronological order of business development, the business procedure of the archive digital editing and studying mode is delineated as "topic matching-material discovery-narrative construction-result presentation." In this framework, "topic matching" pertains to digital storytelling, with a focus on collecting resource context, combing through it, and analyzing topic matching as central tasks. "Material discovery" and "narrative construction" involve storified narratives, with work centers dedicated to the analysis of story elements, context application, organization of plot units, and the analysis and generation of narrative clues and plots. Finally, "result presentation" involves narrative visualization, where discourse construction and digital communication serve as the central components.

4.3 Technical Framework

The "Technical Framework of Archive Digital Editing and Studying Mode" not only explores the necessary technical capabilities for digital editing and studying but also takes into account the existing technical foundation of archive editing and studying, aligning with the principle of "technology for the good." This approach involves selecting technologies that enhance the overall archive editing and studying experience. The technical framework of the archive digital editing and studying mode comprises four main technical modules. Firstly, the technology module for the archive digital editing and studying mode platform is centered on human-computer interaction. It involves functional design aligned with the key operations of "topic matching-material discovery-narrative construction-results presentation," forming the foundation for the operation of the archive digital editing and studying mode.

Secondly, the digital storytelling technology module aims to upgrade archive editing and studying objects. It supports topic matching by constructing specialized databases for archives and databases for archive editing and studying topics.

Thirdly, the storified narrative technology module is directed towards enhancing archive editing and studying methods. It supports material discovery and narrative construction through the creation of "resource-element" dual-layer topic maps and context-embedded archival narrative knowledge graphs, focusing on both static associations and dynamic aggregations.

Lastly, the narrative visualization technology module aims to improve archive editing and studying services. It supports results presentation by developing frameworks for digital products and financial media platforms.

4.4 Application Exploration

The "Application Exploration of Archive Digital Editing and Studying Mode" delves into the specific application scenarios of archive digital editing and studying mode with various types of archival data. The cases discussed primarily stem from practical projects in which the author actively participated. Notably:

Digital Storytelling of Suzhou Silk Archives
Illustrates the construction concept of the Suzhou Silk Archives Special Database and Digital Storytelling Framework.

Emphasizes the significance of archive digital editing and studying mode in advancing the transformation of heritage value.

Storified Narrative of Wu Baokang's Character Archives
Demonstrates the construction of storified narratives, emphasizing plot-driven storytelling and context application.

Highlights the value of archive digital editing and studying mode in preserving the cultural legacy of figures like Wu Baokang.

Narrative Visualization of Anhui University's Historical Photo Archives
Presents an innovative approach to editing and studying photo archives with a focus on narrative visualization.

Showcases the value of archive digital editing and studying mode in supporting the construction of cultural memory within institutions like Anhui University.

These projects represent distinct application scenarios for archive digital editing and studying mode, illustrating its versatility and adaptability across different contexts.

5 Conclusions

This paper introduces three main innovations: a unique research perspective examining archive digital editing and studying from both macroscopic and microscopic viewpoints, an innovative academic viewpoint emphasizing the importance of expanding the digital space, and advancements in technology and methods. It constructs a comprehensive toolkit applicable to various scenarios. However, the paper falls short in practical research and mode validation due to pandemic limitations, and further exploration is needed for broader application scenarios.

References

1. Xiaji, H., Yan, L.: The 'constancy' and 'change' of archive editing and studying in the information age. Arch. Sci. Bull. **4**, 39–44 (2016)
2. Siqi, D., Ying, L.: Research on development strategy of archives editing and studying in the data age. Shanxi Arch. **1**, 112–117 (2020)
3. Zhijie, W.: Research on archive editing and studying work in new technology environment. Beijing Arch. **6**, 25–27 (2019)
4. Xiaoke, F., Yongxian, X., Qiaoling, W.: A new approach to archive editing and studying based on digital humanities. Arch. Sci. Study **5**, 138–142 (2020)
5. Jingyi, Z., Wei, G., Lichao, L., Ying, Z.: Analysis of innovative ideas of editing and studying of photo archives. Arch. Sci. Bull. **4**, 71–76 (2021)
6. Jiming, H., Xing, L., Ye, C.: Research on content structure of archive micro editing and studying in the new media environment. Beijing Arch. **6**, 16–20 (2022)
7. Linxing, Z., Yunping, C.: From the perspective of narration, archival cultural communication: value, mechanisms, and path selection. Arch. Manage. **1**, 36–38 (2021)

8. Battad, Z., White, A., Si, M.: Facilitating information exploration of archival library materials through multi-modal storytelling. In: Cardona-Rivera, R.E., Sullivan, A., Young, R.M. (eds.) ICIDS 2019. LNCS, vol. 11869, pp. 120–127. Springer, Cham (2019). https://doi.org/10.1007/978-3-030-33894-7_13
9. Si, M.: Facilitate knowledge exploration with storytelling. Procedia Comput. Sci. **88**, 224–231 (2016)
10. Yuxue, X.: Research on the visual narrative service of digital archive resources. Arch. Sci. Study **3**, 122–128 (2020)
11. Davidson, A., Peter, H.: Digital storytelling and participatory local heritage through the creation of an online moving image archive: a case-study of Fraserburgh on Film. J. Documentation **78**(2), 389–415 (2022)

An Ontology for the Annotation of the Archives of the Royal Site of San Leucio

Alba Amato(✉) and Angelo Di Falco

Department of Political Science, University of Campania "Luigi Vanvitelli", Caserta, Italy
{alba.amato,angelo.difalco}@unicampania.it

Abstract. Semantic Digital Libraries and Ontologies gained attention across diverse research communities including Cultural Heritage and History. The development of the ontology for these specific domains is meant to support the implementation of intelligent applications such as decision support systems, recommender systems and semantic search. In this paper we present San Leucio ontology, a domain ontology developed as part of RASTA project for the implementation of an automatic storytelling process for Bourbon Royal Sites.

1 Introduction

Semantic Digital Libraries and Ontologies gained attention across diverse research communities including Cultural Heritage and History [8]. The development of the ontology for these specific domains [5] is meant to support the implementation of intelligent applications such as decision support systems, recommender systems and semantic search [6]. Historical documents have undergone digitization and publication on the web through various applications so, in order to enhance search and retrieval operations, portals and digital libraries dedicated to Cultural Heritage have recently embraced Semantic Web technologies [9]. Moreover, it is well-established that this domain is very complex: data are often heterogeneous, semantically rich, and highly interlinked. For this reason, searching and linking them with related contents represents a challenging task [11].

In this paper we present San Leucio ontology, a domain ontology developed as part of RASTA project[1] for the implementation of an automatic storytelling process for Bourbon Royal Sites. In particular the automatic storytelling requires a series of key activities and it is fundamental to define the domain ontologies, which act as a tool to semantically annotate the archival content and define a palette of items to be selected for the composition of stories through the use of an ad hoc storycomposer. This phase will constitute the foundation for understanding and organizing the material available. As second step of the project, the

[1] http://rasta.unicampania.it/.

L. Barolli (Ed.): AINA 2024, LNDECT 204, pp. 37–44, 2024.
https://doi.org/10.1007/978-3-031-57942-4_5

existing digital content relating to the royal Bourbon sites is metadated [4] and the digital material will be subject to semantic annotations using the previously defined domain ontologies, thus feeding the archiving system with relevant metadata and relevant semantic information. Subsequently, we will proceed with the definition of the main threads for the stories, identifying key themes expressed in terms of semantics [3]. These themes will include rituals and the historical-geographical transformation of royal sites. At the same time, a targeted search will be conducted for digital content linked to royal Bourbon sites on the web, including "open" data from sources such as Europeana, Wikipedia, Twitter and other social networks. This data will be subjected to a process of metadating and semantic annotation, thus expanding the base of materials available for the creation of stories. Finally, the storytelling system for Bourbon sites will be activated and made available together with a manual dedicated to the creation and management of stories and related contents. This will allow widespread use of the system, helping to bring the narrative of the Bourbon Royal Sites to an innovative and engaging level.

The paper is organized as follows: in Sect. 2 ontologies proposed in literature for historical domain are shown, in Sect. 3 the Royal Site of San Leucio is described. Section 4 describes the ontology creation and in Sect. 5 Conclusions are drawn.

2 Related Works

Several ontologies have been proposed in literature for the historical and archiving domain. In [12] is presented a method to integrate the fine-grained knowledge of cultural heritage into an ontology, especially when it is necessary to build a fine-grained ontology for particular purposes, such as knowledge-based terminological resources.

In [2] several extensible ontology for concepts and information in cultural heritage and museum documentation are shown. Those are formal ontologies created with the aim of enabling and facilitating the integration, mediation and exchange of heterogeneous information relating to cultural heritage. More precisely, the main objective of those ontological models is to provide a set of semantic definitions useful for transforming heterogeneous data derived from local sources into a coherent set of global information, therefore usable by everyone.

In [10] author propose HIMIKO (Historical Micro Knowledge and Ontology) as a package containing a model and editing system for collecting pieces of historical knowledge scattered in the documents. The model enables help historians to construct document-based Historical Linked Data and brings a new perspective on the digital analysis of historical data.

In [1] authors presented the details of the process of developing a domain ontology for Nigeria history, which is the first of its kind in terms of focus (Nigeria), and perspectives (political, cultural). The NHO is a documentation of Nigerian historical knowledge for meaningful use, which makes the knowledge useful by both humans and software agents. It also provides an electronic archive of Nigerian history in a compact and easily accessible way.

In [7] authors present first results in creating an ontology of historical Finnish occupations, AMMO, that enables selection of groups of people based on their occupation, occupational groups, or socioeconomic class. For interoperability, AMMO is linked to the HISCO international historical occupation classification and to a late 20th century Finnish occupational classification. AMMO will be used as a component in two semantic portals for Finnish war history.

3 Domain Description

The Royal Site of San Leucio is a monumental complex in Caserta, commissioned by Charles of Bourbon, king of Naples and Sicily which is considered, together with the Royal Palace of Caserta and the Vanvitelli Aqueduct, an UNESCO World Heritage Site. The historical archives of San Leucio are very important to preserve the cultural heritage of a society providing insights into the past, including the customs, traditions, and values of previous generations. The documentation mainly includes correspondence between the site administrators and various personalities, volumes of accounts, ledgers, monthly budgets of the hunting branch, and also documents relating to the Military Division and the Majorate of the Count of Caserta. Archiving allows the documentation to be made accessible through specific description tools that represent the complex documentary object of study. One important instrument is the annotation of the archives that are directed to represent semantic information, instead of content. Ontologies offer the means of explicit representation of the meaning of different terms or concepts, together with their relationships. The proposed ontology, also aims at representing the concepts and relationship related to the royal site of San Leucio, in order to annotate the archives, so improving the knowledge of this important Royal Site.

4 Ontology Creation

From an operational point of view, the ontology in question was entirely built using the open-source Protégé software[2]. Protégé is a free open-source editor that allows to simplify the construction and creation of Semantic Web ontologies through the use of specific interfaces and graphic windows for the definition of ontological elements. Protégé integrates within it a semantic reasoner which can be launched in order to verify the consistency of the built ontology and derive any implicit information based on the concepts expressed in the model. The software offers different types of reasoners, among these we chose to use HermiT to validate the constructed ontological models. In general, Protégé allows to define concepts

[2] https://protege.stanford.edu.

and relationships between concepts in the form of classes and properties respectively, as defined by the principles underlying Semantic Web technologies. The construction of an ontology by Protégé occurs through the production of files in OWL format containing the concepts modeled in textual form.

Our project started with the domain analysis, focusing on a comprehensive examination and revision of historical documents. We systematically reviewed prior research pertaining to ontologies and historical documents, specifically seeking insights from studies that employed methodologies similar to ours. Additionally, we conducted an exploration to identify any existing ontologies that could serve as guidelines for developing our initial ontology. The first step was the search for already existing ontologies for reuse. However, we did not find ontologies of the royal Bourbon sites but only generic ontologies containing terms that are also common to our ontologies and which we will reuse by connecting them to ours. The analysis phase played a crucial role in structuring the identified terms into a conceptualization format for the event ontology. To validate and refine our understanding, experts convened to discuss and confirm the accuracy of the identified issues. In this phase we begin to design the overall conceptual structure of the domain, identifying the main concepts of the domain and their properties, looking for the relationships between the various concepts, possibly creating abstract ones, specifying which ones of these they have instances etc.

These discussions were facilitated by subject matter experts in both history and ontology, who contributed with their expertise to substantiate and support the tasks. The collaborative effort ensured a robust foundation for the development of our ontology, drawing on both existing knowledge and expert insights.

The insights gained from the analysis task played a crucial role in structuring the identified terms into a conceptualization format tailored for the event ontology. This involved the thoughtful formulation of enumerated terms, shaping them into a coherent and meaningful framework. Furthermore, we drew upon the knowledge gleaned from existing ontologies, utilizing them as guidelines to inform the development of our own ontology.

In essence, the analysis task served as a foundational step, facilitating the systematic organization and conceptualization of terms for the event ontology in alignment with the objectives of our work.

In Fig. 1 the basic classes that were matched to our domain are shown. In particular are people, political office, factor, spatial thing (location) and time and date. Then, we expanded the ontology by adding some classes like Noble Title and Global Domain that are used to know people involved in the process. Subsequently we add concepts, relations and entities, until we reach the level of detail necessary to satisfy the objectives of the ontology.

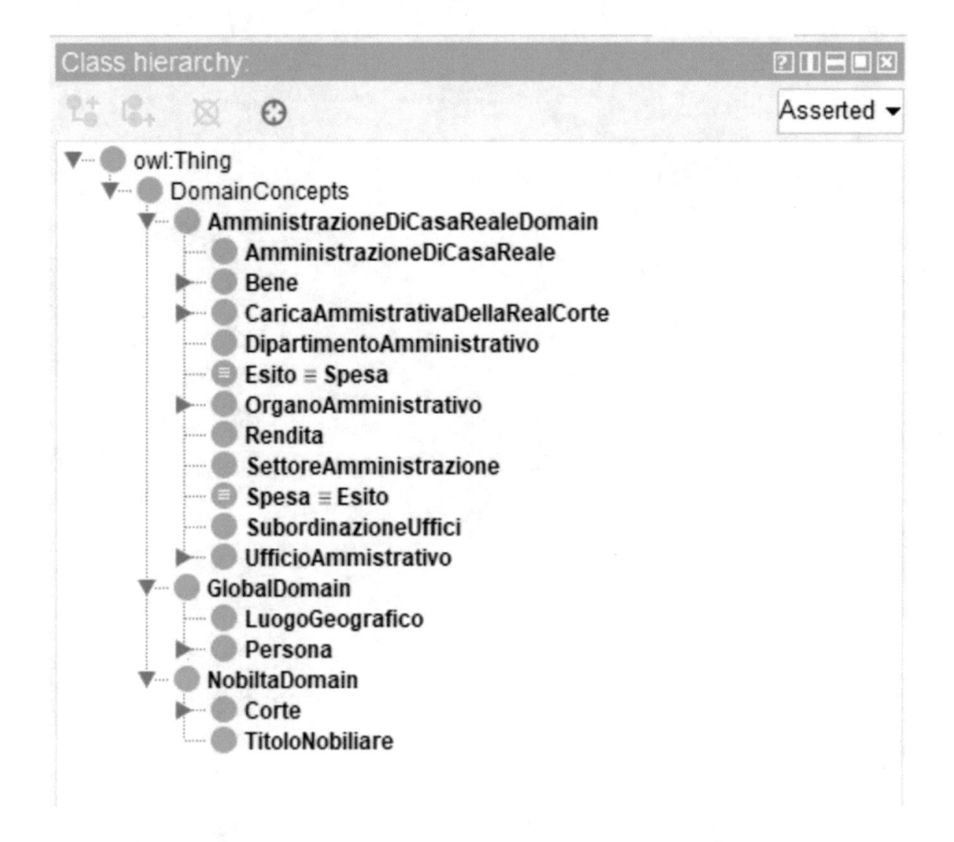

Fig. 1. Classes

In Fig. 2 and in Fig. 3 the terms from the previous step were listed in the form of hierarchical taxonomy. The approach used was the combined one which involves identifying the concepts that are salient and then generalize and specialize. Concepts alone do not provide enough information, so this is important also define the relationships between the objects of the domain.

Fig. 2. Hierarchies

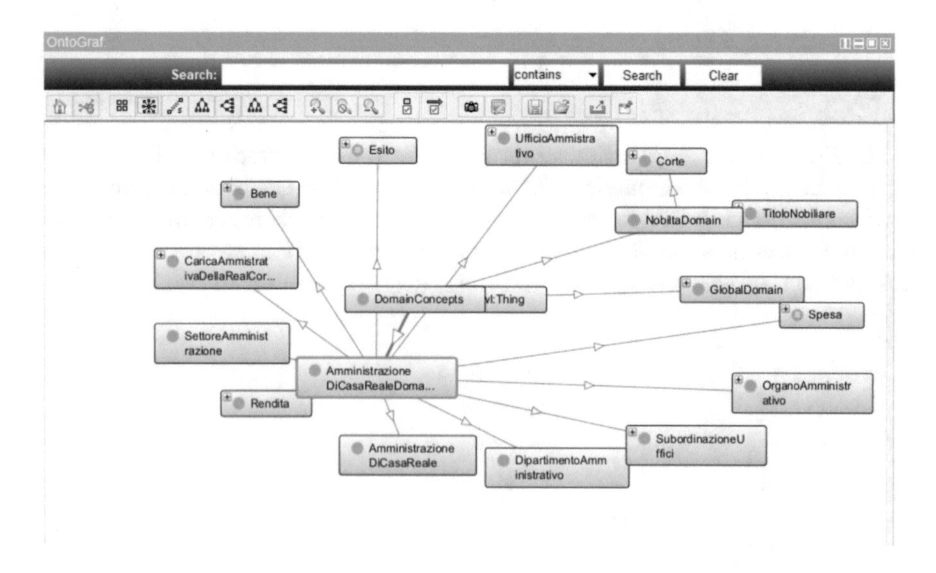

Fig. 3. Part of Domain Ontology Taxonomy

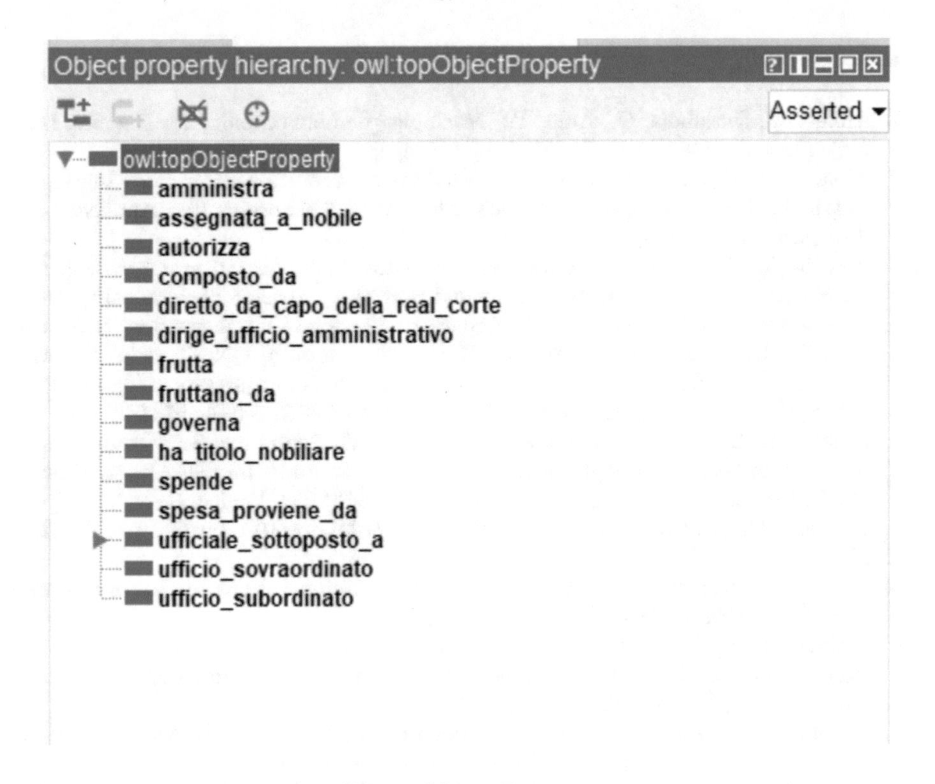

Fig. 4. Object Property

Once the ontology has been developed, we analyzed it to discover any syntactic, logical and semantic inconsistencies between its elements. Often these controls are useful for automatic classification, which lead to the discovery of new concepts based on properties, entities and relationships between classes, as in Fig. 4.

5 Conclusion

Building domain ontologies is a difficult task, especially when the domain experts have a little background on knowledge engineering techniques and lack the skills of domain conceptualization [6]. In this paper we present San Leucio ontology, a domain ontology developed as part of RASTA project for the implementation of an automatic storytelling process for Bourbon Royal Sites. Our future works include using the developed ontology for supporting semantic document retrieval of historical documents.

Acknowledgements. The work described in this paper has been supported by the research project RASTA: Realtà Aumentata e Story-Telling Automatizzato per la valorizzazione di Beni Culturali ed Itinerari; Italian MUR PON Proj. ARS01 00540.

References

1. Afolabi, I., Daramola, O., Adio, T.: Developing domain ontology for Nigerian history. Aust. J. Basic Appl. Sci. **8**(6), 30 (2014)
2. Alba, A., Giuseppe, C.: Comparative analysis of ontologies for archival representation. In: Barolli, L. (ed.) Advanced Information Networking and Applications. Springer, Cham (2024)
3. Amato, A.: Towards the interoperability of metadata for cultural heritage. In: Barolli, L. (ed.) Complex, Intelligent and Software Intensive Systems - Proceedings of the 17th International Conference on Complex, Intelligent and Software Intensive Systems (CISIS-2023), 5–7 July 2023, Toronto, ON, Canada. Lecture Notes on Data Engineering and Communications Technologies, vol. 176, pp. 309–317. Springer, Cham (2023). https://doi.org/10.1007/978-3-031-35734-3_31
4. Amato, A., Branco, D., Venticinque, S., Renda, G., Mataluna, S.: Metadata and semantic annotation of digital heritage assets: a case study. In: IEEE International Conference on Cyber Security and Resilience, CSR 2023, Venice, Italy, 31 July–2 August 2023, pp. 516–522. IEEE (2023). https://doi.org/10.1109/CSR57506.2023.10224935
5. Corda, I.: Ontology-based representation and reasoning about the history of science. Ph.D. thesis (2007)
6. Fatihah, R., Noah, M., Azman, S.: Building an event ontology for historical domain to support semantic document retrieval. Int. J. Adv. Sci. Eng. Inf. Technol. **6**, 1154 (2016). https://doi.org/10.18517/ijaseit.6.6.1634
7. Gasbarra, L., Koho, M., Jokipii, I., Rantala, H., Hyvönen, E.: An ontology of Finnish historical occupations. In: Hitzler, P., et al. (eds.) ESWC 2019. LNCS, vol. 11762, pp. 64–68. Springer, Cham (2019). https://doi.org/10.1007/978-3-030-32327-1_13
8. Hacking, I.: Historical Ontology, pp. 583–600. Springer, Cham (2002). https://doi.org/10.1007/978-94-017-0475-5
9. Hyvönen, E., Lindquist, T., Törnroos, J., Mäkelä, E.: History on the semantic web as linked data: an event gazetteer and timeline for the world war i. In: Proceedings of CIDOC 2012, vol. 2012. International Committee for Documentation, CIDOC, International (2012). Proceeding volume: 2012; CIDOC 2012; Conference date: 01-01-1800
10. Ogawa, J., Ohmukai, I., Nakamura, S., Kitamoto, A.: Collecting pieces of historical knowledge from documents: introduction of HIMIKO (historical micro knowledge and ontology). In: Baillot, A., Tasovac, T., Scholger, W., Vogeler, G. (eds.) Annual International Conference of the Alliance of Digital Humanities Organizations, DH 2022, Graz, Austria, 10–14 July 2023, Conference Abstracts (2023). https://doi.org/10.5281/ZENODO.8107411
11. Pandolfo, L.: STOLE: a reference ontology for historical research documents. In: Bellodi, E., Bonfietti, A. (eds.) Proceedings of the Doctoral Consortium (DC) Co-located with the 14th Conference of the Italian Association for Artificial Intelligence (AI*IA 2015), Ferrara, Italy, 23–24 September 2015. CEUR Workshop Proceedings, vol. 1485, pp. 13–18. CEUR-WS.org (2015). https://ceur-ws.org/Vol-1485/paper3.pdf
12. Wei, T., Chen, Y.: A methodology for building domain ontology of cultural heritage. Digit. Sch. Humanit. **38**(4), 1710–1719 (2023). https://doi.org/10.1093/llc/fqad045

An Architecture for Metadata and Semantic Annotation of Historical Archives

Alba Amato[1(✉)], Dario Branco[2], and Salvatore Venticinque[2]

[1] Università della Campania "Luigi Vanvitelli", Caserta, Italy
alba.amato@unicampania.it

[2] Università della Campania "Luigi Vanvitelli", Aversa, Italy
{dario.branco,salvatore.venticinque}@unicampania.it

Abstract. Historical archives constitute a treasure of humanity in which we can find information from the past, to know where we come from and what events characterized previous populations. The concept of a historical archive, therefore, can be depicted through an enormous quantity of documents and testimonies of inestimable value preserved over the years in safe places more or less accessible to the general public. Over the years and thanks to digital transformation, historical and cultural heritage archives have also had the need to digitize their documentary treasure within more performing digital archives able to manage metadata and annotation. In this paper, we present an architecture for metadata and Semantic Annotation of archives and an open-source software solution.

1 Introduction

Historical archives constitute a treasure of humanity in which we can find information from the past, to know where we come from and what events characterized previous populations.

In the context of digital archives, metadata is the information that must be provided in the electronic document in order to be able to correctly relate it to a set of information so managing and preserving it over time.

Metadata constitute a fundamental part of a digital archive, an environment in which logical locations have more relevance than physical ones and the link between documents and the information on them is consequently inseparable.

Research in metadata has proposed flexible and richer schemata (such as Dublin Core), but only the advent of semantic annotation has brought about a change of paradigm in the description of resources. However, in order to be effective, semantic annotation has to rely on a rich semantic model, enabling metadata to provide a rich description of the resource content. In fact, although the relative simplicity of currently available semantic models benefits processing, interoperability, and sharing, it often prevents metadata from being actually useful [8].

L. Barolli (Ed.): AINA 2024, LNDECT 204, pp. 45–54, 2024.
https://doi.org/10.1007/978-3-031-57942-4_6

In this paper, we present an the integrated utilization of standard and supporting technologies for metadata and semantic annotation of digital archives implemented by open-source software.

2 Related Works

Interpreting metaphors and symbols in historical archives is still too difficult for technology [3]. The problem is not simply to accurately identify objects in works of art, but to tell stories and transfer meanings of an image within the culture and historical context in which the image was created [12]. Many efforts have been made in the Cultural Heritage field to build digital repositories based on standards for technological interoperability and for enriching the semantic content and the interconnection among the archived objects and with other repositories [1].

As stated in [11], the implementation of a common platform poses two challenges: 1) the partners use more than one preservation infrastructure, which requires a consensus for harvesting mechanisms and update workflows; and 2) the nature of the digital objects ranges from manuscripts or other text-based resources to artifacts from various contexts. With respect to information retrieval and in order to implement a uniform object representation, a consistent data basis is obligatory.

A record or a collection in an archive is usually defined according to an *Application Profile*. Application Profiles consist of data elements drawn from one or more namespace schemas combined and optimized for a particular local application. The Encoded Archival Description (EAD[1]) is an XML standard for encoding archival finding aids, maintained by the Technical Subcommittee for Encoded Archival Standards of the Society of American Archivists, in partnership with the Library of Congress. Many EAD elements have been or can be, mapped to content standards such as DACS and ISAD(G) [9] and other structural standards (such as MARC or Dublin Core), increasing the flexibility and interoperability of the data.

Several *Application Profiles*, such as the Europeana Data Model (EDM) and PICO, a standard defined by the Italian Ministry for Cultural Heritage, include elements that allow for the interoperable delivery and presentation of archived content over the web. One for all is the IIIF standard (International Image Interoperability Standard) [10], which is a set of open standards for delivering high-quality, attributed digital objects online at scale. It's also an international community developing and implementing IIIF APIs. The IIIF standard supports the semantic wrapping of delivered images and compliant technologies implemented by *Annotation Servers* which allow defining annotations, storing them, listing, searching, and exporting.

The research project RASTA: Realtà Aumentata e Story-Telling Automatizzato per la valorizzazione di Beni Culturali ed Itinerari is an italian project

[1] https://www.loc.gov/ead/.

aiming at investigate methodologies and techniques for semantic annotation and archiving of digital content, in line with national and European standards and methodologies for automated storytelling, using semantic correlation and logical inference techniques.

The availability of standard APIs for linked information allows for the design and implementation of personalized recommendation systems, which are a relevant service for digital archives that aims at promoting the valorization of their content to visitors [5]. In [1], a prototype development of such a mechanism that is based on IIIF standards and IIIF-compliant open-source technologies is described.

3 Metadata Representation and Semantic Annotation

Focusing on historical documents and archives allows for specialization of the requirement analysis landing on two complementary standards which are the Encoded Archival Description for the archival description and the IIIF standard for the presentation of digital contents.

In Fig. 1 a digital document from the collection of administration of the Royal site of San Leucio, in the province of Caserta, in Italy, has been published by the AtoM[2] (Access to Memory) software. Atom is primarily intended for describing and accessing archives, it acts as a public web-based catalog so that users can search and browse your archives. It uses templates to map Dublin Core and another standard to the EAD application profiles. AtoM represents the access point for the editor that needs to store and describe new items, eventually

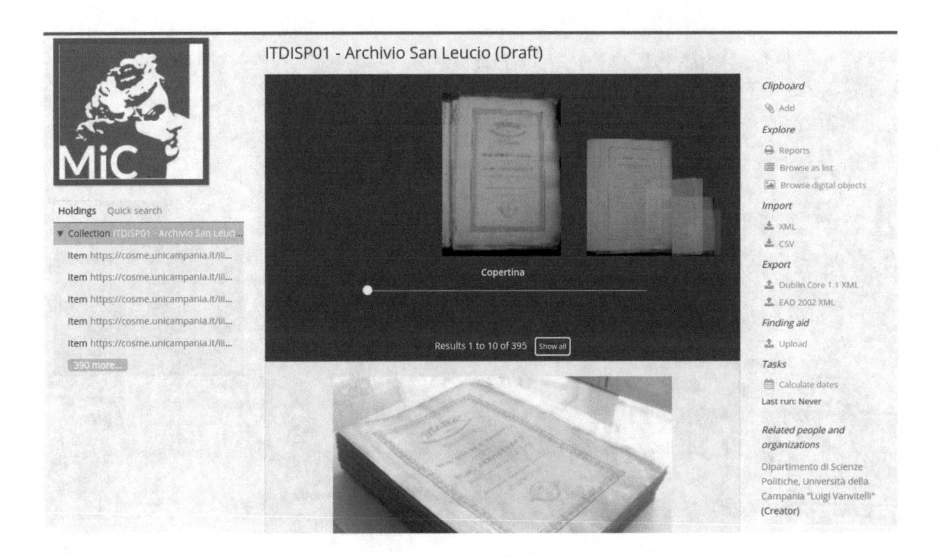

Fig. 1. Atom view

---
[2] https://www.accesstomemory.or.

organized into collections and it is the technology that allows for describing metadata according to a chosen standard.

In the case of a digitized manuscript, the metadata describes the full manuscript and can be slightly specialized for some pages, which usually inherit the same metadata of the whole manuscript. Moreover, such metadata have not been defined to provide information about the content of the manuscript or to describe the semantics of each page. The metadata of the manuscript is shown in Fig. 1 follow the ISAD(G) (General International Standard Archival Description), which is encoded by the EAD XML schema. ISAD(G) is used to create finding aids for archival collections, and includes elements such as title, date of record creation, the structure of the item, conditions of access and use. For this reason, a complementary representation model and a supporting tool are needed to describe the semantics of the digital content.

The IIIF standard is widely used nowadays for the interoperable delivery of high-resolution images through the web [6]. The presentation API allows to description of a digitized manuscript as a sequence of canvas, each one containing the digital version of a page. The manifest is the overall description of the structure and properties of the digital representation of an object in JSON format. It carries the information needed for an IIIF-compliant viewer to present the digitized content to the user, such as a title and other descriptive information about the object. Each manifest describes how to present a single object such as a book, a photograph, or a statue. In Fig. 2 the same manuscript archived in ATOM is shown through the Mirador IIIF viewer[3]. Many providers or aggregators, use to include in their metadata a link to the IIIF manifest of the archived record and integrate an IIIF viewer that retrieves the manifest, and the linked content,

Fig. 2. Mirador view

[3] https://projectmirador.org/.

directly from the data source. On the left side of Fig. 2 the viewer visualizes the metadata included in the manifest. It means that for presentation purpose the IIIF model can include the same metadata we set in ATOM. However a direct mapping between an EAD record and a IIIF record is not possible. In fact, the EAD supports multiple nested levels of description, which can be a fond, sub-fond, collections, and files until the last level of the hierarchy that corresponds to the single items. An IIIF manifest can be used to represent the last level of description of and EAD record. Nevertheless, the IIIF presentation APIs support semantic wrapping through an annotation element. As it is shown in Listing 1, a manifest eventually contains a list of annotation pages. Each *Annotation-Page* can include one or more items, each one defining an area of a canvas and associated to it a tag or a text. In particular, line 10 of Listing 1 defines that the current annotation is used to tag a part of the image, line 13 specifies the value of the tag and in line 17, the tag *target* refers on which canvas and the parameters of the rectangular part that has been annotated.

Listing 1. Example of IIIF Annotation

```
1  {
2   "annotations": [
3   {
4    "id": "https://cosme.unicampania.it/sanleucio",
5    "type": "AnnotationPage",
6    "items": [
7    {
8     "id": "https://cosme.unicampania.it/rasta/sanleucio/index.
         json/canvas/9/annotation/0-tag",
9     "type": "Annotation",
10    "motivation": "tagging",
11    "body": {
12      "type": "TextualBody",
13      "value": "tree",
14      "language": "it",
15      "format": "text/plain"
16          },
17    "target": "https://cosme.unicampania.it/rasta/sanleucio/
         index.json/canvas/9#xywh=3534,2437,67,56"
18       }]}
19      ]
20 }
```

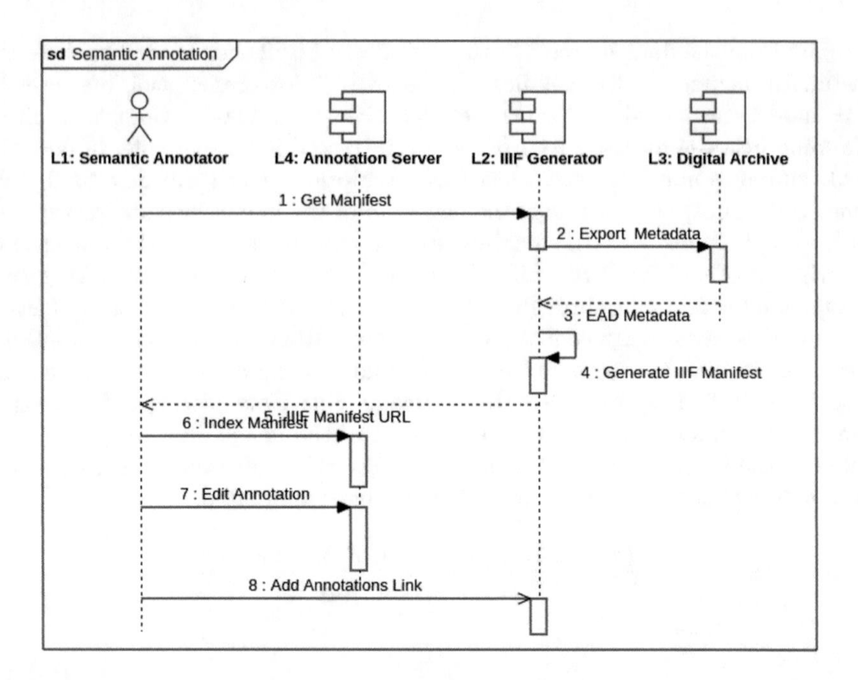

Fig. 3. Semantic Annotation

In Fig. 3 the integrated usage of the EAD data model instantiated into the digital archive and of the IIIF presentation APIs for semantic annotation is described through an UML sequence diagram. The *Semantic Annotator* selects from the archive the manuscript to be annotated and uses an application services to generate from the EAD metadata an IIIF manifest with the archival metadata.

The *Annotation Server* is used to index the manifest and to annotate the contained images. The *Annotation Server* is used as an annotation store embedding into the manifest a link to the annotation list, and eventually to the endpoint of the search service. The archival record and the IIIF manifest can refer to each other in their metadata.

4 Ontology Based Annotation

In Fig. 4 it is shown an example of annotation by *SimpleAnnotationServer*[4], an open-source software. It can act as an IIIF Search API endpoint to allow the annotations to be searched by some query in the IIIF viewer.

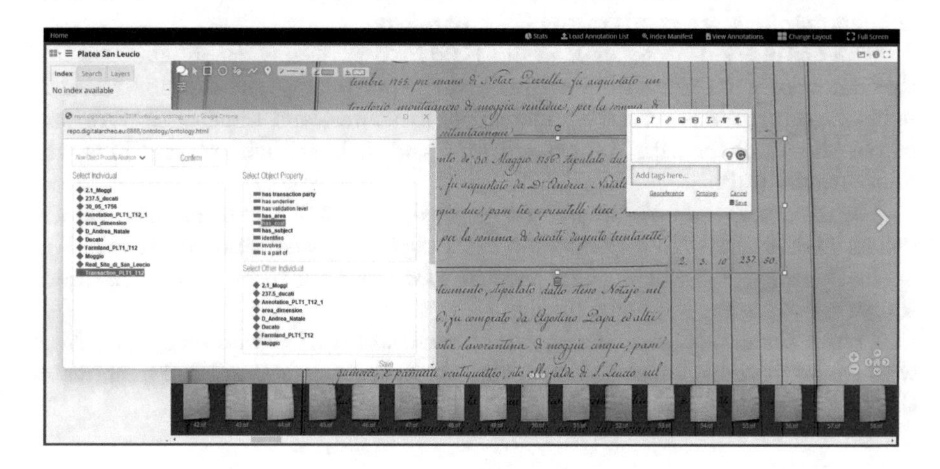

Fig. 4. Ontology based annotation

We extended the *SimpleAnnotationServer* in order to allow the annotation with concepts and individuals of an Ontology [2,4] which is essential to carry out logical reasoning.

The annotation model allows the annotation of a selected area of an image either with an individual or with an entity of the ontology. In the first case, the ontology is populated with the annotation that has an object properties to the selected individual. In the second case, a new individual of the selected class must be created too.

In Fig. 4 an example of ontology-based annotation is shown. The annotated page belongs to an inventory of goods. We can see a table that contains transactions related to the purchase of lands. In the first column, a transaction per row is recorded in handwritten natural language by a notary and describes the seller, the date of the transaction, the farmland, its size, and its cost. The other columns report as numbers the values of size and cost in units and sub-units of that time (1700-1800). Three columns correspond to the units *moggio*, *passi* and *passitelli* used to measure the land extension. The last two columns report the cost in terms of *ducati* and *grana*.

Figure 4 highlights the farmland annotation, which is also selected into the window to define its object property *has_cost*.

[4] https://github.com/glenrobson/SimpleAnnotationServer.

New individuals populate a domain ontology, that uses, for the representation of transactions and of measures (e.g. time, area dimension, and cost) two external ontologies.

The REA Enterprise Ontology has been initially created by William E. McCarthy [7], mainly for modeling of accounting systems. The persons of the domain ontology are also buyers of the REA ontology, and the farmlands of the domain ontology are goods of the REA ontology.

The Ontology of units of Measure (OM) 2.0 models concepts and relations that underlie the formulation of quantitative knowledge in scientific research. It has a strong focus on units, quantities, measures, and dimensions. We populated the ontology with the ancient unit and defined their relationships with international units.

For example, the unit *moggio* corresponds to 3, 387 square meters. We also define multiple units to express that the *moggi* is 30 *passi* and 900 *passitelli.*

In this way, a reasoner could use the OM ontology to infer the measure in square meters from the value in *moggi.*

Figure 5 shows a simplified graph visualization of the ontology populated with the individuals generated from the annotations. It is simplified in the sense that only a subset of entities and individuals are shown.

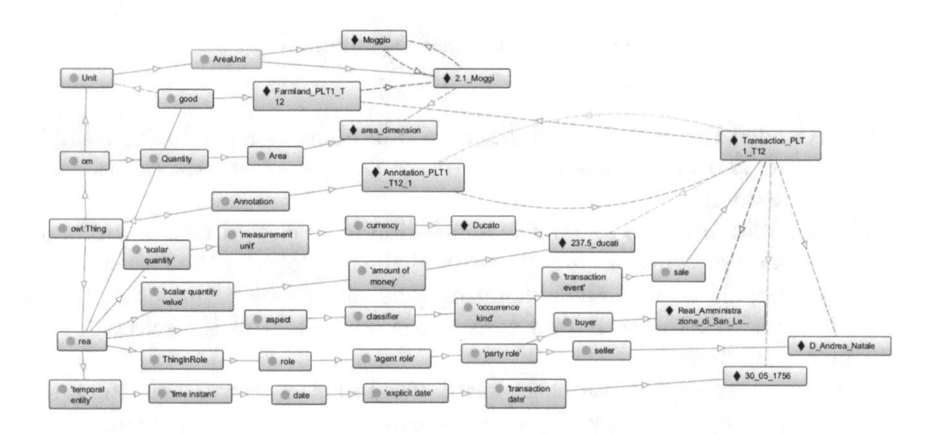

Fig. 5. Populated ontology

5 Conclusions

The increasing availability of digital archives and the adoption of interoperability frameworks allows for developing intelligent services which can work across a large amount of data which are complemented with standard meta-data and linked data valorized by common vocabularies. Nevertheless, linked data limit reasoning capability by expert systems and intelligent agents, which need to extract the semantic from multimedia data. The definition of domain ontology and their usage for image annotations can help to overcome the difficulties relating to the interpretation of handwriting. In this context the proposed solution allow to enrich digitized manuscript with ontology based semantic annotations using interoperability standards for their representation.

Acknowledgments. The work has been supported by ARS01_00540 - RASTA project, funded by the Italian Ministry of Research PNR 2015-2020.

References

1. Amato, A., Aversa, R., Branco, D., Venticinque, S.: Semantic wrap and personalized recommendations for digital archives. In: Barolli, L. (ed.) Complex, Intelligent and Software Intensive Systems - Proceedings of the 17th International Conference on Complex, Intelligent and Software Intensive Systems (CISIS-2023), 5–7 July 2023, Toronto, ON, Canada, Lecture Notes on Data Engineering and Communications Technologies, vol. 176, pp. 299–308. Springer, Cham (2023). https://doi.org/10.1007/978-3-031-35734-3_30
2. Amato, A., Aversa, R., Branco, D., Venticinque, S., Renda, G., Mataluna, S.: Porting of semantically annotated and geo-located images to an interoperability framework. In: Barolli, L. (ed.) Complex, Intelligent and Software Intensive Systems, pp. 508–516. Springer, Cham (2022). https://doi.org/10.1007/978-3-031-08812-4_49
3. Amato, A., Branco, D., Venticinque, S., Renda, G., Mataluna, S.: Metadata and semantic annotation of digital heritage assets: a case study. In: IEEE International Conference on Cyber Security and Resilience, CSR 2023, Venice, Italy, pp. 516–522. IEEE (2023)
4. Amato, A.A.D.: An ontology for the annotation of the archives of the royal site of San Leucio. In: Barolli, L. (ed.) Advanced Information Networking and Applications, pp. 37–44. Springer, Cham (2024)
5. Ambrisi, A., et al.: Intelligent agents for diffused cyber-physical museums. In: Camacho, D., Rosaci, D., Sarné, G.M.L., Versaci, M. (eds.) Intelligent Distributed Computing XIV, pp. 285–295. Springer, Cham (2022)
6. Branco, D., Aversa, R., Venticinque, S.: A tool for creation of virtual exhibits presented as IIIF collections by intelligent agents. In: Barolli, L. (ed.) Advanced Information Networking and Applications, pp. 241–250. Springer, Cham (2023). https://doi.org/10.1007/978-3-031-28694-0_22
7. Geerts, G.L., McCarthy, W.E.: The ontological foundation of rea enterprise information systems. In: Annual Meeting of the American Accounting Association, Philadelphia, PA, vol. 362, pp. 127–150 (2000)
8. Goy, A., et al.: PRiSMHA (providing rich semantic metadata for historical archives). In: Joint Ontology Workshops (2017). https://api.semanticscholar.org/CorpusID:8890884
9. Shepherd, E., Smith, C.: The application of ISAD(G) to the description of archival datasets. J. Soc. Arch. **21**(1), 55–86 (2000)
10. Snydman, S., Sanderson, R., Cramer, T.: (IIIF): a community & technology approach for web-based images. In: Archiving Conference, Archiving 2015, pp. 6–21(6). Society for Imaging Science and Technology (2015)

11. Steiner, E., Koch, C.: A digital archive of cultural heritage objects: standardized metadata and annotation categories. New Rev. Inform. Netw. **20**(1–2), 255–260 (2015)
12. Wu, M., Brandhorst, H., Marinescu, M.C., Lopez, J.M., Hlava, M., Busch, J.: Automated metadata annotation: what is and is not possible with machine learning. Data Intell. **5**(1), 122–138 (2023)

Social Robots and Edge Computing: Integrating Cloud Robotics in Social Interaction

Theodor-Radu Grumeza(✉), Thomas-Andrei Lazăr, and Alexandra-Emilia Fortiş

West University of Timişoara, Faculty of Mathematics and Informatics, Timişoara 300223, Romania
theodor.grumeza@e-uvt.ro

Abstract. In this study, the authors explore the integration of cloud robotics for social interactions using the SoftBank Robotics Pepper robot in relation with an academic environment. The main objective is to enhance social robots ability to interact with humans by providing guidance and information using the Pepper robot as a case study. The initial steps assess Pepper's language processing capabilities, identifying limitations in its vocabulary. The novelty of this research lies in employing Meta LLAMA 2, a large language model, trained on educational content, to enhance audio-to-text coversion. This approach aims to boost Pepper's comprehension and response to human inquiries, marking a step forward in the practical application of cloud-edge robotics in social contexts.

Keywords: Edge computing · Cloud Robotics · LLAMA 2 · Natural Language Processing · Social Robots · Large Language Models

1 Introduction

Cloud-based robotics [1] combines cloud computing with robotic systems, creating a new approach in cloud-edge applications. By transferring control logic from the robot's local hardware to cloud platforms, it facilitates a more effective process for exchanging services and information among a network of agents.

This research is centered around the cloud-edge integration of social robots, employing the SoftBank Robotics Pepper as use-case [2], interaction with humans being one of its key features. The main idea is to place the humanoid robot into a crowded academic environment to test its abilities in offering guidance and other information. An approach is being explored where the robot should help students or visitors, provide them information on services, and assist them in filling out forms or applications, ensuring that they have up-to-date information on policies and procedures. By implementing the application of Natural Language Processing (NLP), the Meta's LLAMA 2 [3], a collection of pretrained and fine-tuned large language models (LLMs), the authors are trying to transform the social robot's behavior, by training more interaction skills and the use of a broader dictionary of words.

L. Barolli (Ed.): AINA 2024, LNDECT 204, pp. 55–64, 2024.
https://doi.org/10.1007/978-3-031-57942-4_7

Understanding the integration of social robots using a Large Language model with cloud edge computing is covered by the literature review in the second section of this research. The third section refers to the implied methods, the system specifications, highlights some open problems that could be of use in implementing a new approach in social robots and, finally, the phases in implementation of the system. The fourth section offers the results of the research, emphasizing the precision, similarity and F1 score which are of use in determining the correctitude of predictions of the trained model from a lexical and semantic point of view. The last section is dedicated to conclusions and future work.

2 Related Work

Recent scientific research [4–6] emphasizes the importance of edge computing, representing an imperative technology in today's intelligent society, enabling stable and efficient artificial intelligence services for the rapidly expanding array of terminal devices and data streams. Edge devices have integrated into numerous fields, such as social robots enhancing interactive experiences in education and customer service [7], advanced robotics in healthcare [8], warehouse management systems employing robots for efficient inventory control [9], smart agricultural robots revolutionizing precision farming and many more. This widespread adoption illustrates the diverse and innovative applications of edge technology in contemporary industries, extending beyond conventional uses like smart homes and autonomous vehicles [10,11].

Cloud-based robotics, as an application of cloud-edge, represents an innovative convergence of cloud computing technologies with robotic systems [12]. This approach wants to enhance the possibility to reconfige robotic applications while reducing their intrinsic complexity and cost. By migrating the control logic from the local robot hardware to cloud-based platforms, it enables a more efficient mechanism for service and information exchange across a network of robots or agents.

Recent advancements in robotics have been driven using machine learning, especially data-driven methods like Deep Learning and Reinforcement Learning [13], which require repetitive task execution by robots in order to gather data. Simulations capable of handling complex environments and robot dynamics could simplify data collection and protect real robots from damage. They also help identify potential issues scenarios in a cloud environment, involving robots before real-world deployment. While simulation tools like Gazebo, V-REP, Webots, or Choregraphe exist, they often struggle with complex environment simulation as presented in the work of Ayala et al. [14].

Large Language Models (LLMs) are advanced AI assistants, excelling in complex reasoning tasks and expert knowledge in various fields, including specialized areas like robotics in social contexts. They are built using auto-regressive transformers and are initially trained on a vast, self-supervised data sets. This is followed by an alignment process with human preferences through methods like

Reinforcement Learning with Human Feedback (RLHF). Despite the simplicity of this methodology, the high computational demands have restricted LLM development to a limited number of entities. Several publicly released pretrained LLMs, such as BLOOM [15] or LLaMa-1 [3]. However, these models do not fully replace closed "product" LLMs like ChatGPT, BARD, and Claude.

3 Proposed Methods

In what follows, it can be stated that the computational framework employed for this study consisted in a computer equipped with an AMD Ryzen 9 7950X CPU, 32GB of DDR4 RAM, operating at a speed of 3200MHz, and a 1TB NVMe SSD. Additionally, an Nvidia RTX 3070 GPU, featuring 8GB of VRAM, was employed for the data training processes. To manage and process the input data from the robot, a setup of two Virtual Machines (VMs) was established. Each VM was configured with 8 virtual CPUs, 16GB of RAM, and 50GB of NL-SAS storage.

The workflow involved capturing the robot's input in a WAV file format on one VM, followed by processing the audio data and subsequently transmitting the output back to the robot. The operating systems chosen for this setup were Ubuntu 22.04 LTS Desktop for the main computer and Ubuntu 22.04 LTS Server for the VMs. The implementation and coding tasks were carried out using Jupyter Notebook.

Regarding interaction with social robots one needs to take into account some open problems:

- How the incorporation of NLP (Natural Language Processing) could improve human-robot interaction?
- How the response generation in social robots could be made more natural and context-aware?
- What methods can be employed for improving the accuracy of language processing in robots without significantly increase the computational costs?

Using only the main board of the robot and a preset of dictionaries will not be sufficient to offer that type of information. Initial approaches we to evaluate the performance of the language processing feature on the Pepper robot which is of most importance when interacting with humans. The primary objective was to assess Pepper's capability in accurately comprehending spoken questions from humans. Observation indicated that the efficiency of language processing on Pepper is somewhat limited due to its constrained vocabulary also presented in [16]. This vocabulary, essentially a database of words and phrases, determines the range of speech that the recognition system can understand; a broader vocabulary allows for more extensive comprehension.

To facilitate this evaluation, a Python script was designed to record the user's voice. This script prompts the robot to record for a set duration, subsequently transferring the audio file to the same directory as the running script. Following this, a second script comes into play, utilizing the Pinecone API as a dynamic

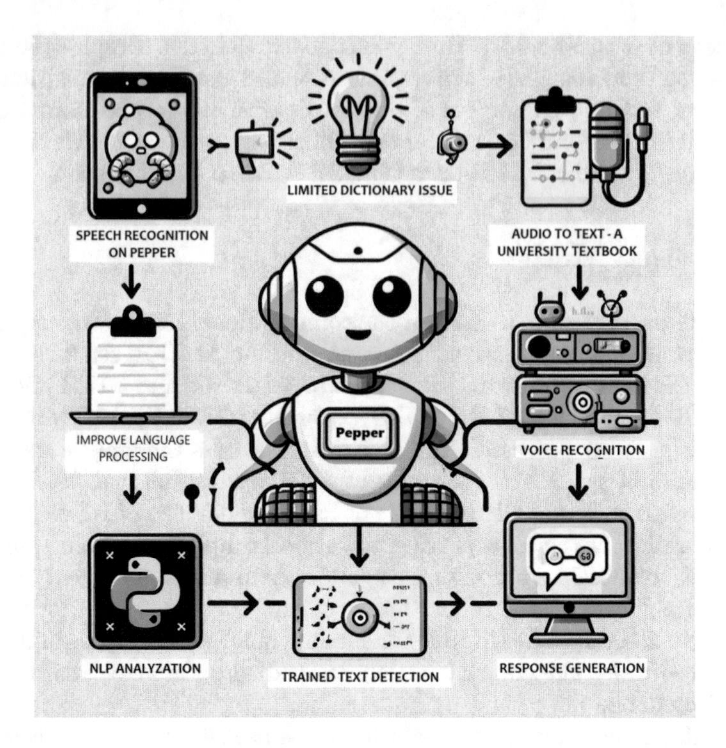

Fig. 1. System diagram

data vector, enabling the model to learn more effectively to extract text from the audio file.

The LLAMA 2 1.7B model was trained on a dataset that includes a PDF file of an academic textbook, such that it will interpret the text extracted from the audio file. The process involves the NLP analysis of the extracted text to identify the user's query. Once it comprehends the question, the NLP generates an appropriate response. This methodology not only tests Pepper's ability to interact with users but also explores the integration of advanced NLP techniques in enhancing robot-human communication (Fig. 1).

3.1 Advanced Library Integration for Improved Robot-Human Interaction

This research explored the integration of cloud-edge technology with SoftBank Robotics' Pepper robot, utilizing a suite of Python libraries like 'langchain', 'pypdf', 'unstructured', `'sentence_transformers'`, `'pinecone-client'`, and `'huggingface_hub'`. These tools were pivotal in augmenting Pepper's functionalities, particularly in processing complex text data, crucial for effective interaction with university students and visitors.

3.2 Customizing Pepper for Interactive Settings

The authors harnessed the 'PyPDFLoader'module to equip Pepper with the capability to parse and analyze text from PDF documents, such as university guidelines and protocols. This feature is of major importance for providing up to date, accurate information to the university community. It sets the foundation for Pepper's advanced comprehension and interpretation abilities relevant to a university context.

3.3 Context-Aware Data Processing Modules

Modules like "RecursiveCharacterTextSplitter" and 'HuggingFaceEmbeddings' were implemented, to enable Pepper to process and understand text in a detailed and nuanced manner. These modules were crucial for handling various tasks, such as responding to student inquiries, assisting in document completion, and navigating information requests.

3.4 Enhanced Interaction Through Advanced NLP Techniques

A significant improvement in Pepper's natural language processing (NLP) capabilities was obtained by incorporating transformer-based models from "HuggingFaceHub" and "SentenceTransformer". This enhancement was essential for tasks like semantic search and automated question answering, facilitating natural and relevant interactions between Pepper and its users.

3.5 Developing a Reliable Question Answering Framework

A sophisticated question-answering system was integrated, evaluated using metrics like `'accuracy_score'`, `precision_score`, `'recall_score'`, `'f1_score'`. This system was instrumental in allowing Pepper to accurately address queries regarding university services, ensuring the dissemination of trustworthy and prompt information.

4 Results

Regarding text classification and machine learning model evaluation, F1 score, precision, and similarity are important metrics, each offering a unique perspective on model accuracy. The F1 score, in particular, is a composite measure that accounts for both precision and recall. Precision refers to the proportion of positive predictions that are correctly identified, while recall is about the proportion of actual positive predictions that the model accurately identifies. The F1 score is calculated as the harmonic mean of precision and recall, with a higher score indicating greater model precision and recall.

The similarity score, on the other hand, assesses the similarities in two text parts. It can be computed using various metrics like cosine similarity, Jaccard similarity, or Levenshtein distance. A higher similarity score suggests greater resemblance between the two text segments.

The F1 score is particularly useful when both precision and recall are vital. Precision is the preferred metric when correctly identifying positive predictions and it is more important than avoiding incorrect negative predictions. For instance, in loan approval models, a high precision score implies a lower likelihood of approving loans to those likely to default.

In a parallel approach, the similarity score is valuable for evaluating text classification models, aiding in assessing the model's ability to accurately identify similar text segments. For example, in product classification, a high similarity score suggests the model is more likely to correctly group products with similar features.

The Fig. 2 serves to demonstrate the method of assessing similarity between two distinct lists of character strings, named `'supposed_answers'` and `‘ai_generated_answers‘`. This assessment is executed through the formation of a new list, entitled `‘similarity.scores‘`. Each entry in this list is assigned a binary value, either 1 or 0. Specifically, a value of 1 is allocated if the corresponding elements in both lists match exactly, indicating a perfect correlation. Conversely, a value of 0 is assigned if the elements differ, signaling a lack of similarity.

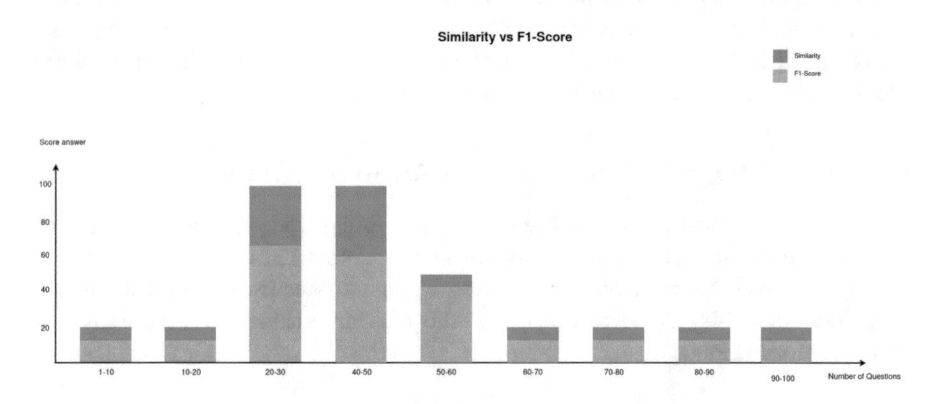

Fig. 2. Similarity assessing method

This binary scoring system is necessary in revealing certain patterns in the model's performance. It becomes apparent that while the model may occasionally fail to produce answers that closely mirror the supposed answers, it is capable of generating true positive responses. This implies that the model can provide accurate and relevant information on specific subjects, even if it does not always align perfectly with the anticipated answers.

To illustrate this point, a specific example is depicted in 3.

Although the answer provided by the robot is succint it touches upon an aspect of the University Senate's role in policy-making process, but without the depth and breadth of the complete answer.

Answer vs Prediction

	Question	Answer provided by the book	Answer provided by the robot
Human interface	What role does the UVT Senate play in shaping university policies?	The UVT Senate is pivotal in shaping university policies. It holds significant responsibilities such as approving academic regulations, endorsing strategic plans, and making key institutional decisions. This role ensures that the university is governed and directed effectively, maintaining its overall vision and mission.	Promoting institutional communication, endorsing strategic plans, and making key institutional decisions

Fig. 3. Answer vs Prediction

This example highlights the AI's capability to grasp the essence of a topic and provide relevant, if not comprehensive, responses. Its performance is reflective of its training on a dataset that comprises various segments or chunks of information. This structure allows the AI to access and relay pertinent information, even if it doesn't always achieve the level of detail or specificity found in the ideal answers (Fig. 4).

The F1 score, as mentioned earlier, reflects the authors' findings with values between 0.02 up to 0.10 in some cases. This indicates that the model, trained on a chunk of a PDF document, accurately generates responses 40% of the time as showed in Fig. 5. This suggests a need for more refined fine-tuning.

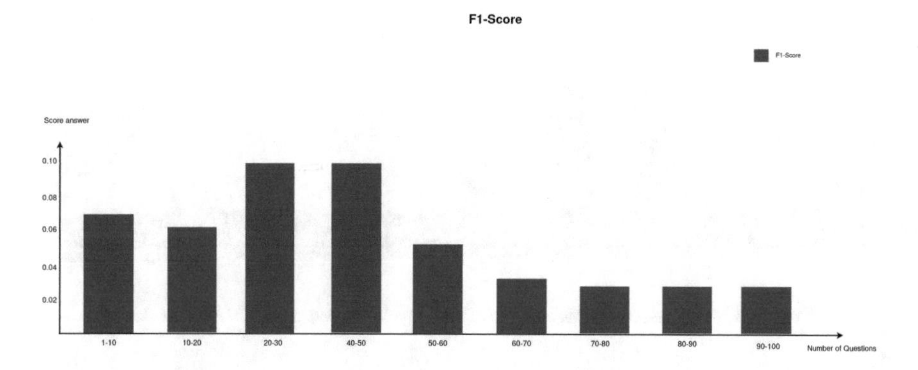

Fig. 4. Precision Score comparison

Pinecone was also utilized as a dynamic data vector, enabling the model to learn more effectively. The benefits of this approach include:

1. Model Suitability for Limited Hardware: Given that the version of Pepper trained for this research is a recent one, with limited processing power, the model provides responses via the cloud, offering somewhat relevant answers based on the results. However, the accuracy of these responses can not be guaranteed.

2. Enhanced Autonomy for Pepper: By integrating a LLMA model, Pepper gains assistance with information processing, though this does not involve physical movement. Taking into consideration that the authors integrated the language model on a computational model, the robot does not have to compute the answer by itself. Making a request is needed in order to give the answer.

3. Optimization for Simplicity: The first node is specifically designed to optimize Pepper's behavior, considering its limited ability to comprehend complex sentences. Taking into account that the robot has a limited dictionary and understanding of the words it was trained in such a manner that improvements audio-to-text extraction were observed.

Precision and similarity are fundamentally the same concept, yet precision incorporates an additional aspect: it examines the grammar of each sentence. In the current study, a method of lexical similarity was employed, involving a comparison in order to identify the words check if they can be found in the primary dataset. Precision aids this process by also verifying the positional arrangement of the words to determine if they are placed in correct ordered and if the spelling is in accordance with the dictionary. Furthermore, the gathered data, as illustrated in Fig. 3, indicates that in cases of similarity, the response is also grammatically correct.

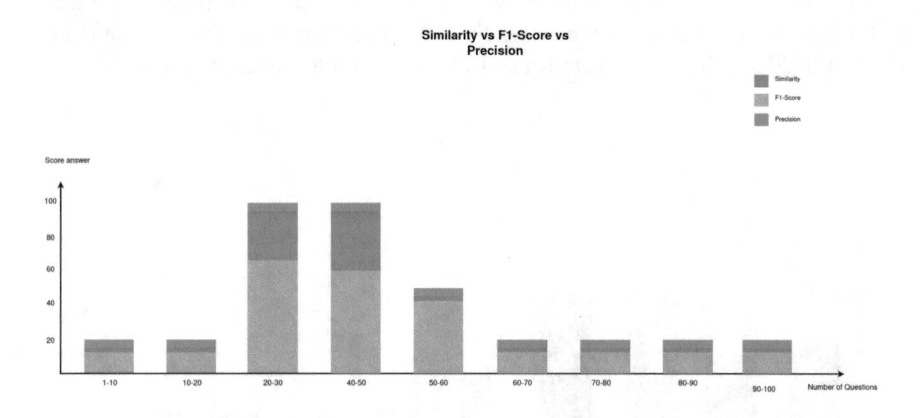

Fig. 5. Comparison between Similarity, F1-Score and Precision score

5 Conclusions and Future Work

5.1 Conclusions

The performed study demonstrates an approach of optimizing AI learning based on LLAMA 2, which is a NLP, and it is focused on learning chunks of information from a document. This method proves to be efficient enough to use the gathered information such that the response could be used as guidance to individuals who require assistance.

Implementing Cloud-Edge and problem segmentation techniques conducted to a better response time in terms of latency. If the entire model was implemented to run on the Pepper's robot hardware system, it would be nearly impossible to achieve such a rapid response. A significant step forward in this study is that success was obtained by using Cloud-Edge technology for responses, enhancing the interactivity of the social robot.

In conclusion, the tests that were conducted have shown that the Pepper robot can be effectively used as a guide demonstrating a considerable ability to accurately respond to questions by offering semantic and lexical correct answers. Given that the pretrained model on data chunks is capable of addressing specific types of questions will enable the robot to guide users by providing certain information.

Furthermore, after integrating and enhancing the speech recognition from a limited dictionary to a much broader but slightly slower one it is significant that the social robot will understand and will be able to interact with more people, not responding to just a limited set of words.

5.2 Future Work

Next steps in this research will be focused on enhancing the interactivity by enabling Natural Language Processing (NLP) commands for the Pepper social robot. This would involve instructing the robot with data associated to the locations of different rooms and guiding it efficiently through the requested direction. Additionally, another goal is to optimize and upgrade the hardware component. Identifying the most suitable variant will improve the training of the model or even replace the current model entirely to achieve superior responses from alternative NLP models.

Starting from the current results, the authors' intention is to perform additional evaluations on typical edge devices, such as integrating Cloud-Edge computing on an Raspberry Pi 4 or NodeMCU ESP32 microcontroller in order to assess their response times.

Acknowledgement. This article was partially supported by the UVT 1000 Develop Fund of the West University of Timişoara.

References

1. Wan, J., Tang, S., Yan, H., Li, D., Wang, S., Vasilakos, A.V.: Current status and open issues: cloud robotics. IEEE Access **4**, 2797–2807 (2016)
2. Pandey, A.K., Gelin, R., Robot, A.: Pepper: the first machine of its kind. IEEE Robot. Autom. Mag. **25**(3), 40–48 (2018)
3. Touvron, H., et al.: Llama 2: open foundation and fine-tuned chat models. arXiv preprint arXiv:2307.09288 (2023)
4. Cao, K., Liu, Y., Meng, G., Sun, Q.: An overview on edge computing research. IEEE Access **8**, 85714–85728 (2020)
5. Shi, W., Cao, J., Zhang, Q., Li, Y., Lanyu, X.: Edge computing: vision and challenges. IEEE Internet Things J. **3**(5), 637–646 (2016)
6. Satyanarayanan, M.: The emergence of edge computing. Computer **50**(1), 30–39 (2017)

7. Elfaki, A.O., et al.: Revolutionizing social robotics: a cloud-based framework for enhancing the intelligence and autonomy of social robots. Robotics **12**(2), 48 (2023)
8. Wan, S., Zonghua, G., Ni, Q.: Cognitive computing and wireless communications on the edge for healthcare service robots. Comput. Commun. **149**, 99–106 (2020)
9. Queralta, J.P., Qingqing, L., Zou, Z., Westerlund, T.: Enhancing autonomy with blockchain and multi-access edge computing in distributed robotic systems. In: 2020 Fifth International Conference on Fog and Mobile Edge Computing (FMEC), pp. 180–187. IEEE (2020)
10. Biswas, A., Wang, H.-C.: Autonomous vehicles enabled by the integration of IoT, edge intelligence, 5G, and blockchain. Sensors **23**(4), 1963 (2023)
11. Nasir, M., Muhammad, K., Ullah, A., Ahmad, J., Baik, S.W., Sajjad, M.: Enabling automation and edge intelligence over resource constraint IoT devices for smart home. Neurocomputing **491**, 494–506 (2022)
12. Dr Subarna Shakya: Survey on cloud based robotics architecture, challenges and applications. J. Ubiquitous Comput. Commun. Technol. **2**(1), 10–18 (2020)
13. Morales, E.F., Murrieta-Cid, R., Becerra, I., Esquivel-Basaldua, M.A.: A survey on deep learning and deep reinforcement learning in robotics with a tutorial on deep reinforcement learning. Intell. Serv. Robot. **14**(5), 773–805 (2021)
14. Ayala, A., Cruz, F., Campos, D., Rubio, R., Fernandes, B., Dazeley, R.: A comparison of humanoid robot simulators: a quantitative approach. In: 2020 Joint IEEE 10th International Conference on Development and Learning and Epigenetic Robotics (ICDL-EpiRob), pp. 1–6. IEEE (2020)
15. BigScience Workshop, Le Scao, T., et al.: Bloom: a 176b-parameter open-access multilingual language model. arXiv preprint arXiv:2211.05100 (2022)
16. Pande, A., Mishra, D.: The synergy between a humanoid robot and whisper: bridging a gap in education. Electronics **12**(19), 3995 (2023)

Fine-Tuned CNN for Clothing Image Classification on Mobile Edge Computing

Diogen Babuc(✉) and Alexandra-Emilia Fortiş

Faculty of Mathematics and Informatics, West University of Timişoara, 300223 Timişoara, Romania
diogen.babuc00@e-uvt.ro

Abstract. This paper explores the use of convolutional neural networks for clothing image classification based on evaluation metrics such as F1-score, accuracy, precision, and recall. The goal is to improve the accuracy of image detection and classification to help individuals choose the appropriate clothing, recycle clothing, and help retailers supervise their inventory and optimize the goods. A comparison between four state-of-the-art models was performed, emphasizing some thresholds for these models (recall greater than 74%, precision above 89%). The ZF NET model has the highest accuracy (over 81%), which shows its effectiveness in identifying and categorizing clothing items. The VGG-16 model has the best F1-score (86.4%), which reflects a good balance between precision and recall. We also propose a fine-tuned convolutional neural network called *Clothes-DAT*, and compare it with the four reference models, in terms of recall, accuracy, and F1-score. Additionally, *Clothes-DAT* has strong generalization ability, being able to handle not only the training data sets but also new data.

Keywords: Convolutional Neural Network · Clothing Image Classification · Performance Evaluation Metrics · Edge Device

1 Introduction

In the landscape of technology, the intersection of machine learning and the fashion industry has become a persuasive area of exploration. Convolutional neural networks (CNN) play a crucial role in clothing image classification [1]. CNN deep learning architectures are designed to understand spatial hierarchies in images. In clothing classification, different parts of an outfit contribute to its overall class. CNN models excel at capturing local features. This allows them to recognize patterns such as textures, colors, and shapes at different levels of generalization [2]. CNN models use parameter sharing through convolutional layers, which reduces the number of parameters compared to fully connected networks. This makes CNN algorithms efficient for image-related tasks, including clothing classification, where there can be a large number of parameters to learn [3].

L. Barolli (Ed.): AINA 2024, LNDECT 204, pp. 65–75, 2024.
https://doi.org/10.1007/978-3-031-57942-4_8

This paper aims to develop a robust program using multiple CNN standards, involving the adjusted one, for efficient and accurate clothing image classification (including object detection). Clothing image detection allows the creation of a personalized shopping experience. Designers can use image classification to gather inspiration, understand market trends, and refine their designs based on favorite styles. People can maintain modesty in dressing by establishing a balance between colors and textures so that they can wear appropriate clothing for any event [4]. In addition, clothing recycling reduces the demand for new textile production, curbs associated manufacturing emissions, and reduces the environmental impact on air quality [5]. Clothing image classification helps retailers and manufacturers manage and organize their inventory effectively, reducing errors and optimizing stock levels [6]. The three main objectives of this paper are:

1. Mobile edge deployment for clothing image classification for concrete occasions;
2. Performance metrics' calculation of four state-of-the-art CNN models (ZF NET, AlexNet, ResNet, and VGG-16);
3. Building the fine-tuned convolutional neural network, called *Clothes-DAT*, with improved statistical results.

The motivation behind this initiative lies in addressing the growing demand for smart applications that integrate with our daily lives.

2 Background Information and Related Works

This section offers a comprehensive examination of relevant prior works. It outlines key methodologies and studies that have shaped the terrain of image classification. It focuses on deep learning methods and performance evaluation metrics. CNN models are significant because of their capacity to handle spatial hierarchies. CNN automatically learns and extracts pertinent features. It also demonstrates translation invariance and effectively classifies variables and visual patterns linked to various clothing types [3]. CNN become a key piece of technology in the field of computer vision for garment recognition and classification [7].

Residual Network (ResNet)
ResNet is a deep convolutional neural network architecture that has been proven to be highly accurate in image classification tasks, including those related to clothing. ResNet allows the gradient to flow more easily through the network by addressing the vanishing gradient issue. ResNet enables the network to learn and represent hierarchical features of varying complexity [8]. This is crucial for clothing image classification because clothing items can have patterns, textures, and details that require the network to learn a hierarchy of features [9].

The authors of [10] stated that ResNet emerges as crucial in addressing the primary concern where the fashion industry seeks innovative solutions to accommodate evolving trends and meet the escalating demand for clothing. The proposed deep learning model, anchored in the pre-trained ResNet, attains an 82.1% accuracy in identifying clothing patterns. It plays a key role in establishing a user-friendly recommendation system. ResNet's significance lies in its ability to change storage automation and organization processes. This tends to align with the goals of the fashion industry, to streamline and improve customer experiences.

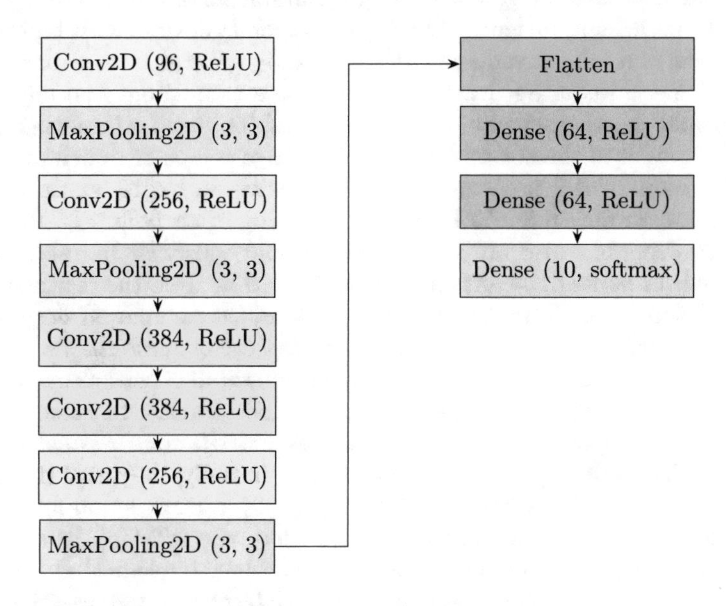

Fig. 1. ZF NET deep learning architecture for clothing image classification on mobile edge computing.

Zeiler-Fergus Network (ZF NET)

ZF NET has been surpassed by later architectures like AlexNet and ResNet, but it still has relevance in certain contexts. ZF NET, compared to architectures such as ResNet, has fewer layers, making it computationally less intensive [11]. This can be beneficial for deployment on mobile devices with limited computational resources. This is a common scenario in edge computing. ZF NET offers a balance between accuracy and computational efficiency. ZF NET's architecture emphasizes capturing spatial hierarchical features through convolutional and pooling layers (see Fig. 1). This can be beneficial for clothing images, where local and global spatial features play a role in classification. The paper [12] aims to showcase the effectiveness of a multi-layer classification technique in enhancing general results by using the ZF NET architecture. This framework allows modification of each layer to incorporate various learning strategies such as Support Vector Machine, K-Nearest Neighbors, Bayesian classifiers, etc.

Visual Geometry Group 16 (VGG-16)

VGG-16 is a convolutional deep learning architecture with some relevant characteristics for the classification of clothing images in mobile edge computing. It consists of repeated blocks of 3×3 convolutional layers with max-pooling layers in between [13]. This simplicity can make it easier to implement and deploy, especially in resource-constrained environments. VGG-16 has been pre-trained on large data sets like ImageNet, making it suitable for transfer learning. This is beneficial for clothing image classification on mobile devices with limited data, as the model can take advantage of the knowledge gained from a diverse set of images. The architecture's flexibility allows for modifications and adjustments based on specific requirements [14]. This adaptability can be important when a model is exposed to the constraints of mobile edge computing environments.

The authors of [15] used multiple CNN models to recycle clothes that are continually discarded in South Korea, contributing to air pollution. Using recycling, 1500 liters of water are saved, which would otherwise be used to create a single pair of pants. The authors categorize and analyze the features of each garment found in the data set. For this purpose, they employed deep learning architectures, *FashionNet* and *Attentive Fashion Grammar Network*, supported by deep learning architecture VGG-16. To encompass all critical elements such as pollution, resource conservation, and enhancing data transfer to the network, the authors integrated the *Deep Learning Clothes Classification System* (DLCCS) approach using both edge and cloud computing. Initially, they used the AlexNet model and devices of the Internet of Things (IoT), a camera and an electronic control device for physical computing, to explore the IoT [16]. In addition to AlexNet and the proposed model, DLCCS, the authors implemented GoogLeNet and ResNet to observe differences in performance (accuracy and precision). Their authentic system performed well, correctly labeling the test images at a rate of 65%.

AlexNet

AlexNet has been pre-trained on large data sets like ImageNet. This makes it suitable for transfer learning. Transfer learning allows the model to increase the knowledge gained from a diverse set of images, even if the target data set (clothing images in this case) is relatively small [17]. AlexNet is not as lightweight as some newer architectures, but its adaptability can be improved through techniques such as model compression and quantization [18]. These techniques help reduce the model size and make it more suitable for deployment on mobile devices with limited computational resources. The algorithm proposed from [19], which uses the AlexNet convolutional neural architecture, presents an optimized sequential algorithm for the classification of clothing images. This paper summarizes traditional methods for image classification, highlighting the importance of classifying clothing images due to the growing popularity of online clothing transactions. Accurate classification aids in data storage and holds the potential for automatic clothing recognition, clothing retrieval, and clothing recommendation [20].

The usage of CNN state-of-the-art models in image classification in the context of edge computing holds paramount significance, as these models are designed to balance accuracy and computational efficiency, enabling real-time processing on resource-constrained edge devices [21].

3 Proposed Method

By increasing the power of machine learning, particularly convolutional neural networks, it becomes possible to create intelligent systems capable of recognizing and classifying clothing items with great precision and accuracy. This can simplify the shopping experience and open the way for personalized recommendations, trend analysis, and more [22].

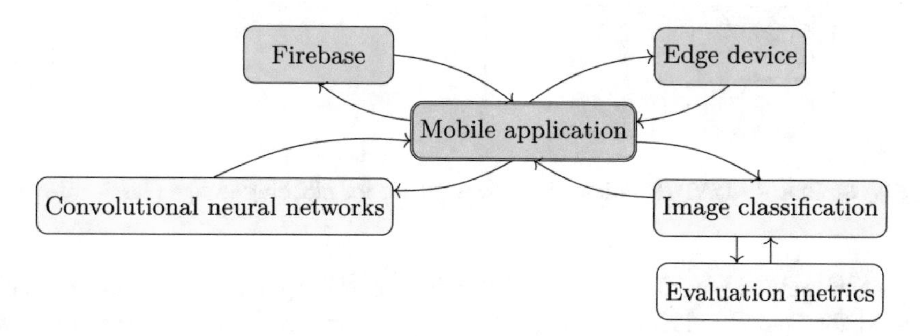

Fig. 2. System context diagram for the mobile application of the clothing classification.

The choice of an edge device, specifically a mobile phone, is a procedure that democratizes access to this technology [23]. By deploying the classification models on a mobile platform, users can utilize their smartphone cameras as sensors to capture new images [24]. This approach improves accessibility and empowers users to make informed decisions, bridging the gap between traditional retail experiences and the digital era. Edge computing reduces the need to send image data to a remote server for processing. This is especially important for applications where low latency is crucial, such as fast automated image classification.

The user interface of the mobile application [25] allows users to capture or upload images (with the camera as a sensor) of clothing for classification (Fig. 2). We used a data set (with 5403 unique pictures) that is publicly accessible on Kaggle[1]. Image preprocessing is performed with basic tasks such as resizing and data formatting before sending the image for classification. The newly obtained format of the data set contains images and the senders' IDs. Subsequently, it constructs ten values for labels and a column for information on whether the cloth is for adults or children. Firebase receives the uploads of captured images for storage, along with any associated metadata, with secure access to the database.

[1] https://www.kaggle.com/datasets/agrigorev/clothing-dataset-full. Clothing data set, last accessed: January 5th, 2023.

The edge device (the edge computing unit) is a local device capable of running deep learning models. Edge's engine runs inference on the edge device using CNN models. Users can also identify a type of cloth with CNN models.

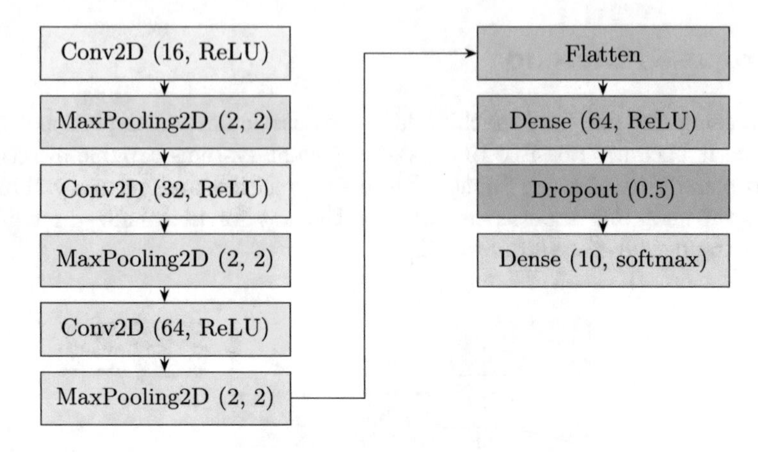

Fig. 3. Clothes-DAT deep learning architecture for clothing image classification.

A customized model, *Clothes-DAT* (Clothes of Diogen and Alexandra Tech), is used for specific clothing categories. This relatively shallow deep model has fewer layers and parameters, making it more resource-efficient and suitable for deployment on edge devices where computational capabilities are constrained. This model often leads to faster inference times compared to deeper models. The suggested model has been improved with a random forest layer and adapted hyperparameters (Fig. 3). In real-time applications, such as clothing image classification for augmented reality or even personalized shopping experiences, a faster inference speed is crucial. Since it might be difficult to maintain modesty when dressing in today's environment, our model strives to balance color and other elements so that people can wear suitable clothing for any occasion or meeting.

4 Results

Performance evaluation metrics are essential to assess the effectiveness of CNN models for clothing image classification. These metrics provide insights into how well the model is performing and help researchers, engineers, and practitioners make informed decisions about model improvements and deployment [2].

Accuracy is a key indicator of the model's overall correctness in categorizing images of clothes. It shows the proportion of cases that were accurately classified as all instances. Recall evaluates the capacity to record all positive occurrences, while precision gauges the correctness of positive predictions. The F1-score strikes a balance between the two [26]. When it comes to clothing image cat-

Table 1. Performance evaluation metrics.

%	ZF NET	AlexNet	VGG-16	ResNet	Clothes-DAT
Recall	76.8	75.2	74.5	75.7	78.0
Precision	90.2	89.5	90.7	89.7	90.6
Accuracy	81.4	78.7	77.5	73.6	84.5
F1-score	86.2	84.2	86.4	85.5	86.8

egorization, recall is important in establishing missing positive instances, while precision is important when proposing incorrect clothing items.

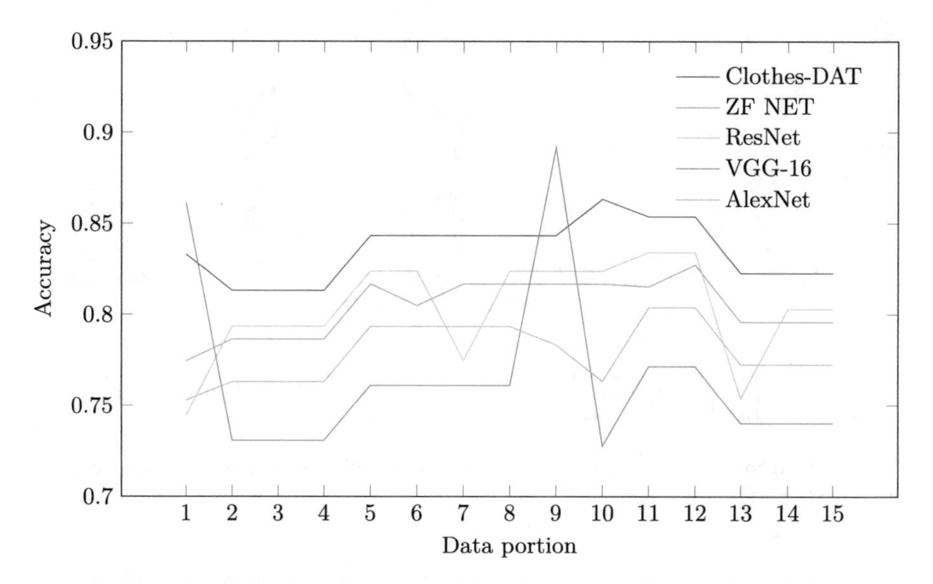

Fig. 4. K-fold cross-validation accuracy for deep learning approaches on 15 data portions.

All the deep learning models evaluate metrics, such as precision, recall, F1-score, accuracy, which are based on ground truth data. The image data is collected and sent through feedback from users on the classification results. Clothes-DAT, with its balanced performance, might be a good choice (compared to AlexNet, VGG-16, etc.) since overall accuracy, precision, and recall are essential for an application (Table 1). This model achieved the most accurate (84.5%). For recall, the proposed model provided the best result (78%), while the most precise model was VGG-16 (with 90.7%). In e-Commerce or fashion-related applications, high recall contributes to a better user experience by ensuring that the system accurately captures and displays a wide range of available clothing options. Excellent precision is important in applications with edge computing, where computational resources may be limited; reducing false positives is essential to avoid unneeded processing and ensure efficient resource utilization. Clothes-DAT also had the best result for the F1 score (86. 8%). The performance of Clothes-DAT suggests that this deep learning architecture might be robust in varied scenarios, but the specific use case should guide the choice.

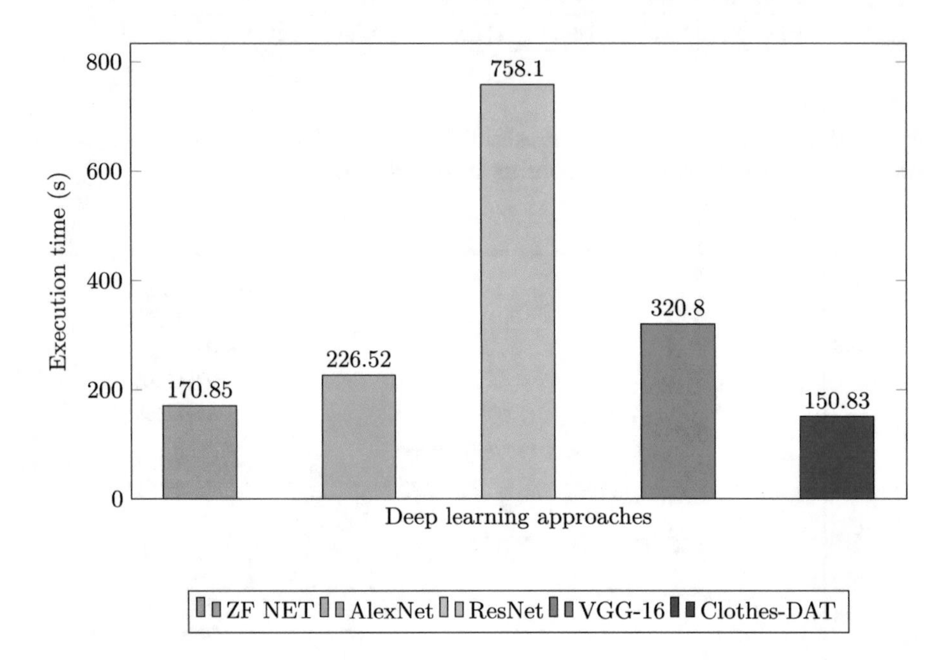

Fig. 5. Execution times for models' construction, training part, classification, and performance evaluation metrics.

Considering the k-fold cross-validation, the most performant model for most folds is Clothes-DAT, with all accuracies above 81%. Given that the model performs well across different portions, it is well-suited for deployment on resource-constrained devices like mobile devices at the edge. The consistent accuracy suggests that the model is efficient and effective, making it a strong candidate for real-time applications. The success of the model across various folds also indicates that the hyperparameters (learning rate, batch size, etc.) have been well tuned. Proper hyperparameter tuning is crucial to achieving optimal model performance. For the first and ninth portions, the most performant model is VGG-16, with an accuracy of around 89%. Also, VGG-16 is the worst deep-learning model for most of the portions (Fig. 4).

The provided execution times represent the computational cost, specifically the time taken for training and evaluation, of different neural network models in the context of clothing image classification. ZF NET model demonstrates relatively efficient training time, suggesting it can be suitable for real-time applications or scenarios where quick model updates are crucial. ResNet has a significantly higher computational cost, which could be attributed to its deeper architecture. It may be more suitable for scenarios where model accuracy is prioritized over training speed. Clothes-DAT stands out as the fastest model among the listed ones (Fig. 5). This could be advantageous for real-time applications on mobile devices, especially when resource constraints are a concern.

5 Conclusions

We performed an analysis for clothing image classification on mobile edge computing and calculated performance evaluation metrics for four state-of-the-art deep learning architectures and a proposed fine-tuned deep learning model, to choose the appropriate outfit for various events.

The proposed model, *Clothes-DAT*, provided the best results for recall (78%), accuracy (84.5%), and F1-score (86.8%). This suggests that it's effective in both minimizing false negatives (good recall) and providing accurate overall predictions. The accuracy of 84.5% indicates that Clothes-DAT performs well in correctly classifying clothing items. This is a key metric for assessing the model's overall competence. VGG-16 furnished the best precision, 90.7%. This means that when it predicts a specific type of clothing, it is very conceivable to be correct. Edge devices often have resource constraints in terms of processing power. CNN models for edge computing can be optimized, using techniques such as model quantization or compression, to fit within these constraints while maintaining acceptable performance. However, by performing image classification on the edge device, only relevant results are transmitted, improving bandwidth efficiency.

Achieving a balance between precision and recall in clothing image classification enhances user satisfaction, supports personalized recommendations, and fosters the effectiveness of downstream applications.

Acknowledgment. This article was supported by the UVT 1000 Develop Fund of the West University of Timisoara.

References

1. Vijayaraj, A., et al.: Deep learning image classification for fashion design. Wireless Commun. Mob. Comput. **2022** (2022)
2. Nayak, A., Shah, J., Kuruvilla, A., Akshaya, J., Sandesh, B.J.: Fine-grained fashion clothing image classification and recommendation. In: 2021 2nd International Conference on Electronics, Communications and Information Technology (CECIT). IEEE (2021)
3. Elleuch, M., Mezghani, A., Khemakhem, M., Kherallah, M.: Clothing classification using deep CNN architecture based on transfer learning. In: Abraham, A., Shandilya, S.K., Garcia-Hernandez, L., Varela, M.L. (eds.) HIS 2019. AISC, vol. 1179, pp. 240–248. Springer, Cham (2021). https://doi.org/10.1007/978-3-030-49336-3_24

4. Kelsey Sherrod Michael: Wearing your heart on your sleeve: the surveillance of women's souls in evangelical Christian modesty culture. Fem. Media Stud. **19**(8), 1129–1143 (2019)
5. Yang, C., Zhang, X.: Environmental protection clothing design and materials based on green design concept. Front. Mater. **10**, 1225289 (2023)
6. Papachristou, E., Chrysopoulos, A., Bilalis, N.: Machine learning for clothing manufacture as a mean to respond quicker and better to the demands of clothing brands: a greek case study. Int. J. Adv. Manuf. Technol. **115**, 691–702 (2021)
7. Park, S., Suh, Y., Lee, J.: Clothing classification using CNN and shopping mall search system. ICIC Express Lett. Part B: Appl. **11**(8), 773–780 (2020)
8. Gao, Z., Han, L.: Clothing image classification based on random erasing and residual network. J. Phys: Conf. Ser. **1634**(1), 012136 (2020)
9. Cychnerski, J., Brzeski, A., Boguszewski, A., Marmolowski, M., Trojanowicz, M.: Clothes detection and classification using convolutional neural networks. In: 2017 22nd IEEE International Conference on Emerging Technologies and Factory Automation (ETFA). IEEE (2017)
10. Singh, M., Dalmia, S., Ranjan, R.K., Singh, A.: Dress pattern classification using ResNet based convolutional neural networks. In: Garg, L., et al. (eds.) ISMS 2021. LNNS, vol. 521, pp. 91–103. Springer, Cham (2023). https://doi.org/10.1007/978-3-031-13150-9_8
11. Seo, Y., Shin, K.: Hierarchical convolutional neural networks for fashion image classification. Expert Syst. Appl. **116**, 328–339 (2019)
12. Willimon, B., Walker, I., Birchfield, S.: A new approach to clothing classification using mid-level layers. In: 2013 IEEE International Conference on Robotics and Automation, pp. 4271–4278. IEEE (2013)
13. Dhariwal, S., Liu, Y., Karali, A., Vlassov, V.: Clothing classification using unsupervised pre-training. In: 2020 Fourth International Conference on Multimedia Computing, Networking and Applications (MCNA), pp. 82–89. IEEE (2020)
14. Mogan, J.N., Lee, C.P., Lim, K.M., Muthu, K.S.: VGG16-MLP: gait recognition with fine-tuned VGG-16 and multilayer perceptron. Appl. Sci. **12**(15), 7639 (2022)
15. Noh, S.-K., et al.: Deep learning system for recycled clothing classification linked to cloud and edge computing. Comput. Intell. Neurosci. **2022** (2022)
16. Shi, W., Cao, J., Zhang, Q., Li, Y., Lanyu, X.: Edge computing: vision and challenges. IEEE Internet Things J. **3**(5), 637–646 (2016)
17. Tan, Z., Yuping, H., Luo, D., Man, H., Liu, K.: The clothing image classification algorithm based on the improved Xception model. Int. J. Comput. Sci. Eng. **23**(3), 214–223 (2020)
18. Mani Raj Paul: Classification of garments from fashion MNIST dataset using AlexNet CNN architecture. EPRA Int. J. Multidisc. Res. (IJMR) **8**(10), 296–299 (2022)
19. Jun, X., Wei, Y., Wang, A., Zhao, H., Lefloch, D.: Analysis of clothing image classification models: a comparison study between traditional machine learning and deep learning models. Fibres Text. Eastern Eur. **30**(5), 66–78 (2022)
20. Li, J., Shi, W., Yang, D.: Clothing image classification with a dragonfly algorithm optimised online sequential extreme learning machine. Fibres Text. Eastern Eur. (2021)
21. Stahl, R., Hoffman, A., Mueller-Gritschneder, D., Gerstlauer, A., Schlichtmann, U.: DeeperThings: fully distributed CNN inference on resource-constrained edge devices. Int. J. Parallel Prog. **49**, 600–624 (2021)
22. Donati, L., Iotti, E., Mordonini, G., Prati, A.: Fashion product classification through deep learning and computer vision. Appl. Sci. **9**(7), 1385 (2019)

23. Li, J., Lv, T.: Deep neural network based computational resource allocation for mobile edge computing. In: 2018 IEEE Globecom Workshops (GC Wkshps). IEEE (2018)
24. Shubathra, S., Kalaivaani, P.C.D., Santhoshkumar, S.: Clothing image recognition based on multiple features using deep neural networks. In: 2020 International Conference on Electronics and Sustainable Communication Systems (ICESC). IEEE (2020)
25. Mach, P., Becvar, Z.: Mobile edge computing: a survey on architecture and computation offloading. IEEE Commun. Surv. Tutor. **19**(3), 1628–1656 (2017)
26. Kayed, M., Anter, A., Mohamed, H.: Classification of garments from fashion MNIST dataset using CNN LeNet-5 architecture. In: 2020 International Conference on Innovative Trends in Communication and Computer Engineering (ITCE), pp. 238–243. IEEE (2020)

Homomorphic Cryptography Authentication Scheme to Eliminate Machine Tools Gaps in Industry 4.0

Shamsher Ullah[1], Jianqiang Li[1(✉)], Farhan Ullah[2,3], Diletta Cacciagrano[2], Muhammad Tanveer Hussain[4], and Victor C. M. Leung[1]

[1] National Engineering Laboratory for Big Data System Computing Technology, Shenzhen University, Shenzhen 518060, People's Republic of China
{shamsher,lijq}@szu.edu.cn, vleung@ieee.org

[2] Division of Computer Science, University of Camerino, 9 Via Madonna Delle Carceri, 62032 Camerino, Italy
farhan@nwpu.edu.cn, diletta.cacciagrano@unicam.it

[3] School of Software, Northwestern Polytechnical University, Xi'an 710072, Shaanxi, People's Republic of China

[4] Department of Mathematics, University of Management and Technology, Lahore, Pakistan
tanveerhussain@umt.edu.pk

Abstract. Cryptographic methods are becoming more vital in the era of Industry 4.0, as they aid in protecting data and personal information. It is now possible to merge information technology, networks, and industrial production into a single system, which has significantly influenced our everyday communication. Cryptographic techniques are the most efficient mechanism of safeguarding against all forms of threats. In order to eliminate the Machine Tools Gaps (MTG) and to maintain security and authenticity in Industry 4.0 for real-world applications, our proposed Homomorphic Cryptography Authentication (HCA) scheme is being utilized. Our scheme protects users from being able to deny transactions while also preventing attacks from attackers. It also offers security and privacy properties like integrity, authenticity, unforgeability, confidentiality, verifiability, and un-traceability.

Keywords: Homomorphic cryptography · Authentication · Homomorphic Encryption · ABE · Industry 4.0

1 Introduction

Industry 4.0 need to protect security and privacy when the users communicate with each other. Numerous Homomorphic Encryption (HE) schemes, such as Paillier [1], ElGmal [2], and Benaloh [3], enable message computations by conducting a similar computation on the ciphers. In any case, none of these methods address the data's authenticity or integrity. Authenticated encryption protects privacy and integrity. HC is comparable

L. Barolli (Ed.): AINA 2024, LNDECT 204, pp. 76–85, 2024.
https://doi.org/10.1007/978-3-031-57942-4_9

to classical cryptography. It secures data transport, processing, and storage. Homomorphism transfers problems across algebraic systems. It allows secure computation delegation between transactional entities. HE encrypts data without decryption. In HE, the message space is always enclosed inside a ring, and the computational model uses arithmetic circuits over this ring [4]. Several security applications, such as electronic voting and Private Information Retrieval (PIR), show that simple homomorphic cryptosystem are unsuitable to analyze extensive transformations on encrypted data. We further enhance their construction in this work by ensuring its authenticity, accuracy and efficiency by using HC techniques.

1.1 Homomorphic Attribute Based Encryption

ABE may be implemented utilizing probabilistic polynomial time algorithms message (m), the attribute space (A), and the class of permitted access policies F, (*Setup*; *KeyGen*; *Enc*; *Dec*).

1.2 Authentication

In authentication, the private key computing from Eqs. 1 and 2 is difficult. Our method is utilized to maintain the validity of the message and to ban eavesdropper attempts. TTP randomly generate x, r ϵ $Z^*{}_p$ and make the user's private data only visible to the specified user [5].

$$d' = g(x - r)mod(p). \tag{1}$$

$$d'' = g\left(rt'\right)mod(p). \tag{2}$$

where d is a private key and x, $r\epsilon Z_p^*$.

1.3 Contribution

Industry 4.0 requires a homomorphic cryptography authenticator. Homomorphism ensures the supplier and buyer's genuineness. Our approach protects vendor and buyer privacy and verifies transactions. The seller-buyer relationship needs stricter confidentiality and integrity standards. Our solution preserves secrecy and authenticity due to homomorphic cryptography. TTP oversees all major system managements (e.g., key exchange, key generation, and so on).

2 Preliminaries

This section defines Industry 4.0, HC for social networks and for Industry 4.0.

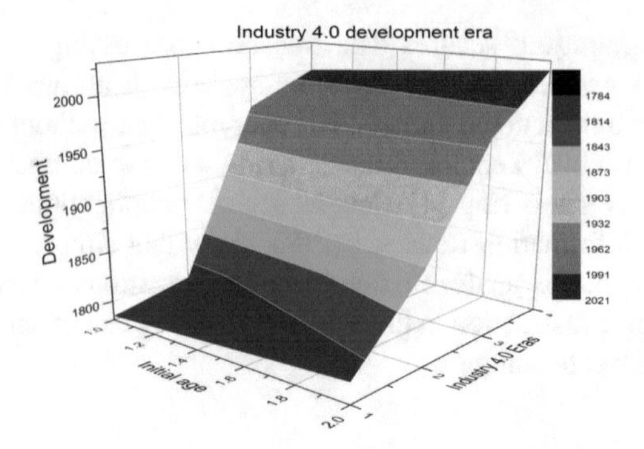

Fig. 1. Industry 4.0 development decades.

2.1 Industry 4.0

In industry 4.0, Social Networks (SN) like Facebook, WhatsApp, Twitter, and Google + are crucial for storing data and connecting with hundreds of millions of consumers globally. Users may change the world's knowledge via this interaction. Once a user enters information into the SN, the user's privacy is forfeited. We employ HE cryptography to preserve privacy [6] (Fig. 1).

2.2 HC for Social Network

The growing use of SNS for communication and information sharing has produced new vulnerabilities and cyber hazards, resulting in privacy breaches on Facebook, Twitter, LinkedIn, etc. SNS providers' policy of sharing users' personal information make maintaining privacy more challenging. Suppliers create risks by offering alternative data-exchange channels (such as API). Traditional cryptography encrypts and decrypts sensitive SNS data. This strategy threatens the privacy of SNS data collected, stored, and shared. Now, modern encryption technologies are used to guarantee data privacy in ubiquitous computing environments via SNS. HC approaches allow the modification and operation of encrypted SNS user data without decoding it. This research aims to use homomorphic cryptography to SNS user data confidentiality. The users of SNS are able to perform homomorphic operations on encrypted data without the need of decoding ciphertexts. These operations are carried out on the encrypted data. Homomorphic operations are those that based on the additive and (or) multiplicative properties of homomorphism over C.T. HC in SNS is made up of HE algorithms, primitives, characteristics, and operations.

2.3 Related Work

This section defines the homomorphic cryptosystem for the Fourth industrial revolution (Industry 4.0). A function $f : A \to B$, where f is a homomorphism on two algebraic objects A; B, keeps the algebraic structure intact, for example group and ring.

1) If the addition operations on A and B are both addition operations, then the homomorphism requirement is satisfied according to Eq. 3.

$$f(a+b) = f(a) + f(b);\ \forall\, a, b \in A. \tag{3}$$

2) It is homomorphic, if the multiplication operations on A and B are both performed on the same data set by Eq. 4.

$$f(ab) = f(a)f(b);\ \forall\, a, b \in A. \tag{4}$$

3) In addition, if the operation on A is addition and the operation on B is multiplication, Eq. 5 satisfies the homomorphism condition.

$$f(a+b) = f(a)f(b);\ \forall\, a, b \in A. \tag{5}$$

Definition 1: Let a group G and H, there is a petty homomorphism $f : G \to H$ given by $f(g) = 1H, \forall\, g \in G$.

Definition 2: Let n be a positive integer. The function g: $Z \to Z_n$ defined by $g(a) = a(mod\ n)$ is a ring homomorphism (Eq. 3).

Definition 3: Let $f : C \to C$ by $f(z) = \bar{z}$, i.e., f is complex conjugation. So, f is a homomorphism of C. It's obviously bijection, therefore it's an isomorphism of C.

1) Homomorphic cryptosystem: A cryptosystem is considered homomorphic if it demonstrates either additive or multiplicative homomorphism, but not both [7]. Certainly, HE approaches may be advantageous in a variety of instances. Other than that, the efficacy of various HE approaches is enough for practical applications. On the other hand, the shortcomings of this family approaches are readily evident. Paillier's and RSA, for instance, only permit one sort of operation, while the majority of other HE algorithms provide a variety of operations.
2) Homomorphic properties:

a) Un-padded RSA: If the modulus and exponent of the RSA Public Key (PK) are m and e, respectively, the encryption of a message y is given by Eq. 6;

$$\sigma(y) = y^e (\text{mod m}). \tag{6}$$

To decrypt C.T σ(y) one makes use of the private key pair $(f; m)$:

$$PT = \sigma(y)^f (\text{mod m}). \tag{7}$$

The homomorphic property;

$$\sigma(y_1).\sigma(y_2) = y_1^e y_2^e (mod\ m) = (y_1 y_2)^e (mod\ m) = \sigma(y_1.y_2). \tag{8}$$

b) El-Gamal [8]**:** Suppose that the PK is (G, r, s, t) in a group G, where $t = s^y$ and y is the secret key, the encryption of a message m is:

$$\sigma(\text{m}) = \left(s^u, m, t^u\right),\ \forall\, u \in \{0, 1, \ldots, r-1\}. \tag{9}$$

The homomorphic property is based on Eq. 10;

$$\sigma(y_1).\sigma(y_2) = \left(s^{u_1}, y_1, t^{u_1}\right)\left(s^{u_2}, y_2, t^{u_2}\right) = \left(s^{u_1+u_2}, (y_1.y_2), t^{u_1+u_2}\right) = \sigma(y_1.y_2). \quad (10)$$

c) **RSA:** RSA and other cryptographic algorithms do computations modulo $n = (p \times q)$, which is the product of two large prime numbers. The encryption technique is simple, needing just a product and a square, but decryption requires exponentiation. If the modulus m and quadratic non-residue y of the PK are identical, then the encryption of a bit b is determined by Eq. 11:

$$\sigma(\mathrm{b}) = \left(y^b, u^2\right)(mod\ m), \forall\ u \in \{0, 1, \ldots, r-1\}. \quad (11)$$

The homomorphic property is;

$$\sigma(\mathrm{b}_1).\sigma(b_2) = \left(y^{b_1}u_1^2 y^{b_2}u_2^2\right) = y^{b_1+b_2(u_1u_2)^2} = \sigma(b_1 \oplus b_2). \quad (12)$$

d) **Benaloh's scheme:** The proposed system is an extension of the Goldwasser-Micali scheme, which handles inputs of $l(p)$ bits. If the Benaloh PK is m, g, and d, then y is encrypted as [9] Eq. 13:

$$\sigma(\mathrm{y}) = g^y u^d(\mathrm{mod\ m}), \ \forall\ u \in \{0, 1, \ldots, r-1\}. \quad (13)$$

The homomorphic property is based on Eq. 14;

$$\sigma(y_1).\sigma(y_2) = \left(g^{y_1}u_1^d)(g^{y_2}u_2^d\right) = g^{y_1+y_2}d = \sigma(y_1+y_2)u^d(mod\ d). \quad (14)$$

e) **Paillier:** Paillier cryptosystem is a homomorphic cryptosystem. If the PK is the modulus m and the base k, then the encryption of a message y is [9], Eq. 15;

$$\sigma(\mathrm{y}) = (k^y u^m)\left(mod\ m^2\right). \quad (15)$$

Upon the multiplication of two plain texts, a homomorphic property is based on Eq. 16.

$$\sigma(y_1).\sigma(y_2) = (k^{y_1}u_1^m)(k^{y_2}u_2^m) = k^{y_1+y_2}(u_1u_2)^m$$

$$= \sigma(y_1+y_2)\left(mod\ m^2\right). \quad (16)$$

3) Attribute-Based Encryption: A sender may encrypt and send a message to an identity without knowing the identity's PK certificate. The ability to execute PK encryption without the need of certificates offers a broad variety of real-world applications. If a user wishes to send an encrypted email to a recipient, such as ****@gmail:com, neither the availability of a Public-Key Infrastructure nor the receiver being online at the time the email is produced is required. All Attribute-Based Encryption systems have the trait of treating identities as a string of characters rather than a single entity. The first viable and secure Attribute-Based Encryption technique was published decades later, by Boneh and Franklin [10]. Their method made novel use of groups for which an efficiently computable bilinear map exists, which was previously undiscovered area.

4) Homomorphism: A homomorphism maintains the operations of two algebraic structures of the same kind (such as two groups, rings, or vector spaces). This means a map $f : A \rightarrow B$ between two sets A, B equipped with the same structure such that, if Δ is an operation of the structure Eq. 17:

$$\mathrm{f}(\mathrm{x}\Delta\mathrm{y}) = \mathrm{f}(\mathrm{x})\Delta\mathrm{f}(\mathrm{y}). \tag{17}$$

for every pair $x, y \in A$.

5) Bilinear Mapping: Let us consider that the following two multiplicative cyclic groups G_1 and G_2 of prime order p. Assume that g is a generator of the function G_1. The following are the characteristics of the bilinear map [6]:
   (i) Bilinearity: $e(g^x, g^y) = e(g, g)^{xy};\ \forall\ g \in G, x, y \in Z_p^*$.
   (ii) Non-degeneracy: Let $g \in G_1$ such that $e(g, g) \neq 1$.
   (iii) Compatibility: It has been created a technique for assessing bilinear mapping that is both efficient and accurate.

6) Computational Assumption: An assumption of computational hardness is a notion that a specific issue cannot be solved effectively in the context of computational complexity theory [6].
   (i) Let g be a primitive root belong to Z_p and t be a nonzero element of Z_p, the hard problem Discrete Logarithm Problem (DLP) Eq. 18:

$$g^x = \mathrm{t}(\mathrm{mod}\,\mathrm{p}). \tag{18}$$

   (ii) Let G be a multiplicative cyclic group of order p, with generator $< g >$. The security hardness is based on decisional Bilinear Diffie-Hellman (g, g^x, g^y, g^z), where $(x, y, z) \in Z_p^*$, Eq. 19:

$$g^z = g^{xy}(\mathrm{mod}\ \mathrm{p}). \tag{19}$$

Privacy in social networks is protecting personal data. It preserves basic rights, assigns property rights, balances consents, and organizes user behavior. Garg et al. [11] proposed a secure, and provable authentication scheme for IoT, enabling Industry 4.0. Provably Secure Authentication (PSA) and Key Agreement Protocol (KAP) are emphasized for IoT. The proposed scheme cannot meet all industry 4.0 security and privacy criteria. Astorga et al., [12] presented identity management in industry 4.0. Protecting edge-server and industrial IoT communications. Proposed scheme protects edge server and industrial IoT devices, but it has the lacks of authentication and can't secure all of Industry 4.0. A Lightweight Key Exchange (LKE) technique was presented by Gaba et al. [13], which prevents unauthorized parties from accessing the industrial network. In this paper, our proposed scheme revokes emotional proxy acts (such as eavesdropping) and protects user privacy and integrity.

3 Proposed Scheme

Our proposed scheme-based HC. It has three parties (sender, receiver and TTP [5]) and multi platforms of industry 4.0. The flow of our scheme is shown in Fig. 2.

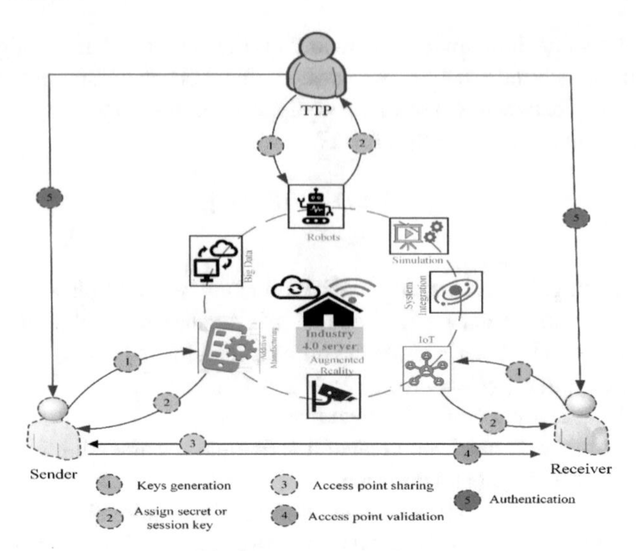

Fig. 2. Our scheme flows.

A. Sender: The sender delivers industry 4.0 data. Industry 4.0 sends to TTP and receiver. Key creation, secret or session key assignments, access point sharing, and access point validation make up the sender. Below are the phases:

(1) Keys setup: In key setup, it takes the security parameter k and produce the output such as p_k and s_k (Algorithm 1).

Algorithm 1	:	Sender keys setup
Input	:	Select k as a security parameter, where $k \in Z_p^*$
Output	:	Sender PK and secret key (S_{p_k}, S_{s_k})
Step 1	:	Generate g_1 with prime order p, where $g_1 \leftarrow g$;
Step 2	:	Compute bilinear map $\hat{e}$ as $\hat{e} = G_1 \rightarrow G_2$;
Step 3	:	Set $R_\tau = \{S_{\tau_1}, S_{\tau_2}, ..., S_{\tau_n}\}$, let $R_\tau = (0,1)^*$;
Step 4	:	Randomly selects $S_x, S_\gamma \epsilon S_\tau$ and generate $S_t{''} \in Z_p$;
Step 5	:	Computes $S_y = \hat{e}(g,g)^{S_\alpha}$ and $S_T = S_{g^\beta}$, where $S_{p_k} = (e, g, S_y, R_T)$ and $S_{s_k} = (x, S_t{''}, 1 \leq j \leq n);\psi$
Step 6	:	Finish

(2) Keys generation: Sender transmits private and PKs to industry manager during key generation. Algorithm (Algorithm 2), assigns the manager the secret or session key for transferring information to the recipient.

Algorithm 2	:	Keys generation
Input	:	Select $(x, \gamma) \in Z_p^*$
Output	:	Output (s_k)
Step 1	:	Compute the set of elements: $\delta = (g)^{(x-\gamma)}$;
Step 2	:	Compute $\delta = g^{\gamma t_j^{-1}}$, where $1 \leq j \leq n$;
Step 3	:	Return $(s_k)_w = (\delta, \forall\, a_j{''} \in w: \delta')$ to the users;
Step 4	:	TTP verify δ & w attributes;
Step 5	:	Finish;ψ

(3) Assign secret or session key: Industry 4.0 servers deliver session or secret keys for all parties' connections. If users choose session-based connections, it assigns one session to all participants. If the session time is up, it re-generates the key. Recursive connectedness. If the user desires a secret-key connection, the industry 4.0 server will assign one. Using the right secret key, the user may transmit and receive. Else it will reject and display the error message $\perp$ (Algorithm 3).

Algorithm 3	:	Secret or session key
Input	:	Input $(\delta, \gamma) \in \delta'$
Output	:	Output (S_k, s_k)
Step 1	:	Compute $\delta = (g)^{(x-\gamma)}$;
Step 2	:	Compute $\delta = g^{\gamma t_j^{-1}}$, where $1 \leq j \leq n$;
Step 3	:	Return $(s_k)_w = (\delta, \forall\, a_j{''} \in w: \delta')$ to the users;
Step 4	:	Return session key $(S_k)_w = (\delta, \forall\, a_j{''} \in w: \delta')$ to the user;
Step 5	:	Else display error message $(\perp)$;
Step 6	:	Finish;

B. Receiver

(1) Keys generation: The receiver communicates R_d and R_p to the industry manager during key generation. The manager is given the secret or session key for sending information.
(2) Assign secret or session key: Industry 4.0 server delivers session S'_{R_p} or secret key s_{R_p} for all receivers' connections.

If users choose session-based connections, it assigns one session to all participants. If the session time is up, it re-generates the key. Recursive connectedness. If the user desires a secret-key connection, the industry 4.0 server will assign one. Using the right secret key, the user may transmit and receive. Else it will reject and display the error message.

(3) Access point sharing: With key generation and secret or session keys, users may sell and buy information. Sender may give required information to receiver, and receiver can receive it using specified keys.

4 Analysis of Our Scheme

The analysis of our scheme in term of security properties such as integrity, authenticity, un-forgeability, confidentiality, verifiability & un-traceability (Table II). The analysis of Pailliar, ECC and HEC based HE and comparison is shown in Fig. 3.

1) Integrity: In integrity the access point of the sender and validation point of the receiver remain constant (Theorem 1).

Theorem 1: The eavesdropper tries to change the original message of the sender and receiver (access point sharing and validation).

Proof: The eavesdroppers are unable to generate the access keys (e.g., secret or session key). Therefore, the integrity of the access point is secure and the integrity property has remained constant.

Theorem 2: How to maintain authenticity property?

Proof: On behalf of a private and PKs, it fulfills the authenticity property and prevents the eavesdropper attacks.

Theorem 3: Let the original users (sender & receiver) are capable to forge.

Proof: The valid private and PKs are used to provides secret and session keys. Due to its own secret or session the communicating parties (sender and receiver) cannot be able to forge.

Theorem 4: If an eavesdropper knows something about the original access point, they can use it?

Proof: The keys setup is based on private and PKs elements (Algorithm 2) respectively. Therefore, the confidentiality property remains constant.

Theorem 5: Can the user (sender & receiver) provides verifiability property?

Proof: Due to the usage of private and public keys (Algorithms 1, 2 & 3). Therefore, our proposed scheme provides verifiability property.

Theorem 6: Let the sender and receiver is un-traceable.

Proof: From (Algorithm 3), it is clear that the value of sender and receiver is based on the secret/session keys. Therefore, un-authorized users cannot trace the original users.

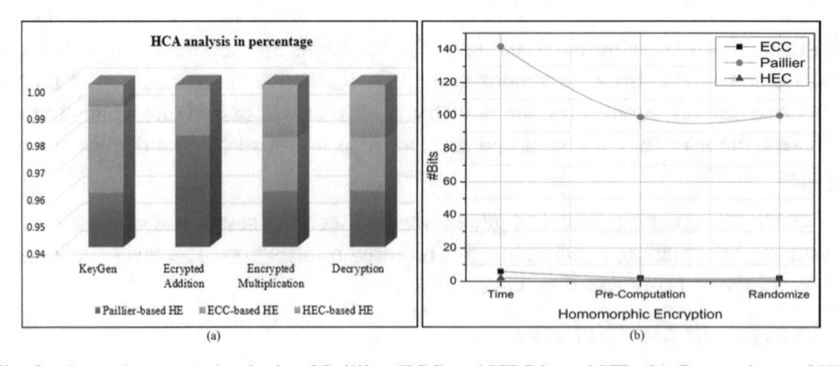

Fig. 3. Our scheme (a) Analysis of Pailliar, ECC and HEC based HE, (b) Comparison of HE.

5 Conclusion and Future Work

The current digital world is built based on the Fourth Industrial Revolution, known as Industry 4.0. In this environment, users of digital technology send enormous volumes of data. The sender and receiver data communication environments suffer from a lack of maintenance due to the Machine Tools Gap (MTG). As a result, every aspect of Industry 4.0, including information technology, networks, industrial production, and so on, must ensure its maintenance in terms of security and privacy. Cryptographic techniques are the most effective means of protecting against attacks. The sender and receiver legitimacy are guaranteed by our proposed scheme. Our proposed scheme provides the security and privacy properties such as integrity, authenticity, un-forgeability, secrecy, verifiability, and un-traceability. In the future, it is openly challenging for researchers to implement HCA for any data trading organization with Industry 4.0.

References

1. Pascal, P.: Public-key cryptosystems based on composite degree residuosity classes. In: Stern, J. (eds.) EUROCRYPT 1999. LNCS, vol. 1592, pp. 223–238. Springer, Heidelberg (1999). https://doi.org/10.1007/3-540-48910-X_16
2. Taher, E.: A public key cryptosystem and a signature scheme based on discrete logarithms. IEEE Trans. Inf. Theory **31**, 469–472 (1985)
3. Josh, B.: Dense probabilistic encryption. In: Proceedings of the Workshop on Selected Areas of Cryptography, pp.120–128 (1994)
4. An, J.H., Bellare, M.: Does encryption with redundancy provide authenticity? In: Pfitzmann, B. (ed.) EUROCRYPT 2001. LNCS, vol. 2045, pp. 512–528. Springer, Heidelberg (2001). https://doi.org/10.1007/3-540-44987-6_31
5. Ullah, S., Li, X.Y., Lan, Z.: A novel trusted third party based signcryption scheme. Multimedia Tools Appl. **79**, 22749–22769 (2020). https://doi.org/10.1007/s11042-020-09027-w
6. Shamsher, U., Lan, Z., Wasif, S.M., Tanveer, H.M.: τ-Access policy: attribute-based encryption scheme for social network based data trading. IEEE China Commun. **18**(8), 183–198 (2021)
7. Ullah, S., Li, X.Y., Lan, Z.: A review of signcryption schemes based on hyper elliptic curve. In: 2017 3rd International Conference on Big Data Computing and Communications (BIGCOM), IEEE, pp. 51–58 (2017)
8. Boruah, D., Saikia, M.: Implementation of ElGamal Elliptic Curve Cryptography over prime field using C. In: International Conference on Information Communication and Embedded Systems (ICICES2014), pp. 1-7. Chennai, India (2014). https://doi.org/10.1109/ICICES.2014.7033751
9. Saikia, M.: A Brief Overview of Homomorphic Cryptosystem and Their Applications (2015). https://doi.org/10.13140/RG.2.1.5062.5360
10. Boneh, D., Franklin, M.: Identity-based encryption from the weil pairing. In: Kilian, J. (eds.) CRYPTO 2001. LNCS, vol. 2139, pp 213–229. Springer, Heidelberg (2001). https://doi.org/10.1007/3-540-44647-8_13
11. Garg, S., Kaur, K., Kaddoum, G., Choo, K.-K.R.: Toward secure and provable authentication for internet of things: realizing industry 4.0. IEEE Internet Things J. 7(5), 4598-4606 (2020). https://doi.org/10.1109/JIOT.2019.2942271
12. Astorga, J., Barceló, M., Urbieta, A., Jacob, E.: How to survive identity management in the industry 4.0 era. IEEE Access (2021). https://doi.org/10.1109/ACCESS.2021.3092203
13. Gaba, G.S., Kumar, G., Monga, H., Kim, T.-H., Liyanage, M., Kumar, P.: Robust and lightweight key exchange (LKE) protocol for industry 4.0. IEEE Access, **8**, 132808–132824 (2020). https://doi.org/10.1109/ACCESS.2020.3010302

Fixed-Parameter Tractability for Branchwidth of the Maximum-Weight Edge-Colored Subgraph Problem

Alessandro Aloisio(✉)

Dipartimento di Scienze Umanistiche e Sociali Internazionali, Università degli Studi Internazionali di Roma, Roma, Italy
alessandro.aloisio@unint.eu

Abstract. A k-edge-coloring of a(n undirected) graph is an assignment of one of k possible colors to each of the edges of the graph such that different colors are assigned to any two adjacent (but different) edges. Given a weight function on colored edges of a graph G, a *maximum-weight k-edge-coloring* of G is a k-edge-coloring of a subgraph of G whose total weight of colored edges is maximum. The MAXIMUM WEIGHT EDGE-COLORED SUBGRAPH asks, for a graph G, an integer k, and a weight function w for k-colored edges of G, to find a maximum-weight k-edge-colored subgraph of G. We propose a fixed-parameter tractable algorithm for the MAXIMUM WEIGHT EDGE-COLORED SUBGRAPH problem with respect to two parameters: the branchwidth of the graph and the number k of colors. This result can be transferred to the treewidth.

1 Introduction

Graph coloring comprises a class of problems in which colors have to be assigned to various components of graphs. The most well-known problems concern vertex and edge colorings [20], which have captured attention since the beginning of graph theory. Indeed, coloring problems form a captivating area of discrete mathematics due to two features: they typically require non-trivial proofs that are often of profound mathematical interest, and they have widespread applications across various fields of computer science.

Given a graph, *edge coloring* [20] aims to assign colors to the edges in such a way that no two adjacent, but different, edges are assigned the same color. The edge coloring problem consists in determining the minimum number of colors required to color all the edges of a graph.

Among the applications of edge coloring, we can mention the following: task assignment to agents with time conflicts [10], frequency allocation in fiber optic networks [22], and link scheduling in sensor networks using the TDMA MAC protocol [26]. These applications are also employed in military contexts [34]. In [10], a game

This work is partially supported by the project 'Soluzioni innovative per il problema della copertura nelle multi-interfacce e relative varianti', UNINT, and by the Italian National Group for Scientific Computation (GNCS-INdAM).

L. Barolli (Ed.): AINA 2024, LNDECT 204, pp. 86–95, 2024.
https://doi.org/10.1007/978-3-031-57942-4_10

theory model is proposed related to graphical games, as seen in [4,6–9]. This model is particularly interesting when considering a decentralized version of the original edge coloring problem. Such an approach implies that some of the positive and negative results, exemplified in [4,6–8], could be extended to the multi-agent model.

In the application example discussed in [26], the authors refer to a theoretical problem applied to a sensor network consisting of a large number of static sensor nodes deployed on-the-fly for unattended operation. By proposing Time Division Multiple Access (TDMA) MAC protocols to avoid problems in the sensor network, they suggest a link scheduling algorithm that resorts to the edge-coloring problem of the sensor networks.

In this paper, we propose an initial study of the MAXIMUM WEIGHT EDGE-COLORED SUBGRAPH with respect to parameterized complexity theory. The problem consists in finding a subgraph that is k-edge-colorable while maximizing the sum of weights that depend both on edges and colors. We formulate a simple dynamic programming algorithm that witnesses that this problem is fixed-parameter tractable with respect to the graph branchwidth, and the number of colors as parameters. We also point out how to transfer these results to the treewidth.

The paper is structured as follows. Section 2, presents preliminaries and a formal definition of the problem. Section 3 covers the related work. Our results on branchwidth are discussed in Sect. 4. The paper concludes in Sect. 5, where we summarize our results and delineate directions for future work.

2 Preliminaries and Problem Statement

We generally consider undirected and finite graphs without loops and multiple edges, formalized as pairs $G = \langle V, E \rangle$ that consist of a finite vertex set V and a set E of edges that are 2-element subsets of V. For a graph denoted by G we permit to denote by $V(G)$, and by $E(G)$, the vertex set, and the edge set of G, respectively. Every edge of a graph G is a 2-element set $\{u, v\} \in E(G)$, where the vertices u and v are called the *endpoints* or *extremes* of the edge. A vertex v is *incident* to an edge e if $e = \{v, u\}$ for some vertex u. Two edges are called *adjacent* if they share a common endpoint. By $\deg_G(u)$ we mean the *degree* of a vertex u, that is, the number of edges of G that u is incident with. By $\delta(G)$ and $\Delta(G)$ we refer to the minimum and maximum degree of the vertices in G, respectively.

Let G be a graph. A subset of edges in G is called a *matching* in G if it does not contain any two adjacent edges. By a *maximum matching* of G we mean a maximum-cardinality matching, that is, a matching of G with a maximum number of edges among all matchings. Let $k \in \mathbb{N}^{\geq 1}$ be a positive integer. A graph G is said to be *k-edge-colorable* if each of its edges can be colored with a color in $[k] = \{1, \ldots, k\}$ such that there are no two adjacent (non-coinciding) edges of the same color. The *chromatic index* of G, denoted by $\chi'(G)$, is the smallest $k \in \mathbb{N}^{\geq 1}$ for which G is k-edge-colorable.

In [36,37], Shannon proved that for any graph G, $\Delta(G) \leq \chi'(G) \leq \left\lfloor \frac{3\Delta(G)}{2} \right\rfloor$ holds. Furthermore, in [37,38], Vizing showed that $\Delta(G) \leq \chi'(G) \leq \Delta(G) + \mu(G)$ holds for any multigraph G, where $\mu(G)$ denotes the maximum multiplicity of an edge in G. When the number of available colors, denoted as k, is less than $\chi'(G)$, there are not enough colors to color all the edges. This leads to another problem, which aims to color as many edges as possible using only k colors.

A subgraph H of G is called *maximum k-edge-colorable* if H is k-edge-colorable and contains the maximum number of edges among all k-edge-colorable subgraphs of G. For $k \geq 0$ and a graph G, let $\nu_k(G) = \max\{|E(H)| : H$ is a k-edge-colorable subgraph of $G\}$. Clearly, a k-edge-colorable subgraph is maximum if it contains exactly $\nu_k(G)$ edges. Please note that $\nu_1(G)$ is the size of a maximum matching of G, denoted simply as $\nu(G)$.

Let $w : E(G) \times [k] \to \mathbb{R}^{>0}$ be a function that assign a positive weight $w(e,c)$ to every couple made of an edge $e = \{u,v\} \in E(G)$ and a color $c \in [k]$. A subgraph H of G is called *maximum weight k-edge-colorable* if H is k-edge-colorable and maximizes $w(H) = \sum_{e \in E(H)} w(e,c)$. We can now formally state the problem tackled by our paper.

MAX-WEIGHT–EDGE-COLORED–SUBGRAPH

Input: A graph $G = \langle V,E \rangle$, $k \in \mathbb{N}^{\geq 1}$, and a colored-edge weight function $w : E(G) \times [k] \to \mathbb{R}^{>0}$.

Solution: A k-edge-coloring c of a subgraph H of G.

Goal: Find a subgraph H of G with k-edge-coloring $c : E(H) \to [k]$ whose total weight $w(H) = \sum_{e \in E(H)} w(e, c(e))$ of its colored edges is maximum.

When the weights are unitary, the problem becomes:.

Maximum Edge-Colorable Subgraph

Instance: A graph $G = \langle V,E \rangle$, and $k \in \mathbb{N}^{\geq 1}$.

Solution: A subgraph H of G that is k-edge-colorable.

Goal: Find a maximum k-edge-colorable subgraph.

The decision version of the previous problem is the following.

Maximum Edge-Colorable Subgraph (decision version)

Instance: A graph $G = \langle V,E \rangle$, and $k,l \in \mathbb{N}^{\geq 1}$.

Question: Is there an k-edge-colorable subgraph of G with $\geq l$ edges?

It is not difficult to prove that the decision version of the maximum k-edge-colorable subgraph problem is NP-complete for every $k \geq 2$. In fact, when G is cubic and $k = 2$, we have that $\nu_2(G) = |V|$ if and only if G contains two edge-disjoint perfect matchings. The latter condition is equivalent to saying that G is 3-edge-colorable, which is an NP-complete problem as Holyer demonstrated in [28].

We study the MAXIMUM WEIGHT EDGE-COLORED SUBGRAPH problem from the perspective of parameterized complexity theory, which provides a better complexity analysis than the classical theory of NP-completeness. We present preliminary results addressing the fixed-parameter tractability (FPT) of this problem with respect to two graph-theoretic parameters. Specifically, we introduce an algorithm that solves our problem in FPT time concerning branchwidth (treewidth) and k. For additional theoretical details, readers are encouraged to refer to [21].

3 Related Work

A survey on edge coloring can be found in [20]. Albertson and Haas investigated the problem in [2] for cubic graphs. In [19], the authors proved that for any cubic multigraph G $\nu_2(G)+\nu_3(G) \geq 2|V(G)|$, and in [31,32], Mkrtchyan et al. showed that for any cubic multigraph G, $\nu_2(G) \leq \frac{|V(G)|+2\nu_3(G)}{4}$. Finally, in [29], it is shown that the sequence ν_k is convex in the class of bipartite multigraphs.

In [23], an approximation algorithm for the problem is presented for each $k \geq 2$. For any fixed value of $k \geq 2$, the algorithms are proven to have specific approximation ratios. In [30], two approximation algorithms for the maximum 2-edge-colorable subgraph and maximum 3-edge-colorable subgraph problems are presented. Some structural properties of maximum k-edge subgraph problem are proved in [19,33].

Galby et al. [25] demonstrated that k-edge coloring is fixed-parameter tractable when parameterized by the number of colors k and the number of vertices of maximum degree. In [27], Grüttemeier et al. proposed a kernel with the parameter k along with the number of edges that need to be deleted from the graph to achieve the maximum edge-colorable subgraph, and presented other results. In [1], Akanksha Agrawal et al. showed FPT results for the maximum edge colorable subgraph problem when the parameters are: k and the solution size, k and the vertex cover number, along with other results. The authors of [10,11] also examined specific fixed-parameter aspects of the maximum edge colorable subgraph problem, forming our study's basis.

4 Graphs with Bounded Branchwidth

We describe here a fixed-parameter tractable (FPT) algorithm for the MAXIMUM WEIGHT EDGE-COLORED SUBGRAPH problem on graphs with bounded *branchwidth*.

Definition 1 [21]. Let $G = \langle V, E\rangle$ be a graph. A *branch decomposition* of a graph G is a pair $\langle T, \phi\rangle$ that consists of a tree T with $|E|$ leaves whose all internal nodes have degree 3, and a bijection $\phi : E \to \mathit{Leaves}(T)$ from the edge set E of G to the leaves of T.

Now suppose that G has at least two edges, that is, that $|E| \geq 2$. Let $\langle T, \phi\rangle$ be a branch decomposition of G. Then each removal of an edge $\{i, j\}$ from T divides this tree into two connected components, and in doing so it partitions the edge set E of G into two subsets X and $E \setminus X$. We denote by $\delta(X)$ the *border (vertices)* of X, and that is, the set of those vertices of G that are both incident to an edge of X and to an edge of $E \setminus X$. The *width* of the branch decomposition $\langle T, \phi\rangle$ is the maximum of the cardinalities

$|\delta(X)|$ of border vertices $\delta(X)$ over all partitions $\langle X, E \setminus X \rangle$ of the edges of G that arise by removal of edges from T.

The *branchwidth* $bw(G)$ of G is 0 if $|E| \leq 1$, and otherwise (if $|E| \geq 2$) it is the minimum width over all branch decomposition of G.

4.1 A Dynamic Programming Algorithm

Consider a graph $G = \langle V, E \rangle$ with a branch decomposition $\langle T, \phi \rangle$ of width h. We now present a dynamic programming algorithm designed to solve the MAXIMUM WEIGHT EDGE-COLORED SUBGRAPH problem for G, leveraging the structure and properties of the branch decomposition $\langle T, \phi \rangle$. To avoid confusion between the vertices of the graph and those of T, we refer to the vertices of T as nodes. We arbitrarily pick an internal node of T, call it the *root* by orienting all tree edges away from the root; we permit to denote this directed tree again by T. While the root of (the now directed tree) T has three children, all other internal nodes of T have precisely two children.

Let $T(i)$ be the subtree induced by the node i of T and all its descendants. We denote by $E_i := \phi^{-1}(Leaves(T(i)))$ the edges of G that occur in the leaves of the subtree $T(i)$ of T at i; by $G(i)$ the subgraph of G induced by the edges in E_i; and by $\delta(i)$ (*border*) the set of those nodes of G that are both incident to an edge of E_i and to an edge of $E \setminus E_i$.

In the core of the algorithm, we compute the optimal value of a constrained version of the problem restricted to the subgraph $G(i)$. The constraints are given by the colors incident on each node in the border $\delta(i)$: namely, for every $u \in \delta(i)$, a set $A(u)$ of edge colors that must be used is given.

Max.-weight edge-colorable subgraph of $G(i)$ with fixed border-edge colors	
Input:	A subgraph $G(i)$ induced by E_i, a border $\delta(i)$, an integer $k \geq 1$, a function $w : E(G) \times [k] \to \mathbb{R}^{>0}$, and a collection $\mathscr{A} = \bigcup_{u \in \delta(i)} \{A(u)\}$, with $A(u) \in [k]$
Solution:	A k-edge coloring c of a subgraph H of $G(i)$ such that the color *incident* to the vertex u in $\delta(i)$ are those in $A(u)$, for every $u \in \delta(i)$.
Goal:	Find a subgraph H of $G(i)$ with k-edge-coloring $c : E(H) \to [k]$ whose total weight $W(i, \mathscr{A}) = \sum_{e \in E(H)} w(e, c(e))$ of its colored edges is maximum.

Here, a color is *incident* to a vertex u if it is used in at least one edge incident to u in $G(i)$. Please note that the branchwidth of G is h, so the number of subsets in $\mathscr{A}$ is less or equal to h.

If i is a leaf corresponding to an edge $e = \{u, v\} \in E$, then there are three different cases.

Case 1: Neither u and v are not connected to some other vertices in $V \setminus \{u, v\}$. Then $\mathscr{A} = \emptyset$, because the border is empty, and $W(i, \mathscr{A}) = \max_{c \in [k]} \{w(e, c)\}$. In fact, edge e is isolated. Therefore we choose the color $c \in [k]$ that maximize the weight $w(e, c)$.

Case 2: Only one of the two vertices u and v is connected with the rest of the graph. We can suppose w.l.o.g. that u is the connected vertices.

Then, $\mathscr{A} = \{A(u)\}$ and

- $W(i,\mathscr{A}) = 0 \quad$ if $A(u) = \emptyset$;
- $W(i,\mathscr{A}) = w(e,c) \quad$ if $A(u) = \{c\}$, with $w \in [k]$
- $W(i,\mathscr{A}) = -\inf \quad$ otherwise.

In fact, the only way to have a color $c \in A(u)$ incident to u is by coloring $e = \{u,v\}$ with c, thereby obtaining a profit of $w(e,c)$ in $G(i)$. Please note that, since there is no way to have two or more colors incident to u in $G(i)$, there are no solutions if $|A(u)| \geq 2$. Thus, we set $W(i,\mathscr{A}) = -\infty$.

Case 3: u and v are connected to the rest of the graph, so, $\mathscr{A} = \{A(u),A(v)\}$ and

- $W(i,\mathscr{A}) = 0 \quad$ if $A(u) = A(v) = \emptyset$;
- $W(i,\mathscr{A}) = w(e,c) \quad$ if $A(u) = A(v) = \{c\}$, with $c \in [k]$;
- $W(i,\mathscr{A}) = -\inf \quad$ otherwise.

In fact, the only way to have a color $c \in [k]$ incident to both u and v, is to color edge $e = \{u,v\}$ with color c, obtaining a value $W(i,\mathscr{A}) = w(e,c)$. Also here, if $|A(u)| \geq 2$ or $|A(v)| \geq 2$, there is no way to have a feasible solution of the constrained problem, therefore we set $W(i,\mathscr{A}) = -\inf$.

We now analyze the case of an internal node i of T, different from *root*, with children j and k. We can compute $W(i,\mathscr{A})$ by using the already computed values $W(j,\mathscr{B})$ and $W(k,\mathscr{C})$. Please note that $G(j)$ and $G(k)$ can share some common vertices, that is $V(G_j) \cap V(G_k) \neq \emptyset$. Moreover, every vertex in $V(G_j) \cap V(G_k)$ must belong to $\delta(j) \cap \delta(k)$ and vice versa, thus $V(G_j) \cap V(G_k) = \delta(j) \cap \delta(k)$.

To compute $W(i,\mathscr{A})$, for a given collection $\mathscr{A}$, we can solve this maximization problem, for every possible color collection $\mathscr{B}$ and $\mathscr{C}$. If there is no solution, we set $W(i,\mathscr{A}) = -\infty$.

$$\begin{aligned} \max \quad & W(j,\mathscr{B}) + W(k,\mathscr{C}) \\ \text{s.t.} \quad & A(u) = B(u) \quad \forall u \in ((\delta(j) \setminus \delta(k)) \\ & A(u) = C(u) \quad \forall u \in (\delta(k) \setminus \delta(j)) \\ & A(u) = B(u) \cup C(u) \quad \forall u \in ((\delta(j) \cap \delta(k)) \cap \delta(i)) \\ & B(u) \cap C(u) = \emptyset \quad \forall u \in (\delta(j) \cap \delta(k)) \end{aligned} \tag{1}$$

In fact, the border of $G(i)$ is a subset of the union of the borders of $G(j)$ and $G(k)$, i.e., $\delta(i) \subseteq \delta(j) \cup \delta(k)$.

To merge the solutions found for $G(j)$ and $G(k)$ successfully, we must ensure the following: for each node u in $\delta(i)$ that is also in $\delta(j)$ but not in $\delta(k)$, the set of colors $A(u)$ equals $B(u)$ (first constraint); for each node u in $\delta(i)$ that is also in $\delta(k)$ but not in $\delta(j)$, the set of colors $A(u)$ equals $C(u)$ (second constraint); for each node u in $\delta(i)$ that also belongs to both $\delta(k)$ and $\delta(j)$, the set of colors $A(u)$ equals $C(u) \cup B(u)$ (third constraint); and there is no color that belongs to both $B(u)$ and $C(u)$ when u is in both $\delta(j)$ and $\delta(k)$ (fourth constraint).

Please note that the fourth constraint prevents a color from being incident to more than one edge on the same node of $G(i)$. In fact, if $c \in C(u) \cup B(u)$, then c colors an edge in $G(j)$ and another edge in $G(k)$, both incident to u.

We now compute the constrained problem for node *root*. Since *root* has exactly three children: j, k, and l, we utilize a slightly different maximization problem. Moreover, as $G(root)$ represents the entire graph G, the border of $G(root)$ is empty, i.e., $\delta(X_{root}) =$

$\emptyset$. This implies that $\mathscr{A} = \emptyset$, and we solve the following problem for every possible collection of $\mathscr{B}$, $\mathscr{C}$, and $\mathscr{D}$.

$$\begin{aligned} \max \quad & W(j,\mathscr{B}) + W(k,\mathscr{C}) + W(l,\mathscr{D}) \\ \text{s.t.} \quad & B(u) \cap C(u) = \emptyset \quad \forall u \in (\delta(j) \cap \delta(k)) \\ & B(u) \cap D(u) = \emptyset \quad \forall u \in (\delta(j) \cap \delta(E_l)) \\ & C(u) \cap D(u) = \emptyset \quad \forall u \in (\delta(k) \cap \delta(E_l)) \end{aligned} \tag{2}$$

In the preceding problem, the first constraint ensures that each vertex u belonging to both borders $\delta(j)$ and $\delta(k)$ must not have an incident color on edges of both $G(j)$ and $G(k)$ that are adjacent via vertex u. The other two constraints also guarantee the same for $G(j)$ and $G(l)$, and $G(k)$ and $G(l)$, respectively.

After computing $W(root, \mathscr{A})$, we can find a solution for the MAXIMUM WEIGHT EDGE-COLORED SUBGRAPH problem by going back from the *root* to the leaves of T, following the chosen optimal values $W(i, \mathscr{A})$, and coloring the edges with respect to the constraints and the optimal values.

We analyze now the time complexity of the algorithm. For a leaf i of T, $G(i)$ is an edge $\{u,v\}$, and there are at most $O(2^{2k})$ possible collections $\mathscr{A} = \{A(u), A(v)\}$. This results in a complexity of $O(2^{2k})$ for an edge.

If i is an internal node, we note that for any two collections $\mathscr{B}$ and $\mathscr{C}$, the collection $\mathscr{A}$ is uniquely determined. Since there are at most h elements in every collection, each chosen among 2^k subsets of $[k]$, we have to solve $O(2^{2hk})$ Problems (1). This leads to a complexity for and internal node i of $O(h \cdot 2^{2hk})$, since solving Problem (1) for two specific collections $\mathscr{B}$ and $\mathscr{C}$ costs $O(h)$ in time .

Finally, for *root* there are three different collections $\mathscr{B}$, $\mathscr{C}$, and $\mathscr{D}$, thus the complexity is $O(h \cdot 2^{3hk})$. It is easy to observe that we can simplify the issue by replacing the root problem with one involving an internal node and another similar problem.

In conclusion, since the number of internal node is at most equal to the leaves of T, which are $|E(G)|$, the time complexity of the algorithm is $O(h \cdot 2^{2hk}|E(G)|)$. We can then state the following:

Theorem 2. *There is a deterministic algorithm that, given an instance $\langle G,k,w \rangle$ of* MAXIMUM WEIGHT EDGE-COLORED SUBGRAPH *and a branch decomposition $\langle T, \phi \rangle$ of G of width h, computes a solution of this problem (that is, a maximum-weight k-edge-colored subgraph of G) in $O(h \cdot 2^{2hk} m)$ time where $m = |E(G)|$.*

As the notions of branchwidth and treewidth of a graph are well-known to be closely related, we can transfer this result to also obtain a similar result for treewidth. For this purpose we use two results from the literature: first, treewidth and branchwidth can be bounded in terms of each other, see Theorem 3 below, and second, there is an FPT-algorithm for obtaining a branch decomposition whose width is at most twice the branchwidth of the considered graph, see Theorem 4 below.

Theorem 3 (Treewidth Versus Branchwidth (Robertson and Seymour *[35]***)).** *For all graphs G with branchwidth $bw(G) > 1$ it holds:*

$$bw(G) \leq tw(G) + 1 \leq \frac{3}{2} bw(G) \,. \tag{3}$$

Theorem 4 (Fast Approx. of Branchwidth (Fomin and Korhonen *[24]*). *There is a deterministic algorithm that, for a given graph G and an integer h, runs in time* $2^{O(h)}n$ *with* $n = |V(G)|$ *and either produces a branch decomposition of G of width* $2h$ *or determines that the branchwidth of G is more than h.*

By bounding the branchwidth of a graph by its treewidth via the first inequality in (3) in Theorem 3 and using the approximation result of a branch decomposition in Theorem 4 when an upper bound on the branchwidth of a graph is given, we can transfer the fixed-parameter tractability result for MAXIMUM WEIGHT EDGE-COLORED SUBGRAPH of Theorem 2 with respect to branchwidth to a fixed-parameter tractability result for this problem with respect to treewidth. In this way we obtain the following corollary.

Corollary 5. *There is a deterministic algorithm that, given an instance* $\langle G,k,w\rangle$ *of* MAXIMUM WEIGHT EDGE-COLORED SUBGRAPH, *and 2-approximate branch decomposition* $\langle T,\phi\rangle$, *computes a solution of this problem (a maximum-weight edge-k-colored subgraph of G) in time* $O(t \cdot 2^{4(t+1)k}m)$, *where and* $t = tw(G)$.

5 Conclusion and Future Work

We propose a preliminary investigation of the MAXIMUM WEIGHT EDGE-COLORED SUBGRAPH problem from the perspective of parameterized complexity theory. We describe a simple FPT-algorithm for the MAXIMUM WEIGHT EDGE-COLORED SUBGRAPH problem with respect to the branchwidth of the instance graph and the number of available colors. We also argued that this result can be transferred to one that uses treewidth instead of branchwidth as a parameter.

We believe that the problem deserves further investigation. In particular, we want to explore whether FPT results can also be obtained for more general forms of graphs, such as those with bounded cliquewidth [10], and follow the ideas presented in [12,13,18]. Another line of research we believe is worth pursuing involves generalizing the model by introducing a budget constraint, as discussed in [3,5,14–18]. This will also include using a non-linear objective function and extending the weight function to the vertices.

References

1. Agrawal, A., Kundu, M., Sahu, A., Saurabh, S., Tale, P.: Parameterized complexity of maximum edge colorable subgraph. Algorithmica **84**(10), 3075–3100 (2022)
2. Albertson, M.O., Haas, R.: The edge chromatic difference sequence of a cubic graph. Disc. Math. **177**(1), 1–8 (1997)
3. Aloisio, A.: Coverage subject to a budget on multi-interface networks with bounded carving-width. In: Advances in Intelligent Systems and Computing (WAINA). LNCS, vol. 1150, pp. 937–946. Springer, Cham (2020). https://doi.org/10.1007/978-3-030-44038-1_85
4. Aloisio, A.: Distance hypergraph polymatrix coordination games. In: Proceedings of the 22nd Conference of Autonomous Agents and Multi-Agent Systems (AAMAS), pp. 2679–2681 (2023)

5. Aloisio, A.: Algorithmic aspects of distributing energy consumption in multi-interface networks. In: Advances in Intelligent Systems and Computing (WAINA). Springer, Cham (2024)
6. Aloisio, A., Flammini, M., Kodric, B., Vinci, C.: Distance polymatrix coordination games. In: Proceedings of the 30th International Joint Conference Artificial Intelligence (IJCAI), pp. 3–9 (2021)
7. Aloisio, A., Flammini, M., Kodric, B., Vinci, C.: Distance polymatrix coordination games (short paper). In: SPIRIT Co-located with 22nd International Conference AIxIA 2023 (CEUR), 7–9 November 2023, Rome, vol. 3585 (2023)
8. Aloisio, A., Flammini, M., Vinci, C.: The impact of selfishness in hypergraph hedonic games. In: Proceedings of the 34th Conference of Artificial Intelligence (AAAI), pp. 1766–1773 (2020)
9. Aloisio, A., Flammini, M., Vinci, C.: Generalized distance polymatrix games. In: Fernau, H., Gaspers, S., Klasing, R. (eds.) Theory and Practice of Computer Science: 49th International Conference on Current Trends in Theory and Practice of Computer Science (SOFSEM 2024), pp. 25–39. Springer, Cham (2024). https://doi.org/10.1007/978-3-031-52113-3_2
10. Aloisio, A., Mkrtchyan, V.: On the fixed-parameter tractability of the maximum 2-edge-colorable subgraph problem. CoRR (2019)
11. Aloisio, A., Mkrtchyan, V.: Algorithmic aspects of the maximum 2-edge-colorable subgraph problem. In: Barolli, L., Woungang, I., Enokido, T. (eds.) Advanced Information Networking and Applications (AINA-2021), vol. 3, pp. 232–241. Springer, Cham (2021). https://doi.org/10.1007/978-3-030-75078-7_24
12. Aloisio, A., Navarra, A.: Balancing energy consumption for the establishment of multi-interface networks. In: Italiano, G.F., Margaria-Steffen, T., Pokorný, J., Quisquater, J.-J., Wattenhofer, R. (eds.) Theory and Practice of Computer Science (SOFSEM 2015). LNCS, vol. 8939, pp. 102–114. Springer, Heidelberg (2015). https://doi.org/10.1007/978-3-662-46078-8_9
13. Aloisio, A., Navarra, A.: Budgeted constrained coverage on bounded carving-width and series-parallel multi-interface networks. Internet of Things **11**, 100259 (2020)
14. Aloisio, A., Navarra, A.: Budgeted constrained coverage on series-parallel multi-interface networks. In: Barolli, L., Amato, F., Moscato, F., Enokido, T., Takizawa, M. (eds.) Advanced Information Networking and Applications (AINA-2020). LNCS, vol. 1151, pp. 458–469. Springer, Cham (2020). https://doi.org/10.1007/978-3-030-44041-1_41
15. Aloisio, A., Navarra, A.: Constrained connectivity in bounded x-width multi-interface networks. Algorithms **13**(2), 31 (2020)
16. Aloisio, A., Navarra, A.: On coverage in multi-interface networks with bounded pathwidth. In: Advances in Intelligent Systems and Computing (WAINA). Springer, Cham (2024)
17. Aloisio, A., Navarra, A., Mostarda, L.: Distributing energy consumption in multi-interface series-parallel networks. In: Barolli, L., Takizawa, M., Xhafa, F., Enokido, T. (eds.) Web, Artificial Intelligence and Network Applications (WAINA-2019). LNCS, vol. 927, pp. 734–744. Springer, Cham (2019). https://doi.org/10.1007/978-3-030-15035-8_71
18. Aloisio, A., Navarra, A., Mostarda, L.: Energy consumption balancing in multi-interface networks. J. Ambient. Intell. Humaniz. Comput. **11**(8), 3209–3219 (2020)
19. Aslanyan, D., Mkrtchyan, V.V., Petrosyan, S.S., Vardanyan, G.N.: On disjoint matchings in cubic graphs: maximum 2-edge-colorable and maximum 3-edge-colorable subgraphs. Disc. Appl. Math. **172**, 12–27 (2014)
20. Cao, Y., Chen, G., Jing, G., Stiebitz, M., Toft, B.: Graph edge coloring: a survey. Graphs Comb. **35**(1), 33–66 (2019)
21. Cygan, M., et al.: Parameterized Algorithms. Springer, Cham (2015). https://doi.org/10.1007/978-3-319-21275-3

22. Erlebach, T., Jansen, K.: The complexity of path coloring and call scheduling. Theoret. Comput. Sci. **255**(1), 33–50 (2001)
23. Feige, U., Ofek, E., Wieder, U.: Approximating maximum edge coloring in multigraphs. In: Approximation Algorithms for Combinatorial Optimization, pp. 108–121 (2002)
24. Fomin, F.V., Korhonen, T.: Fast FPT-approximation of branchwidth. In: Proceedings of the 54th Annual ACM SIGACT Symposium on Theory of Computing (STOC 2022), pp. 886–899. Association for Computing Machinery, New York (2022)
25. Galby, E., Lima, P.T., Paulusma, D., Ries, B.: On the parameterized complexity of k-edge colouring. arXiv preprint arXiv:1901.01861 (2019)
26. Gandham, S., Dawande, M., Prakash, R.: Link scheduling in sensor networks: distributed edge coloring revisited. In: Proceedings of the IEEE 24th Annual Joint Conference of the IEEE Computer and Communications Societies, vol. 4, pp. 2492–2501 (2005)
27. Grüttemeier, N., Komusiewicz, C., Morawietz, N.: Maximum edge-colorable subgraph and strong triadic closure parameterized by distance to low-degree graphs. In: Proceedings of the 17th Scandinavian Symposium and Workshops on Algorithm Theory (SWAT), vol. 162, pp. 26:1–26:17 (2020)
28. Holyer, I.: The np-completeness of edge-coloring. SIAM J. Comput. **10**(4), 718–720 (1981)
29. Karapetyan, L., Mkrtchyan, V.: On maximum k-edge-colorable subgraphs of bipartite graphs. Disc. Appl. Math. **257**, 226–232 (2019)
30. Kosowski, A.: Approximating the maximum 2- and 3-edge-colorable subgraph problems. Disc. Appl. Math. **157**(17), 3593–3600 (2009). Sixth International Conference on Graphs and Optimization 2007
31. Mkrtchyan, V.V., Petrosyan, S.S., Vardanyan, G.N.: On disjoint matchings in cubic graphs. Disc. Math. **310**(10), 1588–1613 (2010)
32. Mkrtchyan, V.V., Petrosyan, S.S., Vardanyan, G.N.: Corrigendum to "on disjoint matchings in cubic graphs" [discrete math. 310 (2010) 1588-1613]. Disc. Math. **313**(21), 2381 (2013)
33. Mkrtchyan, V.V., Steffen, E.: Maximum Δ-edge-colorable subgraphs of class II graphs. J. Graph Theory **70**(4), 473–482 (2012)
34. Perucci, A., Autili, M., Tivoli, M., Aloisio, A., Inverardi, P.: Distributed composition of highly-collaborative services and sensors in tactical domains. In: Proceedings of 6th International Conference in Software Engineering for Defence Applications (2020)
35. Robertson, N., Seymour, P.: Graph minors. x. obstructions to tree-decomposition. J. Combinator. Theory Ser. B **52**(2), 153–190 (1991)
36. Shannon, C.E.: A theorem on colouring the lines of a network. J. Math. Phys. **28**(1–4), 148–152 (1949)
37. Stiebitz, M., Scheide, D., Toft, B., Favrholdt, L.M.: Graph Edge Colouring. John Wiley and Sons (2012)
38. Vizing, V.G.: On an estimate of the chromatic class of a p-graph. Diskret. Analiz **3**, 25–30 (1964)

On Coverage in Multi-Interface Networks with Bounded Pathwidth

Alessandro Aloisio[1(✉)] and Alfredo Navarra[2]

[1] Dipartimento di Scienze Umanistiche e Sociali Internazionali, Università degli Studi Internazionali di Roma, Roma, Italy
alessandro.aloisio@unint.eu

[2] Dipartimento di Matematica e Informatica, Università degli Studi di Perugia, Perugia, Italy
alfredo.navarra@unipg.it

Abstract. In dealing with diverse devices equipped with multiple communication interfaces, a significant challenge arises in selectively activating a subset of interfaces on each device to establish effective communication connections. The goal is to ensure that devices at the endpoints of any connection share at least one active interface, forming the foundation of an extensively studied model known in the literature as "Multi-Interface" networks. We explore the latest variation wherein each interface is linked to a cost and a profit, with two as a limit to the possible active interfaces per device. Recognizing the NP-hard nature of this problem, we narrow our focus to graphs with bounded pathwidth, where the challenge persists. Subsequently, by exploiting dynamic programming techniques, we provide two pseudo-polynomial time-optimal algorithms.

1 Introduction

Nowadays, we are surrounded by compact yet powerful devices, exploited for a diverse array of applications. Furthermore, heterogeneous devices have the capability to interact with each other through various protocols and connecting interfaces

In this paper, we explore networks of heterogeneous devices that can establish connections by means of different communication interfaces. The selection of the most profitable interface to establish a connection depends on various factors, especially when considering battery-powered devices. Priority is given to energy consumption issues to maximize the network's lifespan, crucial in critical realms like the tactical one [28].

Indeed, the depletion of batteries in even a single device can lead to network disconnection and subsequent undesirable failures. This issue can be mitigated by implementing appropriate solutions during the selection of network connections based on specific requirements. This introduces complex and natural optimization problems that need to consider multiple parameters simultaneously.

This work is partially supported by the project 'Soluzioni innovative per il problema della copertura nelle multi-interfacce e relative varianti', UNINT, and by the Italian National Group for Scientific Computation (GNCS-INdAM).

L. Barolli (Ed.): AINA 2024, LNDECT 204, pp. 96–105, 2024.
https://doi.org/10.1007/978-3-031-57942-4_11

In a broader context, a network comprising heterogeneous devices (in particular with respect to the available interfaces) can be represented as an undirected graph, denoted as $G = (V, E)$, where V represents the set of devices and E represents the set of potential connections. Each device $u \in V$ is associated with a set of available interfaces $W(u)$. The comprehensive set of all possible interfaces in the network is determined by $\bigcup_{u \in V} W(u)$, with its cardinality denoted as k. A connection is considered established when the devices at the endpoints of the corresponding edge activate at least one common interface. When a node $u \in V$ activates an interface α, it consumes energy $c(\alpha)$ to maintain α as active.

1.1 Related Work

Over the past decade, extensive research has been conducted on multi-interface networks, which model various network optimization problems [17,19]. Among the versions already investigated, one of the earliest is the so-called *Coverage* problem, which involves determining the most cost-effective method to establish all the connections specified by the input graph G. Further insights into this matter can be found in [3,10,13–15,21,23].

The problem of *Connectivity*, as discussed in [25], involves determining the cheapest way to ensure the overall connectivity of the network, i.e., the goal is to establish the least expensive set of connections so that every pair of nodes in G is connected through a path created by these connections. Both Connectivity and Coverage have been also examined in terms of minimizing the maximum cost incurred on a single node, as in [21]. The problem of *Cheapest Path* is explored in [25]. It can be seen as a generalization of the classic Shortest Path problem. Additionally, traditional problems like Maximum Matching [24] or Flow [16,20] have been subjects of investigation.

1.2 Our Results

In this paper, our focus lies on the Coverage problem, which is subject to a budget constraint denoted by b for the overall cost and a maximum limit denoted by q on the number of interfaces each device can activate. Specifically, our primary emphasis is on the scenario where $q = 2$. As per [10], we refer to this problem as *CMI(b,q)*, and we specifically investigate the *unbounded case*, wherein there is no restriction on the number k of available interfaces across the entire network. Moreover, each interface is associated with a profit, introducing an incentive aspect.

The computational complexity of the resulting version of the Coverage derives from [10]. However, even in cases where feasibility is ensured and $q = 2$, in [1,11,12] the authors showed that the problem is *NP*-hard through a polynomial time transformation from the classic Knapsack problem [26,27]. Due to its complexity, our attention turns to graphs [9,22] with bounded *pathwidth*, which model the underlying topology of the network and exhibit *NP*-hardness even when $q = 2$.

In order to deal with the *CMI(2,b)* problem on graphs with bounded pathwidth, we introduce two pseudo-polynomial time dynamic programming algorithms that provide optimal solutions.

2 Preliminaries

For a graph G, we denote by V its node set and by E its edge set. Unless otherwise stated, the graph $G = (V,E)$ representing the network is assumed to be undirected, connected, and without multiple edges and loops. Let $N(u) \subset V$ be the neighbours of a node $u \in V$. Let $[k] = \{1,\dots,k\}$, for every positive integer k. When considering the network graph G, we simply denote the number of its nodes and edges by n and m, respectively. A global assignment of the interfaces to the nodes in V is given in terms of an appropriate interface assignment function W, as follows.

Definition 1. *A function* $W : V \rightarrow 2^{[k]}$ *is said to* cover *graph* G *if* $W(u) \cap W(v) \neq \emptyset$*, for each* $(u,v) \in E$.

The considered *CMI(q,b)* optimisation problem is formulated as follows.

CMI(q,b): Coverage in Multi-Interface Networks	
***Input*:**	A graph $G = (V,E)$, an allocation of available interfaces $W : V \rightarrow 2^{[k]}$ covering graph G, an interface cost function $c\colon [k] \rightarrow \mathbb{N}_{>0}$, two integers $q,b \geq 1$, two profit functions $p\colon [k] \rightarrow \mathbb{N}_{>0}$ and $p\colon [k]^2 \rightarrow \mathbb{N}_{\geq 0}$.
***Solution*:**	If possible, an allocation of active interfaces $W_A : V \rightarrow 2^{[k]}$ covering G such that for all $u \in V$, $W_A(u) \subseteq W(u)$ and $\lvert W_A(u)\rvert \leq q$; with $c(W_A) = \sum_{u\in V}\sum_{\alpha\in W_A(u)} c(\alpha) \leq b$. Otherwise, a negative answer.
***Goal*:**	Maximize the total profit $p(W_A) = \sum_{u\in V}\sum_{\alpha\in W_A(u)} p(\alpha) + \sum_{(u,v)\in E}\sum_{\alpha\in(W_A(u)\cap W_A(v))} p(\alpha,\alpha)$.

Actually, *CMI* (q,∞) in the absence of profit functions was examined in [10]. The research demonstrated that the problem is *NP*-hard, even when considering unitary costs. This finding can be generalized to *CMI(q,b)*, even in scenarios where feasibility is ensured. The subsequent theorem establishes this extension by exploiting a polynomial time transformation from the well-known knapsack problem.

Theorem 1 **([11]).** *CMI(q,b)* is NP-hard, even when the input instance admits a feasible solution and q=2.

2.1 Graphs with Bounded Pathwidth

In this section, we revise a formal definition of the *pathwidth* of a graph.

Definition 1 **([22]).** A path decomposition of a graph $G = (V,E)$ is a set $\mathscr{P} = (X_1,\dots,X_r)$ of subsets of V, that is $X_i \subseteq V$ for each $i \in [r]$, called *bags*, such that: (i) for every $u \in V$ there exists $i \in [r]$ with $u \in X_i$; (ii) for every $(u,v) \in E$, there exists $i \in [r]$ with $(u,v) \in X_i$; (iii) for every three bags X_i, X_j, and X_k, with $i \leq j \leq k$, it holds that $X_i \cap X_k \subseteq X_j$.

The *width* of a path decomposition $\mathscr{P}$ is determined by the largest number of vertices in any bag of $\mathscr{P}$ minus one, expressed as $\max_{i\in[r]} |X_i| - 1$.

The *pathwidth* of a graph G corresponds to the minimum width across all possible path decompositions of G. To prevent confusion, we will refer to the elements of $\mathscr{P}$ as nodes and the elements of V as vertices in the rest of the paper.

As highlighted in [22], a significant characteristic of path decomposition is the *pathwidth separator* property, which we utilize in our algorithm. This property asserts that for any three nodes X_i, X_j, and X_k, where X_j is positioned between X_i and X_k, every path in G connecting a vertex in $X_i \setminus X_j$ to a vertex in $X_k \setminus X_j$ must include a vertex from X_j. This implies that the node X_j serves as a separator, isolating the vertices in $X_i \setminus X_j$ from those in $X_k \setminus X_j$. Our algorithm works with a particular kind of path decomposition called *nice*, which is helpful in designing dynamic programming algorithms. Furthermore, there exists a linear time algorithm that transforms every path decomposition into a nice path decomposition, preserving the same width, [22].

Definition 2 **([22]).** A path decomposition of a graph $G = (V, E)$ is *nice* if $|X_1| = |X_r| = 1$, and for every $i \in \{1, 2, \ldots, r-1\}$ there is a vertex $v \in V$, such that either $X_{i+1} = X_i \cup \{v\}$ *(introduce node)*, or $X_{i+1} = X_i \setminus \{v\}$ *(forget node)*.

Using the definition before, it turns out that the number of nodes in a *nice* path decomposition is $2|V| + 1$. This happens because of property (iii) in Definition 1, which states that every vertex $v \in V$ is part of a series of bags in a row.

3 A Dynamic Programming Algorithm via Costs

In this section, we introduce a pseudo-polynomial-time algorithm in b that optimally solves *CMI(2,b)* when the underlying graph has a limited pathwidth. The algorithm utilizes dynamic programming techniques, making use of the previously outlined nice path decomposition, with a particular focus on the pathwidth separator property.

Given an instance of *CMI(2,b)*, first of all we identify, a path decomposition for the graph $G(V, E)$ having a width of h, which can be computed in linear time [18] when the pathwidth is limited. Subsequently, we calculate a nice path decomposition of $G(V, E)$, which is a process still achievable within linear time [22]. Let $G(X_i)$ represent the subgraph induced by the vertices in $\bigcup_{j=1}^{i} X_j$ in this setting. Additionally, consider $f(X_i, \mathscr{A}, d)$ as the maximum profit of *CMI(2,b)* on $G(X_i)$, where $\mathscr{A}$ denotes a collection of $|X_i|$ subsets $A(u)$ taken from the available interfaces $W(u)$ for $u \in X_i$, which adhere to the following constraints:

- for each vertex u in X_i, the *active* interfaces are those in $A(u)$, where $|A(u)| \leq 2$;
- $c(W_A) = d$, that is the cost related to the subgraph $G(X_i)$ equals d.

By leveraging the pathwidth separator property, the algorithm's core calculates, at each node X_i in the path decomposition $\mathscr{P}$, the values $f(X_i, \mathscr{A}, d)$ for every possible collection $\mathscr{A}$ and integer $0 \leq d \leq b$. The algorithm initiates at X_1 and concludes at X_r. It is evident that the constrained version of *CMI(2,b)* may not always be solvable. In such instances, we set $f(X_i, \mathscr{A}, d) = -\infty$. Given that X_1 contains only one vertex u, resulting in no edges in $G(X_1)$ and $\mathscr{A} = \{A(u)\}$, the following holds:

- $f(X_1, \mathscr{A}, d) = p(A(u))$ if $|A(u)| \subseteq W(u)$, $|A(u)| \leq 2$, and $c(A(u)) = d$;
- $f(X_1, \mathscr{A}, d) = -\infty$ otherwise.

Basically, for X_1, we activate all feasible subsets of up to two interfaces for u, incurring a total cost of d. Notably, given that $G(X_1)$ comprises only one vertex, these partial solutions are essential for constructing an optimal solution for the entire graph, mainly when G consists of more than one vertex.

In every *introduce node* $X_{i+1} = X_i \cup \{v\}$, the computation of the value $f(X_{i+1}, \mathscr{A}, d)$ involves solving the constrained maximization problem shown in (1). This problem is defined for a particular collection $\mathscr{A}$ of active interface sets and a specific total cost d. Importantly, the solution relies on the values $f(X_i, \mathscr{B}, d_1)$ that were previously computed in the preceding node X_i.

$$
\begin{aligned}
\max \quad & f(X_i, \mathscr{B}, d_1) + p(A(v)) + \sum_{u \in (X_i \cap N(v))} \sum_{\alpha \in A(u) \cap A(v)} p(\alpha, \alpha) \\
\text{s.t.} \quad & B(u) = A(u), \quad \forall u \in X_i \\
& A(v) \cap B(u) \neq \emptyset, \quad \forall u \in X_i \cap N(v) \\
& d_1 + c(A(v)) = d
\end{aligned}
\tag{1}
$$

In Problem (1), the updated collection $\mathscr{A}$ is identical to $\mathscr{B}$ for every vertex, except for the newly added one, v (first constraint). The vertex v can establish communication with all its neighboring vertices u in $G(X_{i+1})$ by ensuring the presence of at least one common active interface (second constraint). The total cost is exactly d (third constraint).

The objective function combines the previously computed optimal value $f(X_i, \mathscr{B}, d_1)$ with the profit derived from activating the interfaces $A(v)$ in the new vertex v, as well as the profit associated with the edges connecting v to other vertices in $G(X_{i+1})$.

In each *forget node* $X_{i+1} = X_i \setminus \{v\}$, we compute $f(X_{i+1}, \mathscr{A}, d)$ by solving the following constrained maximization problem. Here, $\mathscr{A}$ is a specific collection of interface subsets $A(u)$ for each vertex $u \in X_i$, with each subset having at most 2 elements. The parameter d represents a specific cost.

$$
\begin{aligned}
\max \quad & f(X_i, \mathscr{A} \cup B(v), d) \\
\text{s.t.} \quad & A(u) \cap B(v) \neq \emptyset, \quad \forall u \in X_{i+1} \cap N(v) \\
& B(v) \subseteq W(v)
\end{aligned}
\tag{2}
$$

Indeed, the value $f(X_{i+1}, \mathscr{A}, d)$ essentially represents the maximum value of $f(X_i, \mathscr{A} \cup B(v), d)$ for every conceivable subset of active interfaces $B(v)$. This subset must be compatible with each $A(u)$, signifying that $A(u) \cap B(v) \neq \emptyset$ for every vertex u adjacent to v. Furthermore, it's crucial that the cost d remains consistent for both $G(X_i)$ and $G(X_{i+1})$.

When the algorithm ends, the optimal solution for *CMI(2,b)* is represented by the maximum value $f(X_r, \mathscr{A}, d)$, for every possible collection of interface subsets $A(u) \subseteq W(u)$, where $|A(u)| \leq 2$, for the sole vertex $u \in X_r$, and for every possible cost $d \in [b]$.

After finding the maximum profit for Problem 2, we can construct an optimal solution for *CMI(2,b)* by applying standard methods of dynamic programming.

To complete this section, we proceed to determine the algorithm's complexity. During each *introduce node* step, we tackle at most $b\left(k+\binom{k}{2}\right)^{h+1}$ problems, specifically those outlined in Eq. (1). These problems pertain to every possible subset of active interfaces $A(u)$, with $|A(u)| \leq 2$, and every positive integer $d \in [b]$. Additionally, this encompasses every vertex $u \in X_{i+1}$ within a limit of $h+1$.

At every *forget node*, the complexity remains upper bounded by $b\left(k+\binom{k}{2}\right)^{h+1}$. This complexity arises from considering all subsets of active interfaces $A(u)$, where $u \in X_{i+1}$, the various possible subsets of the active interface $B(v)$ for the forget vertex v, and all positive integers d less than or equal to b.

In conclusion, since the computational time to solve Problem (1) or Problem (2) is $O(h)$, and since there are $O(n)$ nodes (bags) in a path decomposition, the time complexity of the dynamic programming algorithm is $O\left(n \cdot b \cdot h\left(k+\binom{k}{2}\right)^{h+1}\right) = O\left(n \cdot b \cdot h \cdot \left(\frac{k^2}{2}\right)^{h+1}\right)$. We can then state the following:

Theorem 2. There is a deterministic algorithm that, given an instance of *CMI(2,b)* with G as the underlying graph, a path decomposition $\mathscr{P}$ of width h for G, computes an optimal solution of *CMI(2,b)* in $O\left(n \cdot b \cdot h \cdot \left(\frac{k^2}{2}\right)^{h+1}\right)$ time.

4 Dynamic Programming via Profits

In this section, when the pathwidth is limited, we present an alternative pseudo-polynomial time algorithm for *CMI(2,b)*, still relying on the dynamic programming technique. The key distinction from the previous algorithm lies in its focus on profits rather than costs. As outlined in the preceding section, the initial phase of the algorithm involves computing a path decomposition and subsequently a nice path decomposition for a given instance of *CMI(2,b)*. Both steps are accomplished in linear time in terms of the size of the instance [18,22] when the pathwidth is limited. Let $\mathscr{P} = (X_1, \ldots, X_r)$ represent the obtained nice path decomposition with a width equal to h. Here, again, $G(X_i)$ refers to the subgraph induced by the vertices in $\bigcup_{j=1}^{i} X_j$. Let $\rho(X_i, \mathscr{A}, r)$ denote the minimum cost of *CMI(2,b)* on $G(X_i)$, where $\mathscr{A}$ constitutes a collection of $|X_i|$ subsets $A(u)$ originating from the available interfaces $W(u)$, with $u \in X_i$, that satisfies the following constraints:

- for each vertex u in X_i, the *active* interfaces are those in $A(u)$, where $|A(u)| \leq 2$;
- $p(W_A) = r$, that is the profit related to the subgraph $G(X_i)$ equals r.

By utilizing the pathwidth separator characteristic, the central part of the algorithm calculates, at each node X_i of $\mathscr{P}$, the values $\rho(X_i, \mathscr{A}, r)$ for every conceivable collection $\mathscr{A}$ and for every integer $0 \leq r \leq \mu$, where μ represents an upper bound on the optimal value of the *CMI(2,b)* instance.[1] The algorithm starts at X_1, and ends at X_r. Given that

[1] E.g., μ might be the sum of the profits over all the available interfaces, both on the vertices and the edges.

the constrained version of *CMI(2,b)* may not always be solvable, we set $\rho(X_i,\mathscr{A},d) = +\infty$ in such cases. In the first node X_1, the following conditions hold because X_1 contains only one vertex u and no edges:

- $\rho(X_1,\mathscr{A},r) = c(A(u))$ if $|A(u)| \subseteq W(u)$, $|A(u)| \leq 2$, and $p(A(u)) = r$;
- $\rho(X_1,\mathscr{A},r) = +\infty$ otherwise.

Substantially, for X_1 we activate all the possible subsets of at most two interfaces available for u, with a total profit equal to r.

In every *introduce node* step $X_{i+1} = X_i \cup \{v\}$, the computation of the value $\rho(X_{i+1},\mathscr{A},r)$ for a given collection $\mathscr{A}$ of active interface sets and a specific total profit r involves solving the following constrained minimization problem. This problem utilizes the values $f(X_i,\mathscr{B},r_1)$ already calculated in the preceding node X_i:

$$
\begin{aligned}
\min \quad & \rho(X_i,\mathscr{B},r_1) + c(A(v)) \\
\text{s.t.} \quad & B(u) = A(u), \quad \forall u \in X_i \\
& A(v) \cap B(u) \neq \emptyset, \quad \forall u \in X_i \cap N(v) \\
& r_1 + p(A(v)) + \sum_{u \in N(v) \cap X_i} \sum_{\alpha \in A(u) \cap A(v)} p(\alpha,\alpha) = r
\end{aligned}
\tag{3}
$$

In Problem (3), the updated collection $\mathscr{A}$ is identical to $\mathscr{B}$ for every vertex, except for the newly added one, v (first constraint). The vertex v can establish communication with all its neighboring vertices u in $G(X_{i+1})$ by ensuring the presence of at least one common active interface (second constraint). The total profit is exactly r (third constraint). The objective function combines the previously computed optimal value $\rho(X_i,\mathscr{B},r_1)$ with the cost of the interfaces $A(v)$ activated in the new vertex v.

In any *forget node* $X_{i+1} = X_i \setminus \{v\}$, the value $\rho(X_{i+1},\mathscr{A},r)$, where $\mathscr{A}$ represents a specific collection of interface subsets $A(u)$, with $u \in X_i$, and $|A(u)| \leq 2$, and a specific profit r, is computed by solving the following constrained minimisation problem:

$$
\begin{aligned}
\min \quad & \rho(X_i,\mathscr{A} \cup B(v),r) \\
\text{s.t.} \quad & A(u) \cap B(v) \neq \emptyset, \quad \forall u \in X_{i+1} \cap N(v) \\
& B(v) \subseteq W(v)
\end{aligned}
\tag{4}
$$

In essence, $\rho(X_{i+1},\mathscr{A},r)$ represents the minimum value among $\rho(X_i,\mathscr{A} \cup B(v),r)$ for every feasible subset of active interfaces $B(v)$. Feasibility here implies compatibility with all $A(u)$, ensuring that $A(u) \cap B(v) \neq \emptyset$ for every u adjacent to v. Notably, the profit r in $G(X_{i+1})$ remains consistent with that of $G(X_i)$.

At the conclusion of the algorithm, the optimal value of *CMI(2,b)* corresponds to the maximum r such that $\rho(X_r,\mathscr{A},r) \leq b$, considering all possible collections of interface subsets $A(u) \subseteq W(u)$, where $|A(u)| \leq 2$, for the singular vertex $u \in X_r$. After finding the minimum cost for Problem 4, we can construct an optimal solution for *CMI(2,b)* by applying standard methods of dynamic programming.

We can now finalize this section by determining the algorithm's complexity. During each *introduce node* operation, we solve at most $(\mu+1)\left(k+\binom{k}{2}\right)^{h+1}$ problems (3), one for every possible subset of active interfaces $A(u)$, with $|A(u)| \leq 2$, for every positive integers r less than or equal to μ, and for every $u \in X_{i+1}$ that are at most $h+1$.

At any *forget node*, the complexity is also at most $(\mu+1)\left(k+\binom{k}{2}\right)^{h+1}$, considering all subsets of active interfaces $A(u)$ for every $u \in X_{i+1}$, all possible subsets of active interfaces $B(v)$ for the forget vertex v, and all positive integers r less than or equal to μ. In conclusion, since every Problem (3) and any Problem (4) can be solved in $O(h)$ time, and there are $O(n)$ nodes (bags) in a path decomposition, the time complexity of the dynamic programming algorithm is $O\left(n\mu(k+\binom{k}{2})^{h+1}\right) = O\left(n \cdot \mu \cdot h \cdot \left(\frac{k^2}{2}\right)^{h+1}\right)$. We can then state the following:

Theorem 3. There is a deterministic algorithm that, given an instance of *CMI(2,b)* with G as the underlying graph, a path decomposition $\mathscr{P}$ of width h for G, and an upper bound μ of the optimal value, computes an optimal solution of *CMI(2,b)* in $O\left(n \cdot \mu \cdot h \cdot \left(\frac{k^2}{2}\right)^{h+1}\right)$ time.

5 Concluding Remarks

We have explored a novel variant of the extensively studied Coverage problem within the topic of multi-interface networks. Specifically, our attention is directed toward situations where individual nodes can activate a maximum of two interfaces. The fundamental topology is composed of graphs featuring a constrained pathwidth, which is one of the common parameters used in the fixed parameter complexity theory.

A possible way of investigation could be studying budgeted constrained coverage problems with respect to other parameters (e.g. treewidth and cliquewidth). Another line of research which we thing is worth to be investigated is thinking at each vertex as an agent that gets a utility from the profit functions where she is directly involved or from profits on edges that are at some distance [2,4–9].

References

1. Aloisio, A.: Coverage subject to a budget on multi-interface networks with bounded carving-width. In: Barolli, L., Amato, F., Moscato, F., Enokido, T., Takizawa, M. (eds.) WAINA 2020. AISC, vol. 1150, pp. 937–946. Springer, Cham (2020). https://doi.org/10.1007/978-3-030-44038-1_85
2. Aloisio, A.: Distance hypergraph polymatrix coordination games. In: Proceedings of the 22nd Conference Autonomous Agents and Multi-Agent Systems (AAMAS), pp. 2679–2681 (2023)
3. Aloisio, A.: Algorithmic aspects of distributing energy consumption in multi-interface networks. In: WAINA, Advances in Intelligent Systems and Computing, pp. 114–123. Springer (2024)
4. Aloisio, A.: Fixed-parameter tractability for branchwidth of the maximum-weight edge-colored subgraph problem, pp. 86–95. Springer (2024)

5. Aloisio, A., Flammini, M., Kodric, B., Vinci, C.: Distance polymatrix coordination games. In: Proceedings of the 30th International Joint Conference Artificial Intelligence (IJCAI), pp. 3–9 (2021)
6. Aloisio, A., Flammini, M., Kodric, B., Vinci, C.: Distance polymatrix coordination games (short paper). In: SPIRIT co-located with 22nd International Conference AIxIA 2023, November 7-9th, 2023, Rome, Italy, vol. 3585 (2023)
7. Aloisio, A., Flammini, M., Vinci, C.: The impact of selfishness in hypergraph hedonic games. In: Proceedings of the 34th Conference Artificial Intelligence (AAAI), pp. 1766–1773 (2020)
8. Aloisio, A., Flammini, M., Vinci, C.: Generalized distance polymatrix games. In: Proceedings of 49th International Conference Current Trends in Theory & Practice of Computer Science (SOFSEM). Springer, Cham (2024)
9. Aloisio, A., Mkrtchyan, V.: Algorithmic aspects of the maximum 2-edge-colorable subgraph problem. In: Barolli, L., Woungang, I., Enokido, T. (eds.) AINA 2021. LNNS, vol. 227, pp. 232–241. Springer, Cham (2021). https://doi.org/10.1007/978-3-030-75078-7_24
10. Aloisio, A., Navarra, A.: Balancing energy consumption for the establishment of multi-interface networks. In: Proceedings 41st International Conference on Current Trends in Theory and Practice of Computer Science, (SOFSEM), vol. 8939, pp. 102–114 (2015)
11. Aloisio, A., Navarra, A.: Budgeted constrained coverage on bounded carving-width and series-parallel multi-interface networks. Internet of Things **11**, 100259 (2020)
12. Aloisio, A., Navarra, A.: Budgeted constrained coverage on series-parallel multi-interface networks. In: Advanced Information Networking and Applications - Proceedings of the 34th International Conference on Advanced Information Networking and Applications, AINA-2020, Caserta, Italy, 15–17 April, vol. 1151, pp. 458–469 (2020)
13. Aloisio, A., Navarra, A.: Constrained connectivity in bounded X-width multi-interface networks. Algorithms **13**(2), 31 (2020)
14. Aloisio, A., Navarra, A., Mostarda, L.: Distributing energy consumption in multi-interface series-parallel networks. In: Barolli, L., Takizawa, M., Xhafa, F., Enokido, T. (eds.) WAINA 2019. AISC, vol. 927, pp. 734–744. Springer, Cham (2019). https://doi.org/10.1007/978-3-030-15035-8_71
15. Aloisio, A., Navarra, A., Mostarda, L.: Energy consumption balancing in multi-interface networks. J. Ambient. Intell. Humaniz. Comput. **11**(8), 3209–3219 (2020)
16. Audrito, G., Bertossi, A., Navarra, A., Pinotti, C.: Maximizing the overall end-user satisfaction of data broadcast in wireless mesh networks. J. Discrete Alg. **45**, 14–25 (2017)
17. Caporuscio, M., Charlet, D., Issarny, V., Navarra, A.: Energetic performance of service-oriented multi-radio networks: issues and perspectives. In: Proceedings 6th International Workshop on Software and Performance (WOSP), pp. 42–45. ACM (2007)
18. Cattell, K., Dinneen, M.J., Fellows, M.R.: A simple linear-time algorithm for finding path-decompositions of small width. Inf. Process. Lett. **57**(4), 197–203 (1996)
19. D'Angelo, G., Di Stefano, G., Navarra, A.: Multi-interface wireless networks: complexity and algorithms. In: Ibrahiem, S.R., El Emary, M.M. (eds.) Wireless Sensor Networks: From Theory to Applications, pp. 119–155. CRC Press, Taylor & Francis Group (2013)
20. D'Angelo, G., Di Stefano, G., Navarra, A.: Flow problems in multi-interface networks. IEEE Trans. Comput. **63**, 361–374 (2014)
21. D'Angelo, G., Stefano, G.D., Navarra, A.: Minimize the maximum duty in multi-interface networks. Algorithmica **63**(1–2), 274–295 (2012)
22. Fomin, F.V., Kaski, P.: Exact exponential algorithms. Commun. ACM **56**(3), 80–88 (2013)
23. Klasing, R., Kosowski, A., Navarra, A.: Cost minimization in wireless networks with a bounded and unbounded number of interfaces. Networks **53**(3), 266–275 (2009)
24. Kosowski, A., Navarra, A., Pajak, D., Pinotti, C.: Maximum matching in multi-interface networks. Theoret. Comput. Sci. **507**, 52–60 (2013)

25. Kosowski, A., Navarra, A., Pinotti, M.: Exploiting multi-interface networks: connectivity and cheapest paths. Wireless Netw. **16**(4), 1063–1073 (2010)
26. Martello, S., Toth, P.: Knapsack Problems. Biddles Ltd., Guildford (1990)
27. Navarra, A., Pinotti, C.M.: Online knapsack of unknown capacity: how to optimize energy consumption in smartphones. Theor. Comput. Sci. **697**, 98–109 (2017)
28. Perucci, A., Autili, M., Tivoli, M., Aloisio, A., Inverardi, P.: Distributed composition of highly-collaborative services and sensors in tactical domains. In: Ciancarini, P., Mazzara, M., Messina, A., Sillitti, A., Succi, G. (eds.) SEDA 2018. AISC, vol. 925, pp. 232–244. Springer, Cham (2020). https://doi.org/10.1007/978-3-030-14687-0_21

Sensitivity Analysis of Performability Model to Evaluate PBFT Systems

Marco Marcozzi[1,2](✉), Antinisca Di Marco[3], and Leonardo Mostarda[4]

[1] Computer Science Division, University of Camerino, Camerino, Italy
marco.marcozzi@unicam.it
[2] Institute of Data Science and Digital Technologies, Vilnius University, Vilnius, Lithuania
[3] Department of Information Engineering, Computer Science and Mathematics, University of L'Aquila, L'Aquila, Italy
antinisca.dimarco@univaq.it
[4] Department of Mathematics and Computer Science, University of Perugia, Perugia, Italy
leonardo.mostarda@unipg.it

Abstract. The evaluation of performance and availability of consensus protocols used in distributed networks is a crucial part in the development of tailored solutions for diversified applications. When building an analytical model for performability evaluation, the sensitivity analysis of the model parameters is a necessary step to determine the output variance at the variation of the input parameters. The sensitivity analysis performed in this work shows that the input parameters of the analyzed performability model have a highly non-linear effect on the considered performability metrics, determining a large variance on the output for small variation of the inputs.

1 Introduction

The importance of distributed networks, and in particular Distributed Ledger Technology (DLT), e.g. blockchain, surges from the diversified field of application, with solutions in widespread areas of industry, finance, cloud computing, entertainment, and healthcare, among others [1,9,15]. The need of developing tailored solutions for specific business or institutional applications lead to an increase in researching the fundamental and most effective mechanisms in DLT, with regard to the consensus protocol adopted by a specific solution [8]. Briefly, a consensus protocol is a set of algorithms and procedures established in a distributed computer network, as it might be a DLT platform, to agree on certain tasks assigned to the network, e.g. transactions to be checked and registered in the local memory of each computer participating in the network. The consensus protocol and its performance are vital in a decentralized setting, since there is no central authority enforcing policies, therefore the consensus protocol ensures the correctness and safety of operation in the network [6,13,16,17]. As an example, DLT seeks to be a viable solution also for systems where devices have limited

L. Barolli (Ed.): AINA 2024, LNDECT 204, pp. 106–113, 2024.
https://doi.org/10.1007/978-3-031-57942-4_12

computational power, i.e., for Internet of Things (IoT) environments (as [12]), and a PBFT consensus protocol could fit the requirement for such systems.

Jointly, the need of more efficient and performing systems also made rise the efforts to test the characteristics of a DLT project, with attention on the cost-benefit analysis for what concerns the time and resources spent in assessing performance and limitations of certain settings, even before implementing the planned system itself. More specifically, a prolific literature has been established on the assessment and evaluation of performance and availability of consensus protocols for DLT, for which some efforts in literature review have been made, finding that common techniques involve simulation, benchmarking, and analytical modeling of the systems under evaluation [4,14].

This work presents the sensitivity analysis of the parameters involved in the definition of an analytical model [10] developed to assess the performance and availability (performability) of consensus protocols used in DLT, that are based on a Practical Byzatine Fault-Tolerant (PBFT) scheme to reach agreement in a network, while accounting for the presence of possibly malicious participants. Here, the definition of "performability" used to assess simultaneously performance and availability analysis is from [11], who states that "performability relates directly to system effectiveness and is a proper generalization of both performance and reliability."

The rest of this paper presents the analytical model (Sect. 2) developed to assess five performability metrics – availability, blocking probability, throughput, mean queue length, and latency –, while reporting the sensitivity analysis performed to enquire which of these metrics, and how much, are affected to variations of input parameters, characteristic of the presented model (Sect. 3). Follows a conclusion summarizing the findings, along with possible extensions and future developments on this work (Sect. 4).

2 Performability Model

The analytical model presented in [10] is based on continuous-time Markov chains and it employs a queueing theory approach to find goal metrics. The main objective of [10] is to find the steady-state solution for the model, such that – once obtained the probabilities of the system to be in any possible state of the model – the performability metrics can be obtained as a consequence of the analysis of the limiting probabilities, i.e. the steady-state solution of the Markov chain(s).

As for other Byzantine systems using unsigned messages, the model requires the number of participants (servers), N, to be $N \geq 4$. Indeed, if $N < 4$ the problem has no solution [7].

Since the formalism connected with queueing theory assigns a Poisson statistic to the processes involved, there are rates describing these processes and they can be considered as free parameters of the system, along with the number of servers, the percentage of Byzantine participants, and the buffer size. For instance, servers break-down independently at rate ξ and they can be repaired individually at rate η. It implies that, if there are, for example, $n \in [0, N]$ servers, the resultant break-down rate is $n\xi$, while the repair rate remains η. The jobs to be performed by the system, in the form of messages/transactions bundled in blocks of transactions, are checked to prove the validity of the submitted transactions. Then, after the validity of each transaction is confirmed, the block is committed by all the honest nodes, each in their own memory. Whether a block of transactions can not be served immediately (because the system is busy), the transactions contained are stored in a finite memory buffer (with size $J \geq 1$), until the memory buffer is not saturated. After saturation, the system rejects incoming transactions, that are lost.

Note that break-downs and incoming transactions are modelled as a Poisson process with rates η and λ, while repair time and service/processing time are exponentially distributed with rates ξ and μ. This is because, while using Markov chains, the two distributions are modelling two related processes – a Poisson point process, or the time between two Poisson events (which is distributed exponentially) [2].

Figure 1 presents the state diagram for the proposed model. Nodes colored in red represent the area in which consensus is not reached, while nodes in green are those for which the system is available. The three indices, h, f, j, reported in each node are indicating states that the system may occupy.

As mentioned earlier in this section, the free parameters of the system are various and they describe different aspects of the modelled system. The total number of nodes is $N \geq 4$. These are divided into honest $h \leq H$ and Byzantine nodes $f \leq F$, where $H \in [0, N]$ is the maximum number of honest nodes and $F \in [0, N]$ the maximum amount of Byzantine nodes, such that $H + F = N$. The buffer size is $J \geq 1$ and it determines the maximum number of jobs $j \in [0, J]$ in the system. In this context, it is assumed that N, H, h, F, f, J, j are positive integers. Therefore, for simplicity of exposition, when dealing with divisions, the *ceiling* $\lceil \cdot \rceil$ and *floor* $\lfloor \cdot \rfloor$ functions are implicitly applied, accordingly. On the other hand, all the rates connected with a process – ξ, η, λ, μ, and μ_t – are positive real numbers. The latter, μ_t, is the rate indicating the process of losing a block of transactions, due to the internal time-out defined by the system. For instance, in the eventuality that there are not enough messages agreeing on a job before the time-out occurs, this job is not committed by the servers that received it and it is lost. This possibility may happen when the mean service time $1/\mu < 1/\mu_t$ or when there are more Byzantine nodes than the maximum tolerated ratio. For instance, in a BFT system, there may be present servers that are not acting accordingly to the rules set by the protocol – called Byzantine servers – either maliciously, either because of malfunctioning. Indeed, a system with N servers can tolerate up to $F < N/3$ Byzantine participants, regardless if they are acting deliberately in contrast with the network or they are unresponsive. In this formulation, however, unresponsive nodes include also honest nodes that are broken-down servers, and not exclusively Byzantine actors.

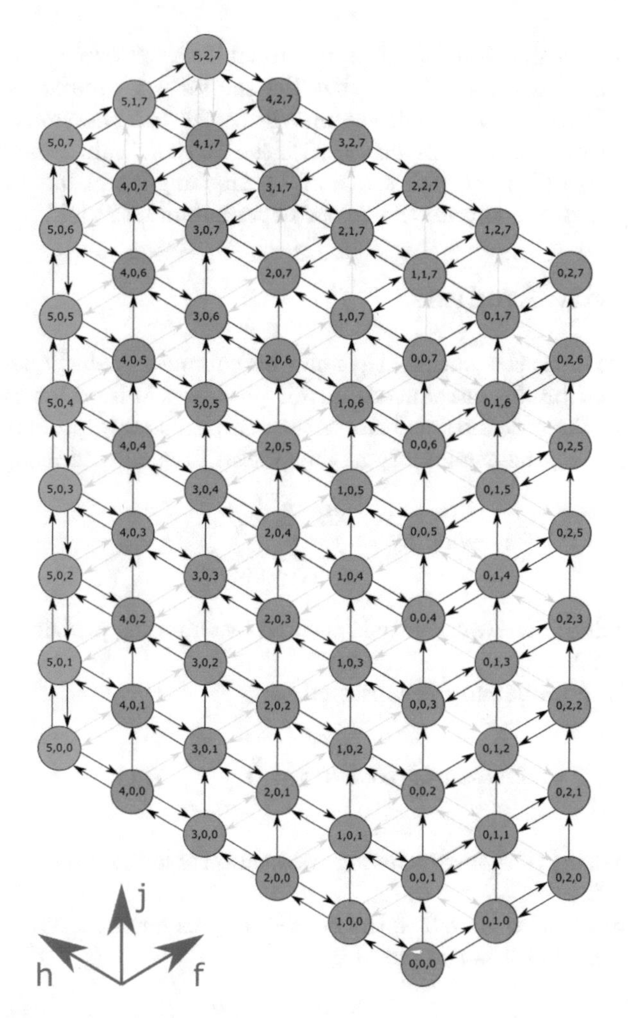

Fig. 1. Depiction of a PBFT state diagram with $H = 5, F = 2$ (hence, $N = 7$), and $J = 7$. The states in the diagram have indices (h, f, j), where $h \in [0, H]$, $f \in [0, F]$, and $j \in [0, J]$. From [10].

At the beginning of this section it was declared that the model is used to compute important metrics regarding the performability of the system. To achieve this goal, the steady-state probabilities for each state in the Markov chain have to be determined. The limiting distribution of the probabilities of the states, the vector $\boldsymbol{P}$, is obtained by writing the balance equations of the Markov chain, which describes the balance of the incoming rates and the outgoing rates for each possible state in the system. With the help of the state diagram in Fig. 1, the balance equations for the system are determined. Although, it is in this instance out of scope to determine the balance equations and solve them, for which it is suggested to check the full procedure in [10].

The stationary distribution of state probabilities $\boldsymbol{P}$ can be determined in several ways, but the main idea is that the the matrix equation $\mathbf{Q}\boldsymbol{P} = 0$ has to be solved. Here, $\mathbf{Q}$ the coefficient matrix of the balance equations, i.e. the stochastic transition matrix associated with the Markov chain. Since the solution to be found belong to the null space of $\mathbf{Q}$, the singular value decomposition method is applied to compute the vector of probabilities $\boldsymbol{P}$ [2].

3 Sensitivity Analysis

In the Introduction, the metrics that can be computed using queueing theory where listed and briefly introduced. In this part, a definition for each metric is presented, and these definitions involve the computed state probabilities $P_{h,f,j}$.

In this context, the availability of the system is calculated using

$$availability = \sum_{j=0}^{J} \sum_{f=0}^{F} \sum_{h>2N/3}^{H} P_{h,f,j}, \tag{1}$$

where the condition on availability is ensured by summing probabilities of states for which $h > 2N/3$.

The blocking probability is expressed as

$$blocking\_probability = \sum_{f=0}^{F} \sum_{h=0}^{H} P_{h,f,J}, \tag{2}$$

which tells that if $j > J$, any incoming transaction is lost because of full memory buffer.

Throughput can be viewed as the number of jobs being served by the system in the unit time, and it is computed by

$$throughput = \mu \sum_{j=1}^{J} \sum_{f=0}^{F} \sum_{h>2N/3}^{H} P_{h,f,j}, \tag{3}$$

where $h > 2N/3$ indicates that the is available, i.e. their number h is greater or equal than the quorum.

The mean queue length can be determined by enumerating

$$mean\_queue\_length = \sum_{j=0}^{J} \sum_{f=0}^{F} \sum_{h=0}^{H} j\, P_{h,f,j}, \tag{4}$$

and it expresses the mean number of transactions/jobs in the buffer waiting to be served by the system.

Lastly, the latency is

$$latency = \frac{mean\_queue\_length}{throughput}, \tag{5}$$

and this quantity measures the average time that elapses from when a transaction enters the system until it is served.

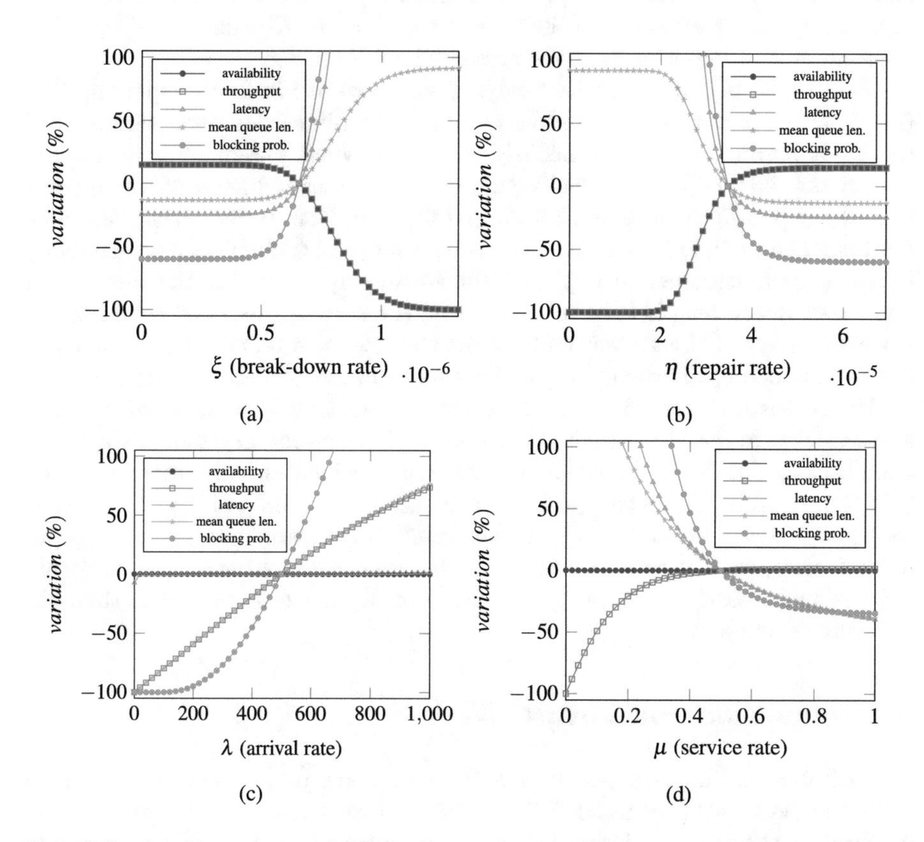

Fig. 2. Sensitivity analysis, where the parameters under investigations are ξ, η, λ, and μ.

While a detailed study of the effect of the number of nodes N, the ratio of honest (h) and Byzantine servers (f) has been presented in [10], the actual magnitude of the effects given by the other free parameters (rates of the arrival, service, break-down and repair processes) has not been properly estimated.

To address this aspect, the following sensitivity analysis studies the percentage variation of the performability metrics by changing one parameter at the time. Indeed, for each parameter, the value from a studied case is used as reference, then by varying the considered parameter and keeping fixed the others, the output variation on the performability metrics can be analysed.

In this work, ranges for each variable are taken such that the reference value is the median of the definition interval, i.e. $\xi \in \left[0, 2\,\xi^{\text{ref}}\right]$, $\eta \in \left[0, 2\,\eta^{\text{ref}}\right]$, $\lambda \in \left[0, 2\,\lambda^{\text{ref}}\right]$, and $\mu \in \left[0, 2\,\mu^{\text{ref}}\right]$. Note that the reference values for the parameters are chosen to be reflecting the results from a benchmark [5] on Tendermint [3]. Since they replicate the results from a benchmark, those reference values for the parameters appear to be reasonable. These reference values are: $\xi^{\text{ref}} = 6.593 \cdot 10^{-7}$, $\eta^{\text{ref}} = 3.47 \cdot 10^{-5}$, $\lambda^{\text{ref}} = 500$, and $\mu^{\text{ref}} = 0.5$. Although there are also other parameters, i.e. $J = 4096$, $N = 64$, $H = 54$, $K = 3000$, and $\mu_t = 0.2$, these are not of importance in the presented analysis.

The results of the sensitivity analysis, as shown in Fig. 2, are reporting that (most of) the performability metrics are highly non-linear for small variations of the input parameters, simultaneously identifying what parameters and metrics are related. For instance, break-down and repair rates influence all the metrics in a highly non-linear fashion (Fig. 2a and Fig. 2b). Conversely, arrival rate does not influence availability and latency, while it influences the blocking probability in a seemingly-exponential way, and almost linearly determine the throughput and mean queue length (Fig. 2c). Similarly, the service rate does not influence the availability of the system and – after the reference point – its throughput, but it responds not linearly for the other performability metrics (Fig. 2d).

Nonetheless, the trend of the performability metrics in relation with arrival and service rate is easily explainable if considering that for a system in statistical equilibrium, if $\mu \gg \lambda$, the effective throughput coincides with the arrival rate (Fig. 2c), resulting in no change for latency and establishing an "orderly" trend for other metrics. Instead, if $\mu \leq \lambda$, the system is saturated and the throughput follows (damped) the service rate, while the other metrics (except for availability) are non-linearly decreasing, passing from higher to lower values than the reference, when $\mu \gg \lambda$.

4 Conclusion and Future Work

Analytical modeling is a powerful tool to evaluate performability metrics of consensus protocols used in DLT. The sensitivity analysis on the free parameters involved in the performability model is a necessary step to better understand connections among parameters and their effect on the results. In this work, the dependence of the performability metrics on the inspected parameters is highly non-linear, with large variations for the value of each metric for slight variation of the input parameter. The main finding can be summarized as a warning in the reliability of the attempted approach to assess performability through analytical modeling, if a well-defined interval of application (hence of input parameters) is not provided. Indeed, researchers have to focus on the definition of accurate input, i.e., parameters must have low variance, when trying to make predictions on the system under development.

Possible future works involve the concrete implementation of a PBFT-based distributed network to test the accuracy and reliability of the prediction obtained by using the analytical model.

References

1. Antal, C., Cioara, T., Anghel, I., Antal, M., Salomie, I.: Distributed ledger technology review and decentralized applications development guidelines. Future Internet **13**(3), 62 (2021)
2. Bolch, G., Greiner, S., De Meer, H., Trivedi, K.S.: Queueing Networks and Markov Chains: Modeling and Performance Evaluation with Computer Science Applications. Wiley, Hoboken (2006)
3. Buchman, E., Kwon, J., Milosevic, Z.: The latest gossip on BFT consensus. arXiv preprint arXiv:1807.04938 (2018)
4. Fan, C., Ghaemi, S., Khazaei, H., Musilek, P.: Performance evaluation of blockchain systems: a systematic survey. IEEE Access **8**, 126927–126950 (2020)
5. Fu, W.-K., et al.: Soteria: A provably compliant user right manager using a novel two-layer blockchain technology. In: 2020 IEEE Infrastructure Conference, pp. 1–10. IEEE (2020)
6. Ismail, L., Materwala, H.: A review of blockchain architecture and consensus protocols: use cases, challenges, and solutions. Symmetry **11**(10), 1198 (2019)
7. Lamport, L., Shostak, R., Pease, M.: The byzantine generals problem. ACM Trans. Program. Lang. Syst. **4**(3), 382–401 (1982)
8. Lashkari, B., Musilek, P.: A comprehensive review of blockchain consensus mechanisms. IEEE Access **9**, 43620–43652 (2021)
9. Macrinici, D., Cartofeanu, C., Gao, S.: Smart contract applications within blockchain technology: a systematic mapping study. Telematics Inform. **35**(8), 2337–2354 (2018)
10. Marcozzi, M., Mostarda, L.: Analytical model for performability evaluation of Practical Byzantine Fault-Tolerant systems. Expert Syst. Appl. **238**, 121838 (2024)
11. Meyer. On evaluating the performability of degradable computing systems. IEEE Trans. Comput. **100**(8), 720–731 (1980)
12. Russello, G., Mostarda, L., Dulay, N.: A policy-based publish/subscribe middleware for sense-and-react applications. J. Syst. Softw. **84**(4), 638–654 (2011)
13. Sankar, L.S., Sindhu, M., Sethumadhavan, M.: Survey of consensus protocols on blockchain applications. In: 2017 4th International Conference on Advanced Computing and Communication Systems (ICACCS), pp. 1–5. IEEE (2017)
14. Smetanin, S., Ometov, A., Komarov, M., Masek, P., Koucheryavy, Y.: Blockchain evaluation approaches: state-of-the-art and future perspective. Sensors (Basel, Switzerland) **20**(12), 1–20 (2020)
15. Tasatanattakool, P., Techapanupreeda, C.: Blockchain: challenges and applications. In: 2018 International Conference on Information Networking (ICOIN), pp. 473–475. IEEE (2018)
16. Xiao, Y., Zhang, N., Lou, W., Hou, Y.T.: A survey of distributed consensus protocols for blockchain networks. IEEE Commun. Surv. Tutorials **22**(2), 1432–1465 (2020)
17. Zhang, S., Lee, J.-H.: Analysis of the main consensus protocols of blockchain. ICT Express **6**(2), 93–97 (2020)

Algorithmic Aspects of Distributing Energy Consumption in Multi-interface Networks

Alessandro Aloisio(✉)

Dipartimento di Scienze Umanistiche e Sociali Internazionali, Universitá degli Studi Internazionali di Roma, Roma, Italy
alessandro.aloisio@unint.eu

Abstract. In contemporary communication networks, diverse devices with multiple interfaces enable the establishment of connections by selectively activating interfaces. This scenario forms the basis of the well-explored model known as Multi-Interface networks. This paper investigates a variation where each device is restricted to activating a fixed number p of its available interfaces, focusing specifically on the Coverage problem. Given a network $G = (V, E)$, with nodes representing devices and edges representing potential connections, the goal is to activate at most p interfaces at each node to establish all specified connections. A connection is formed when the two endpoints share a common interface. The challenge is to maintain a balanced consumption among devices, represented by parameter p. Recent findings have proven the problem to be *NP*-hard, even in the basic case of $p = 2$. Our investigation persists with the case of $p = 2$ in graphs with limited branchwidth, presenting an optimal resolution algorithm, which shows the problem to be fixed parameter tractable with respect to the branchwidth and the total number of available interfaces. We also show that this result can be transposed to the treewidth.

1 Introduction

As technology advances, powerful devices become widely available, facilitating communication among heterogeneous devices using various protocols and interfaces. This paper focuses on networks of diverse devices using different communication interfaces to establish connections. Despite devices often having multiple interfaces like Bluetooth, Wi-Fi, 4G, and 5G, their full potential is rarely utilized. Selecting optimal interfaces depends on factors such as interface availability, communication bandwidth, energy consumption costs, and the device's surroundings. For instance, an experimental study in [27] explores choosing between Bluetooth and Wi-Fi based on energy consumption for data transmission among smartphones. Considering the portable nature of devices, managing energy consumption is crucial for prolonging network lifetime and preventing device failures due to battery drainage.

This introduces optimization challenges in networks made of heterogeneous devices. Each device has a set of interfaces, and connection establishment relies on

This work is partially supported by the project 'Soluzioni innovative per il problema della copertura nelle multi-interfacce e relative varianti' - UNINT, and by the Italian National Group for Scientific Computation (GNCS-INdAM).

L. Barolli (Ed.): AINA 2024, LNDECT 204, pp. 114–123, 2024.
https://doi.org/10.1007/978-3-031-57942-4_13

shared active interfaces among links. Activating an interface consumes energy and provides bandwidth with neighbours sharing the same interface.

Formally, a network of devices is described by a graph $G = (V, E)$, with V as the set of devices and E as the set of potential connections based on device distance and shared interfaces. Each device $v \in V$ is associated with a set of available interfaces $W(v)$. The collection of all possible interfaces in the network is determined by $\bigcup_{v \in V} W(v)$, denoted by k. A connection is established when the endpoints of a corresponding edge share at least one active interface. If an interface α is activated at node u, it incurs energy consumption $c(\alpha)$ to maintain α as active, providing a maximum communication bandwidth with all neighbors sharing interface α.

This paper focuses on the *Coverage* problem, aiming to establish all connections in a graph G at the lowest cost, regardless of the interface. The objective is to ensure a common active interface at the endpoints of each edge, minimizing the overall activation cost in the network. The study tackles the coverage version where the number of interfaces that a device can activate is limited to p, [12,14,15]. This additional constraint aims to control the energy spent by individual devices, offering a nuanced perspective on optimizing overall cost in the presence of constraints [9].

The proposed model, enhances energy balance in nodes, contributing to the network's longevity, which is vital in military tactical networks [30].

1.1 Related Work

Extensive research on multi-interface networks has occurred in recent years, emphasizing the benefits of utilizing multiple interfaces per device in various contexts. This research revisits fundamental problems in standard network optimization, especially in routing and network connectivity (e.g., [17,19,21,23,24]). Combinatorial problems in multi-interface wireless networks have been explored since [18], with subsequent investigations into the Coverage problem and its variants in papers like [1,9,10,22,28].

Connectivity issues have been addressed in [16,29]. The problem consists in finding the cheapest way to ensure the connectivity of the entire network. In other words, it aims to find at each node a subset of the available interfaces that must be activated in order to guarantee a path between every pair of nodes in G while minimizing the overall cost of the interfaces activated among the whole network.

Further insights into the Coverage model can be found in [11–14], where the authors expanded the model by introducing two profit functions, one for the devices and another for the links, incentivizing the activation of interfaces within the network. This novel perspective results in improved connections while keeping the overall consumed energy limited.

1.2 Our Results

In this paper, we delve into the *CMI(2)* (Coverage with Maximum Interface 2) problem, where each node can activate at most 2 interfaces. Following [9], we consider the *unbounded case*, where there is no restriction on the number k of interfaces available across the network. Although the general problem is known to be NP-hard, we explore specific graph topologies, particularly graphs with bounded *branchwidth* [25], extending the investigation beyond previously studied structures [9]. Graphs with limited branchwidth represent a standard class of networks that has garnered significant interest from researchers over the last few decades. Additionally, the *fixed parameter tractability theory* [20] has analysed this parameter (branchwidth) for numerous classic optimization problems. This type of analysis aims to provide a better understanding of the complexity of these problems.

We present an optimal algorithm that produces a solution with a time complexity demonstrating that our problem is fixed-parameter tractable with respect to the branchwidth [20] and the number of total interfaces. Additionally, we prove that *CMI(2)* is also fixed-parameter tractable with respect to the treewidth [20] and the number of available interfaces.

1.3 Outline

The next section provides definitions and notation

in order to formally describe the *CMI(p)* problem. Section 3 is the core of the paper, containing the formal definition of the branchwidth and the resolution algorithm that show *CMI(2)* to be fixed-parameter tractable. Finally, Sect. 4 contains conclusive remarks and a discussion of interesting open problems.

2 Preliminaries

For a graph G, we denote by V its vertex set, by E its edge set. Unless otherwise stated, the graph $G = (V, E)$ representing the network is assumed to be undirected, connected, and without multiple edges and loops. When considering the network graph G, we simply denote the number of its vertices and edges by n and m, respectively. A global assignment of the interfaces to the nodes in V is given in terms of an appropriate interface assignment function W, as follows.

Definition 1. A function $W\colon V \rightarrow 2^{\{1,\ldots,k\}}$ is said to *cover* graph G if $W(u) \cap W(v) \neq \emptyset$, for each $\{u, v\} \in E$.

The cost of activating an interface for a vertex is assumed to be identical for all vertices and given by a cost function $c\colon \{1, \ldots, k\} \rightarrow \mathbb{R}_{>0}$, i.e., the cost of interface α is denoted as $c(\alpha)$. The considered *CMI(p)* optimization problem is formulated as follows.

CMI(p): Coverage in Multi-Interface Networks	
Input:	A graph $G = (V, E)$, an allocation of available interfaces $W: V \rightarrow 2^{\{1,\ldots,k\}}$ covering graph G, an interface cost function $c\colon \{1, \ldots, k\} \rightarrow \mathbb{R}_{>0}$, an integer $p \geq 1$
Solution:	If possible, an allocation of active interfaces $W_A: V \rightarrow 2^{\{1,\ldots,k\}}$ covering G such that for all $v \in V$, $W_A(v) \subseteq W(v)$ and $\lvert W_A(v)\rvert \leq p$; Otherwise, a negative answer
Goal:	Minimize the total cost of the active interfaces, $c(W_A) = \sum_{v \in V} \sum_{\alpha \in W_A(v)} c(\alpha)$

It is worth noting that we can explore two variations of the aforementioned problem: the cost function c may range over $\mathbb{R}_{>0}$, or $c(\alpha) = 1$, for every $\alpha \in \{1, \ldots, k\}$ (*unit cost case*). In both instances, we presume $k \geq 2$, as the case $k = 1$ has a straightforward and unique solution (all nodes must activate their sole interface).

In [9], it was demonstrated that *CMI(p)* is a specific instance of the broader *CMI*(∞) problem (refer to [21,28]), where each node is limited to activating at most p interfaces. Notably, the basic variant with $p = 2$ proves to be generally more challenging than *CMI*(∞). However, certain graph classes exhibit more manageable characteristics. For instance, in trees and complete graphs, *CMI*(∞) has been established as *APX*-hard and not approximable within $O(\log k)$, respectively, while *CMI(2)* is solvable in polynomial time.

3 Graphs with Bounded Branchwidth

We describe here an algorithm for *CMI(2)* on graphs with bounded *branchwidth.*

Definition 2 ([20]). A *branch decomposition* of a graph $G = (V, E)$ is a pair (T, χ), where:

- T is a tree with $|E|$ leaves whose all internal nodes have degree three;
- χ is a bijection from E to the leaves of T.

Each edge $\{i, j\}$ of T divides the tree into two components, so it divides E into two parts X and $E \setminus X$. We denote by $\gamma(X)$ (*border*) the set of those vertices of G that are both incident to an edge of X and to an edge of $E \setminus X$. The *width* of a decomposition (T, χ) is the maximum value on the edges of T. The *branchwidth* of G, denoted by $bw(G)$, is the minimum width over all branch decompositions of G. We put $bw(G) = 0$ if $|E| = 1$, and $bw(G) = 0$ if $|E| = 0$.

3.1 An Optimal Algorithm

Let $G(V, E)$ be a graph with a branch decomposition (T, χ) of width h. We arbitrarily choose an internal edge $\{l, r\}$ of T, where l and r are not leaves of T. Subsequently, we divide T into two subtrees, L and R, by removing the edge $\{l, r\}$ and orient their edges from l and r to their respective leaves. In doing so, l and r become their roots.

We can now formulate an algorithm that efficiently solves *CMI(2)* for the graph G. This algorithm capitalizes on the structural properties inherent in the branch decomposition (T, χ). For clarity, we will refer to the vertices of T (belonging to subtrees L and R) as *nodes* to avoid potential confusion with the vertices of the original graph.

As mentioned earlier, we will assume that L and R are directed with the roots l and r, respectively, and that every edge is directed from the root towards the leaves. Denote by $L(i)$ the subtree induced by an internal node i of L and all its descendants. Similarly, denote by $R(i)$ the subtree induced by an internal node i of R and all its descendants. For $L(i)$ (or for $R(i)$), denote also by X_i its respective leaves, by $G(i)$ the subgraph of G induced by the edges in X_i, and by $\gamma(X_i)$ (*border*) the set of those vertices of G that are both incident to an edge of X_i and to an edge of $E \setminus X_i$. In the core of the algorithm, we compute the optimal value of a constrained version of the problem restricted to the subgraph $G(i)$. The constraint is given by the interfaces active on each node in the border $\gamma(X_i)$. In particular, we compute $\varphi(i, F)$, which is the optimum value of *CMI(2)* on $G(i)$, where F is a family of subsets $A(u)$ of interfaces defined as:

- F contains $|\gamma(X_i)|$ sets, one for every vertex in $\gamma(X_i)$;
- the interfaces in $A(u)$ must belong to $W(u)$;
- the interfaces active on vertex u are those in $A(u)$, for every $u \in \gamma(X_i)$;
- the cardinality of $A(u)$ is at most 2, fo every $u \in X_i$

Please note that the maximum number of subsets in F is h since the branchwidth of G is h.

Case: leaf.
If i is a leaf corresponding to an edge $\{u, v\} \in E$, then there are three different cases. In the first case, neither u nor v is connected to any other nodes in $V \setminus \{u, v\}$; thus, $F = \emptyset$ because the border is empty. In the second case, either u or v is in the border of i, but not both. Therefore, either $F = \{A(u)\}$ or $F = \{A(v)\}$. In the last case, both u and v are in the border, implying $F = \{A(u), A(v)\}$. Therefore, to compute $\varphi(i, F)$ for a leaf, we can solve the following minimization problem.

$$
\begin{aligned}
\min \quad & 2c(\alpha) + \sum_{u \in \gamma(X_i)} \left(\sum_{\beta \in A(u) \setminus \{\alpha\}} c(\beta) \right) \\
\text{s.t.} \quad & \alpha \in A(u) \qquad \forall u \in \gamma(X_i) \\
& \alpha \in W(u) \cap W(v)
\end{aligned} \tag{1}
$$

In practice, to establish communication along the edge $\{u, v\}$, we must select a common interface $\alpha \in W(u) \cap W(v)$ with a total cost of $2c(\alpha)$. Subsequently, we calculate the cost contributed by the other interfaces in F, excluding the double counting of the interface α. The first constraint ensures that α must be active on every vertex u of the border and contained in the corresponding set $A(u)$. If Problem (1) has no feasible solution, we set $\varphi(i, F) = +\infty$.

Case: internal node.
We now analyze the case of an internal node i in L (or R) with children j and k. We can compute $\varphi(i, F)$ using the previously computed values $\varphi(j, F_j)$ and $\varphi(k, F_k)$. It's important to note that $G(j)$ and $G(k)$ may share some common vertices, meaning $\gamma(X_j) \cap \gamma(X_k) \neq \emptyset$. The following constrained minimization problem solves $\varphi(i, F)$ for a particular collection F, where $F_j = \bigcup_{u \in \gamma(X_j)} \{B(u)\}$ and $F_k = \bigcup_{u \in \gamma(X_k)} \{C(u)\}$.

$$
\begin{aligned}
\min \quad & \varphi(j, F_j) + \varphi(k, F_k) - \sum_{u \in \gamma(X_j) \cap \gamma(X_k)} \sum_{\alpha \in B(u) \cap C(u)} c(\alpha) \\
\text{s.t.} \quad & A(u) = B(u) \quad \forall u \in \gamma(X_j) \setminus \gamma(X_k) \\
& A(u) = C(u) \quad \forall u \in \gamma(X_k) \setminus \gamma(X_j) \\
& A(u) = B(u) \cup C(u) \quad \forall u \in (\gamma(X_j) \cap \gamma(X_k)) \cap \gamma(X_i) \\
& |B(u) \cup C(u)| \leq 2 \quad \forall u \in \gamma(X_j) \cap \gamma(X_k) \\
& A(u) \subseteq W(u) \quad \forall u \in \gamma(X_i)
\end{aligned}
\tag{2}
$$

In fact, the border of $G(i)$ is a subset of the union of the borders of $G(j)$ and $G(k)$, that is $\gamma(X_i) \subseteq \gamma(X_j) \cup \gamma(X_k)$. Then, in order to merge the solutions computed for $G(j)$ and $G(k)$, we must guarantee that:

- the total cost is given by $\varphi(j, F_j) + \varphi(k, F_k)$ minus the cost given by the interfaces in $B(u) \cap C(u)$, for every $u \in \gamma(X_j) \cap \gamma(X_k)$. This prevents counting the cost of an interface more than once for a specific vertex;
- for each vertex u in $\gamma(X_i)$ that is also in $\gamma(X_j)$ but not in $\gamma(X_k)$, the set of active interfaces $A(u)$ equals $B(u)$ (first constraint);
- for each vertex u in $\gamma(X_i)$ that is also in $\gamma(X_k)$ but not in $\gamma(X_j)$, the set of active interfaces $A(u)$ equals $C(u)$ (second constraint);
- for each vertex u in $\gamma(X_i)$ that is also in $\gamma(X_k)$ and in $\gamma(X_j)$, the set of active interfaces $A(u)$ equals $C(u) \cup B(u)$ (third constraint);
- the number of active interfaces in each vertex belonging both in $\gamma(X_j)$ and $\gamma(X_k)$ must be at most 2 (fourth constraint);
- the set $A(u)$ must belong to the interfaces present on u (last constraint).

Case: union of the subtrees L and R.
We now consider the case where we merge every solution computed for the trees L and R to determine an optimal solution for T and, consequently, for *CMI(2)* in graph G. In this case, we only need to check if two solutions for $G(L)$ and $G(R)$ are compatible, that it the number of interfaces does not exceed the limit.

$$
\begin{aligned}
\min \quad & \varphi(j, F_L) + \varphi(k, F_R) - \sum_{\alpha \in B(u) \cap C(u)} c(\alpha) \\
\text{s.t.} \quad & |B(u) \cup C(u)| \leq 2 \quad \forall u \in \gamma(X_L) = \gamma(X_R) \\
& B(u) \cup C(u) \subseteq W(u) \quad \forall u \in \gamma(X_L) = \gamma(X_R)
\end{aligned}
\tag{3}
$$

In fact, $G(L)$ and $G(R)$ have the same borders, i.e., $\gamma(X_L) = \gamma(X_R)$. This means that for every $u \in \gamma(X_L) = \gamma(X_R)$, we must guarantee that the number of active interfaces on it is less than or equal to 2 (first constraint). Moreover, we must subtract once the cost for the interfaces in $B(u) \cap C(U)$, where $u \in \gamma(X_L) = \gamma(X_R)$, from the objective function, since it is already counted in $\varphi(k, F_R)$ and in $\varphi(j, F_L)$.

After finding the minimum cost for Problem 3, we can construct an optimal solution for *CMI(2)* by applying standard methods of dynamic programming.

We analyse now the time complexity of the algorithm.

If i is a leaf of T, since the graph $G(i)$ is only made up of an edge $\{u, v\}$, there are at most $(k^2 + k)^2/4$ ways for the collections $F = \{A(u), A(v)\}$. This leads to a time complexity of $O(k^4)$. Please note that we assume every subset in F is non-null, as we must guarantee communication on every edge.

If i is an internal node, the number of Problems (2) that we need to solve is determined by all possible combinations for the two collections F_j and F_k, each having at most h sets. In fact, there is only one way to determine the family F_i when F_j and F_k are fixed. This implies a time complexity of $O\left(\left(\frac{k^2}{2}\right)^{2h} h\right)$ for every internal node, as we require $O(h)$ computational time to solve a single Problem (2).

It is easy to check that the time complexity for the last case, the union of trees L and R, is upper-bounded by that of every internal node.

In conclusion, since the number of internal nodes of T is upper bounded by the number of the leaves, which are $m = |E|$, the time complexity of the algorithm is $O\left(\left(\frac{k^2}{2}\right)^{2h} h \cdot m\right)$.

We can finally state the following:

Theorem 1. Given an instance of *CMI(2)* with underlying graph G, and a branch decomposition (T, χ) of width h for G, it is possible to find an optimal solution in $O\left(\left(\frac{k^2}{2}\right)^{2h} h \cdot m\right)$ time.

Given the well-established connection between the concepts of branchwidth and treewidth in graph theory, we can extend this outcome to derive a comparable result for treewidth as well. Indeed, the relationships between these two parameters for every graph G with branchwidth $h > 1$ are captured by the following inequalities [20]:

$$\mathrm{bw}(G) \leq \mathrm{tw}(G) + 1 \leq \frac{3}{2}\mathrm{bw}(G) \tag{4}$$

where bw and tw state for branchwidth and treewidth, respectively. To obtain our final result, Theorem 3, we utilize the following theorem regarding a fast algorithm for finding a branch decomposition.

Theorem 2 ([26]). There is a deterministic algorithm that, for a given graph G and an integer h, runs in time $2^{O(h)}n$ and either produces a branch decomposition of G of width $2h$ or determines that the branchwidth of G is more than h.

By using Equation (4) and Theorem (2), we can write the following result.

Theorem 3. Given an instance of *CMI(2)* with G as the underlying graph, and a 2-approximate branch decomposition (T, χ), it is possible to find an optimal solution in $O\left(\left(\frac{k^2}{2}\right)^{4(t+1)} t \cdot m\right)$ time, where t is the treewidth of G.

4 Concluding Remarks

In this study, we investigate a variant of the well-explored Coverage problem within the domain of multi-interface networks. This problem involves identifying the most cost-effective way to establish all connections outlined by an input graph. This is achieved by activating suitable subsets of interfaces at network nodes. The modification to the original model introduces an additional constraint, limiting each node to activate at most p interfaces. We specifically focus on the case with $p = 2$ while considering graphs with bounded branchwidth and treewidth. Although the problem is generally known to be *NP*-hard, we present an optimal fixed-parameter tractable algorithm that demonstrates *CMI(2)* to be tractable with respect to the branchwidth (or treewidth) and the number of available interfaces.

A potential avenue for further investigation is to study *CMI(2)* in relation to other parameters, such as local treewidth and cliquewidth, with the aim of deriving both positive and negative results.

Furthermore, while *NP*-hardness is established for general graphs with a maximum degree of 4, and the problem is optimally solvable in polynomial time for a maximum degree of 2, the case of graphs with a maximum degree of 3 remains to be investigated.

Another line of research that we believe is worth pursuing involves analysing the multi-interface coverage problem through the lens of game theory. By doing so, the problem becomes decentralized, with devices functioning as agents [2,4,6], striving to maximize their utility (e.g., connections), all while keeping overall energy consumption under control. In this context, one could employ one of the models presented in [3,5,7,8] and investigate the degradation of social welfare with respect to different standard metrics (e.g., the price of anarchy and stability).

References

1. Aloisio, A.: Coverage subject to a budget on multi-interface networks with bounded carving-width. In: Barolli, L., Amato, F., Moscato, F., Enokido, T., Takizawa, M. (eds.) WAINA 2020. AISC, vol. 1150, pp. 937–946. Springer, Cham (2020). https://doi.org/10.1007/978-3-030-44038-1_85
2. Aloisio, A.: Distance hypergraph polymatrix coordination games. In: Proceedings of the 22nd Conference on Autonomous Agents and Multi-Agent Systems (AAMAS), pp. 2679–2681 (2023)
3. Aloisio, A.: Fixed-parameter tractability for branchwidth of the maximum-weight edge-colored subgraph problem. In: WAINA, Advances in Intelligent Systems and Computing (2024)
4. Aloisio, A., Flammini, M., Kodric, B., Vinci, C.: Distance polymatrix coordination games. In: Proceedings of the 30th International Joint Conference on Artificial Intelligent (IJCAI), pp. 3–9 (2021)
5. Aloisio, A., Flammini, M., Kodric, B., Vinci, C.: Distance polymatrix coordination games (short paper). In: SPIRIT co-located with 22nd International Conf. AIxIA 2023, November 7-9th, 2023, Rome, Italy, vol. 3585 (2023)
6. Aloisio, A., Flammini, M., Vinci, C.: the impact of selfishness in hypergraph hedonic games. In: Proceedings of the 34th Conference Artificial Intelligence (AAAI), pp. 1766–1773 (2020)

7. Aloisio, A., Flammini, M., Vinci, C.: Generalized distance polymatrix games. In: Fernau, H., Gaspers, S., Klasing, R. (eds.) SOFSEM 2024. LNCS, vol. 14519, pp. 25–39. Springer, Cham (2024). https://doi.org/10.1007/978-3-031-52113-3_2
8. Aloisio, A., Mkrtchyan, V.: Algorithmic aspects of the maximum 2-edge-colorable subgraph problem. In: Barolli, L., Woungang, I., Enokido, T. (eds.) AINA 2021. LNNS, vol. 227, pp. 232–241. Springer, Cham (2021). https://doi.org/10.1007/978-3-030-75078-7_24
9. Aloisio, A., Navarra, A.: Balancing energy consumption for the establishment of multi-interface networks. In: Italiano, G.F., Margaria-Steffen, T., Pokorný, J., Quisquater, J.-J., Wattenhofer, R. (eds.) SOFSEM 2015. LNCS, vol. 8939, pp. 102–114. Springer, Heidelberg (2015). https://doi.org/10.1007/978-3-662-46078-8_9
10. Aloisio, A., Navarra, A.: Budgeted constrained coverage on bounded carving-width and series-parallel multi-interface networks. Internet Things **11**, 100,259 (2020)
11. Aloisio, A., Navarra, A.: Budgeted constrained coverage on series-parallel multi-interface networks. In: Barolli, L., Amato, F., Moscato, F., Enokido, T., Takizawa, M. (eds.) AINA 2020. AISC, vol. 1151, pp. 458–469. Springer, Cham (2020). https://doi.org/10.1007/978-3-030-44041-1_41
12. Aloisio, A., Navarra, A.: Constrained connectivity in bounded X-width multi-interface networks. Algorithms **13**(2), 31 (2020)
13. Aloisio, A., Navarra, A.: On coverage in multi-interface networks with bounded pathwidth. In: WAINA, Advances in Intelligent Systems and Computing (2024)
14. Aloisio, A., Navarra, A., Mostarda, L.: Distributing energy consumption in multi-interface series-parallel networks. In: Barolli, L., Takizawa, M., Xhafa, F., Enokido, T. (eds.) WAINA 2019. AISC, vol. 927, pp. 734–744. Springer, Cham (2019). https://doi.org/10.1007/978-3-030-15035-8_71
15. Aloisio, A., Navarra, A., Mostarda, L.: Energy consumption balancing in multi-interface networks. J. Ambient. Intell. Humaniz. Comput. **11**(8), 3209–3219 (2020)
16. Athanassopoulos, S., Caragiannis, I., Kaklamanis, C., Papaioannou, E.: Energy-efficient communication in multi-interface wireless networks. Theory Comput. Syst. **52**, 285–296 (2013)
17. Bahl, P., Adya, A., Padhye, J., Walman, A.: Reconsidering wireless systems with multiple radios. SIGCOMM Comput. Commun. Rev. **34**(5), 39–46 (2004)
18. Caporuscio, M., Charlet, D., Issarny, V., Navarra, A.: Energetic performance of service-oriented multi-radio networks: issues and perspectives. In: Proceedings of the 6th International Workshop on Software and Performance (WOSP), pp. 42–45. ACM (2007)
19. Cavalcanti, D., Gossain, H., Agrawal, D.: Connectivity in multi-radio, multi-channel heterogeneous ad hoc networks. In: Proceedings of the 16th International Symposium on Personal, Indoor and Mobile Radio Communications (PIMRC), pp. 1322–1326. IEEE (2005)
20. Cygan, M., et al.: Parameterized Algorithms. Springer, Cham (2015). https://doi.org/10.1007/978-3-319-21275-3
21. D'Angelo, G., Di Stefano, G., Navarra, A.: Multi-interface wireless networks: complexity and algorithms. In: Ibrahiem, S.R., El Emary, M.M. (ed.) Wireless Sensor Networks: From Theory to Applications, pp. 119–155. CRC Press, Taylor & Francis Group (2013)
22. D'Angelo, G., Stefano, G.D., Navarra, A.: Minimize the maximum duty in multi-interface networks. Algorithmica **63**(1–2), 274–295 (2012)
23. Draves, R., Padhye, J., Zill, B.: Routing in multi-radio, multi-hop wireless mesh networks. In: Proceedings of the 10th International Conference on Mobile Computing and Networking (MobiCom), pp. 114–128. ACM (2004)
24. Faragó, A., Basagni, S.: The effect of multi-radio nodes on network connectivity—a graph theoretic analysis. In: Proceedings of the 19th International Symposium on Personal, Indoor and Mobile Radio Communications, (PIMRC), pp. 1–5. IEEE (2008)

25. Fomin, F.V., Kaski, P.: Exact exponential algorithms. Commun. ACM **56**(3), 80–88 (2013)
26. Fomin, F.V., Korhonen, T.: Fast FPT-approximation of branchwidth. In: Proceedings of the 54th Annual ACM SIGACT Symposium on Theory of Computing, STOC 2022, New York, NY, USA, pp. 886–899. Association for Computing Machinery (2022)
27. Friedman, R., Kogan, A., Krivolapov, Y.: On power and throughput tradeoffs of WiFi and Bluetooth in smartphones. In: Proceedings of the 30th International Conference on Computer Communications (INFOCOM), pp. 900–908. IEEE (2011)
28. Klasing, R., Kosowski, A., Navarra, A.: Cost minimization in wireless networks with a bounded and unbounded number of interfaces. Networks **53**(3), 266–275 (2009)
29. Kosowski, A., Navarra, A., Pinotti, M.: Exploiting multi-interface networks: connectivity and cheapest paths. Wireless Netw. **16**(4), 1063–1073 (2010)
30. Perucci, A., Autili, M., Tivoli, M., Aloisio, A., Inverardi, P.: Distributed composition of highly-collaborative services and sensors in tactical domains. In: Ciancarini, P., Mazzara, M., Messina, A., Sillitti, A., Succi, G. (eds.) SEDA 2018. AISC, vol. 925, pp. 232–244. Springer, Cham (2020). https://doi.org/10.1007/978-3-030-14687-0_21

Blockchain and Financial Services a Study of the Applications of Distributed Ledger Technology (DLT) in Financial Services

Ramiz Salama[1(✉)], Diletta Cacciagrano[2], and Fadi Al-Turjman[3,4]

[1] Department of Computer Engineering, AI and Robotics Institute, Research Center for AI and IoT, Near East University, Nicosia, Mersin 10, Turkey
ramiz.salama@neu.edu.tr
[2] Computer Science, University of Camerino, Camerino, Italy
[3] Artificial Intelligence, Software, and Information Systems Engineering Departments, AI and Robotics Institute, Near East University, Nicosia, Mersin10, Turkey
[4] Research Center for AI and IoT, Faculty of Engineering, University of Kyrenia, Kyrenia, Mersin10, Turkey

Abstract. Untrusted parties can agree on a database's status without the need for an intermediary thanks to blockchain technology. A blockchain could eliminate the need for banks to supply some financial services like payments and securitization by offering a ledger that is not managed by anyone. Furthermore, blockchain enables the deployment of "smart contracts," which are self-executing contracts written on the network that can automate a variety of manual tasks, including will distribution and compliance as well as claim processing. "Distributed ledger technology (DLT)," a relative of blockchain, may assist companies in enforcing more stringent regulations and policies on data exchange and cooperation for use cases that don't require a high degree of decentralization but could benefit from improved coordination.

Keywords: Blockchain · Financial Services · Information Security · Technology · Banking

1 Introduction

In the last ten years, blockchain technology has garnered a lot of attention from investors and financial professionals, extending beyond the niche market of Bitcoin aficionados. "It's worse than tulip bulbs" JP Morgan Chase CEO Jamie Dimon said in September 2017, alluding to the 17th-century Dutch tulip market frenzy. There will not be a joyful conclusion. The result will be someone's death. A similar opinion was voiced by Lloyd Blankfein, senior chairman of Goldman Sachs, who stated, "Something that changes 20% [overnight] does not feel like a currency". It functions as a con game strategy. Notwithstanding the gloom, it is still unclear if blockchain and decentralized ledger technology (DLT) will disrupt or replace certain aspects of the financial industry [1]. AI and machine learning enablers have been ingrained in our daily lives and have had a big

L. Barolli (Ed.): AINA 2024, LNDECT 204, pp. 124–135, 2024.
https://doi.org/10.1007/978-3-031-57942-4_14

impact on the internet of things explosive growth. Prof. DUX, an AI learning facilitator, is one of the pioneering efforts in this sector [2]. It's a new kind of AI facilitator designed to help students study at their own pace and with the highest caliber possible across a wide range of subjects.

2 Relevant Work

The literature that is now available indicates that a large number of studies have looked at the importance, characteristics, and implementation issues of blockchain generally in the fields of finance, banking, auditing, and accounting. Coverage. According to Liu et al. (2019), blockchain is a cutting-edge technique for storing, analyzing, and documenting data and business transactions. It has the power to change the structure of the accounting industry. Dai and Vasarhelyi looked at how blockchain technology might be used to create an accounting system that is transparent, real-time, and auditable in their 2017 study. These skills will make accounting more sophisticated and reliable. Blockchain technology is essential for accounting in current stage of economic development because of its speed and security in document management, claims (Kwilinski, 2019).Blockchain technology can help prevent accounting fraud, which is more difficult for management to commit when internal controls are lax (Rückeshäuser, N., 2017). Like any other technology, this one has its share of obstacles. The security benefits of the blockchain in accounting aren't entirely available or dependable, according to Coyne and McMickle (2017). It's currently under development, so a more improved version should be available shortly [3, 4]. Furthermore, it is thought that blockchain might affect audit procedures and efficiency. (Brender and Gauthier, 2021). More businesses are utilizing blockchain technology or intend to incorporate it into their everyday operations. It was difficult for the auditors to audit businesses that had adopted blockchain technology because there isn't currently a standard for blockchain audits. The auditors stressed how crucial it is to guarantee the reliability and validity of the data that is stored on the blockchain. Many financial industries have studied blockchain technology. According to Chen and Bellavitis (2020), decentralized businesses can be established through the use of blockchain technology. The possibility for transparent, international, decentralized financial services is one of blockchain technology's main advantages for the financial industry. There are benefits, drawbacks, and limitations to decentralized finance [5–10].

If intermediaries are eliminated from essential services offered by banks, like these, distributed ledger technology (DLT) and blockchain might radically revolutionize the financial sector.

1. ***Compensation***

Trillions of dollars are being lost due to an antiquated system that has delayed payouts and high fees. If you are employed in San Francisco and would like to send money to your family in London, you may have to pay a wire transfer fee of $25 fixed rate plus additional fees totaling 7%. You pay exchange rate expenses in addition to a cut to your bank and the receiving bank. It can take a week for your family's bank to even record the transaction (Fig. 1).

Fig. 1. In January 2014, there were just over 50,000 verified Bitcoin transactions daily; by August 2022, that figure had risen to over 249,000. Using Blockchain.com as a reference

Because banks make so much money processing payments, they have little motivation to raise expenses. For example, payments income from cross-border B2B and C2B transactions is expected to reach $175 billion in 2020. Cryptocurrencies like bitcoin and ether, which anybody can use to send and receive money, are built on open blockchains. (Ethereum and Bitcoin, correspondingly). Everyone may send money swiftly, affordably, and globally thanks to public blockchains, which do not require reliable third parties to validate transactions [11].

Examples of how payments have been enhanced via blockchain

The amount of transactions for cryptocurrencies like bitcoin and ether has increased over the past few years, mainly in a positive direction, even though fiat currencies (like the US dollar) won't ever fully replace them when it comes to payments. To be exact, the Ethereum network created history in 2020 when it settled $1 trillion worth of transactions in a single year. Blockchain technology is being used by some businesses to enhance business-to-business payments in developing nations. BitPesa, a blockchain-based payment system used in Kenya, Nigeria, and Uganda, among other nations, is one example. The company apparently seen a 20% increase in million-dollar transactions month over month (Fig. 2).

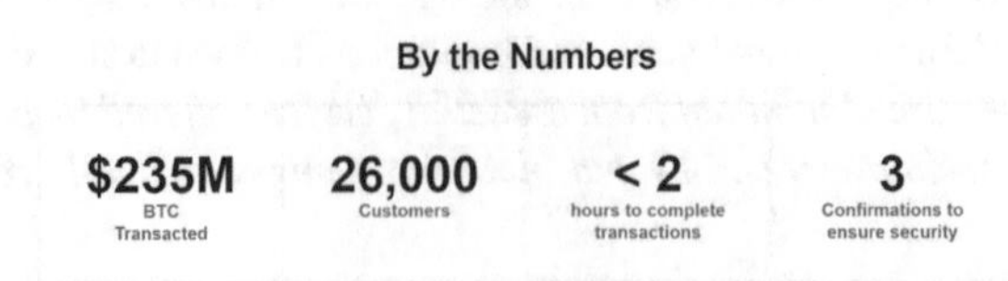

Fig. 2. Blockchain technology is being used by some businesses to enhance business-to-business payments in emerging economies

Remittances from Sub-Saharan Africa, where sending money is most expensive globally, are also largely handled via BitPesa. Transfer expenses in the area have been dramatically reduced by BitPesa and other cryptocurrency payment systems. The capacity of companies to accept cryptocurrencies as payment is another area of interest for blockchain corporations. For example, there are numerous interfaces between e-commerce platforms like Shopify and WooCommerce and BitPay, a company that assists businesses in accepting and storing bitcoin payments.It committed to supporting payments made with DoorDash and Uber Eats in 2022. BitPay Card owners can use their BitPay Wallet to buy an Uber Eats or DoorDash gift card, or they can use their BitPay Wallet to pay for meals when ordering from a website that accepts direct bitcoin payments, such as Menufy or Takeaway.com. Customers can now use MasterCard's banQi app to pay at foreign points of sale in addition to all Via Varejo stores, thanks to a partnership between AirFox and MasterCard [12, 13].

3 System for Clearance and Settlements

Our financial infrastructure's development has much to do with the three-day settlement time of a typical bank transfer, as previously mentioned. The consumer is not the only one affected. For the banks themselves, moving money around the globe presents logistical challenges. These days, a straightforward bank transfer from one account to another has to go via a convoluted network of middlemen, including custodial services and correspondent banks, before it can reach its final destination. The two bank balances need to be compared over a broad network of dealers, funds, asset managers, and other players in the global financial system [14] (Fig. 3).

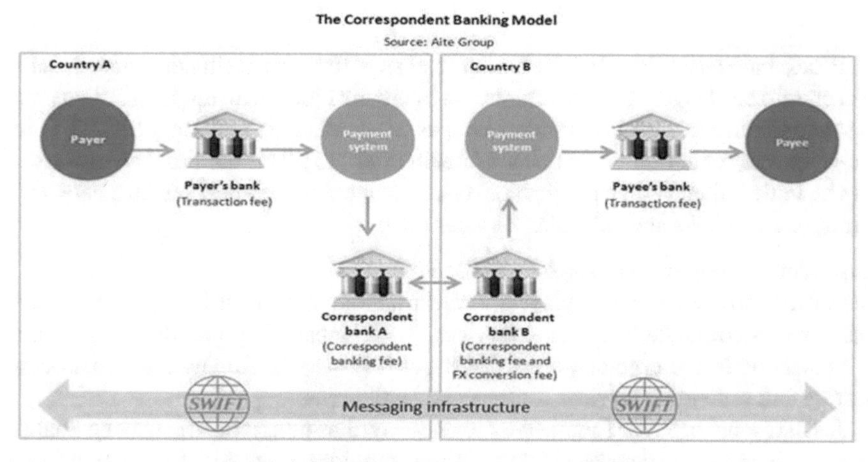

Fig. 3. The Model of Correspondent Banking

Since UniCredit Banca and Wells Fargo do not now have a functional financial relationship, they must look through the SWIFT network for a correspondent bank that can settle the transaction for a charge and has connections to both companies. Since each correspondent bank keeps separate ledgers at both sending and receiving banks, these ledgers must be reconciled at the end of each day.

Money is never moved via the centralized SWIFT protocol; only payment orders are exchanged. After then, a network of intermediaries processes the actual money. Every intermediary drives up the cost of the deal and boosts the possibility of a failure (Fig. 4).

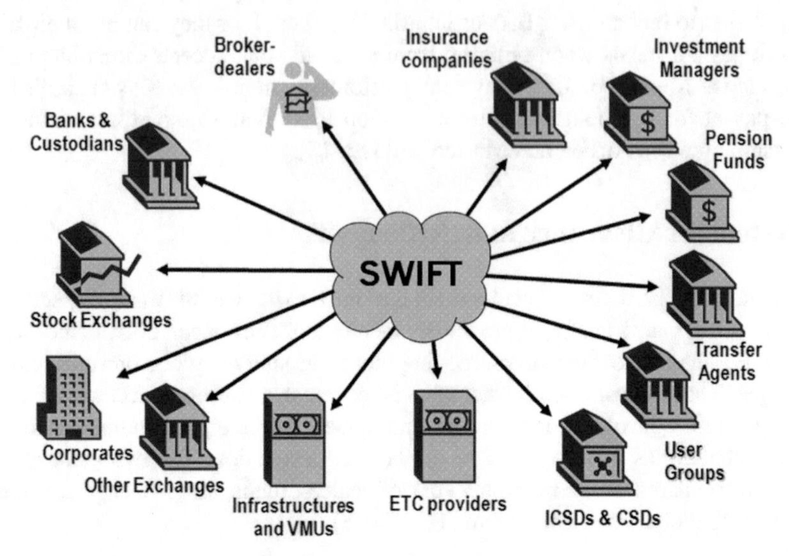

Fig. 4. The centralized SWIFT protocol

Blockchain technology has the ability to change the current situation since it acts as a decentralized "ledger" of transactions. An interbank blockchain might maintain public and transparent records of all transactions, as an alternative to utilizing SWIFT to reconcile the ledgers of individual financial institutions [15]. This implies that transactions may be settled directly on a public blockchain as opposed to depending on a network of correspondent banks and custodian services [16].

Illustrations of improved blockchain transactions

One of the most well-known players in clearing and settlement is Ripple, a business that provides corporate blockchain services. The company is primarily recognized for its connection to the cryptocurrency XRP, but it is also developing blockchain-based clearing and settlement services for financial institutions.

Just like emails, SWIFT messages are one-way, thus transactions cannot be finalized unless both parties have inspected them. Ripple provides banks with a quicker, two-way communication protocol that seamlessly interacts with the bank's current databases and ledgers to enable real-time messaging and settlement. More than 300 individuals from more than 40 countries have signed up with Ripple to test out its blockchain network thus far (Fig. 5).

Additionally, ripple enables quicker cross-border transaction resolution. A traditional bank transaction would need that both traders have local currency accounts in the countries where they intend to receive their money if a trader in Mexico wanted to send money to their counterpart in the US. This condition is dropped by ripple. By utilizing Mexican pesos to purchase XRP tokens on the exchange, the Mexican trader can swiftly

International Payment Infrastructure Costs

Global Average Cost: 20.9 bps on payment volume

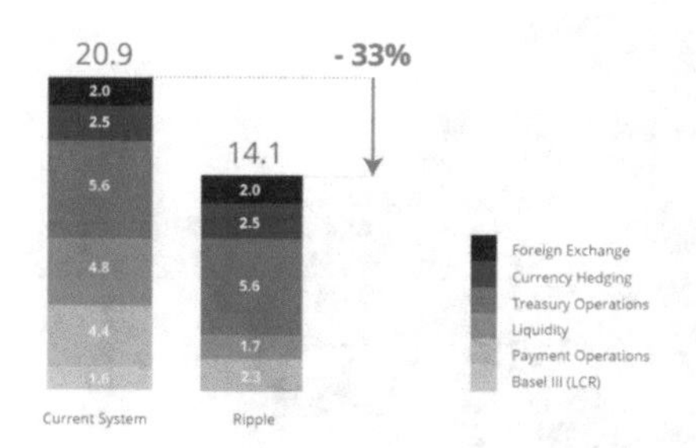

Fig. 5. International Payment Infrastructure Costs

pay their American counterpart. These XRP tokens can be exchanged for dollars at the US store. Furthermore, according to Ripple, the entire transaction can be completed in a single second. Another significant business developing distributed ledger technology for banks is R3. It states that the goal is to develop a "new operating system for financial markets". In May 2017, it received $107 million from a consortium of financial institutions that included HSBC, Bank of America, and Merrill Lynch. It has lost some important members, but it has also lost some other important members, including as Goldman Sachs, which left because it wanted more operational control over the system.

4 Obtaining Funds

The process of raising venture capital is difficult. In the hopes of making a little cash, entrepreneurs prepare decks, sit through long partner meetings, and tolerate protracted equity and valuation disputes. Some companies raise money using initial coin offers (ICOs), which are supported by open-source blockchains like Ethereum and Bitcoin. During an ICO, projects trade tokens or coins for cash. (Frequently expressed in ether or bitcoin). The success of the blockchain project is, in theory, correlated with the token's value. Token investing provides a direct means for investors to wager on value and usage. Blockchain businesses can sell tokens to the general public directly through initial coin offerings (ICOs), eschewing the conventional financing method. Before releasing a workable product, a few well-known initial coin offerings (ICOs) raised hundreds of millions or even billions of dollars. Filecoin, a startup that holds blockchain data, raised $257 million in 2017, in contrast to EOS, which is constructing a "global computer," which raised roughly $4 billion during its year-long initial coin offering (ICO). However, since then, the EOS blockchain has struggled due to problems like a declining user base and important developers quitting the project. Since 2017, ICOs have also had issues. Even though use was still very common in 2018, ICO financing took a severe turn as the bubble broke in the middle of the year (Fig. 6).

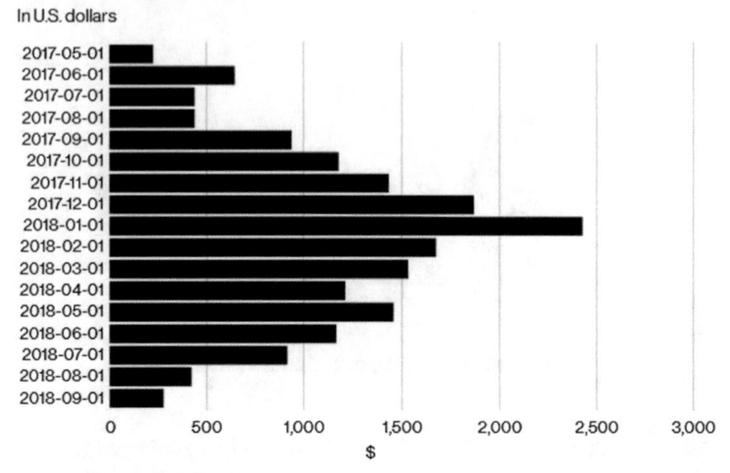

Fig. 6. ICO Monthly ICO Funding

Recently, regulators have expressed concern over initial coin offerings (ICOs) due to their rigorous observation of transactions and enforcement of infractions. For instance, in February 2021, the SEC filed charges against people who they claimed were involved in digital asset fraud, which included unregistered initial coin offerings. According to the SEC, the people used two initial coin offerings (ICOs) to defraud investors of over \$11 million. At the same time, Initial Coin Offerings (ICOs) signal a fundamental shift in the way businesses finance their expansion. The primary advantage of initial coin offerings (ICOs) is their worldwide and online nature, which allows firms to reach a far wider pool of potential investors. You are no longer restricted to wealthy people, organizations, and other people who can give the government proof that they are reliable investors. Companies can obtain funding instantly through initial coin offerings (ICOs). A token's value is established instantly upon sale on a global 24-h market. Ten years is the period of time for businesses that receive venture capital funding. One is the democratization of everything. Instances of how blockchain technology has enhanced fundraising.

While most initial coin offerings (ICOs) have gone to blockchain initiatives in the pre-revenue stage, an increasing number of digital businesses are beginning to structure their business models around a decentralized paradigm.

Telegram, a messaging software, raised \$1.7 billion through an initial coin offering (ICO).The ICO aims to bootstrap a payment system on top of the messaging network by providing tokens to users. The next Facebook, Google, and Amazon, according to proponents of blockchain technology, will be developed around decentralized protocols and launched through an initial coin offering (ICO), directly affecting margins in investment banking.

In this area, a large number of blockchain firms with great potential have emerged. Companies such as CoinList, which emerged from a collaboration between Protocol Labs and AngelList, are facilitating the mainstreaming of digital assets by assisting blockchain companies in organizing legitimate and compliant initial coin offerings (ICOs). For example, a Coin-List ICO helped the decentralized banking network Ondo banking raise $10 million. CoinList's monthly trade volume hit $1 billion in 2021 (Fig. 7).

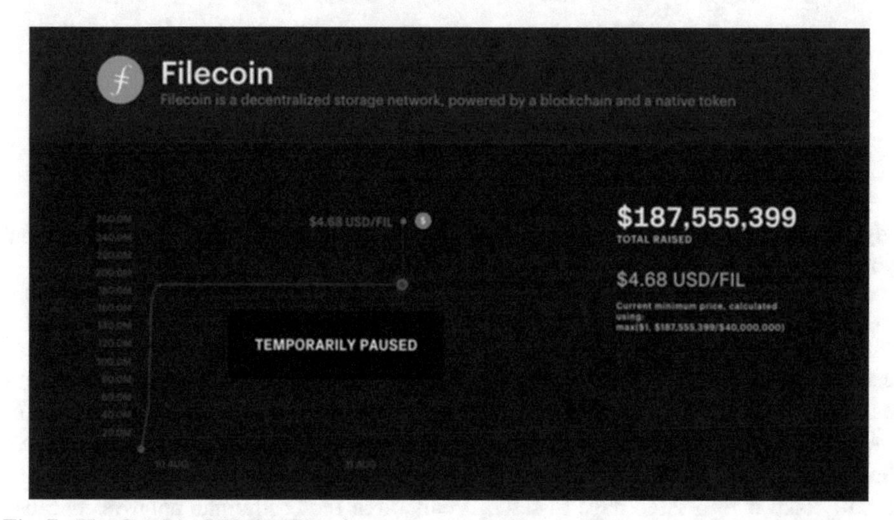

Fig. 7. The CoinList ICO for Filecoin was so popular that, within an hour of its launch, servers were overloaded. In the end, Filecoin's ICO raised almost $257 million.

With a streamlined API, CoinList has created a bank-grade compliance mechanism that blockchain companies may use to support projects that require everything from investor accreditation to due diligence. Though its platform is intended for blockchain firms, CoinList's focus on lowering the administrative and regulatory burden connected with financing is highlighted by the public marketplaces. Currently, investment banks are experimenting with automation in an effort to reduce the tens of thousands of labor hours required for each initial public offering.

5 Investments

You need a method to track the owners of assets like stocks, loans, and commodities when you buy or sell them. By using a sophisticated network of brokers, exchanges, clearing-houses, central security repositories, and custodian banks, today's financial markets do this. The foundation of these multiple parties is an antiquated paper ownership structure that is sluggish, imprecise, and susceptible to fraud. Let's say you wish to purchase Apple stock. Using a stock exchange, which connects you with a vendor, you might place an order. The ownership certificate for a share used to need to be paid for, therefore you had to do that. Trying to complete this transaction electronically makes things much more difficult. The day-to-day administration of the assets, such as processing dividends, exchanging certificates, and keeping records, is not something we wish to handle. We thus give the shares to custodian banks for storage. The custodians themselves must rely on a reliable third party to hold onto all of the paper certificates because buyers and sellers don't always rely on the same custodian banks (Fig. 8).

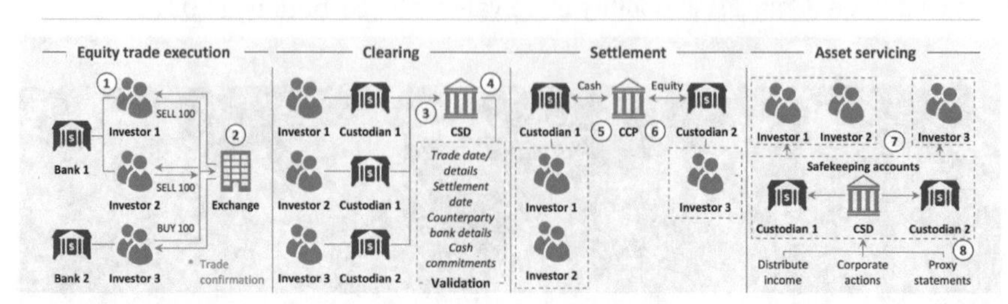

Fig. 8. Order clearing and settlement on an exchange involves a number of middlemen and potential sites of failure.

This basically means that a number of different third parties are involved in the process of processing your request to buy or sell something. Because each participant keeps a separate ledger with their own interpretation of the facts, it is challenging to transfer ownership. This system is inefficient and imprecise at the same time. As everyone's books need to be updated and reconciled at the end of the day, securities transactions usually take one to three days to settle. Deals often require human approval because there are so many stakeholders involved. Every party charges a price. By generating a decentralized database of distinct digital assets, blockchain technology has the potential to significantly alter the financial markets. By substituting cryptographic tokens for assets "off-chain", the rights to an asset can be transferred via a distributed ledger. While Bitcoin and Ethereum have solely used digital assets to achieve this, new blockchain companies are investigating ways to tokenize tangible assets like gold, stocks, and real estate (Fig. 9).

There's plenty of room for disruption. The top four US custodial banks, State Street, BNY Mellon, Citi, and JP Morgan, are together responsible for managing assets worth over $12 trillion. Profits are not derived from the small percentages of normally charged fees, but rather from the amount of the assets. Due to the complete elimination of middlemen like custodial institutions by blockchain technology, tokenized securities have the potential to reduce asset exchange costs. Tokenized securities can also function like programmable equities by using smart contracts to execute buybacks or dividend distributions with just a few lines of additional code. Finally, but just as importantly, storing physical assets on blockchain could increase the number of global markets [20].

Instances where blockchain technology has enhanced securities procedures
The transfer of financial products valued at trillions of dollars to the blockchain is an endeavor being undertaken by multiple blockchain technology businesses. For example, the Polymesh blockchain network, run by 14 authorized financial institutions, made it possible to validate new blocks before they were added to the chain. Polymath introduced Polymesh in 2021. Since then, Polymath has added Huobi and the investment firm

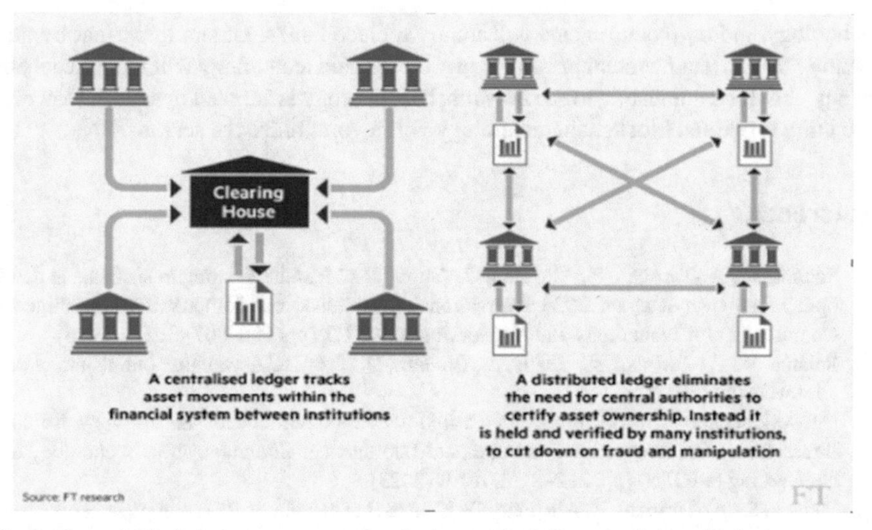

Fig. 9. Several blockchain startups are developing methods for tokenizing physical assets such as gold, real estate, and equities

Greentrail as node operators. Financial institutions are not sitting idly by in the interim. By April 2023, the Australian Stock Exchange plans to replace its current bookkeeping, clearing, and settlements system with a blockchain-based substitute created by Digital Asset Holdings. The recent appointment of a head of group digital assets by the London Stock Exchange indicates that it is also pursuing blockchain solutions in the UK [17, 18].

6 Conclusion

It is noteworthy that the application of blockchain technology in the financial services industry will benefit all stakeholders, including financial institutions and the general public. By fostering collaboration among financial institutions and standardizing procedures throughout the sector, blockchain technology has the potential to lower the costs of financial services and provide more equitable access to previously marginalized populations. By building confidence between third-party entities and enabling information transfer through previously unheard-of techniques, blockchain can assist collaboration between financial institutions and enhance the banking industry's function as a trusted middleman. The banking industry has a historic chance to quickly upgrade through the integration of public and private blockchains. The banking sector can use blockchain technology more extensively to provide more effective and secure products and services to both new and existing customers, and the legal ambiguities in the field can be resolved through a combination of appropriate governmental regulation and partnerships between the public and private sectors.

Disruption takes time because blockchain technology is still very much in its infancy and hasn't been extensively tested or matured. Devoted believers believe that blockchain technology and cryptocurrencies will totally replace banks. Others think that by augmenting the current financial infrastructure, blockchain technology will increase its efficiency. It will be interesting to see how much technology is adopted by banks. However, one thing is certain: blockchain technology will revolutionize the sector.

References

1. Salama, R., Al-Turjman, F., Altrjman, C., Gupta, R.: Machine learning in sustainable development–an overview. In: 2023 International Conference on Computational Intelligence, Communication Technology and Networking (CICTN), (pp. 806–807). IEEE (2023)
2. Salama, R., Al-Turjman, F., Bhatla, S., Gautam, D.: Prof.DUX available online.https://dux.aiiot.website/ (2023)
3. Network security, trust & privacy in a wired wireless environments–an overview. In: 2023 International Conference on Computational Intelligence, Communication Technology and Networking (CICTN) (pp. 812–816). IEEE (2023)
4. Salama, R., Al-Turjman, F., Altrjman, C., Kumar, S., Chaudhary, P.: A comprehensive survey of Block-chain powered cybersecurity-a survey. In: 2023 International Conference on Computational Intelligence, Communication Technology and Networking (CICTN), (pp. 774–777). IEEE (2023)
5. Salama, R., Al-Turjman, F., Bordoloi, D., Yadav, S.P.: Wireless sensor networks and green networking for 6G communication-an overview. In: 2023 International Conference on Computational Intelligence, Communication Technology and Networking (CICTN), pp. (830–834). IEEE (2023)
6. Salama, R., Al-Turjman, F., Bhatia, S., Yadav, S.P.: Social engineering attack types and prevention techniques-a survey. In: 2023 International Conference on Computational Intelligence, Communication Technology and Networking (CICTN), (pp. 817–820). IEEE (2023)
7. Salama, R., Altrjman, C., Al-Turjman, F.: Smart grid applications and blockchain technology in the AI era. NEU J. Artifi. Intell. Internet Things **1**(1), 59–63 (2023)
8. Salama, R., Alturjman, S., Al-Turjman, F.: Internet of things and AI in smart grid applications. NEU J. Artifi Intell. Internet Things **1**(1), 44–58 (2023)
9. Salama, R., Altrjman, C., Al-Turjman, F.: A survey of machine learning (ML) in Sustainable Systems. NEU J. Artifi. Intell. Internet Things **2**(3) (2023)
10. Salama, R., Altrjman, C., Al-Turjman, F.: A Survey of Machine Learning Methods for Network Planning. NEU J. Artifi. Intell. Internet Things, **2**(3) (2023)
11. Salama, R., Altrjman, C., Al-Turjman, F.: A survey of the architectures and protocols for wireless sensor networks and wireless multimedia sensor networks. NEU J. Artifi. Intell. Internet Things, **2**(3) (2023)
12. Al-Turjman, F., Salama, R., Altrjman, C.: Overview of IoT solutions for sustainable transportation systems. NEU J. Artifi. Intell. Internet Things, **2**(3) (2023)
13. Salama, R., Altrjman, C., Al-Turjman, F.: An overview of the internet of things (IoT) and machine to machine (M2M) communications. NEU J. Artifi. Intell. Internet Things, **2**(3) (2023)
14. Salama, R., Al-Turjman, F., Altrjman, C., Bordoloi, D.: The use of machine learning (ML) in sustainable systems-an overview. In: 2023 International Conference on Computational Intelligence, Communication Technology and Networking (CICTN), (pp. 821–824). IEEE (2023)
15. Al-Turjman, F., Salama, R.: Cyber security in mobile social networks. In: Security in IoT Social Networks, pp. 55–81. Academic Press (2021)

16. Al-Turjman, F., Salama, R.: Security in social networks. In: Security in IoT Social Networks, pp. 1–27. Academic Press (2021)
17. Salama, R., Al-Turjman, F.: AI in block-chain towards realizing cyber security. In: 2022 International Conference on Artificial Intelligence in Everything (AIE), pp. 471–475. IEEE (2022)
18. Al-Turjman, F., Salama, R.: An overview about the cyberattacks in grid and like systems. In: Smart Grid in IoT-Enabled Spaces, pp. 233–247 (2020)

A Personalized Parking Guidance Service Based on Computer Vision Technology for Large Car Parks

Zhi-Wei Kao and Chi-Yi Lin(✉)

Department of Computer Science and Information Engineering, Tamkang University, New Taipei City, Taiwan
chiyilin@mail.tku.edu.tw

Abstract. In recent years, many car parks have adopted parking management systems, leveraging digital technology to enhance user convenience. This work in progress explores how large communities, public institutions, or shopping malls with thousands of parking spaces can provide more detailed "end-to-end parking guidance services" based on the existing parking management system to further improve service quality. Specifically, we assign a dedicated parking space for each vehicle and provide personalized guidance on the displays in the parking lot. In this way, drivers are freed from the task of finding available parking spots; as long as drivers follow the exclusive guidance, their vehicles can be parked smoothly. The difference between this work and existing literature or practices lies in our use of cameras installed in the parking lot to dynamically track vehicles with computer vision technology, allowing vehicles to enjoy exclusive parking guidance services without any additional equipment on their side.

1 Introduction

With the advancement of latest technologies, in recent years the wave of digital transformation has swept across all industries, such as retail, healthcare, manufacturing, and even construction [1]. Taking Taiwan as an example, many newly developed areas have witnessed the construction of large-scale residential communities, with individual community units numbering close to two thousand households. In order to provide a more convenient lifestyle for the vast number of community residents, community developers must invest efforts in realizing smart building solutions, smart homes, property management, visitor management, parking management, and other smart services [2].

In terms of parking management, the challenges faced by small and large communities are vastly different. Residents in small communities typically have their own fixed parking spaces, and due to the limited number of spaces, temporary parking for visitors is generally not allowed, making the management issues relatively straightforward. However, large communities with thousands of parking spaces, along with retail spaces and large supermarkets on lower floors and in basement areas, need to consider providing temporary parking for customers visiting the shopping areas. In addition, there is also the parking demand from friends and family visiting the community residents.

L. Barolli (Ed.): AINA 2024, LNDECT 204, pp. 136–144, 2024.
https://doi.org/10.1007/978-3-031-57942-4_15

For customers or visitors arriving by car, entering the underground parking of a large community can be like entering a massive maze. Without proper guidance, it can cause significant inconvenience for them. Therefore, guiding customers or visitors' vehicles to available parking spaces is a problem that large community parking lots need to address.

Similar issues arise in parking lots of public institutions and large shopping centers. For instance, the underground car park at Taipei City Hall has over 2,000 spaces. Despite having lane-specific vacancy indicators, drivers rely on red or green overhead lights upon entering a lane to assess availability. In large shopping centers, unclear signage and unfamiliar layouts may result in customers parking far from the entrance, leading to potential complaints.

To address the aforementioned issues and align with the trend of digital transformation, this study aims to propose a more personalized service for parking guidance in large parking lots. More precisely, our goal is to provide an end-to-end guidance service for each vehicle entering the parking lot, starting from the moment when the vehicle passes through the entrance gate and gradually guiding it to a specific parking space assigned by the parking management system. The parking space assignment logic can be configured by administrators based on specific considerations or customized through a parking reservation app, allowing drivers to express preferences, such as proximity to a specific pedestrian exit. By doing so, drivers no longer need to feel troubled about finding parking spaces. They can smoothly complete the parking process by following the personalized guidance within the parking lot.

The rest of this paper is organized as follows. In Sect. 2 we briefly introduce the existing parking management systems, which serve as the foundation of our work. Section 3 explains the proposed parking guidance service and its flow of operation. In Sect. 4, we show the prototype implementation and present preliminary results. Finally, the conclusions and future work are discussed in Sect. 5.

2 Existing Parking Management Systems

The current indoor parking space management system operates based on two main components. The first involves determining the usage status of parking spaces, which can be achieved by assessing changes in distance measured by ultrasonic sensors or by using cameras for direct image recognition to detect the presence of parked vehicles within parking spaces. The second component, once the status of parking spaces is known, involves changing the color of indicator lights above the parking spaces to signal to drivers from a distance whether the spaces are available. Take Fig. 1(a) as an example, green lights indicate that the parking spaces under them are available, while red lights mean those spaces beneath them are occupied. In terms of guidance indications, existing systems typically operate on a lane-by-lane basis, displaying the current availability of parking spaces in each lane. Figure 1(b) illustrates an example indicating that there are 7 available spaces remaining in that lane.

Fig. 1. A car park in Taipei City with parking space indicator lights and lane-based parking space availability display [3].

Currently, practical implementations use both ultrasonic sensors and video cameras to detect vehicle presence in parking spaces. I-View Communication Inc.'s ParkGuide parking guidance system [4] shown in Fig. 2 is an example, employing a dual-sensing approach. In well-lit conditions, it uses image recognition to identify the license plate number of parked vehicles, recording it in the database. In low-light conditions, ultrasonic sensors determine space occupancy. The system includes a car-finding kiosk for users to locate their parked vehicles. ParkGuide also uses colored lights above parking spaces, such as blue for VIP spots. If a non-VIP vehicle occupies a VIP space, the on-site system issues an immediate voice message instructing the driver to move.

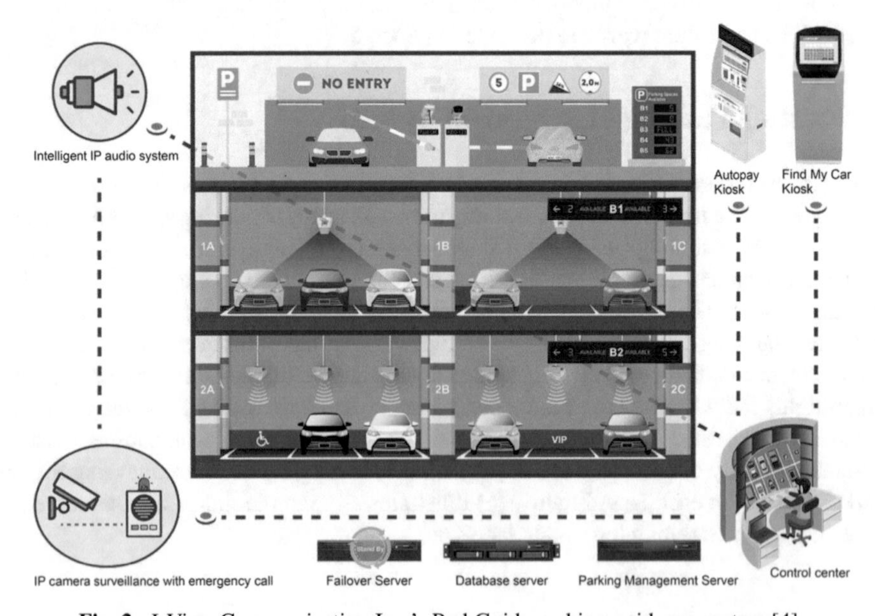

Fig. 2. I-View Communication Inc.'s ParkGuide parking guidance system [4].

In practice, currently available parking guidance systems indeed provide drivers with real-time information about the usage status of parking spaces. However, based on our observation, existing solutions have yet to achieve the functionality of *personalized guidance services* within parking lots. Even services like ParkGuide's VIP parking service merely inform VIP vehicle owners of the designated parking space number through a display screen, using special-colored parking lights (such as blue) for differentiation, without providing directional guidance for VIP vehicles while driving. It is also common to see people dispute over parking spaces in the news, and if parking management systems can proactively allocate parking spaces for incoming vehicles, it can prevent situations where people compete for parking spots. Furthermore, in parking lots with high traffic flow, traffic congestion in the lanes is prone to occur. Achieving personalized guidance for each vehicle could dynamically alter the driving routes of cars, allowing them to bypass congested lane segments and making the parking process more seamless.

3 The Proposed Parking Guidance System

The system architecture of the proposed end-to-end parking guidance system is illustrated in Fig. 3. There are five types of components in the system: 1) entrance and exit gates and cameras, 2) lane cameras, 3) parking space cameras with indicator lights, 4) guidance displays, and 5) backend server. Components of types 1, 3, and 5 are commonly found in commercially available parking management systems, providing license plate recognition during entry and exit, as well as vehicle parking location identification, which also based on license plate recognition. What distinguishes our research from conventional parking management systems is the additional placement of cameras in the lanes (components of type 2), allowing for dynamic tracking of vehicles within the parking facility. Subsequently, based on the dynamic positions of vehicles, personalized parking guidance instructions are provided on the guidance display (components of type 4). Note that the guidance displays are to be installed at each intersection within the parking facility. The display will present drivers with information including the license

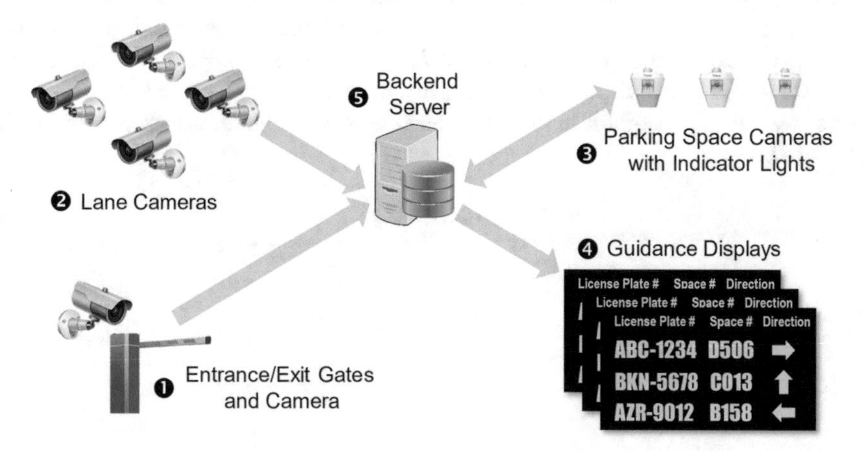

Fig. 3. System architecture of the proposed parking guidance system.

plate number, designated parking space number, and directions to the designated parking space.

We illustrate the system operation process with Fig. 4. First, when a vehicle passes through the entrance gate, the system captures the license plate number through license plate recognition and assigns a specific parking space for the vehicle, as well as planning the driving route to that parking space within the parking facility. As the vehicle moves along the lane, the lane camera streams the captured footage to the backend server, enabling dynamic tracking of the vehicle using computer vision techniques. The backend server, likewise, employs license plate recognition to obtain the license plate number and estimates the vehicle's position while determining its direction of travel. Finally, based on the vehicle's latest status, a personalized guidance instruction is presented on the guidance display.

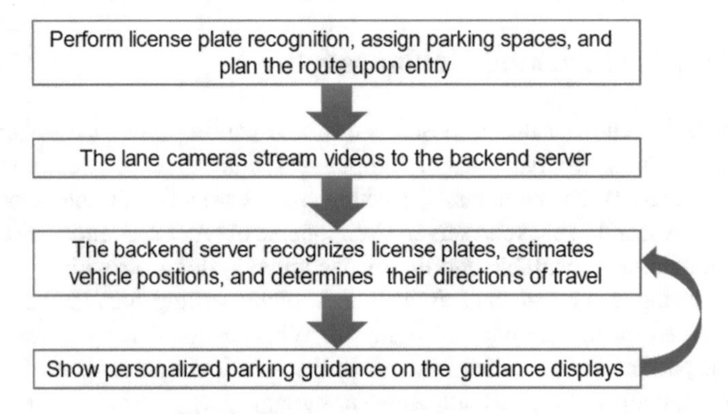

Fig. 4. The system operation process.

As for the estimation of vehicle position in the lane, the method we employ follows the basic principles of photo imaging, as depicted in Fig. 5. The real object on the left has a height of $\boldsymbol{H}$ and is at a distance $\boldsymbol{D}$ from the camera lens. On the right, the green frame represents the camera, with a distance $\boldsymbol{d}$ as the camera's focal length, and the pixel height of the captured object is denoted as $\boldsymbol{h}$. It is obvious that $\boldsymbol{H}/\boldsymbol{D} = \boldsymbol{h}/\boldsymbol{d}$. Since the object we would like to capture is the license plate, we already have the value of $\boldsymbol{H}$. As long as we have the height $\boldsymbol{h}$ in the image and the focal length $\boldsymbol{d}$, we are able to get the value of $\boldsymbol{D}$:

$$\boldsymbol{D} = \frac{\boldsymbol{H}}{\boldsymbol{h}} \times \boldsymbol{d} \tag{1}$$

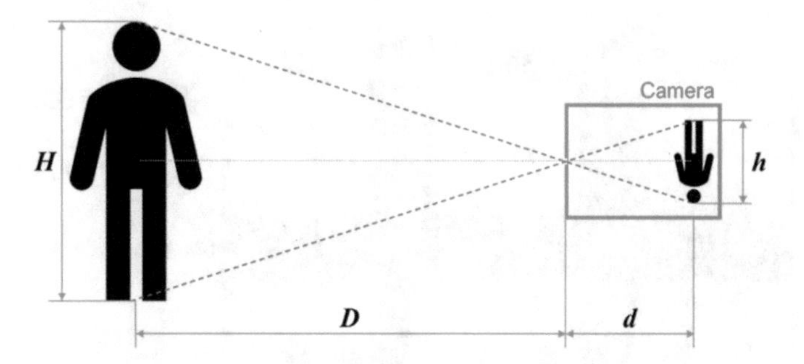

Fig. 5. The estimation of object distance from the camera.

4 Prototype Implementation

In this section, we will illustrate the implementation of our prototype system and present preliminary results to showcase the functionality of the ongoing work.

4.1 Prototype Overview

As described in Sect. 3, the focus of this study lies in utilizing computer vision technology to estimate the real-time location of vehicles in a parking lot, with the aim of providing personalized parking guidance. Therefore, in our prototype, we primarily implemented lane cameras and a backend server, and show the parking guidance information on the screen of the backend server directly.

In the lane camera setup, we utilize a Raspberry Pi 4 development board connected to a Logitech StreamCAM with 1080p resolution for real-time video capture. Using the open-source MJPG-Streamer [5] package, the real-time video is streamed to the backend server with for processing. Once the backend server receives the lane video data, the system applies the YOLOv8 model [6] to draw bounding boxes around the detected license plate positions in the images. Subsequently, the license plate images undergo text recognition using the EasyOCR [7] package. Simultaneously, the system estimates the distance of vehicles from the camera based on Eq. (1), providing a measure of the current vehicle position. Finally, considering the real-time vehicle positions, the system generates appropriate routing guidance.

At this stage, we conducted functional testing in a small-scale simulated parking lot (as illustrated in Fig. 6) to verify the operations of the program logic on the backend server can function correctly. In this parking lot, there are only three lanes with lane cameras and two intersections deployed with guidance displays. The numbers labeled along the lanes denote the parking space number. As for the labels A to H, they represent specific points within the parking lot, enabling the system to assign names to lane segments. For example, the lane area between the two guidance displays in the figure can be labeled as segment DE.

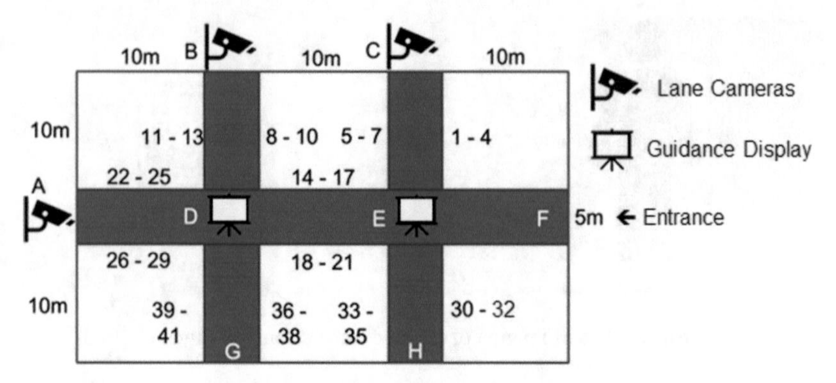

Fig. 6. The small-scale simulated parking lot.

4.2 Preliminary Results

In our implementation, we chose YOLOv8s (small model size) as the model for license plate bounding box selection and utilized a license plate dataset from Kaggle [8] for model training. We selected 192 images from the dataset and added flipped versions of some images, resulting in a total of 301 images, with 243 images for the training set, 38 for the validation set, and 20 for the test set. The training process utilized a batch size of 16 and ran for 25 epochs. The test results are satisfactory, with each image processed in just 10–13 ms, thanks to the Nvidia GeForce RTX3080 graphics card installed in our backend server.

Regarding license plate text recognition, our tests revealed that EasyOCR takes approximately 80–100 ms to process an image with a resolution of 720p. By adjusting the resolution to 152 $\times$ 64, the processing time can be reduced to 10 ms. Therefore, we first use YOLOv8s to select license plate regions, crop the images, and adjust the resolution before sending them to EasyOCR for processing, significantly improving efficiency. Finally, by associating the bounding box positions with the license plate text, distance estimations are performed to determine in which lane segment the vehicle is located.

In terms of establishing the route to the designated parking space, our objective is to enable the system to dynamically adapt routes according to the congestion level of each lane segment. Nevertheless, this functionality is currently under development, and in the present phase, we are performing functional testing with static routes. Figure 7 illustrates the static routing table for the simulated area. Below, we explain how to use this routing table with the example of a vehicle currently at point F, with the designated parking space in segment DG (i.e., destination is G). Since the vehicle is currently at point F, it queries the **Fdst** dictionary, where the sixth element indicates that if the destination is G, the vehicle must first move forward to E. Assuming the vehicle has reached point E, it then queries the **Edst** dictionary, where the sixth element indicates that if the destination is G, its next stop must be D. When the vehicle reaches D, it now queries the **Ddst** dictionary where the sixth element indicates that if it wants to go to G, its next stop is G.

```
#routing table
self.Adst = {'B':['D'], 'C':['D'], 'D':['D'], 'E':['D'], 'F':['D'], 'G':['D'], 'H':['D']}
self.Bdst = {'A':['D'], 'C':['D'], 'D':['D'], 'E':['D'], 'F':['D'], 'G':['D'], 'H':['D']}
self.Cdst = {'A':['E'], 'B':['E'], 'D':['E'], 'E':['E'], 'F':['E'], 'G':['E'], 'H':['E']}
self.Ddst = {'A':['A'], 'B':['B'], 'C':['E'], 'E':['E'], 'F':['E'], 'G':['G'], 'H':['E']}
self.Edst = {'A':['D'], 'B':['D'], 'C':['C'], 'D':['D'], 'F':['F'], 'G':['D'], 'H':['H']}
self.Fdst = {'A':['E'], 'B':['E'], 'C':['E'], 'D':['E'], 'E':['E'], 'G':['E'], 'H':['E']}
self.Gdst = {'A':['D'], 'B':['D'], 'C':['D'], 'D':['D'], 'E':['D'], 'F':['D'], 'H':['D']}
self.Hdst = {'A':['E'], 'B':['E'], 'C':['E'], 'D':['E'], 'E':['E'], 'F':['E'], 'G':['E']}
```

Fig. 7. Static routing table in our current implementation.

Indeed, since drivers are not aware of the specific locations of labels A to H within the parking lot, personalized parking guidance cannot rely on a ‘next stop’ approach. As we have deployed guidance displays at the intersections of the lanes, the information shown on these displays must inform drivers whether to proceed straight, turn right, or turn left at the current intersection. In light of this, we create an additional table indicating the directional relationships between adjacent label positions (as illustrated in Fig. 8). Subsequently, the system can generate arrows at each intersection, guiding drivers in the correct direction toward the destination. Finally, we can seamlessly integrate all information, providing personalized guidance to each vehicle’s driver via the guidance displays along the route to the designated parking space.

```
#dirction table N S W E
self.Adt = {'D':['E']}
self.Bdt = {'D':['S']}
self.Cdt = {'E':['S']}
self.Ddt = {'A':['W'], 'B':['N'], 'E':['E'], 'G':['S']}
self.Edt = {'D':['W'], 'C':['N'], 'F':['E'], 'H':['S']}
self.Fdt = {'E':['W']}
self.Gdt = {'D':['N']}
self.Hdt = {'E':['N']}
```

Fig. 8. Direction table showing directional relationships between adjacent label positions.

5 Conclusions and Future Work

In this study, we propose a personalized parking guidance service tailored for large parking lots. By pre-assigning parking spaces and providing guidance, we eliminate the hassle for drivers to manually search for available parking spots. Leveraging computer vision technology, our parking management system can determine the vehicle’s real-time location within the lane segments. Guidance displays positioned at lane intersections then provide individualized parking instructions for each vehicle. For drivers, this means a convenient parking experience without the need for reliance on any device within the vehicle, such as the driver’s smartphone.

The prototype system we have currently implemented is not yet fully mature, and there are still some technical details to be addressed, such as vehicle object tracking across multiple cameras and dynamically adapting parking routes based on the congestion level of lane segments. For vehicle object tracking, we plan to use *Simple Online and Realtime Tracking* (SORT) [9] based solutions such as *ByteTrack* [10] and *BoT-SORT* [11]. For dynamic route planning, *A* search algorithm* based methods such as *Hybrid A* with Jump Point Search* [12] could be an option.

Acknowledgments. This research is supported in part by the National Science and Technology Council, Taiwan, R.O.C., under grant number NSTC 112-2221-E-032-015.

References

1. Kraus, S., Jones, P., Kailer, N., Weinmann, A., Chaparro-Banegas, N., Roig-Tierno, N.: Digital transformation: an overview of the current state of the art of research. SAGE Open **11**, 3 (2021)
2. Verma, A., Prakash, S., Srivastava, V., Kumar, A., Mukhopadhyay, S.C.: Sensing, controlling, and IoT infrastructure in smart building: a review. IEEE Sens. J. **19**, 9036–9046 (2019)
3. Taipei City Government: Department of Transport Annual Report 2016. https://www-ws.gov.taipei/001/Upload/390/relfile/17914/6867/72878f71-034f-4bbc-8682-f8223bde1177.pdf
4. I-View Communication Inc.: ParkGuide® Parking Guidance System. https://www.i-view.com.tw/parkguide-parking-guidance-system/
5. mjpg-streamer. https://github.com/jacksonliam/mjpg-streamer
6. Ultralytics YOLOv8. https://github.com/ultralytics/ultralytics
7. EasyOCR. https://github.com/JaidedAI/EasyOCR
8. Kaggle. https://www.kaggle.com/datasets/andrewmvd/car-plate-detection
9. Bewley, A., Ge, Z., Ott, L., Ramos, F., Upcroft, B.: Simple online and realtime tracking. In: 2016 IEEE International Conference on Image Processing (ICIP), pp. 3464–3468 (2016)
10. Zhang, Y., et al.: ByteTrack: multi-object tracking by associating every detection box. In: Computer Vision – ECCV 2022, pp. 1–21. Springer Nature Switzerland (2022)
11. Aharon, N., Orfaig, R., Bobrovsky, B.-Z.: BoT-SORT: robust associations multi-pedestrian tracking. arXiv preprint arXiv:2206.14651 (2022)
12. Qin, Z., Chen, X., Hu, M., Chen, L., Fan, J.: A novel path planning methodology for automated valet parking based on directional graph search and geometry curve. Robot. Auton. Syst.Auton. Syst. **132**, 103606 (2020)

Mathematical Modelling of COVID-19 Using ODEs

Dharmendra Prasad Mahato[1](✉) and Radha Rani[2]

[1] Department of Computer Science and Engineering,
National Institute of Technology Hamirpur,
Hamirpur 177 005, Himachal Pradesh, India
dpm@nith.ac.in

[2] School of Computing Science and Engineering, Galgotias University,
Plot No. 2, Sector 17-A Yamuna Expressway, Opposite Buddha International Circuit,
Gautam Buddh Nagar, Greater Noida 203201, Uttar Pradesh, India
radharani8288@gmail.com

Abstract. COVID-19 is the disease caused by the SARS-CoV-2 coronavirus. Globally, as of 6:32pm CEST, 19 September 2023, there have been 772 838 745 confirmed cases of COVID-19, including 6 988 679 deaths, reported to WHO. The number of confirmed cases still are being seen. In this paper, we present a prediction model baased on Ordinary Differential Equations. The prediction model takes the help fom the *susceptible-exposed-infected-recovered* (*SEIR*) family of compartmental models. The *SEIR* is a type of epidemiological models. In this paper we also focus on the reinfection rate among the people from the virus SARS-CoV-2 after they have recovered.

Keywords: COVID-19 · pandemic · virus · Ordinary Differential Equation

1 Introduction

Modeling of infectious diseases like corona virus SARS-COV-2 is a challenging task among the researchers. Most of the mathematical prediction models of the COVID-19 epidemic are based on the Ordinary Differential Equations (ODEs). This type of modeling generally is based on *SEIR* model. In this paper, our model also focuses on the cases of people who have already recovered and again have got infected from the virus. In the literature survey, we rarely find this case has been taken for the prediction.

The motivation of this work is to focus on the accurate prediction of the outbreak of the epidemic like COVID-19 with the help of mathematical model [3,6]. In this field, many mathematical models have been proposed [1,2,4]. These models generally focus on the factors like infection, migration, population size change, and other practical factors. In the literature, we find rare research papers

L. Barolli (Ed.): AINA 2024, LNDECT 204, pp. 145–156, 2024.
https://doi.org/10.1007/978-3-031-57942-4_16

describing the models for the infections among the people considering reinfection after getting recovered. In this paper, we try to model the prediction by considering the reinfection among the people who have already recovered from these infections.

1.1 Compartmental Modelling

Our model is based on the concept of compartmental modelling which is widely used as a method of presenting a mathematical modelling of infectious diseases. In this method, the population is presented as subdivided different compartments. The person who has got the infection is called an individuals in the model. The origin of compartmental modeling is in the early 20th century [2,5]. This model generally takes the help of Ordinary Differential Equation (ODEs) are stochastic in nature.

1.2 SEIR Model

In the *SEIR* model, the rate of the susceptible individuals who get exposed from COVID-19, is given as

$$\lambda_1(t)S(t) = \beta S(t)I(t)/N \tag{1}$$

where notations and variables used in the paper are shown in Table 1. The Fig. 1 and Fig. 2 show the basic model of SER and $SEIR$ respectively.

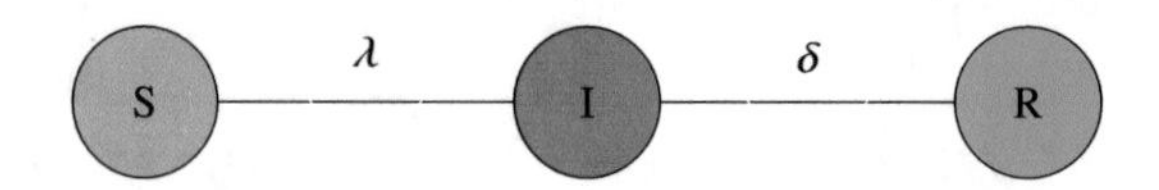

Fig. 1. *SER* Model

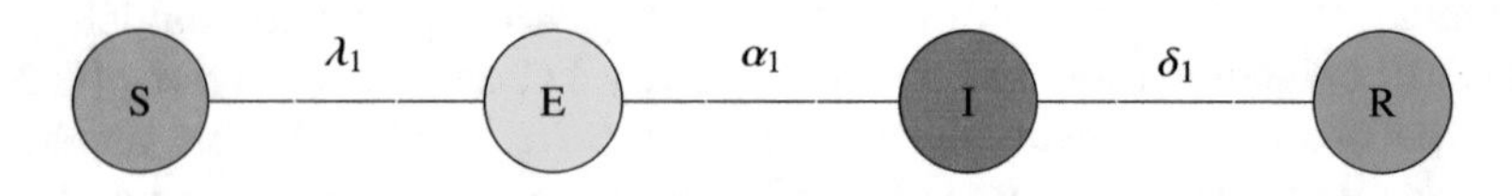

Fig. 2. *SEIR* Model

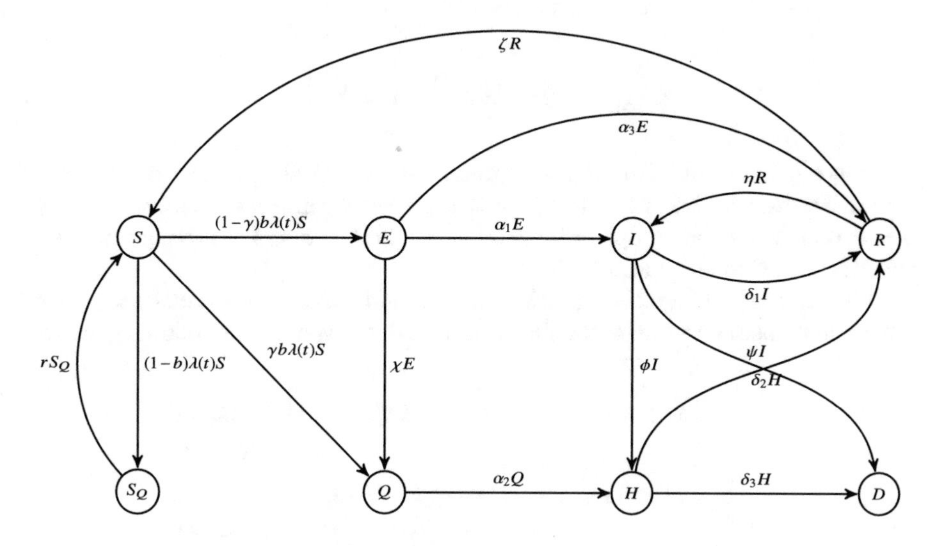

Fig. 3. A transmission diagram of COVID-19 disease

2 Mathematical Modelling for COVID-19

Here we descibe the mathematical model of COVID-19 epidemic with the help of Ordinary Differential Equations (ODEs). The model takes the initiative help from the *SEIR*-type of compartment model as discussed in [2]. When we see *SEIR* model shown in Fig. 2, this is a simple building model but if we extend the model incorporating the measures like quarantine and isolation and even cases of reinfection after the individuals have been recovered, the model looks like as shown in Fig. 3. The notations, parameters and variables used in the shown models have been represented in Table 1.

As we see in Fig. 3, the individuals Q who are quarantined will progress to the next stages like infectious and isolated stage (H) with the rate α_2 and non-quarantined individuals will progress to infectious and isolated stage (I) having the rate α_1. The infected individuals will have to face isolated stage (H) at the rate ϕ. Also the individuals in the isolated stage will recover at the rate of δ_2 and will move to the stage (R). If the individuals from stage (I) will recover with rate δ_1, they will move from (I) stage to the stage (R). If the individuals are going to isolated stage from this stage with the rate of ϕ, they will move to the stage (H). The critical cases of individuals are also losing their lives at the rate of ψ, then the stage will be (D) and the individuals who are losing their lives from the stage (H) will go to the stage (D) with the rate of δ_3.

How our model is different from other models is that we have assumed that some recovered people also have a bad luck to have been infected again with the rate of χ and they will move to again to the stage (S).

Now, if the case of improper isolation is considered, then the new infections [2] are being produced at the rate $\lambda(t)S(t)$ with

$$\lambda(t) = \beta\left[\frac{I(t) + (1-\rho)H(t)}{N}\right] \tag{2}$$

where fraction of reduction is represented as $\rho \in [0, 1]$ in the transmission rate of isolated individuals. Here, $\rho = 1$ represents complete effective isolation, $\rho = 0$ represents the complete ineffective, while $0 < \rho < 1$ represents partially effective isolation as shown in Fig. 3.

The ordinary differential equation (ODE) model of COVID-19 disease considering the transmission diagram shown in Fig. 2 is given by the following system

$$\begin{aligned}
S' &= \mu N + \zeta R + r.S_Q - (1-\gamma)b\lambda(t)S - (1-b)\lambda(t)S \\
&\quad -\gamma b\lambda(t)S - \mu S, \\
E' &= (1-\gamma)b\lambda(t)S - \chi E - \alpha_1 E - \alpha_3 E - \mu E, \\
Q' &= (1-b)\lambda(t)S + \chi E - \alpha_2 Q - \mu Q, \\
I' &= \alpha_1 E + \eta R - \delta_1 I - \mu I - \phi I, \\
H' &= \alpha_2 Q + \phi I - \delta_2 H - \delta_3 H - \mu H, \\
R' &= \alpha_3 E + \delta_1 I + \delta_2 H - \eta R - \zeta R
\end{aligned} \tag{3}$$

In this paper, if $\gamma = 0$, $b = 1$ and $\lambda(t) = \beta\left[\frac{I(t)+(1-\rho)H(t)}{N}\right]$, then model 3 will reduce to as

$$\begin{aligned}
S' &= \mu N + \zeta R + rS_Q - \beta.S.\frac{I + (1-\rho)H}{N} - \mu S, \\
E' &= \beta.S.\frac{I + (1-\rho)H}{N} - (\chi + \alpha_1 + \alpha_3 + \mu)E, \\
Q' &= \chi E - (\alpha_2 + \mu)Q, \\
I' &= \alpha_1 E + \eta R - (\delta_1 + \mu + \phi)I, \\
H' &= \alpha_2 Q + \phi I - (\delta_2 + \delta_3 + \mu)H, \\
R' &= \alpha_3 E + \delta_1 I + \delta_2 H - \eta R - \zeta R \\
D' &= \delta_3.H + \psi.I - \mu.D
\end{aligned} \tag{4}$$

3 The General Model

The general model of the disease begins with the initial model which is *SEIR* model which is generally derived when quarantine and isolation stages are added in the model. Here we assume that the stages described in the model are exponentially distributed.

When we model the characteristics of any epidemic, we generally need to define the incubation period which is defined as the time between the exposure to the outbreak and the time of display of the symptoms. In the literature, the incubation period has been taken as the period of period which ranges from 1 to 12.5 days whose median values are estimated as the duration between 5 to 6 days. In some cases, the incubation period may be as long as 14 to 24 days. But the condition becomes very dangerous if the incubation rate increases [2]. Even a two-week incubation period is enough to make a virus spread worldwide, even at any level of society.

3.1 The Model with Exponentially Distributed Disease Stages

All the notations and variables used in the model are shown and defined in Table 1. The rate of removal of the disease can be expressed as $-\dot{P}_i(s)$ $(i = E, I)$. These functions of the time period in the stages discussed below show to have important properties.

$$p_i(0) = 1, \quad -\dot{P}_i(s) \leq 0, \quad \int_0^\infty p_i(s)ds < \infty, \quad i = E, I \tag{5}$$

Let us assume $k(s)$ as the probabilities that the exposed or infectious individuals at stage s and $l(s)$ represent the exposed or infectious individuals who have not yet been quarantined or isolated at stage s. Then, $1 - k(s)$ or $\bar{k}(s)$ represent the probabilities that the individuals who are exposed or infectious before reaching stage s and $1 - l(s)$ or $\bar{l}(s)$ represent the individuals who are exposed or infectious and have been quarantined or isolated before reaching stage s. This paper assumes that $k(0) = 1$ and $l(0) = 1$. Therefore, $\dot{k}(s) \leq 0$ and $\dot{l}(s) = 1$.

The paper assumes that the probability of surviving natural death is $e^{-\mu t}$. Here, In this paper, let us assume that $S(0) > 0$ are the numbers of initial susceptible individuals and $R(0) > 0$ are the numbers of initial removed individuals. In the model, $E_0(t)e^{-\mu t}$ represents the numbers of individuals who are initially exposed and $I_0(t)e^{-\mu t}$ represents the numbers of individuals who are infectious. $Q_0(t)e^{-\mu t}$ and $I_0(t)e^{-\mu t}$ represent the numbers of individuals who are quarantined and isolated respectively.

Let us assume that P_E and P_I use exponential distribution, then $\tilde{P}_E$ and $\tilde{P}_I$ with mean stage duration α and δ can be expressed as

$$\tilde{P}_E(s) = e^{-\alpha s}, \quad \tilde{P}_I(s) = e^{-\delta s} \tag{6}$$

Here the individuals who are survived from the quarantine stage can be expressed as

Table 1. Definitions of used symbols

Symbol		Definitions
$S(t)$	$\triangleq$	The number of susceptible individuals at the time t
$S_Q(t)$	$\triangleq$	The number of susceptible individuals who are quarantined at the time t
$E(t)$	$\triangleq$	The number of exposed but not yet infectious individuals at the time t
$Q(t)$	$\triangleq$	The number of quarantined or exposed individuals at the time t
$I(t)$	$\triangleq$	The number of susceptible individuals at the time t
$H(t)$	$\triangleq$	The number of isolated or infectious individuals at the time t
$R(t)$	$\triangleq$	The number of recovered individuals at the time t
$D(t)$	$\triangleq$	The number of deaths at the time t
N	$\triangleq$	The total population size (constant)
$C(t)$	$\triangleq$	The number of cumulative new infections at the time t
r	$\triangleq$	The rate of returning from S_Q to S
$\lambda(t)$	$\triangleq$	The force of infection at the time t
β	$\triangleq$	The transmission coefficient
α, α_1	$\triangleq$	The rate at which non-quarantined individuals become infectious
α_2	$\triangleq$	The rate at which quarantined individuals become infectious
α_3	$\triangleq$	The recovery rate of the exposed cases. This parameter is not directly measurable from pure observations and requires lab-based experiments
δ, δ_1	$\triangleq$	The rate at which non-isolated individuals become recovered
δ_2	$\triangleq$	The rate at which isolated individuals become recovered
δ_3	$\triangleq$	The rate at which isolated individuals lose their life
ψ	$\triangleq$	The rate at which infected but not-quarantined individuals lose their life
μ	$\triangleq$	Natural death rate
χ	$\triangleq$	Rate of quarantine
ϕ	$\triangleq$	Rate of isolation
η	$\triangleq$	Rate at which the recovered individuals become again infected
ζ	$\triangleq$	Rate of returning from the recovered group to the susceptible group. This happens for the cases that the body does not develop lifetime immunity
ρ	$\triangleq$	Isolation efficiency ($0 \leq \rho \leq 1$)
b	$\triangleq$	Fraction of contacts infected ($b = 1$ in this paper)
γ	$\triangleq$	Fraction of infected & quarantined at time of exposure ($\gamma = 0$ in this paper)
$p_i(s)$, $P_i(s)$	$\triangleq$	Probability that disease stage i lasts longer than s time units ($i = E, I$)
$k(s)$	$\triangleq$	Probability of not being quarantined at stage age s
$l(s)$	$\triangleq$	Probability of not being isolated at stage age s
T_E	$\triangleq$	Mean of $p_E(s) = e^{-\alpha s}$, $T_E = \frac{1}{\alpha}$
T_I	$\triangleq$	Mean of $p_I(s) = e^{-\delta s}$, $T_E = \frac{1}{\delta}$
$\mathcal{M}_i(s)$, M	$\triangleq$	Expected remaining sojourn at age s: $\int_0^\infty P_i(t\|s)dt(i = E, I)$, $M = \mathcal{M}(0)$
$\mathcal{T}_E$	$\triangleq$	Probability of surviving & becoming infectious: $\int_0^\infty [-\dot{P}(s)]e^{-\mu s}dt$
$\mathcal{T}_{E_k}$	$\triangleq$	"Quarantine-adjusted" probability (similar to $\mathcal{T}_E$): $\int_0^\infty [-\dot{P}_E(s)k(s)]e^{-\mu s}dt$
$\mathcal{T}_I$	$\triangleq$	Probability an infectious person survives and recovers: $\int_0^\infty [-\dot{P}_I(s)]e^{-\mu s}dt$
$\mathcal{T}_{I_I}$	$\triangleq$	"Isolation-adjusted" probability (similar to $\mathcal{T}_I$): $\int_0^\infty [-\dot{P}_I(s)I(s)]e^{-\mu s}dt$
$\mathcal{D}_E$	$\triangleq$	Mean time in exposed stage (adjusted by death):$\int_0^\infty [P_E(s)]e^{-\mu s}dt$
$\mathcal{D}_{E_k}$	$\triangleq$	"Quarantine-adjusted" mean time in exposed stage: $\int_0^\infty [P_E(s)k(s)]e^{-\mu s}dt$
$\mathcal{D}_I$	$\triangleq$	Mean time in infectious stage (adjusted by death):$\int_0^\infty [P_I(s)]e^{-\mu s}dt$
$\mathcal{D}_{I_I}$	$\triangleq$	"Isolation-adjusted" mean time in infectious stage: $\int_0^\infty [P_I(s)I(s)]e^{-\mu s}dt$

$$k(s) = e^{-(\chi+\alpha_1+\alpha_3)s} \quad l(s) = e^{-(\phi+\delta_1)s} \tag{7}$$

where χ, α_1, α_3, ϕ, and δ_1 are constant.
From Eq. 6 and Eq. 7, we have

$$\begin{aligned} E_0(t) &= E(0)e^{-(\chi+\alpha_1+\alpha_3)t}, \\ I_0(t) &= I(0)e^{-(\phi+\delta_1+\psi)t} \end{aligned} \tag{8}$$

$$\begin{aligned} Q_0(t) &= Q(0)e^{-\alpha_2 t}, \\ H_0(t) &= H(0)(t)e^{-(\delta_2+\delta_3)t}, \\ R_0(t) &= R(0)(t)e^{-(\zeta+\eta)t} \\ D_0(t) &= D(0)e^{-\mu t} \end{aligned} \tag{9}$$

where $E(0)$, $I(0)$, $Q(0)$, $H(0)$ and $R(0)$ are constant and represent the numbers of individuals at the stage E, I, Q, H and R when $t = 0$.

Now, if the individuals who are initially infected and have moved into the I, Q, H, and R stages respectively and are still alive at time t, are represented as $\tilde{I}$, $\tilde{Q}$, $\tilde{H}$, $\tilde{R}$, then From Eq. 6 and 7, we have

$$\tilde{I}(t) = \int_0^t \alpha E(0)e^{-(\chi+\alpha_1+\alpha_3+\mu)\tau} e^{-(\phi+\delta_1+\mu)(t-\tau)} d\tau \tag{10}$$

Considering the force of infection $\lambda(t)$ as depicted in Eq. 2 and used in Eq. 3 and 4, the number of individuals who became exposed at some time $s \in (0, t)$ and are still alive is given as

$$E(t) = \int_0^t \lambda(s)S(s)P_E(t-s)k(t-s)e^{-\mu(t-s)}ds + E_0(t)e^{-\mu t} \tag{11}$$

When Eq. 11 is differentiated, we get

$$\begin{aligned} \frac{d(E(t))}{dt} &= \int_0^t \lambda(s)S(s)\dot{P}_E(t-s)k(t-s)e^{-\mu(t-s)}ds \\ &+ \int_0^t \lambda(s)S(s)P_E(t-s)\dot{k}(t-s)e^{-\mu(t-s)}ds + \lambda(t)S(t) - \mu E(t) + \frac{d(E_0(t))}{dt}e^{-\mu t} \end{aligned} \tag{12}$$

In Eq. 12, the first term provides input for the I and the second term provides input for the Q. Therefore,

$$\begin{aligned} I(t) = &\int_0^t \int_0^\tau \lambda(s)S(s)\big[-\dot{P}_E(t-s)k(t-s)\big].P_I(t-\tau)l(t-\tau).e^{-\mu(t-s)}dsd\tau \\ &+ I_0(t)e^{-\mu t} + \tilde{I}(t) \end{aligned} \tag{13}$$

$$\begin{aligned}
Q(t) &= \int_0^t \int_0^\tau \lambda(s)S(s)\big[\\
&\quad - P_E(\tau-s)\dot{k}(\tau-s)\big].P_E(t-\tau|\tau-s) \times e^{-\mu(t-s)}dsd\tau + Q_0(t)e^{-\mu t} + \tilde{Q}_0(t) \\
&= \int_0^t \lambda(s)S(s)P_E(t-s)\bar{k}(t-s)e^{-\mu(t-s)}ds + Q_0(t)e^{-\mu t} + \tilde{Q}_0(t)
\end{aligned} \tag{14}$$

$$\begin{aligned}
H(t) = &\int_0^t \int_0^u \int_0^\tau \lambda(s)S(s)\big[-\dot{P}_E(\tau-s)k(\tau-s)\big].P_I(u-\tau)\dot{l}(u-\tau) \\
&\times P_I(t-u|u-\tau)e^{-\mu(t-s)}dsd\tau du + \int_0^t \int_0^\tau \lambda(s)S(s)\big[\\
&- \dot{P}_E(\tau-s)\bar{k}(\tau-s)\big].P_I(t-\tau).e^{-\mu(t-s)}dsd\tau + H_0(t)e^{-\mu t} + \tilde{H}_0(t)
\end{aligned} \tag{15}$$

$$\begin{aligned}
R(t) &= \int_0^t \int_0^u \int_0^\tau \lambda(s)S(s)\big[-\dot{P}_E(\tau-s)\big].\big[P_I(u-\tau)\big]e^{-\mu(t-s)}dsd\tau du + R_0(t)e^{-\mu t} + \tilde{R}_0(t) \\
&= \int_0^t \int_0^\tau \lambda(s)S(s)\big[-\dot{P}_E(\tau-s)\big]e^{-\mu(t-s)} \int_\tau^t \big[P_I(u-\tau)\big]dsd\tau du + R_0(t)e^{-\mu t} + \tilde{R}_0(t) \\
&= \int_0^t \int_0^\tau \lambda(s)S(s)\big[-\dot{P}_E(\tau-s)\big[1-P_I(u-\tau)\big]e^{-\mu(t-s)}dsd\tau + R_0(t)e^{-\mu t} + \tilde{R}_0(t)
\end{aligned} \tag{16}$$

We have the following integral equation model if we combine these equation:

$$S(t) = \int_0^t \mu N e^{-\mu(t-s)}ds - \int_0^t \lambda(s)S(s)e^{-\mu(t-s)}ds + R_0.e^{-\zeta(t)} + S_0.e^{-\mu(t)} + S_Q.e^{-r(t)} \tag{17}$$

$$E(t) = \int_0^t \lambda(s)S(s)P_E(t-s)k(t-s)e^{-\mu(t-s)}ds + E_0(0)e^{-(\alpha_1+\alpha_3+\xi+\mu)t} \tag{18}$$

$$\begin{aligned}
Q(t) = &\int_0^t \int_0^\tau \lambda(s)S(s)\big[\\
&- P_E(\tau-s)\dot{k}(\tau-s)\big].P_E(t-\tau|\tau-s) \times e^{-\mu(t-s)}dsd\tau + Q(0)e^{-(\alpha_2+\mu)t}
\end{aligned} \tag{19}$$

$$\begin{aligned}
I(t) = &\int_0^t \int_0^\tau \lambda(s)S(s)\big[-\dot{P}_E(t-s)k(t-s)\big].P_I(t-\tau)l(t-\tau).e^{-\mu(t-s)}dsd\tau \\
&+ I(0)e^{-(\phi+\delta_1+\mu)t}
\end{aligned} \tag{20}$$

$$H(t) = \int_0^t \int_0^u \int_0^\tau \lambda(s)S(s)\big[-\dot{P}_E(\tau-s)k(\tau-s)\big].P_I(u-\tau)\dot{l}(u-\tau) \\ \times P_I(t-u|u-\tau)e^{-\mu(t-s)}dsd\tau du + \int_0^t \int_0^\tau \lambda(s)S(s)\big[\\ -\dot{P}_E(\tau-s)\bar{k}(\tau-s)\big].P_I(t-\tau).e^{-\mu(t-s)}dsd\tau + H(0)e^{-(\delta_2+\delta_3+\mu)t} \quad (21)$$

$$R(t) = \int_0^t \int_0^\tau \lambda(s)S(s)\big[-\dot{P}_E(\tau-s)\big[1-P_I(u-\tau)\big]e^{-\mu(t-s)}dsd\tau + R(0)e^{-(\zeta+\eta+mu)t} \quad (22)$$

$$D(t) = \int_0^t \int_0^\tau \lambda(s)S(s)\big[-\dot{P}_E(\tau-s)\big[1-P_I(u-\tau)\big]e^{-\mu(t-s)}dsd\tau + D(0)e^{-(\mu)t} \quad (23)$$

Here $\tilde{X}(t) = X_0(t)e^{-\mu t} + \tilde{X}_0(t)$ $(X = Q, I, H, R)$. And $\tilde{X}(t) \to 0$ as $t \to \infty$.

3.2 The Reduced Model

This section reduces the Eqs. 17-23 to the model as

$$\begin{aligned} S' &= \mu N + \zeta R + rS_Q - \beta.S.\frac{I+(1-\rho)H}{N} - \mu S, \\ E' &= \beta.S.\frac{I+(1-\rho)H}{N} - (\chi+\alpha_1+\alpha_3+\mu)E, \\ Q' &= \chi E - (\alpha_2+\mu)Q, \\ I' &= \alpha_1 E + \eta R - (\delta_1+\mu+\phi)I, \\ H' &= \alpha_2 Q + \phi I - (\delta_2+\delta_3+\mu)H, \\ R' &= \alpha_3 E + \delta_1 I + \delta_2 H - \eta R - \zeta R \\ D' &= \delta_3.H + \psi.I - \mu.D \end{aligned} \quad (24)$$

Here we can see that model 4 is same as model 24.

4 Simulation and Result Analysis

Simulations of these proposed mathematical model was carried out in MATLAB. From the simulation results, we see that the cumulative number $C(t)$ of new infections increases with increasing ϕ when $\chi = 0.1$ (see Fig. 4a) and increasing χ when $\phi = 0$ (see Fig. 4b). Figure 5a and Fig. 5b show that the cumulative number $C(t)$ of new infections increases with increasing χ when $\phi = 0$ (see Fig. 5a) and increasing ϕ when $\chi = 0.1$ (see Fig. 5b). Figure 6a, Fig. 6b and Fig. 7 show the numerical simulations of the EDM (Exponential Distribution Model) under no control. The number of cumulative infections $C(t)$ as well as other disease variables S(t), E(t), I(t), Q(t), $R(t)$ and $H(t)$ are plotted when $\chi = 0$ and $\phi = 0$ (see Fig. 6a), when $\chi = 0.8$ and $\phi = 0$ (see Fig. 6b) and $\chi = 0$ and $\phi = 0.1$ (see Fig. 7).

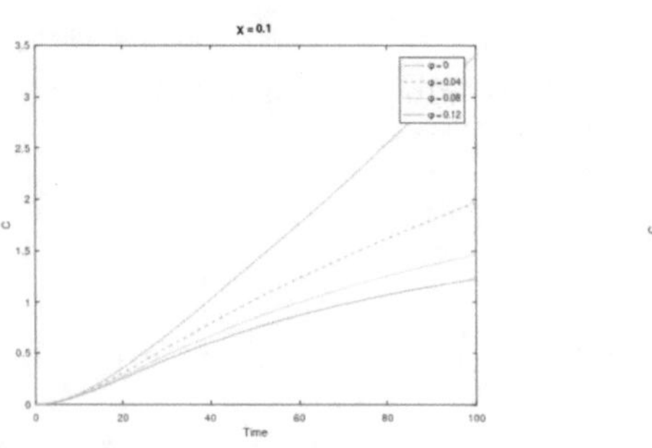

(a) The new infections cumulative number $C(t)$ decreases with the increasing ϕ when $\chi = 0.1$

First subfigure

(b) The new infections cumulative number $C(t)$ decreases with the increasing χ when $\phi = 0$

Second subfigure

Fig. 4. The new infections cumulative number $C(t)$

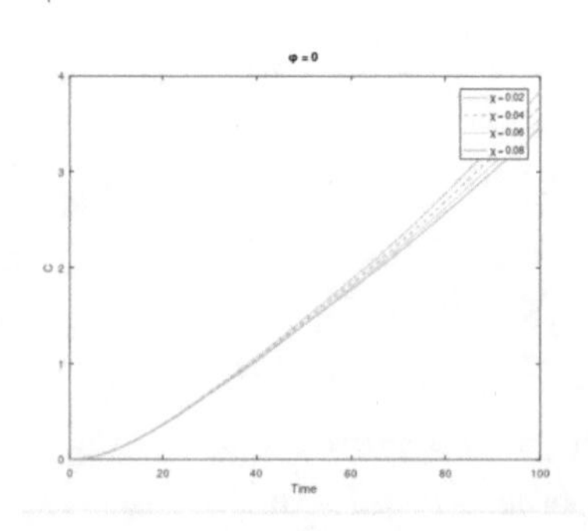

(a) The new infections cumulative number $C(t)$ decreases with the increasing χ when $\phi = 0$

First subfigure

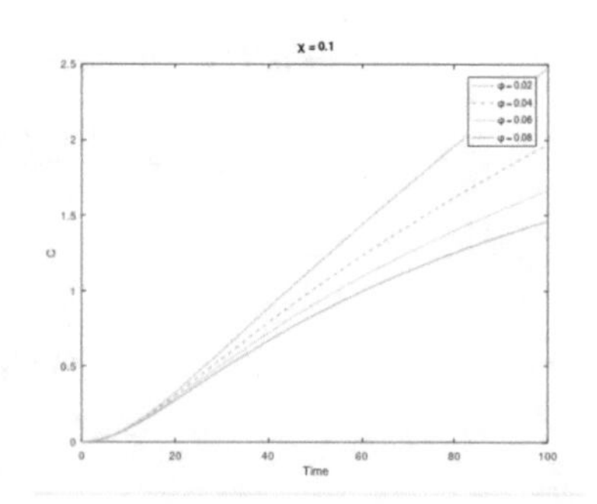

(b) The new infections cumulative number $C(t)$ decreases with the increasing ϕ when $\chi = 0.1$

Second subfigure

Fig. 5. The new infections cumulative number $C(t)$

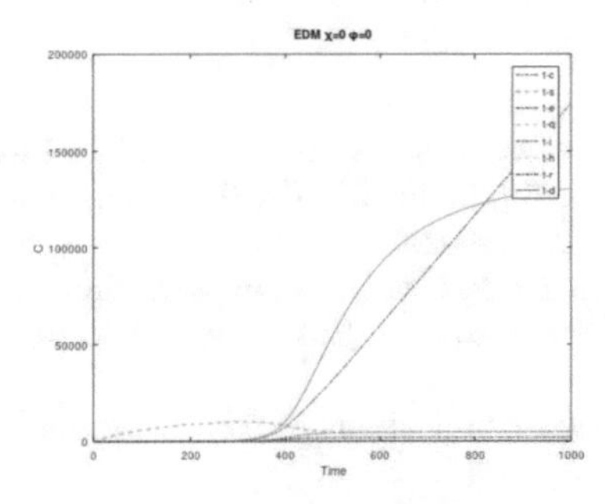

(a) Numerical simulations of the EDM (Exponential Distribution Model) under no control. The number of cumulative infections $C(t)$ as well as other disease variables $S(t)$, $E(t)$, $I(t)$, $Q(t)$, $R(t)$ and $H(t)$ are plotted when $\chi = 0$ and $\phi = 0$

First subfigure

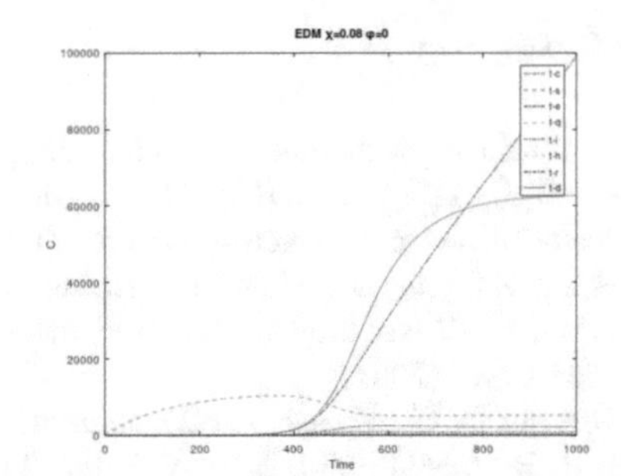

(b) Numerical simulations of the EDM (Exponential Distribution Model) under no control. The number of cumulative infections $C(t)$ as well as other disease variables $S(t)$, $E(t)$, $I(t)$, $Q(t)$, $R(t)$ and $H(t)$ are plotted when $\chi = 0.08$ and $\phi = 0$

Second subfigure

Fig. 6. Numerical simulations of the EDM (Exponential Distribution Model) under no control

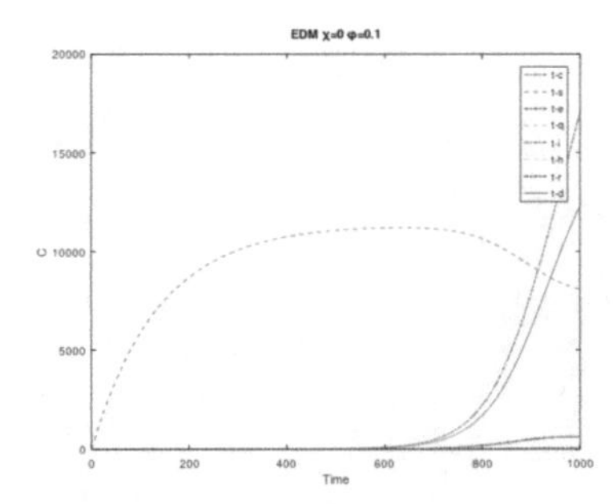

Fig. 7. Numerical simulations of the EDM (Exponential Distribution Model) under no control. The number of cumulative infections $C(t)$ as well as other disease variables $S(t)$, $E(t)$, $I(t)$, $Q(t)$, $R(t)$ and $H(t)$ are plotted when $\chi = 0$ and $\phi = 0.1$

5 Conclusions and Future Scopes

The paper focuses on the prediction model of COVID 19 epidemic with the help of Ordinary Differential Equations(ODEs). In the paper, the novelty is that we have also considered the cases of reinfections after the recovery from the disease.

In future, we will try to model the epidemic with the help of stochastic differential equations and also check if we can use Gaussian Distribution in the model.

References

1. Brauer, F.: Mathematical epidemiology: past, present, and future. Infect. Dis. Modell. **2**(2), 113–127 (2017). https://doi.org/10.1016/j.idm.2017.02.001, http://www.sciencedirect.com/science/article/pii/S2468042716300367
2. Feng, Z., Xu, D., Zhao, H.: Epidemiological models with non-exponentially distributed disease stages and applications to disease control. Bull. Math. Biol. **69**(5), 1511–1536 (2007)
3. Grassly, N.C., Fraser, C.: Mathematical models of infectious disease transmission. Nat. Rev. Microbiol. **6**(6), 477–487 (2008)
4. Hethcote, H.W.: The basic epidemiology models: models, expressions for R0, parameter estimation, and applications. In: Mathematical Understanding of Infectious Disease Dynamics, World Scientific, pp. 1–61 (2009)
5. Kermack, W.O., McKendrick, A.G.: A contribution to the mathematical theory of epidemics. Proc. Royal Soc. London Ser. A Contain. Pap. Math. Phys. Char. **115**(772), 700–721 (1927)
6. Sameni, R.: Mathematical modeling of epidemic diseases; a case study of the COVID-19 coronavirus. arXiv preprint arXiv:2003.11371

Analyzing Monitoring and Controlling Techniques for Water Optimization Used in Precision Irrigation

Rajni Goyal[1(✉)], Amar Nath[1], Utkarsh Niranjan[1], and Rajdeep Niyogi[2]

[1] SLIET, Longowal, Sangrur, Punjab, India
{rajni_pcs2103,amarnath,utkarsh}@sliet.ac.in
[2] IIT Roorkee, Roorkee 247667, India
rajdeep.niyogi@cs.iitr.ac.in

Abstract. Agriculture is the primary food source, pivotal in meeting the global population's food and nutritional requirements. With the current surge in population, the challenge of feeding such a vast number becomes increasingly daunting. Smart and precision farming is a crucial solution to address this challenge effectively. Given that farming inherently relies on water, optimizing its usage is imperative. This paper thoroughly reviews various techniques for optimizing water to ensure judicious water use. The document delves into diverse strategies, encompassing crop, soil, and weather-based management and monitoring approaches geared towards efficiently utilizing water resources. Additionally, the paper explores the application of machine learning and deep learning techniques in optimizing water usage, providing a detailed analysis of the results. Serving as a comprehensive guide, this paper illuminates the multitude of tools, techniques, and methodologies that form the foundation of smart and precision farming practices.

1 Introduction

Agriculture is the backbone of the world economy and food chain, not only because it ensures the majority of the population's food supply but also because a significant portion of the globe's population directly depends on agriculture for their livelihood. The absence of agriculture would have profound and multifaceted impacts on our lives, affecting everything from food availability to employment, economy, environment, and cultural practices. There has been exponential growth in population over the last few decades (close to 10 billion by 2050), causing a rapid escalation of food demand [1]. To produce enough food to feed the world in the future, we need enormous efforts, which can only be achieved through sustainable and intelligent agriculture. One cutting-edge method that helps farmers increase crop yield while utilizing the fewest resources possible is precision agriculture. It makes smart use of water, herbicides, and fertilizers. The plant, not the soil, is fed via precision farming. It lowers expenditures while providing water and nutrients directly to the roots. Adequate water is crucial for cultivating healthy crops. We must be smart about water use so there's

L. Barolli (Ed.): AINA 2024, LNDECT 204, pp. 157–168, 2024.
https://doi.org/10.1007/978-3-031-57942-4_17

enough for the future. Precision irrigation is an intelligent solution to this problem. It helps give plants the right amount of water by keeping an eye on them. IoT (Internet of Things) and ML (Machine Learning) are vital in precision irrigation. This paper explores diverse approaches to promote wise water utilization in agricultural practices. The primary contributions of this study encompass:

- A comprehensive examination of current tools and techniques to monitor and control precision irrigation strategies.
- A comparative analysis of different soil monitoring devices is structured in tabular format.
- An in-depth exploration of machine learning and deep learning techniques applied in precision irrigation, outlining the advantages derived from existing literature and presented in tabular form.
- The identification of future trends and the articulation of existing gaps that warrant attention.

2 Precision Irrigation

A sufficient water supply is necessary for the plant's impressive growth. Lack of water also impacts crop output, and too much water in the roots damages the roots and makes it difficult for plants to breathe. Therefore, the plants should receive the appropriate amount of water. Moreover, excessive usage of water is also costly economically and environmentally. Smart irrigation helps farmers decide how much water should be given to plants. Several parameters help decide about irrigation, like moisture level, the temperature of the soil, environmental factors, humidity, and acidity in the soil [3]. Sensors are placed inside the turf of trees, accurately determining soil moisture content, and according to the humidity in the soil, the water pump can be turned on. Field data is collected and sent to the cloud, and the water pump is turned on/off automatically.

The farm's irrigation schedule is determined by soil moisture sensor readings for each region. The micro-level analysis, scheduling, and effective actuation of irrigation provide optimum crop development and 100% water consumption efficiency. Farmers may control the irrigation system from a smartphone application built for smartphones. This irrigation system is based on information collected from field-installed temperature, humidity, soil moisture, and ultrasonic sensors. Farmers can plan their irrigation systems with the aid of satellite-derived weather information. The local small weather station provides temperature and rainfall data to the weather-based irrigation systems, and a controller controls the irrigation. Three fundamental factors are the foundation of precision irrigation: real-time monitoring and data collecting, data analysis and decision-making, and treatment based on those conclusions. Real-time monitoring and data collection are the most crucial parameters and can be achieved using the Internet of Things (IoT).

3 Monitoring and Controlling in Precision Irrigation

Through the use of wireless sensor networks and IoT, monitoring and controlling precision irrigation entails the gathering of data that shows the status of the soil, plant, and weather in real time to irrigate the plant. Various sensors, actuators, and UAVs are deployed in the fields to gather real-time information and collect data. Fields are monitored in real-time in smart irrigation. Monitoring in precision irrigation can be done through the soil, weather, and plant-based [2].

3.1 Soil-Based Irrigation Monitoring

The soil is observed to determine how much water is required for a specific plant. Many factors, including soil temperature, moisture, salinity, potential hydrogen (pH), and soil water absorption capacity, can decide how much water the soil needs. The list of numerous soil-based monitoring metrics and tools employed are presented in Table 1.

Table 1. Various Soil Monitoring Parameters [2,5]

Ref.	Parameter	Measuring Device	Results
[3]	Soil Moisture	VH400 EC Sensor DS200 TDR Probe	30% increase in water saving
[6]	Soil Moisture	Soil Moisture Sensor	50% reduction in groundwater usage
[7]	Soil Moisture	Soil Moisture Sensor	Improved water saving
[8]	Soil Moisture	Soil Moisture Sensor	58.8% water saving
[9]	Soil Moisture	TDR Probe	water and power conservation
[10]	Salinity	EC Measuring Device	Improved water saving
[11]	pH	PH Meter	Saves time, water and tell nutrients needed for soil

One of the most critical factors in precision irrigation is soil moisture. The moisture content of the soil determines how much water the plant needs. Water volume and plant water uptake can both be measured. The soil moisture level can be measured using various direct and indirect techniques. Gravimetric sampling is the direct method to measure the soil moisture and is used by [12] and found the easiest way. Indirect techniques like electromagnetic property, heat conductivity, neutron count, water potential, and electrical resistance also help to measure the moisture level of the soil [13] and are used by various researchers. Figure 1 represents the various methods of detecting soil moisture and the merits and demerits of each technique.

With the development of technology, satellite, aerial, and ground-based **soil moisture** sensors are becoming more popular in irrigation systems. Sensors can be placed in the field at various depths to comprehend the irrigation schedule, and moisture content can be determined. The performance of the soil-moisture estimation depends heavily on the number of sensors used and the best location for sensor deployment. [14] placed 42 soil moisture sensors at different depths and collected soil moisture data in Alberta, Canada. It is found that optimally placed sensors improve soil moisture-estimation performance. For controlling soil moisture at various depths and the volumetric water contents of soil within a given depth range, [15] developed an Arduino-based automatic system. The study offers a more effective irrigation system for green infrastructure to regulate the moisture distribution in the soil's subsurface.

Soil temperature has a significant effect on the water intake of the plant. An increase in temperature from 14 to 26 $^{\circ}C$, increases the water intake by 30% [16]. Solar radiation, the amount of vegetation present, soil color, and various other elements, such as soil moisture content, significantly impact soil temperature. Maintaining an appropriate soil moisture level also contributes to keeping the soil at an ideal temperature for plant growth. As water viscosity reduces with temperature and more water permeates the soil, soil temperature also hurts soil moisture content. Additionally, the temperature and moisture level of the soil are key factors in how quickly crops and plants contract diseases. According to a study, between the temperatures of 19 °C and 20 °C, plants are entirely immune to disease and only exhibit minimal symptoms. Like this, soil moisture between 30 °C and 33 °C percent is very conducive to disease formation in plants,

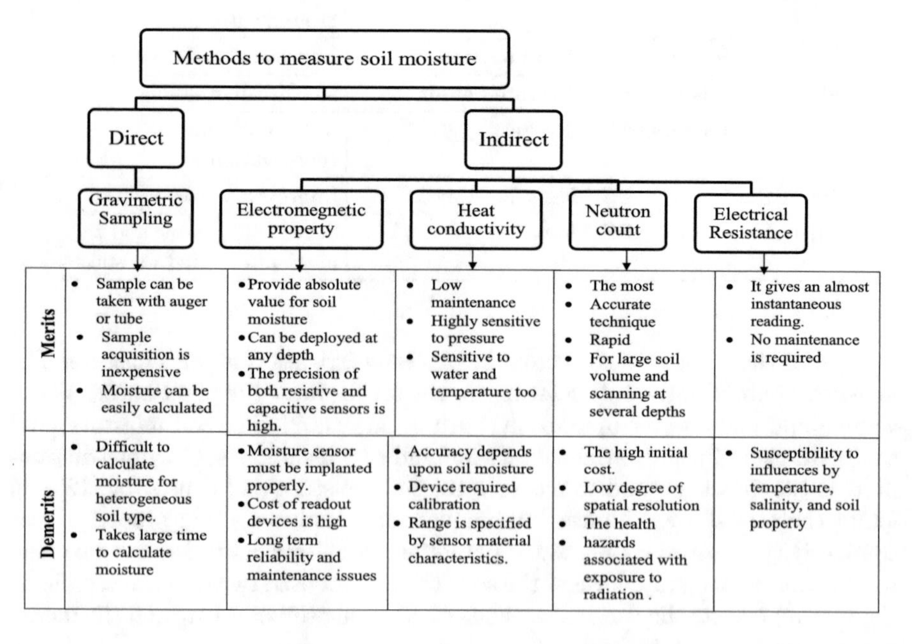

Fig. 1. Soil moisture monitoring methods [13,18,19]

while soil moisture between 13% and 14% percent demonstrates limited disease development in plants [17].

The pH and salinity of the soil also have a significant effect on the watering schedule. Soil nutrients are monitored using pH sensors to ensure that seeds germinate properly. The pH impacts the rate and speed of chemical reactions and the method by which substances are absorbed. The plant's nutrition, output, and growth suffer when the pH level is low. However, production increases when pH levels are ideal. While some crops can grow in soil that has a pH of between 6 and 6.5, others do better in soils that are more acidic or alkaline [20].

Role of IoT devices in soil moisture measuring: The various soil parameters are measured by deploying IoT devices in the fields [21]. The deployed moisture sensors and cost-effective smart devices for irrigation based on a neural network provide better assistance and are remotely monitoring enabled using Message Queuing Telemetry Transport) (MQTT) and HTTP protocols. The system is portable, lightweight, and user-friendly. Various researchers have used different IoT devices to measure soil moisture, temperature and pH level. The details are in the Table 2.

Table 2. IoT devices used to measure soil parameters

Reference	IoT Device Used	Work Done
[22]	Zigbee Wireless Network	Home garden irrigation with integration into the smart home control system.
[23]	Moisture sensors, temperature sensors pressure regulators, molecule sensors, ADC; Zigbee,	Automation with various sensors and wireless communication.
[42]	Arduino, Raspberry Pi3	Robotic model for soil moisture and temperature detection; Automated watering based on data.
[24]	Neural network sensor	Improved reliability through a virtual sensor predicting real-time humidity, temperature, and moisture values.
[25]	Solar panels, sensors, GSM module	Solar-powered smart irrigation system with real-time status alerts to farmers

3.2 Weather-Based Irrigation Monitoring

Monitoring soil data alone is insufficient for the optimum amount of water in the plant's roots, so weather monitoring is also essential. Weather-based irrigation monitoring includes real-time estimating humidity, wind speed, wind direction, reference evapotranspiration (ETo), temperature, solar radiations, and precipitation. Weather parameters have a significant effect on the overall yield of agriculture. Strong winds cause water loss to crops by shredding leaves, lodging crops, and abrasively damaging crops [2,4]. In some situations, soil and plant parameters are impossible, and then weather parameters approximate the irrigation schedule [26]. Atmospheric temperature, for the optimum growth of the plant, is considered 30 °C [4]. High temperature and humidity affect plant growth [2,4]. Considerable research has already been done on smart irrigation based on weather data monitoring.

The [21] presents an open-source technology-based smart system that uses soil and weather data to predict the irrigation schedule. The entire system has been created and implemented on a small scale. Data was sent over the cloud using web services. A web-based system for information visualization and decision support provides real-time insights by analyzing sensor data and weather forecast information. Similarly, [41] employed weather and soil data along with fertility meters and PH meters deployed on the field to measure the proportion of the soil's primary constituents, such as potassium, phosphorous, and nitrogen. This allowed for the determination of the soil's fertility. Evapotranspiration (ET) has a significant impact on irrigation in agriculture. It involves water's simultaneous evaporation and transpiration from the soil, canopies, and water bodies into the atmosphere. Evapotranspiration (ET) based watering scheduling controllers have been proven to be more advantageous in terms of cost, size, and labor needed for irrigation, according to research done in Florida by [27] irrigation. Compared to scheduled procedures, irrigation systems based on ET consume much less water.

Weather monitoring relies heavily on wireless sensor networks (WSN) and gateway nodes. These transmit data from the many sensors placed in the fields to the cloud for additional processing. The work [28] presented an automated irrigation system that uses a GPRS module and wireless sensor network to conserve water. The soil temperature and moisture sensor is placed across the system's wireless network. Water savings of up to 90% compared to the agricultural zone's traditional irrigation procedures were attained.

3.3 Plant-Based Irrigation Monitoring

Irrigation can be scheduled based on plant data. Optical sensors are used to determine the water stress status of the plant. Sensors can be deployed close to the plants or fixed on drones, UAVs, etc., to capture images. These high-definition images help to determine crop health status and nutrient requirements [29] captured images of tea leaves and monitored the plant. The system helped in water conservation and reduced soil erosion by determining the optimum amount of fertilizers needed in the soil. In the study by Dhillon et al. [30], they proposed a system that involves measuring and rigorously testing a thermal infrared (IR) sensor along with relevant environmental variables, including air temperature, photosynthetically active radiation (PAR), wind speed, and relative humidity. The experimentation was conducted in almond, walnut, and grape crops. The daily crop water stress index was calculated using the field data from these tests (CWSI). As a result of offering daily CWSI readings, the results indicated that this technology could be employed as an irrigation management tool. In a greenhouse experiment, [36] analyzed plant health status through the Normalized Difference Vegetation Index (NDVI). Two methods of spectral sensing instruments are used to monitor health status, which can be combined with soil moisture sensing for optimum irrigation. The water level in the plant was also determined by [37] using a leaf water meter sensor(LWM). Photon attenuation through the leaf serves as the basis for the sensor. The sensor has been tested on various plants that have experienced drought and re-watering. LWM is a sensitive, non-destructive, user-friendly technology with promise as a precise irrigation scheduling tool. Readers refer to Table 3 for a more detailed description of various literature on precision irrigation monitoring strategies and machine learning techniques.

4 Role of Machine Learning in Precision Irrigation

Artificial intelligence's machine learning field enables computers to learn without being explicitly programmed. Machine learning models provide intelligent decision support for wise and sustainable use of resources. Machine learning models are being applied to predict the water requirement of the field to achieve water saving. Moreover, It can solve various complicated issues related to irrigation systems. [31] proposed a system based on farmers' knowledge and experience with crops and a fuzzy logic-based decision support system. The proposed system's water-saving was more than the single and multi-threshold-based irrigation scheduling. [32] provided an overview of different recent techniques of irrigation systems in agriculture using IoT and AI. The work by [33] constructed a project using IoT and ML to tackle irrigation challenges by utilizing a variety of sensors, including the temperature, humidity, and pH sensors, the bolt IOT module, and the Raspberry Pi or Arduino module-controlled pressure sensor.. Using the Global System for Mobile Communication (GSM), [34] suggests an intelligent irrigation system that assists farmers in watering their agricultural areas. This system sends out acknowledgment signals describing the task's status, including the soil's humidity level, the ambient temperature, and the motor's status concerning the main power supply or solar power. A fuzzy logic controller is employed to calculate input parameters (such as soil moisture, temperature, and humidity) and provide motor status outputs.

[35] thoroughly analyzes different machine-learning techniques for irrigation prediction. Flexible online learning (OL) framework with author enhancements for irrigation decisions backed by soil property predictions and analysis. Four distinct predictive algorithms were used to forecast soil humidity: deep artificial neural networks, random forests, extreme gradient boosting, and auto-regressive moving averages. All of these techniques were applied in the Hadoop/Spark environment.

Table 3. Summary of field monitoring strategies and machine learning techniques used for irrigation scheduling in recent studies.

Ref.	ML/DL Technique Used	Description	Monitoring Strategy Used			Benefits
			Soil	Weather	Plant	
[38]	PCA, K-means, Clustering, GMM	A mathematical programme is created that determines irrigation schedule based on human-induced irrigation instinct and online weather data.	N	Y	N	• Scalable • Saves money • Minimize water pollution • Preserve water resources especially in drought regions
[39]	SVM	A smart irrigation system that adjusts the quantity of the water automatically based on sensor data.	Y	N	N	•Adjust water quantity automatically • Can be used for home gardening • Saves water
[40]	Linear Regression	The model provides sustainable solution for automating, controlling and monitoring the irrigation process. The model predicts the daily requirement of water-based on sensor data	Y	N	N	• Reduce effort and energy • Saves time • Remote access through mobile app • Avoid wastage of water.
[43]	KNN, DT, SVM, Logistic Regression	A machine learning model based on IoT and wireless sensor network to predict the amount of water and fertilizers needed by plants	Y	N	N	• Remote access • Optimal usage of water and fertilizers • Saves water and fertilizers
[51]	SVM & Bagging	Smart irrigation system to predict the irrigation requirement is based on real-time weather and soil moisture data	Y	Y	N	• Low-cost model • Conserves water • Saves plant nutrients • Reduce labor with 90%
[44]	DT, Random Forest, ANN, SVM	A machine learning model to predict the irrigation time based on soil and weather data. Data is sent to the cloud, and can be accessed using a mobile app remotely.	Y	Y	N	Up to 60% of water saving
[45]	ANN	A sustainable and smart irrigation system based on estimation of ETo using daily data on weather.	Y	Y	N	• Increase irrigation efficiency • Reduce labor cost • Saves water and electricity
[46]	K-means clustering	Based on the physical attributes, special clusters of irrigation network are made and K-means clustering the algorithm is applied. The clustering of the model provided better and easier irrigation making	Y	Y	N	• Economical • Approachable
[47]	Hidden Markov	The system used weather, soil and plant data and compare it with the threshold values to predict the irrigation. The model detected the plant disease condition	Y	Y	Y	• Remote access • Early disease detection • Conserves water
[48]	Mask R-CNN, NN	Automatic water reorganization from aerial footage using UAV captured images	N	N	N	• Reduce time • Conserves water and energy • Reduce labor cost
[49]	RNN	An autonomous irrigation system to optimize the crop yield by predict the water need of plant at a depth of 20 cm	Y	Y	N	Optimizes crop yield Saves the water
[50]	MLR, KNN, DT, RF	A smart system for scheduling the irrigation process by predicting rainfall from the weather station. RMSE values for different models were calculated and Random Forest is considered best by giving minimum RMSE	N	Y	N	• Helps to reduce the usage of ground water • Saves water for future use

The Global Positioning System (GPS) and Radial Function Network (RFN)-based hybrid remote-controlled device was designed to operate the irrigation system, forecast the temperature, regulate the air pressure, and decrease the humidity in water. The system helps to reduce water use, cost, and energy. [34] reviewed the health monitoring and IoT solutions available in the agricultural sector. Monitoring soil health, crop health, IoT-based smart irrigation, and real-time weather forecasting are important focused areas where automation can be used. The IoT-based crop and soil monitoring, automated irrigation system, and real-time weather forecasts can help decrease wastage and increase crop production through efficient fertilizer, herbicides, and water use.

5 Future Trends

The pillars that support "smart" farming include sensors, robots, drones, weather satellites, intelligent software algorithms, and robots. While sensors and weather satellites supply vast amounts of information beneficial to adapting the irrigation, fertilization, and pesticide spraying process and define the ideal time for sowing seeds and harvesting, robots and drones make work time more efficient. However, these technologies present numerous difficulties. Building a universal platform for all types of crops and developing affordable technology is the most significant difficulty. But it's expected that adopting IoT in smart farming would grow agriculture proportionally. Some of the future trends of smart farming include (Table 4):

Table 4. Future Trends of Smart Farming

Trend	Description
Universal Platform	Farmers traditionally view their entire farm as a single unit, but IoT in agriculture requires consideration of different soil types and varying fertilizer and water requirements. There is a need for a universal platform for all types of crops and soil [20]
Precision Fertigation	While precision irrigation has been extensively researched, there is a need for more research on precision fertigation, where optimal amounts of fertilizers and pesticides are provided [20]
Decentralization of Data	Concerns about centralized cloud databases in terms of security have led to the exploration of decentralized approaches like federated learning. This approach involves training machine learning algorithms on dispersed edge devices or servers without transmitting local data samples
Adopting Technology in Developing Nations	Digitalization and intelligent agriculture methods are less prevalent in developing nations, particularly in Africa and some parts of Asia, due to infrastructure challenges. Research is needed to develop technology suitable for these regions to improve sustainable precision irrigation [2]

6 Conclusion

The gradual integration of the Internet of Things (IoT) has become indispensable for effectively managing agricultural resources. Among these resources, water holds a central position as the primary component of agricultural activities. The judicious and efficient water use is pivotal in fostering optimal crop growth. This comprehensive review focuses on the nuanced exploration of water optimization in precision farming, providing valuable insights into crop monitoring and strategies for meticulous irrigation control. The document critically evaluates different soil moisture measurement techniques, offering a comparative analysis to discern their effectiveness. Furthermore, the paper succinctly synthesizes recent advancements in machine-learning techniques specifically applied to precision irrigation and monitoring within the agricultural context. This information will assist researchers and agricultural experts in understanding the optimal utilization of water and the application of machine learning algorithms for water management.

References

1. Alam, A., Biswas, S., Satpati, L.: Population dynamics and its impact: a historical perspective. In: Population, Sanitation and Health: A Geographical Study Towards Sustainability, pp. 3–15. Springer, Cham (2023)
2. Abioye, E.A., et al.: A review on monitoring and advanced control strategies for precision irrigation. Comput. Electron. Agricult. **173**, 105441 (2020)
3. Liao, R., Zhang, S., Zhang, X., Wang, M., Huarui, W., Zhangzhong, L.: Development of smart irrigation systems based on real-time soil moisture data in a greenhouse: proof of concept. Agric. Water Manag. **245**, 106632 (2021)
4. Singh, D.K., Sobti, R.: Long-range real-time monitoring strategy for Precision Irrigation in urban and rural farming in society 5.0. Comput. Indust. Eng **167**, 107997 (2022)
5. Bwambale, E., Abagale, F.K., Anornu, G.K.: Smart irrigation monitoring and control strategies for improving water use efficiency in precision agriculture: a review. Agricult. Water Manag. **260**, 107324 (2022)
6. Wheeler, W.D., Chappell, M., van Iersel, M., Thomas, P.: Implementation of soil moisture sensor based automated irrigation in woody ornamental production. J. Environ. Horticult. **38**(1), 1–7 (2020)
7. Millán, S., Casadesús, J., Campillo, C., Moñino, M.J., Prieto, M.H.: Using soil moisture sensors for automated irrigation scheduling in a plum crop. Water **11**(10), 2061 (2019)
8. Sui, R.: Irrigation scheduling using soil moisture sensors. J. Agric. Sci **10**, 1 (2017)
9. Krishnan, R.S., Golden Julie, E., Harold Robinson, Y., Raja, S., Kumar, R., Thong, P.H.: Fuzzy logic based smart irrigation system using internet of things. J. Clean. Prod. **252**, 119902 (2020)
10. Masina, M., Calone, R., Barbanti, L., Mazzotti, C., Lamberti, A., Speranza, M.: Smart water and soil-salinity management in agro-wetlands. Environ. Eng. Manag. J **18**, 10 (2019)
11. Kumawat, S., Bhamare, M., Nagare, A., Kapadnis, A.: Sensor based automatic irrigation system and soil pH detection using image processing. Int. Res. J. Eng. Technol **4**, 3673–3675 (2017)
12. Rowe, R. O. S. I. A.: Soil moisture. Biosyst. Eng. (2018)
13. Richards, L.A.: Methods of measuring soil moisture tension. Soil Science **68**, 1 (1949)

14. Orouskhani, E., Sahoo, S.R., Agyeman, B.T., Bo, S., Liu, J.: Impact of sensor placement in soil water estimation: a real-case study. arXiv preprint arXiv:2203.06548 (2022)
15. Zhu, H.-H., Huang, Y.-X., Huang, H., Garg, A., Mei, G.-X., Song, H.-H.: Development and evaluation of Arduino-based automatic irrigation system for regulation of soil moisture. Int. J. Geosynth. Ground Eng. **8**(1), 1–9 (2022)
16. Kuiper, P.J.C.: Water uptake of higher plants as affected by root temperature. No. 64-4. Veenman (1964)
17. Sawan, Z.M.: Climatic variables: evaporation, sunshine, relative humidity, soil and air temperature and its adverse effects on cotton production. Inf. Process. Agricult. **5**(1), 134–148 (2018)
18. Schmugge, T.J., Jackson, T.J., McKim, H.L.: Survey of methods for soil moisture determination. Water Resour. Res. **16**(6), 961–979 (1980)
19. Su, S.L., Singh, D.N., Baghini, M.S.: A critical review of soil moisture measurement. Measurement **54**, 92–105 (2014)
20. Sinha, B.B., Dhanalakshmi, R.: Recent advancements and challenges of Internet of Things in smart agriculture: a survey. Future Gen. Comput. Syst. **126**, 169–184 (2022)
21. Goap, A., Sharma, D., Krishna Shukla, A., Rama Krishna, C.: An IoT based smart irrigation management system using machine learning and open source technologies. Comput. Electron. Agricult. **155**, 41–49 (2018)
22. Al-Ali, A.-R., Qasaimeh, M., Al-Mardini, M., Radder, S., Zualkernan, I.A.: ZigBee-based irrigation system for home gardens. In: 2015 International Conference on Communications, Signal Processing, and their Applications (ICCSPA 2015), pp. 1-5. IEEE (2015)
23. Varatharajalu, K., Ramprabu, J.: Wireless irrigation system via phone call and SMS. Int. J. Eng. Adv. Technol **8**, 397–401 (2018)
24. Sami, M., et al.: A deep learning-based sensor modeling for smart irrigation system. Agronomy **12**(1), 212 (2022)
25. Ali, S., et al.: Solar powered smart irrigation system. Pak. J. Eng. Technol. **5**(1), 49–55 (2022)
26. White, S.C., Raine, S.R.: A grower guide to plant based sensing for irrigation scheduling (2008)
27. Davis, S.L., Dukes, M.D.: Irrigation scheduling performance by evapotranspiration-based controllers. Agric. Water Manag. **98**(1), 19–28 (2010)
28. Gutiérrez, J., Villa-Medina, J.F., Nieto-Garibay, A., Ángel Porta-Gándara, M.: Automated irrigation system using a wireless sensor network and GPRS module. IEEE Trans. Instrument. Measur. **63**(1), 166–176 (2013)
29. Jia, X., Huang, Y., Wang, Y., Sun, D.: Research on water and fertilizer irrigation system of tea plantation. Int. J. Distrib. Sens. Netw. **15**(3), 1550147719840182 (2019)
30. Dhillon, R., Francisco, R.O.J.O., Roach, J., Upadhyaya, S., Delwiche, M.: A continuous leaf monitoring system for precision irrigation management in orchard crops. Tarım Makinaları Bilimi Dergisi **10**(4), 267–272 (2014)
31. Viani, F., Bertolli, M., Salucci, M., Polo, A.: Low-cost wireless monitoring and decision support for water saving in agriculture. IEEE Sens. J. **17**(13), 4299–4309 (2017)
32. Ullah, R., et al.: EEWMP: an IoT-based energy-efficient water management platform for smart irrigation. Scientific Program. **2021**, 1–9 (2021)
33. Kanade, P., Prasad, J.P.: Arduino based machine learning and IOT Smart Irrigation System. Int. J. Soft Comput. Eng. **10**(4), 1–5 (2021)

34. Pandey, A.K., Mukherjee, A.: A review on advances in IoT-based technologies for smart agricultural system. Internet of Things Analyt. Agricult. **3**, 29–44 (2022)
35. El Mezouari, A., El Fazziki, A., Sadgal, M.: Hadoop-Spark framework for machine learning-based smart irrigation planning. SN Comput. Sci. **3**(1), 1–10 (2022)
36. Lozoya, C., Eyzaguirre, E., Espinoza, J., Montes-Fonseca, S.L., Rosas-Perez, G.: Spectral vegetation index sensor evaluation for greenhouse precision agriculture. In: 2019 IEEE Sensors, pp. 1–4. IEEE (2019)
37. Cecilia, B., et al.: On-line monitoring of plant water status: validation of a novel sensor based on photon attenuation of radiation through the leaf. Sci. Total Environ. **817**, 152881 (2022)
38. Kılkış, Ş: Sustainable development of energy, water and environment systems index for Southeast European cities. J. Clean. Prod. **130**, 222–23 (2016)
39. Suzuki, Y., Ibayashi, H., Mineno, H.: An SVM based irrigation control system for home gardening. In: 2013 IEEE 2nd Global Conference on Consumer Electronics (GCCE), pp. 365–366. IEEE (2013)
40. Kumar, A., Surendra, A., Mohan, H., Muthu Valliappan, K., Kirthika, N.: Internet of things based smart irrigation using regression algorithm. In: 2017 International Conference on Intelligent Computing, Instrumentation and Control Technologies (ICICICT), pp. 1652–1657. IEEE (2017)
41. Kumar, G.: Research paper on water irrigation by using wireless sensor network. Int. J. Sci. Res. Eng. Technol. 3–4 (2014)
42. Shekhar, Y., Dagur, E., Mishra, S., Sankaranarayanan, S.: Intelligent IoT-based automated irrigation system. Int. J. Appl. Eng. Res. **12**(18), 7306–7320 (2017)
43. Poornima, D., Arulselvi, G.: Implementation of precision soil and water conservation agriculture (PSWCA) through machine learning, cloud-enabled IoT integration and wireless sensor network. Eur. J. Molecul. Clin. Med. **7**, 3 (2020)
44. Glória, A., Cardoso, J., Sebastião, P.: Sustainable irrigation system for farming supported by machine learning and real-time sensor data. Sensors **21**(9), 3079 (2021)
45. Subathra, M.S.P., Blessing, C.J., Thomas George, S., Thomas, A., Dhibak Raj, A., Ewards, V.: Automated intelligent wireless drip irrigation using ANN techniques. In: Peter, J.D., Alavi, A.H., Javadi, B. (eds.) Advances in Big Data and Cloud Computing: Proceedings of ICBDCC18, pp. 555–568. Springer, Singapore (2019). https://doi.org/10.1007/978-981-13-1882-5_49
46. Monem, M.J., Hashemi, S.M.: Spatial clustering of irrigation networks using K-means method (Case study of Ghazvin irrigation network). Iran-Water Resour. Res. **7**(1), 38–46 (2010)
47. Yashaswini, L.S., Vani, H.U., Sinchana, H.N., Kumar, N.: Smart automated irrigation system with disease prediction. In: 2017 IEEE International Conference on Power, Control, Signals and Instrumentation Engineering (ICPCSI), pp. 422–427. IEEE (2017)
48. Albuquerque, C.K.G., Polimante, S., Torre-Neto, A., Prati, R.C.: Water spray detection for smart irrigation systems with mask r-cnn and UAV footage. In: 2020 IEEE International Workshop on Metrology for Agriculture and Forestry (MetroAgriFor), pp. 236–240. IEEE (2020)
49. Anuşlu, T.: Smart precision agriculture with autonomous irrigation system using rnn-based techniques (2017)
50. Kumar, S., Mishra, S., Khanna, P.: Precision sugarcane monitoring using SVM classifier. Procedia Comput. Sci. **122**, 881–887 (2017)
51. Ramya, S., Swetha, A.M., Doraipandian, M.: IoT framework for smart irrigation using machine learning technique. J. Comput. Sci. **16**(3), 355–363 (2020)

Addressing Security Challenges in Copyright Management Applications: The Blockchain Perspective

Nour El Madhoun[1,2,3](✉), Badis Hammi[4], Saad El Jaouhari[1], Djamel Mesbah[1,3,5], and Elsi Ahmadieh[6]

[1] LISITE Laboratory, ISEP, 10 Rue de Vanves, 92130 Issy-les-Moulineaux, France
{nour.el-madhoun,saad.el-jaouhari,djamel.mesbah}@isep.fr

[2] Sorbonne Université, CNRS, LIP6, 4 place Jussieu, 75005 Paris, France
nour.el_madhoun@sorbonne-universite.fr

[3] Université Paris-Saclay, CNRS, Laboratoire Interdisciplinaire des Sciences du Numérique, 91190 Gif-sur-Yvette, France
{nour.el-madhoun, djamel.mesbah}@universite-paris-saclay.fr

[4] SAMOVAR, Télécom SudParis, Institut Polytechnique de Paris, Palaiseau, France
badis.hammi@telecom-sudparis.eu

[5] Adservio Group, T. Franklin, 100 101 Terr. Boieldieu Ét. 9, 92800 Puteaux, France
djamel.mesbah@adservio.fr

[6] Lebanese University, Faculty of Technology, Department Communications and Computer Networks Engineering, Beirut, Lebanon

Abstract. Today, the field of copyright management faces several security vulnerabilities, making it a target for various types of rights violations, including piracy and unauthorized use of creative works. In this context, blockchain technology has emerged as a robust solution in this sector as it offers enhanced security, trust for creators, and increased transparency. In this paper, we first present an overview of four main applications in the field of copyright management: (1) management of licenses and royalties, (2) market for artworks, (3) registration and protection of works, and (4) proof of anteriority. We then analyze in details the critical security vulnerabilities of these applications. Subsequently, we explain how blockchain technology can be used to mitigate these vulnerabilities. Finally, we discuss possible methods of preventing common blockchain-based attacks in these copyright management applications.

Keywords: Artwork · Attack · Blockchain Technology · Copyright · NFT · Proof of Anteriority · Proof of Ownership · Royalties · Security

1 Introduction

The copyright management sector is currently facing a series of security vulnerabilities that pose significant risks to the protection and enforcement of copyrights. These issues make the sector particularly exposed to various types of infringements, such as piracy and unauthorized use of creative works. As a result, these

L. Barolli (Ed.): AINA 2024, LNDECT 204, pp. 169–182, 2024.
https://doi.org/10.1007/978-3-031-57942-4_18

threats seriously compromise the legitimate rights of authors and the originality of their works, while endangering the integrity of artistic and intellectual creation [1,2].

Blockchain technology has recently emerged as a key and robust solution to a variety of security challenges in several sectors, particularly in the copyright management sector. An essential element of this technology is the smart contract which represent decentralized and autonomous programs that is automatically executed on the blockchain when predefined conditions are met [3–7]. Indeed, thanks to the intrinsic properties of the blockchain technology such as decentralization, immutability, and transparency, along with the principle of smart contracts, several significant advantages are provided. These include enhanced copyright security, protection against unauthorized modifications, complete traceability of operations, and boosted confidence among creators and industry professionals [8–10].

We summarize the contributions of this paper as follows. First, we provide an overview of the four main copyright management applications in Sect. 2. Then, we analyze Sect. 3 the critical security vulnerabilities in these applications. Next, we highlight how the blockchain technology could counter these security vulnerabilities in Sect. 4. Finally, we discuss in Sect. 5 the different strategies to prevent the most common attacks on blockchain technology in the context of the discussed copyright management applications.

2 Overview of the Main Copyright Management Applications

Blockchain technology can ensure the security of numerous applications in the field of copyright management. This paper focuses on the four main applications in the latter field: (1) management of licenses and royalties, (2) market for artworks, (3) registration and protection of works, and (4) proof of anteriority. In this section, we present an overview of how these applications are conventionally implemented, excluding the integration of blockchain technology.

2.1 Management of Licenses and Royalties

The management of licenses and royalties is traditionally a complex and centralized process. It involves contractual agreements between the creators of works such as authors, musicians, artists, and so on, and entities wishing to use these works. These contracts define the conditions of use, covering aspects like the duration of the license, the geographical areas concerned, and the types of authorized use (e.g., broadcasting, reproduction, public display, and so on). Typically, the transactions and rights management are handled by intermediaries, such as copyright agencies or collective management organizations. They are responsible for negotiating the terms of licenses, collecting royalties from users (e.g., publishers, broadcasters, or online platforms), and redistributing them to the rights-holders based on the established agreements. The systems used to manage these processes are based on centralized databases that record details of works, contracts, uses, and financial transactions. They aim to accurately track the use of works in order to calculate the royalties due. Additionally, they seek to efficiently manage licensing agreements and payments while ensuring that creators receive fair compensation for the use of their works [11,12].

2.2 Market for Artworks

In the traditional market for artworks, transaction management and authenticity verification are primarily ensured by art galleries, auction houses, and online platforms. These intermediaries have a key role in connecting artists with buyers and facilitating the sale and purchase of works. Indeed, verifying the provenance and authenticity of works, a task carried out by field experts, is essential for buyers and sellers to ensure the value and legitimacy of the exchanged works. However, this centralized model inherently presents complex aspects. For instance, reliance on experts for verification procedures can limit the process's speed and accessibility. Furthermore, centralized structures may introduce additional costs for artists and buyers due to commissions and management fees. Finally, centralizing these operations involves logistical challenges such as secure storage and transportation of works, as well as managing the confidential information of customers and transaction details [13].

2.3 Registration and Protection of Works

The registration and protection of works are traditionally managed by specialized intellectual property organizations. These organizations provide a legal framework for the registration of works, enabling creators to obtain official recognition and proof of their ownership. The registration process involves creators submitting detailed information about their work, including the title, a full description, and often a copy or an extract of the work. Once the work is registered, it is added to a public register, which offers transparency and facilitates the resolution of potential copyright disputes. This system represents a crucial element in the protection of creators' rights, enabling them to legally assert their rights in case of violation. However, its effectiveness can vary depending on the jurisdiction and the particular nature of the work. Moreover, although registration offers some proof of anteriority, demonstrating the work's existence at a specific point in time can pose challenges. This proof is indeed essential in cases where the originality or the priority of creation of the work is under dispute. In such situations, creators may be required to provide additional evidence, such as drafts, correspondence, or testimonials to substantiate their claims, thereby highlighting the importance of documenting and safeguarding reliable proof of the creation of the work (see Sect. 3.3) [14,15].

2.4 Proof of Anteriority

The proof of anteriority often involves depositing creations with specialized organizations, or using testimonials and dated documents. This process aims to establish the existence of a work at a given point in time, especially in the case of copyright disputes. In order to prove anteriority, creators can produce various types of documents, such as drafts, correspondence, or recordings, that attest to the date of creation of the work. Sometimes, it is also possible to rely on testimonials or public records to strengthen ownership claims. These methods of proof are intended to provide legal certainty regarding the originality and ownership of the work, but the quality and reliability of any provided proof are crucial. The proofs must be clear, coherent, and convincing to withstand contestations and divergent interpretations that can arise in legal proceedings. Although these methods are widely used due to their effectiveness in providing legal proof, they require meticulous attention to the documentation and preservation of relevant evidence to effectively support copyright claims (see Sect. 3.4) [16,17].

3 Security Vulnerabilities in the Main Copyright Management Applications

In this section, we analyze the security of the four main applications of copyright management introduced in the previous Sect. 2. We present the security vulnerabilities identified in these applications as follows:

3.1 Vulnerabilities on the Management of Licenses and Royalties

Centralized data management: the management of licenses and royalties traditionally relies on centralized databases managed by intermediaries or institutions. These databases store all data related to copyrights, transactions, and royalty payments. Such centralization creates single points of failure where all information and associated operations can be vulnerable, potentially resulting in data loss, leaks of confidential information, or service interruptions if the central system is compromised (e.g., following a cyberattack). Additionally, reliance on a centralized entity can lead to bottleneck problems where delays in data processing affect the speed and efficiency of transactions. Finally, centralization can also limit system resilience and adaptability in response to rapid market changes or regulatory requirements [18,19].

Lack of transparency in royalty calculation and distribution: the processes for calculating and distributing royalties due to creators are often opaque in the classic systems for management of licenses and royalties. In general, both creators and rights-holders have limited visibility of how royalties are calculated, particularly regarding the actual use of their works. This opacity can result from the complexity of licensing agreements, the use of undisclosed calculation formulas, or the absence of detailed information on sales and uses of works. This situation can lead to a lack of trust for creators, who may feel underpaid or poorly informed about the exploitation of their works. Moreover, in the absence of transparency, errors in the distribution of royalties, whether accidental or fraudulent, are difficult to detect and correct, which can lead to disputes and prolonged conflicts between the involved parties [2,20].

Delays in payment distribution: are often due to the complexity of the administrative processes involved in collecting, calculating, and distributing royalties. Centralized systems may require multiple verification and approval steps, prolonging the time between the collection of royalties and their distribution to creators. In addition, reconciling accounts and processing financial transactions across different institutions and jurisdictions can add further layers of complexity and delays. These delays can have a negative impact on creators, particularly independent artists or small publishers, for whom these payments may represent a significant proportion of their revenues [2,21,22].
Challenges in tracking all instances of work usage: conventional systems managing licenses and royalties encounter complexity in tracking and recording the use of works across various platforms and formats, particularly in the current digital context where works can be easily copied, shared, and distributed. Although these systems are often equipped with tracking mechanisms, they are not always able to thoroughly detect and document each instance of use, especially unauthorized or informal uses, such as those on social media, streaming platforms, or via illegal downloads. This shortcoming makes it difficult for right-holders to receive all the royalties due to them and might results in major financial losses. Furthermore, the limited ability of these systems to provide an accurate and comprehensive tracking of the use of works hampers the transparency and fairness of the distribution of royalties, posing significant challenges for both creators and distributors [23,24].

3.2 Vulnerabilities Related to the Market for Artworks

Authenticity and provenance of artworks: the authenticity and provenance of artworks represent a fundamental challenge in the traditional market for artworks. Authenticating an artwork involves verifying that it was actually created by the artist in question, and provenance concerns the history of the ownership of that artwork. Indeed, authentication and provenance often rely on physical documents such as certificates of authenticity or sales histories, which are subject to errors, omissions, or even falsifications. This creates risks for buyers, who may acquire artworks that are either inauthentic or have uncertain histories. The difficulty of reliably tracing the complete history of an artwork can also affect its market value and perceived legitimacy [25,26].
Counterfeiting and unauthorized duplication: with the advent of modern reproduction technologies, the creation of high-quality reproductions/copies of artworks has become increasingly seamless, often presenting a challenge in distinguishing them from the original works. This poses a problem for buyers who may end up with a counterfeit, and also for artists and creators who may see their rights and potential revenue compromised by such illegal reproductions. In addition, unauthorized duplication can saturate the market, reducing the perceived value of the originals. This vulnerability particularly affects the market for digital artworks, where the copying and distribution of digital files is relatively easy to achieve without quality loss [27,28].

3.3 Vulnerabilities on the Registration and Protection of Works

Complexity in proving ownership and anteriority: determining proof of ownership and anteriority of a work can be complex in the traditional systems of registration and protection of works. To prove ownership, creators must often provide substantial evidence of their efforts to create the work, which may include drafts, correspondence, or other forms of documentation. The proof of anteriority, which involves demonstrating that a work was created at a specific point in time, can be even more difficult to establish. It generally requires tangible evidence, such as dated recordings or testimonials. Such evidence can be hard to preserve reliably over long periods and can be subject to contestation, especially if the documents are altered or lost [29,30].

Lack of universal registers between different countries: each country has its own system for the registration and protection of works, with distinct standards, procedures, and legal requirements. This disparity creates many challenges for creators seeking to protect their works internationally. Differences in registration systems can lead to inconsistencies in copyright recognition and complicate the protection of works across national borders. These differences can limit the effectiveness of copyright protection in a globalized context where works are easily accessible and distributed around the world [31].

3.4 Vulnerabilities Related to the Proof of Anteriority

Dependence on physical proofs that can be altered or lost: the dependence of the proof of anteriority on several documents or physical objects such as manuscripts, drawings, recordings, or testimonies presents several vulnerabilities. Firstly, physical proofs can be altered, intentionally or accidentally, undermining their reliability. Modifications or falsifications can be made to the original documents, making it difficult to determine the authentic state of the work at any given time. Secondly, physical proofs are subject to deterioration and loss. Over time, documents can become damaged or illegible, and there is always a risk of loss due to natural catastrophes, accidents, or negligence. These factors compromise the ability of such evidence to reliably and enduringly establish the date of creation of a work [32,33].

Difficulty in establishing incontestable proof of anteriority: in order to consider a proof of anteriority as incontestable, it needs to be not only accurate and reliable but also recognized and accepted by all the involved parties, including in a judicial context. Traditional methods of proving anteriority, such as legal deposits or testimonials, can be subject to contestation and differing interpretations. For example, questions may arise concerning the authenticity of presented documents, the credibility of witnesses, or the integrity of recordings. Additionally, in an international context, different countries may have varying standards and practices for the proof of anteriority, complicating the mutual recognition of such proofs. This situation can lead to prolonged litigation and judicial uncertainty, making it difficult to defend copyrights on an international stage [34,35].

4 Securing Copyright Management Applications with Blockchain Technology

In this section, we present how blockchain technology can contribute in addressing the ten security vulnerabilities presented in the previous Sect. 3:

Addressing centralized data management: blockchain technology offers a solution to this vulnerability thanks to its decentralized and distributed architecture where data relating to copyrights, transactions, and payment of royalties are no longer stored in a centralized database, but are distributed across a network of blockchain nodes. Each node in the network holds a copy of the entire blockchain, guaranteeing data availability and integrity even in the case of failure of one or more nodes. This approach eliminates single points of failure and significantly reduces the risk of data loss, leaks of confidential information, and service interruptions due to cyberattacks. In addition, decentralization facilitates greater system agility and adaptability to rapid market and regulatory changes, while resolving the problems of bottlenecks associated with centralization and thus improving the speed and efficiency of copyright transactions [36,37].

Addressing the lack of transparency in royalty calculation and distribution: the use of smart contracts on the blockchain provides a solution to this vulnerability. These contracts, once programmed and deployed on the blockchain, enable the automation of the process of calculating and distributing royalties in a reliable and transparent way. Each transaction or use of a work is immutably recorded on the blockchain, providing complete and real-time visibility on the actual use of the works. This traceability makes it possible to precisely calculate the royalties due according to the specific terms of each license agreement, thus eliminating the uncertainties and errors associated with opaque calculation formulas. In addition, thanks to the use of smart contracts, payments can be automatically initiated once the terms of the contract are met, ensuring a rapid and equitable distribution of royalties to creators. This approach promotes a greater degree of transparency for all parties involved and contributes to strengthening trust between creators, rights-holders, and users, while simplifying administration and reducing the possibility of disputes relating to the distribution of royalties [37].

Addressing delays in payment distribution: the decentralized architecture of blockchain technology facilitates direct transactions and automates payment processes through the use of smart contracts. These contracts aim to eliminate the intermediate verification and approval steps usually associated with centralized systems, enabling fast and efficient distribution of royalties to creators. Moreover, blockchain technology ensures transparent reconciliation of accounts in real time by streamlining processing times for financial transactions between different institutions and jurisdictions. This method accelerates the flow of revenue to artists and publishers, particularly those who are independent or small-scale, while improving their financial stability [38].

Addressing the tracking of instances of work usage: the combination of blockchain technology with advanced analysis tools significantly improves the traceability of works across a multitude of digital platforms. This enables real-time tracking of the use of works, whether shared on social networks, distributed via streaming platforms, or downloaded illegally. Each instance of use can be recorded on the blockchain, providing a detailed and unalterable history. This enhanced traceability assures that right-holders receive fair compensation for each use of their work, while minimizing financial losses due to unauthorized use. This method increases transparency and fairness in the distribution of royalties, effectively addressing the major challenges faced by both creators and distributors in the current digital environment [24,39].
Addressing the authenticity and provenance of artworks: in the market for artworks, blockchain technology can be used to create an immutable digital register that ensures the authenticity and traceability of the provenance of artworks [40]. Each artwork registered on the blockchain is identified by a unique cryptographic identifier linked to detailed data on its origin, artist, history of ownership, and journey through the market. These details are permanently and transparently stored on the blockchain, making any falsification nearly impossible. This allows buyers to reliably and transparently verify the authenticity and provenance of an artwork, while significantly reducing the risk of acquiring inauthentic artworks or those with an uncertain history [41].
Addressing counterfeiting and unauthorized duplication: in order to prevent counterfeiting and unauthorized duplication in the market for digital artworks, the use of NFTs (Non-Fungible Tokens) based on blockchain technology offers an innovative and effective solution. Each NFT is a unique digital token associated with a specific artwork and serves as a certificate of ownership and digital authenticity. This uniqueness ensures that even if copies of the artwork exist, only the holder of the NFT owns the original and authenticated version. Indeed, NFTs enable transparent tracking of ownership and transactions, while making any unauthorized reproduction readily identifiable and traceable. The use of NFTs helps to preserve the value of originals and protect the rights and potential revenues of artists and creators [42].
Addressing the complexity in proving ownership and anteriority: blockchain technology simplifies proof of ownership and anteriority by offering a timestamped, immutable registration system. When a work is registered on the blockchain, it receives a unique timestamp certifying its creation date. This information is permanently stored and cannot be altered, providing undeniable proof of the anteriority of the work. In addition, the identity of the creator can be directly linked to this blockchain entry, establishing a clear proof of ownership that is challenging to contest. This system also reduces the need to maintain physical proofs that are susceptible to alteration or loss and streamlines the validation process in the case of copyright disputes [43].
Addressing the lack of universal registers between different countries: blockchain technology operates as a universal ledger for the registration

and protection of works, transcending national borders. Each work registered on the blockchain can be viewed from any country, enabling international recognition and protection of copyrights. This uniformity of registration helps to resolve any inconsistencies due to different national legal systems and registration procedures. Consequently, creators can benefit from a more homogeneous protection of their works around the world, making it easier to manage copyrights in a globalized context where digital works can easily cross borders [44].

Addressing the dependence on physical proofs: blockchain technology offers a digital solution for storing all proofs of anteriority in a cryptographically secure form, where documents, records, or any other type of proof are digitized and stored as transactions in blocks. Each block is cryptographically linked to the previous one, forming an immutable, tamper-resistant chain [45]. Consequently, any modification made to a record is immediately detectable. Additionally, blockchain technology ensures the durability of all proofs because even if physical copies are lost or damaged, their digital versions remain intact and verifiable on the chain [46,47].

Addressing the difficulty in establishing incontestable proof of anteriority: the transparency and immutability of blockchain technology enable reliable and widely recognized proof of anteriority to be established. In fact, when a proof of anteriority is recorded on the blockchain, it is timestamped and becomes accessible to all the involved parties. This timestamp provides indisputable proof of the existence of the work at a given point in time. Moreover, due to the decentralized nature of blockchain, these records are independent of any central authority, which reinforces their credibility and acceptance in a judicial context. Finally, thanks to the distributed nature of blockchain, these proofs of anteriority are recognized internationally, facilitating the defense of copyright on the world stage and reducing the risk of protracted litigation [46,47].

5 Preventing the Most Common Blockchain Attacks in Copyright Management Applications

In this section, we discuss the various prevention strategies that can be adapted to counter the most common attacks on blockchain technology (51% attack, sybil attack, routing attack, double spending attack and smart contract vulnerabilities [7,8,48,49]) in the copyright management applications if the blockchain technology is adopted.

Preventing the 51% attack: a 51% attack occurs when a malicious actor takes control of more than 50% of a blockchain network's computing power, enabling him to manipulate the blockchain, perform potential double-spending, or censor and rewrite transactions. Using Proof of Stake (PoS) blockchains or alternative consensus algorithms such as Proof of Authority (PoA) may make such control economically or logistically unfeasible. Additionally, implementing extra security protocols and redundant validation, such as cross-validation by independent nodes, can strengthen blockchain integrity. The increased decentralization of the network, with a wide and diverse distribution of nodes, can also reduce the probability of domination by a single group. Indeed, the introduction of strict rules for block creation and transaction validation adds an extra layer of security. Moreover, the use of hybrid blockchain networks, combining the characteristics of public and private chains, can offer additional validation and enhanced security. All these joint methods guarantee effective protection against manipulation and attacks, thereby ensuring the reliability and transparency of transactions and registrations in the field of copyright management [50, 51].

Preventing the sybil attack: in a sybil attack, a malicious actor creates multiple false identities to influence or disrupt the network [52]. To remedy such an attack, it is necessary to adopt a robust system of authentication and verification of nodes, such as the use of consensus mechanisms that require some form of proof of identity or economic participation, as in PoS or PoA systems. The establishment of lists of approved nodes or the verification of participants by cryptographic methods, such as digital signatures, can also prevent malicious actors from creating multiple falsified identities. Indeed, continuously monitoring the network must be maintained to quickly detect and isolate any suspicious nodes. Moreover, the use of hybrid networks, which combine features of public and private blockchains, can increase security by restricting access to trusted nodes. These measures, when applied consistently, improve the resilience and reliability of blockchain-based copyright management applications against Sybil attacks [53–55].

Preventing the routing attack: a routing attack, where a malicious entity intercepts or modifies network traffic between blockchain nodes to disrupt or monitor communications, can be prevented by employing end-to-end encryption. This ensures that the transmitted data remain secure and unreadable to unauthorized parties. In addition, the implementation of secure network protocols, such as Transport Layer Security (TLS), for communications between nodes can help prevent data interception. The use of anomaly detection mechanisms is also a solution for monitoring and detecting suspicious activity or unusual traffic patterns that could indicate a routing attack. Furthermore, diversifying data transmission paths and decentralizing network infrastructure can reduce reliance on specific paths, thus minimizing the risk of interception. These strategies enhance the security of blockchain networks in copyright management by ensuring data confidentiality, integrity, and protection against malicious interference[56, 57].

Preventing the double spending attack: a double-spending attack, where an attacker spends the same cryptocurrency or token twice by altering blockchain transaction history, can be prevented with robust consensus mechanisms like Proof of Work (PoW) or PoS. These mechanisms effectively guarantee that only one version of the truth (the longest or most valid chain) is accepted on the network. The implementation of real-time transaction verifications and multiple confirmations for each transaction can also significantly reduce the risk of double spending. This means that a transaction is only considered valid once it has been confirmed by a sufficient number of nodes on the network. Additionally, the constant monitoring of the network is crucial for detecting anomalies and double-spending attempts at an early stage, where network nodes must be able to detect and reject fraudulent transactions, thus preventing them from being recorded in the blockchain. Finally, in systems based on the mechanism PoS, malicious parties risk losing their stake (the tokens they have staked) if they attempt to carry out double-spending attacks, adding an extra layer of deterrence [58].

Preventing smart contract vulnerabilities: smart contract vulnerabilities, often due to code flaws, logic errors, or unexpected interactions with other contracts, require rigorous prevention by examining and auditing the code by blockchain security experts. This involves static and dynamic code analysis, checking for known vulnerabilities, and evaluating the logic of the contracts to identify potential flaws. It is also important to implement good development practices, like defensive programming, extensive unit testing, and security mechanisms such as transaction locks and limits, to reinforce smart contract robustness. Additionally, implementing procedures for managing security updates and corrections is effective, as it ensures rapid updates or corrections to smart contracts upon the discovery of vulnerabilities, without compromising operational continuity or data security. Finally, in order to prevent vulnerabilities right from the design phase, developers should be trained in optimal security practices for smart contract creation. All the aforementioned methods aim to ensure that blockchain-based copyright management applications are safeguarded against smart contract vulnerabilities, thereby securing the integrity, security, and reliability of transactions and records in these systems [5,59,60].

6 Conclusion

Blockchain technology represents a transformative force for addressing the security challenges inherent in the copyright management sector. In this paper, we examined the various security vulnerabilities in the four main copyright management applications and illustrated the effectiveness of blockchain technology in addressing these issues. We also discussed the various strategies appropriate for mitigating the most common attacks on blockchain technology in the context of copyright management applications if these latter are blockchain-based.

References

1. Zhang, Z.: Security, trust and risk in digital rights management ecosystem. In: 2010 International Conference on High Performance Computing & Simulation, pp. 498–503. IEEE (2010)
2. Savelyev, A.: Copyright in the blockchain era: promises and challenges. Comput. Law Secur. Rev. **34**(3), 550–561 (2018)
3. El Madhoun, N., Hatin, J., Bertin, E.: A decision tree for building it applications. Ann. Telecommun. **76**(3), 131–144 (2021)

4. Wilson, K.B., Karg, A., Ghaderi, H.: Prospecting non-fungible tokens in the digital economy: stakeholders and ecosystem, risk and opportunity. Bus. Horiz. **65**(5), 657–670 (2022)
5. Sayeed, S., Marco-Gisbert, H., Caira, T.: Smart contract: attacks and protections. IEEE Access **8**, 24416–24427 (2020)
6. Khan, Z.A., Namin, A.S.: Ethereum smart contracts: vulnerabilities and their classifications. In: 2020 IEEE International Conference on Big Data (Big Data), pp. 1–10. IEEE (2020)
7. Daimi, K., Dionysiou, I., El Madhoun, N.: Principles and Practice of Blockchains. Springer, Cham (2022). https://doi.org/10.1007/978-3-031-10507-4
8. Kaushik, S., El Madhoun, N.: Analysis of blockchain security: classic attacks, cybercrime and penetration testing. In: MobiSecServ 2023 (The Eighth International Conference On Mobile and Secure Services). IEEE (2023)
9. Ding, Y., Yang, L., Shi, W., Duan, X.: The digital copyright management system based on blockchain. In: 2019 IEEE 2nd International Conference on Computer and Communication Engineering Technology (CCET), pp. 63–68. IEEE (2019)
10. Zhao, S., O'Mahony, D.: Bmcprotector: a blockchain and smart contract based application for music copyright protection. In: Proceedings of the 2018 International Conference on Blockchain Technology and Application, pp. 1–5 (2018)
11. Powers, J.B.: Patents and royalties, pp. 129–150. Privatization and public universities. Indiana University Press, Bloomington (2006)
12. Cadavid, J.A.: The origin and purpose of legal protection for the integrity of copyright metadata. IIC-Int. Rev. Intellect. Property Competition Law **54**(8), 1179–1202 (2023)
13. Jeong, S.Y.E.: Value of NFTs in the digital art sector and its market research. Sotheby's Institute of Art-New York (2022)
14. Sprigman, C.: Reform (aliz) ing copyright. Intellectual Property Law and History, pp. 277–360. Routledge, London (2017)
15. Bamakan, S.M.H., Nezhadsistani, N., Bodaghi, O., Qu, Q.: Patents and intellectual property assets as non-fungible tokens; key technologies and challenges. Sci. Rep. **12**(1), 2178 (2022)
16. Pouillard, V.: Intellectual Property Rights and Country-of-Origin Labels in the Luxury Industry. The Oxford Handbook of Luxury Business, Oxford University Press, Oxford (2020)
17. Lamare, F., Portelli, A.: The use of records to manage risks associated with the decommissioning of nuclear facilities. IN: Proceedings of the 29th European Safety and Reliability Conference, European Safety and Reliability Conference (ESREL), pp. 3874–3881 (2019)
18. Towse, R.: Economics of copyright collecting societies and digital rights: is there a case for a centralised digital copyright exchange? Revi. Econ. Res. Copyright Issues **9**(2), 3–30 (2012)
19. Tushnet, R.: All of this has happened before and all of this will happen again: innovation in copyright licensing. Berkeley Technol. Law J. **29**(3), 1447–1488 (2015)
20. Kossecki, P., Akin, O.: Valuation of copyrights to audiovisual works: transparency practices of the copyright management organizations in the european union. Ekonomia i Prawo. Econ. Law **20**(3), 543–571 (2021)
21. Kuo, A.S.Y.: Professional realities of the subtitling industry: the subtitlers' perspective. In: Piñero, R.B., Cintas, J.D. (eds.) Audiovisual translation in a global context, pp. 163–191. Springer, London (2015). https://doi.org/10.1057/9781137552891_10

22. Papadopoulos, T.: The economics of copyright, parallel imports and piracy in the music recording industry. PhD Thesis, Victoria University (2002)
23. Megías, D., Kuribayashi, M., Qureshi, A.: Survey on decentralized fingerprinting solutions: copyright protection through piracy tracing. Computers **9**(2), 26 (2020)
24. Zhu, P., Hu, J., Li, X., Zhu, Q.: Using blockchain technology to enhance the traceability of original achievements. IEEE Trans. Eng. Manag. **70**, 1693–1707 (2021)
25. Oliveri, V., Porter, G., Davies, C., James, P.: Art crime: the challenges of provenance, law and ethics. Museum Manag. Curatorship **37**(2), 179–195 (2022)
26. Luzan, A.: Art provenance yesterday, today, and tomorrow with a particular focus on blockchain technology. Università Ca'Foscari Venezia (2023)
27. Benhamou, F., Ginsburgh, V.: Copies of artworks: the case of paintings and prints. Handb. Econ. Art Culture **1**, 253–283 (2006)
28. O'Dwyer, R.: Limited edition: producing artificial scarcity for digital art on the blockchain and its implications for the cultural industries. Convergence **26**(4), 874–894 (2020)
29. Swan, M.: Blockchain: Blueprint for a New Economy. O'Reilly Media, Inc., Sebastopol (2015)
30. Negrão Chuba, T., Simões Pazelli, G.: A new panorama for intellectual property: the benefits and challenges of blockchain (2023). Available at SSRN 4579686
31. Van Gompel, S.: Copyright formalities in the internet age: filters of protection or facilitators of licensing. Berkeley Technol. Law J. **28**(3), 1425–1458 (2013)
32. Strong, W.S.: The copyright book: a practical guide (2014)
33. Letai, P.: Don't think twice, it's all right: towards a new copyright protection system. In: International Interdisciplinary Business-Economics Advancement Conference, p. 105 (2014)
34. Derrida, J.: Copy, archive, signature: a conversation on photography (2010)
35. Fong, D.: Tales of the tape: the ontological, discursive, and ethical lives of literary audio artifacts. Simon Fraser University (2019)
36. Mehta, D., Tanwar, S., Bodkhe, U., Shukla, A., Kumar, N.: Blockchain-based royalty contract transactions scheme for industry 4.0 supply-chain management. Inf. Process. Manage. **58**(4), 102586 (2021)
37. Liu, J.: Blockchain copyright exchange-a prototype. Buff. L. Rev. HeinOnline **69**, 1021 (2021)
38. Pech, S.: Copyright unchained: how blockchain technology can change the administration and distribution of copyright protected works. Nw. J. Tech. Intell. Prop. HeinOnline **18**, 1 (2020)
39. Jiang, T., Sui, A., Lin, W., Han, P.: Research on the application of blockchain in copyright protection. In: 2020 International Conference on Culture-oriented Science & Technology (ICCST), pp. 616–621. IEEE (2020)
40. Hammi, B., Zeadally, S., Perez, A.J.: Non-fungible tokens: a review. IEEE Internet Things Mag. **6**(1), 46–50 (2023)
41. Zeilinger, M.: Digital art as 'monetised graphics': enforcing intellectual property on the blockchain. Philos. Technol. **31**(1), 15–41 (2018)
42. Guadamuz, A.: The treachery of images: non-fungible tokens and copyright. J. Intellect. Property Law Pract. **16**(12), 1367–1385 (2021)
43. Quinn, J., Connolly, B.: Distributed ledger technology and property registers: displacement or status quo. Law, Innovation Technol. **13**(2), 377–397 (2021)
44. Shakhnazarov, B.A.: Lex registrum as a system of regulation of cross-border relations aimed at protection of intellectual property implemented by means of blockchain technology. Kutafin Law Rev. **9**(2), 195–226 (2022)

45. Hammi, B., Zeadally, S.: Software supply-chain security: issues and countermeasures. Computer **56**(7), 54–66 (2023)
46. Bell, T.W.: Copyrights, privacy, and the blockchain. Ohio NUL Rev. HeinOnline **42**, 439 (2015)
47. Pradeep, A.S.E., Amor, R., Yiu, T.W.: Blockchain improving trust in BIM data exchange: a case study on BIMchain. In: Construction Research Congress 2020, American Society of Civil Engineers Reston, VA, pp. 1174–1183 (2020)
48. Nicolas, K., Wang, Y., Giakos, G.C., Wei, B., Shen, H.: Blockchain system defensive overview for double-spend and selfish mining attacks: a systematic approach. IEEE Access **9**, 3838–3857 (2020)
49. Aggarwal, S., Kumar, N.: Attacks on blockchain. Adv. Comput. **121**, 399–410 (2021)
50. Jing, N., Liu, Q., Sugumaran, V.: A blockchain-based code copyright management system. Inf. Process. Manage. **58**(3), 102518 (2021)
51. Qureshi, A., Megías Jiménez, D.: Blockchain-based multimedia content protection: review and open challenges. Appl. Sci. MDPI **11**(1), 1 (2020)
52. Hammi, B., Idir, Y.M., Zeadally, S., Khatoun, R., Nebhen, J.: Is it really easy to detect sybil attacks in c-its environments: a position paper. IEEE Trans. Intell. Transp. Syst. **23**(10), 18273–18287 (2022)
53. Otte, P., de Vos, M., Pouwelse, J.: Trustchain: a sybil-resistant scalable blockchain. Futur. Gener. Comput. Syst. **107**, 770–780 (2020)
54. Iqbal, M., Matulevičius, R.: Exploring sybil and double-spending risks in blockchain systems. IEEE Access **9**, 76153–76177 (2021)
55. Rajabi, T., et al.: Feasibility analysis for sybil attacks in shard-based permissionless blockchains. Distrib. Ledger Technol. Res. Pract. **2**, 1–21 (2023)
56. Ramezan, G., Leung, C., et al.: A blockchain-based contractual routing protocol for the internet of things using smart contracts. Wirel. Commun. Mob. Comput. **2018** (2018)
57. Wen, Y., Lu, F., Liu, Y., Huang, X.: Attacks and countermeasures on blockchains: a survey from layering perspective. Comput. Netw. **191**, 107978 (2021)
58. Chen, X., Yang, A., Weng, J., Tong, Y., Huang, C., Li, T.: A blockchain-based copyright protection scheme with proactive defense. IEEE Trans. Serv. Comput. (2023)
59. Kongmanee, J., Kijsanayothin, P., Hewett, R.: Securing smart contracts in blockchain. IOn: 2019 34th IEEE/ACM International Conference on Automated Software Engineering Workshop (ASEW), pp. 69–76. IEEE (2019)
60. Zhang, Y., et al.: A decentralized model for spatial data digital rights management. ISPRS Int. J. Geo-Inf. **9**(2), 84 (2020)

A Hybrid Deep Reinforcement Learning Routing Method Under Dynamic and Complex Traffic with Software Defined Networking

Ziyang Zhang(✉), Lin Guan, and Qinggang Meng

Loughborough University, Loughborough, UK
{Z.Zhang,L.Guan,Q.Meng}@lboro.ac.uk

Abstract. Software-defined networking (SDN) routing based on reinforcement learning (RL) is a very promising research topic in recent years, achieving better solutions comparing to traditional network routing based on mathematical models. However, with continuous increase of network complexity and scale, the RL methods show a slow convergence speed and insufficient adaptability to network changes. This leads to the major drawbacks of existing RL algorithms in modern large-scale networks, especially with complexity and dynamics features. Therefore, this paper proposed a novel RL method based on pre-trained data called PRLR, a pre-trained reinforcement learning based SDN routing method, which can effectively improve the QoS of SDN routing and improve the convergence speed of reinforcement learning. The experimental results demonstrate that our proposed PRLR outperforms the benchmarking methods in terms of multiple metrics, such as network delays, bandwidth availability, goodput ratio, as well as the convergence efficiency and works well in dynamic routing topologies.

1 Introduction

Software-Defined Networking (SDN) has been recognized as a dynamic, manageable, cost-effective, and adaptable next-generation networking paradigm. In SDN, the data plane and the control plane are separated, and the network device as the data plane only performs data forwarding, allowing the data plane controller to set forwarding decision rules. Unlike other networks, the control plane understands and controls the routing paths of the entire network. However, in the face of the mutation of the network environment, the routing optimization speed of traditional mathematical methods and commonly used machine learning algorithms (ML) of the software-defined network will slow down. Therefore, proposing a new routing optimization method becomes emergently important to achieve effective quality of service (QoS) in network performance.

L. Barolli (Ed.): AINA 2024, LNDECT 204, pp. 183–192, 2024.
https://doi.org/10.1007/978-3-031-57942-4_19

The major drawback of the current routing protocols is that the main routing algorithms account for a small proportion, and most of them are used to maintain topology and neighbor relationships [1]. As a centralized routing protocol, the SDN routing protocol avoids the time and steps it takes to discover and maintain routing paths. This also means that traditional static routing protocols are not suitable for SDN routing. The use of ML for intelligent data classification and decision making is one of the trends in the web [2]. In this trend, the combination of SDN and ML is becoming an important method of making networks more reliable and secure. Therefore, machine learning can be used for routing algorithms for SDN [3].

Reinforcement learning (RL) is a type of machine learning technique that can be used for intelligent optimization. SDN routing turns the traditional complex routing problem into a relatively simple optimization problem, which is what RL is good at [4]. Today, RL becomes very popular in SDN routing because of its good routing performance in delay, bandwidth and packet loss parts [5].

However, RL-based SDN routing encounters challenges due to its extended duration and numerous iterations needed to achieve optimal performance, often leading to suboptimal outcomes during the learning phase. Existing algorithms typically emphasize post-learning performance, overlooking the necessity to accelerate RL learning. Addressing this, a hybrid strategy is proposed, enhancing RL convergence efficiency while preserving effectiveness. Empirical simulations indicate this method effectively improves RL-based SDN routing by leveraging constrained data.

The original contributions of this paper are summarized as follows: 1) Using constrained reinforcement learning methods to improve SDN routing. 2) Improving the convergence speed of existing RL based SDN routing methods. 3) From the simulation results, the proposed PRLR achieves better network performance (e.g. lower latency, higher bandwidth, lower packet loss and convergence efficiency) than traditional RL methods even if the constrained data did not show the best routing paths. 4) The proposed method PRLR has higher convergence speed comparing to existing mathematical methods and typical RL methods.

The paper is organised as follows: Sect. 2 describes the related work and demonstrates the advantages of the proposed method, Sect. 3 details the proposed constrained RL based SDN routing method, Sect. 4 summarises the results and shows performance validation, and Sect. 5 point out the conclusions and future work.

2 Related Work

Unlike traditional networks, SDN routing is centralized, with all nodes depending on the control plane for directing packet forwarding. This eliminates the need for flooding or similar methods to ascertain the entire network topology [6]. Consequently, the SDN routing algorithm can efficiently analyze and determine the best path for each source-to-destination route.

Caris et al. in [7] proposes a method to improve the performance of SDN routing using Open Shortest Path First (OSPF). In this method, the entire network topology is divided into sub-domains. Between the sub-domain nodes, they run OSPF as the routing protocol, maintaining a best next hop for each node to each destination node [8]. However, as the number of network routers increases, the amount of traffic used to maintain the routing table becomes enormous, which degrades network performance.

Dobrijevic et al. in [9] first used the Ant Colony Optimisation (ACO) algorithm for SDN routing. Gao et al. in [10] proposes a method called CACO-RSP. The method reduces the energy cost of routing path selection calculations. Similar to other ACO approaches, due to the limitations of ACO these methods consume significant computing resources when operating in large SDN topologies. This means that these methods are not suitable for large SDN networks.

Fernández-Fernández et al. in [11] proposes a MODA approach based on the SPEA2 for traffic control. Assefa et al. in [12] has proposed a method called HyMER. This method uses supervised learning methods to eventually output routing path choices when the network state is the input. However, this algorithm is only applicable to static topologies and requires a supervised algorithm to relearn when the network state changes.

Kim et al. in [13] proposes a routing method based on Q-learning which focused on avoid network congestion. Chavula et al. in [14] proposes a method for using Q-learning on network traffic management. This approach allows the network using multipath routing, however, the network latency and jitter also become higher.

Pre-trained RL methods, though popular in machine learning, are less common in SDN routing. The proposed PRLR method involves pre-training with network simulation data to guide the RL algorithm in SDN. This technique boosts QoS by reducing learning time and increasing convergence speed, making the network more adaptable to changes, improving bandwidth, and reducing latency and packet loss.

3 Proposed Method

In the proposed PRLR, pre-trained data from an SDN simulator is utilized to train the reinforcement learning algorithm before deployment in real-world settings. This data, obtained using a greedy algorithm, is designed to ensure accuracy in routing path selection, focusing on parameters like latency, packet loss, or bandwidth, depending on the network's state.

PRLR employs Deep Q-Networks (DQN) as its reinforcement learning method. Being off-policy, DQN can store and learn from past decisions, allowing for reduced learning frequency to decrease network latency. Unlike other methods, in PRLR, the pre-trained data guides the reinforcement learning outcomes, as the greedy method alone may not yield the best routing paths. Over time, the influence of pre-trained data diminishes, ensuring that the reinforcement learning method discovers the most effective results.

In reinforcement learning part of PRLR, the *state* is the node location of the packet, the *action* is the next hop towards the destination, and a *reward* is received when the packet reaches the destination or follows a routing path suggested by the pre-trained data.

The proposed algorithm PRLR is affected by several parameters, which are learning rate α, epsilon reward decay γ and greedy value. α is the learning rate to determine how much of the error is to be learned this time. γ is the attenuation value of the future reward. Epsilon greedy value is the probability of making decisions according to the Q value. Through experience drawn from simulation results, in this method, learning rate α is selected as 0.1, epsilon reward decay γ is selected as 0.8 and epsilon greedy value of DQN is selected as 0.7. Which can make the performance of network routing selection good enough and make sure it works well when the network state changes.

The Bellman equation for reward update in PRLR considers two factors: the discount factor F and pre-trained data. The discount factor, adjustable based on network conditions like delay, bandwidth, and packet loss rate, reflects the quality of each link in the Deep Q networks. Pre-trained data also affects the reinforcement learning reward function, guiding the selection of routing paths.

$$Q(s,a) = Q(s,a) + \alpha(1-t)[r + \gamma * F * max_{a'} Q(s^{'}, a^{'}) - Q(s,a)] + r * t * m \quad (1)$$

Formula (1) presents the Bellman equation used by PRLR to update the Q value. In this equation, *Q(s,a)* denotes the Q value of a potential action a in the current state s. The term s represents the subsequent state following the execution of action a in state s, while $a^{'}$ signifies a potential action in the next state $s^{'}$. The reward value r is 1 when a reward is obtained and 0 otherwise. The variable t, representing learning time, starts at 0.5 and decreases each time a data packet reaches its destination, eventually reaching 0, signifying the diminishing influence of pre-training data over time. The matching value m is 1 when the network state and the next hop align with pre-training data, and 0 in other cases.

Several distinct discount factors are considered, the first being delay. Since routing path selection hinges on the network's overall delay, the delay experienced in forwarding each routing packet is crucial. The delay discount factor can be calculated using the following formula:

$$DF = \frac{Delay_{value}}{Delay_{max}} = \frac{Delay_{max} - Delay(t)}{Delay_{max}} \quad (2)$$

From the Formula (2), DF means delay factor, which can influence γ value of DQN maintenance. $Delay_{max}$ is maximum tolerable delay, which is set as 100ms. $Delay(t)$ is current point to point delay as time t. From this formula, the larger point to point delay is, the smaller delay factor it should be. If the point-to-point delay is larger than maximum tolerable delay, the DF value will become zero to reduce the importance of this routing path.

The second discount factor is about available bandwidth. Bandwidth affects the speed of data packet transmission in the network, thereby affecting the network routing performance. Unlike the delay, the whole bandwidth of the routing path depends on minimum bandwidth of the whole routing path. The discount factor of bandwidth can get by using the formula as follow:

$$BF = \frac{Bandwidth_{value}}{Bandwidth_{max}} = \frac{DataTransferred(t)}{Bandwidth_{max}} \tag{3}$$

From the Formula (3), BF is bandwidth factor, which can influence */gamma* value of DQN maintenance as well as DF. $Bandwidth_{max}$ is maximum delay of the whole routing path, which is set as 100MB. $DataTransferred(t)$ is the data transferred in time t which can represent the bandwidth that has been used.

The third discount factor, packet loss, reflects the reliability of each network routing step. It's measured as the percentage of packets that fail to be forwarded. We use 'good put' as a discount factor, with the delay discount factor derived using the following formula:

$$PF = \frac{Packet_{received}}{Packet_{sent}} = \frac{Packet_{sent} - Packet_{lost}}{Packet_{sent}} \tag{4}$$

Formula (4) shows PF as the delay factor affecting γ in DQN maintenance. $Packet_{sent}$ tracks packets sent, updating the loss rate, while $Packet_{received}$ counts received packets. This data helps the controller monitor and refresh the loss rate every 100 packets. Routes with over 10% loss are bypassed if alternatives exist, with the rate resetting after timeouts or network changes. Links failing five times consecutively are marked unachievable.

In order to test the performance of the proposed method, the network topology needs to be established on the simulator. The topology is like follows (Fig. 1):

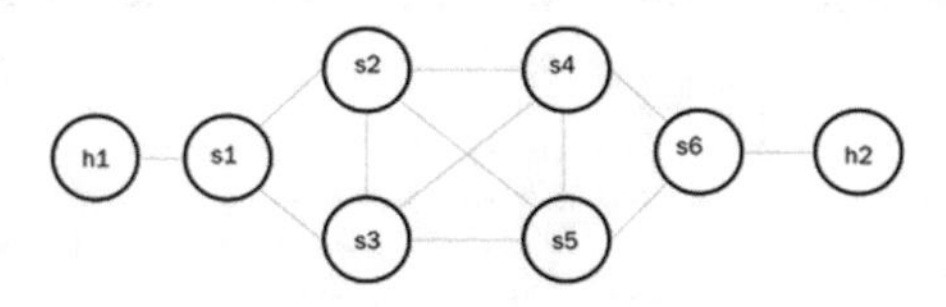

Fig. 1. SDN topology

This paper demonstrates that the proposed method maintains satisfactory network performance even when the greedy algorithm offers insufficient guidance or the network state changes. To illustrate this capability, two different topologies for each discount factor are constructed. The first set of topologies, deemed basic, are those in which the greedy algorithm achieves optimal network performance. The second set comprises topologies where the greedy algorithm fails to identify the best routing path. This setup is used to demonstrate that the proposed method can still perform adequately, even when the pre-training data is less than ideal.

4 Simulation Results and Performance Evaluations

4.1 Comparison Data

This paper demonstrates the benefits of the proposed method through three distinct simulation types, each highlighting a discount factor that impacts routing paths. For each factor, two different topologies are used. The first topology assesses whether the proposed method converges faster than traditional SDN with effective guidance. The second examines if the method can still achieve satisfactory network performance under less effective guidance. To evaluate the performance of the proposed method, comparison data are presented as a control group. This includes the network performance of the greedy algorithm and traditional DQN, serving as a benchmark to establish that the proposed method not only performs well but also converges more rapidly than conventional SDN approaches.

4.2 Routing Performance of Delay

This part comprises two simulations, illustrated in Fig. 2, which shows how the proposed method, with the greedy algorithm's effective guidance, attains optimal routing faster than traditional reinforcement learning methods.

Figure 2 plots pings number from node h1 to h2 against the associated delay. To elucidate the learning process, both DQN and the proposed method update their Q values simultaneously. In each, a delay over 100ms classifies a point-to-point link as unreachable. From the result, the first two time of DQN method did not have data, because it used too much time to find the next hop. While DQN requires eight learning cycles to identify the optimal path, the proposed method, assisted by the greedy algorithm, reaches peak performance in just three attempts, outperforming the conventional DQN routing approach.

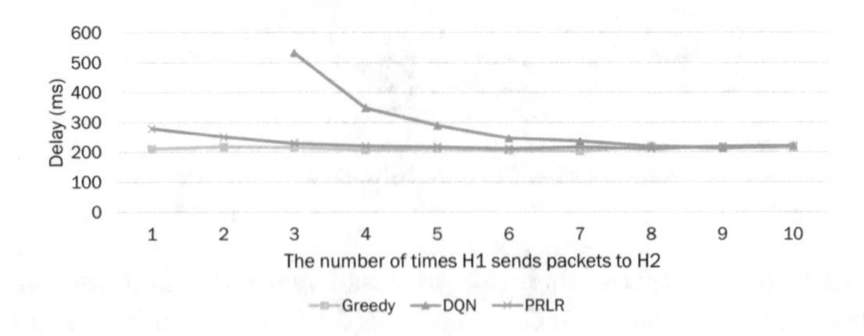

Fig. 2. Delay between different routing methods with best guidance

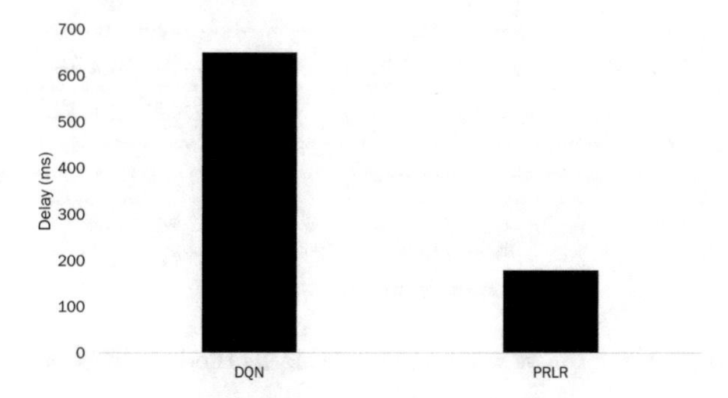

Fig. 3. Time to reach the best routing performance

Figure 3 reveals that while DQN takes over six hundred milliseconds to find the best routing path, the proposed method only needs about two hundred milliseconds, demonstrating significantly faster achievement of optimal routing performance.

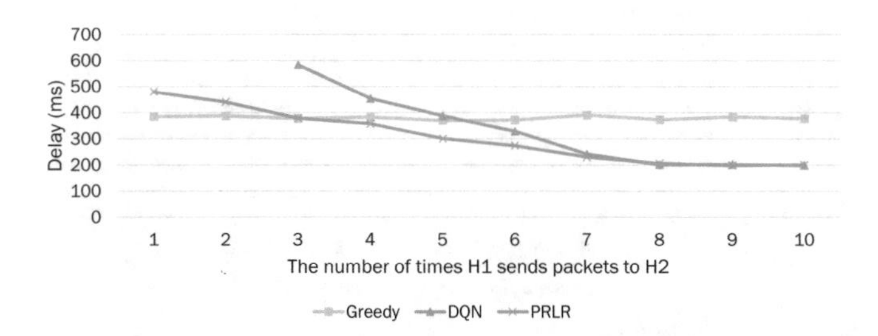

Fig. 4. Delay between different routing methods without best guidance

Figure 4 demonstrates that the proposed method swiftly matches and then surpasses the greedy algorithm's performance, achieving the best routing results after a few learning steps. It outperforms the greedy algorithm more quickly and, unlike DQN's random initial hop selection, shows superior early performance. This aligns with the initial delays observed with DQN in Fig. 2.

4.3 Routing Performance of Bandwidth

This section explores two scenarios. The first assesses whether our method converges more quickly than conventional SDN routing methods under effective mathematical guidance. Similar to the previous delay experiment, the greedy algorithm achieves optimal routing, as shown in Fig. 5. In this figure, which presents the number of pings from node h1 to h2 and the available bandwidth, the greedy algorithm, used as the training method, demonstrates superior routing performance. Our method outperforms traditional reinforcement learning methods in speed to optimal performance, especially with effective guidance.

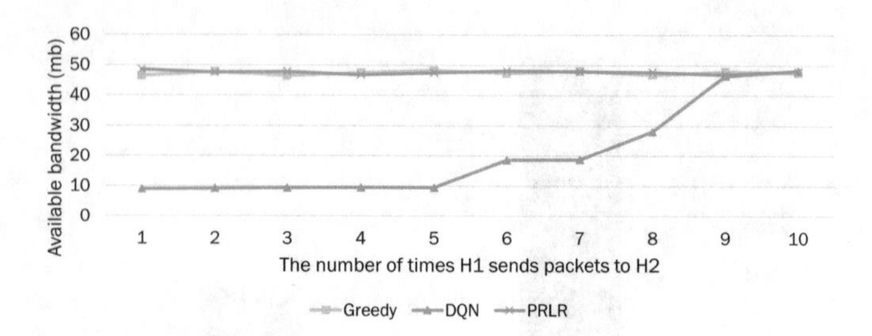

Fig. 5. Bandwidth between different routing methods with best guidance

Figure 6 shows that while the mathematical approach does not always pinpoint the best routing path, both the proposed method and traditional reinforcement learning eventually reach optimal performance. Despite less effective guidance, the proposed approach attains peak performance more swiftly than standard reinforcement learning methods.

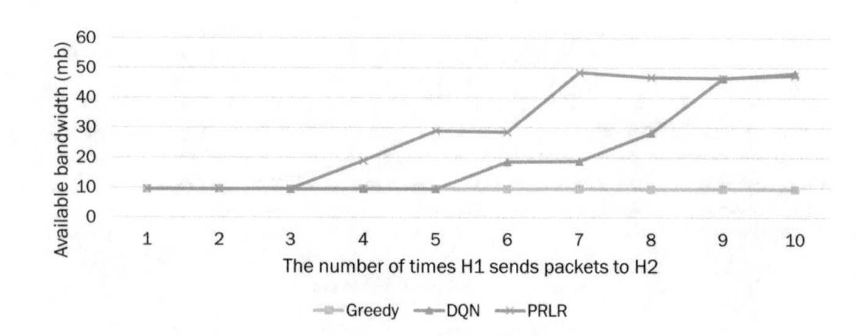

Fig. 6. Bandwidth between different routing methods without best guidance

4.4 Routing Performance of Packet Loss Rate

Figure 7 presents a comparison where the horizontal axis indicates how often node h1 pings h2, and the vertical axis shows the routing path's goodput. Here, the greedy method leads to the best network routing performance, helping the proposed method achieve optimal results quickly, unlike the slower traditional reinforcement learning methods.

The second part of the experiment, shown in Fig. 8, tests the proposed method with less effective mathematical guidance. In this scenario, both the proposed method and traditional reinforcement learning methods eventually reach peak routing performance, but the proposed method does so more rapidly, demonstrating its capacity to surpass typical reinforcement learning approaches even under suboptimal guidance.

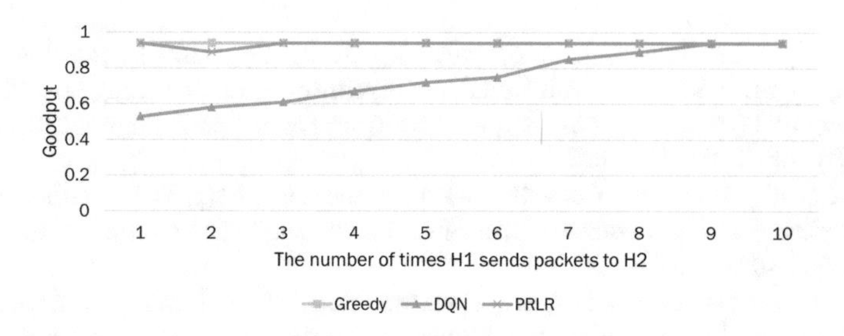

Fig. 7. Goodput between different routing methods with best guidance

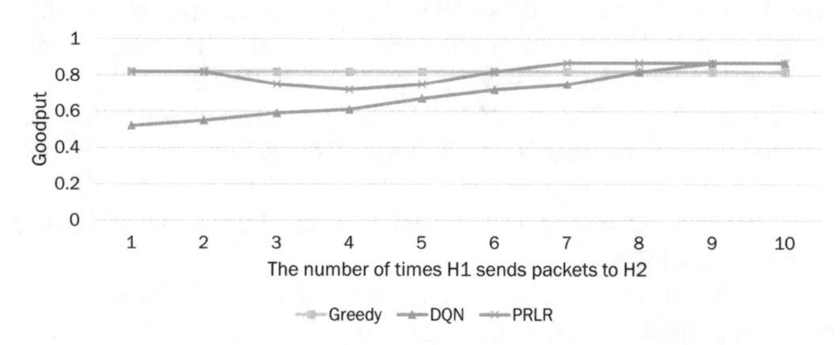

Fig. 8. Goodput between different routing methods without best guidance

5 Conclusions and Future Work

In this paper, a novel RL based hybrid SDN routing algorithm named as PRLR has been proposed and implemented. This algorithm is based on a pre-trained reinforcement learning routing method, and it overcomes the slow convergence drawbacks of traditional RL method. This hybrid method runs a greedy algorithm in simulation to get pre-trained data, then use pre-trained data to guide the reinforcement learning algorithms. From empirical simulation results, the proposed PRLR improves the network performance without absolutely correct guidance of pre-trained data and works well in dynamic routing topologies.

In the future work, the proposed algorithm will be further developed to deal with large-scale network environments. More complex mathematical algorithms and deep reinforcement learning algorithms in larger-scale network environments can be used to further demonstrate the advantages of the proposed pre-training based deep reinforcement learning algorithm.

References

1. Moy, J.: "RFC2328: OSPF Version 2." (1998)
2. Fadlullah, Z.Md., et al.: State-of-the-art deep learning: evolving machine intelligence toward tomorrow's intelligent network traffic control systems. IEEE Commun. Surv. Tutor. **19**(4), 2432–2455 (2017)

3. Swain, P., Kamalia, U., Bhandarkar, R., Modi, T.: CoDRL: intelligent packet routing in SDN using convolutional deep reinforcement learning. In: 2019 IEEE International Conference on Advanced Networks and Telecommunications Systems (ANTS), pp. 1–6. IEEE (2019)
4. Xie, J., et al.: A survey of machine learning techniques applied to software defined networking (SDN): research issues and challenges. IEEE Commun. Surv. Tutor. **21**(1), 393–430 (2018)
5. Amin, R., Rojas, E., Aqdus, A., Ramzan, S., Casillas-Perez, D., Arco, J.M.: A survey on machine learning techniques for routing optimization in SDN. IEEE Access (2021)
6. Zhang, H., Yan, J.: Performance of SDN routing in comparison with legacy routing protocols. In: 2015 International Conference on Cyber-Enabled Distributed Computing and Knowledge Discovery, pp. 491–494. IEEE (2015)
7. Caria, M., Jukan, A., Hoffmann, M.: SDN partitioning: a centralized control plane for distributed routing protocols. IEEE Trans. Netw. Serv. Manage. **13**(3), 381–393 (2016)
8. Moy, J.T.: OSPF: Anatomy of an Internet Routing Protocol. Addison-Wesley Professional, Reading (1998)
9. Dobrijevic, O., Santl, M., Matijasevic, M.: Ant colony optimization for QoE-centric flow routing in software-defined networks. In: 2015 11th International Conference on Network and Service Management (CNSM), pp. 274–278. IEEE (2015)
10. Gao, Y., Cheng, L., Sang, L., Yang, D.: Spectrum sharing for LTE and WiFi coexistence using decision tree and game theory. In: 2016 IEEE Wireless Communications and Networking Conference, pp. 1–6. IEEE (2016)
11. Fernández-Fernández, A., Cervelló-Pastor, C., Ochoa-Aday, L.: A multi-objective routing strategy for QoS and energy awareness in software-defined networks. IEEE Commun. Lett. **21**(11), 2416–2419 (2017)
12. Assefa, B.G., Ozkasap, O.: HyMER: a hybrid machine learning framework for energy efficient routing in SDN. arXiv preprint arXiv:1909.08074 (2019)
13. Kim, S., Son, J., Talukder, A., Hong, C.S.: Congestion prevention mechanism based on Q-leaning for efficient routing in SDN. In: 2016 International Conference on Information Networking (ICOIN), pp. 124–128. IEEE (2016)
14. Chavula, J., Densmore, M., Suleman, H.: Using SDN and reinforcement learning for traffic engineering in UbuntuNet Alliance. In: 2016 International Conference on Advances in Computing and Communication Engineering (ICACCE), pp. 349–355. IEEE (2016)

Improving ML/DL Solutions for Anomaly Detection in IoT Environments

Nouredine Tamani(✉), Saad El-Jaouhari, Abdul-Qadir Khan, and Bastien Pauchet

Institut Supérieur d'Electronique de Paris (Isep), Issy-les-Moulineaux, France
{nouredine.tamani,saad.el-jaouhari}@isep.fr,
abdul-qadir.khan@ext.isep.fr, bastien.pauchet@orange.fr

Abstract. As part of the evolution toward an era of Web 3.0, the Internet of Things (IoT) bridges physical smart devices to digital world to enhance services for consumer convenience. However, the rapid increase of IoT devices led also to the inheritance of security, privacy, and trust problems, already well-known in traditional networks, making IoT devices even more vulnerable. To be able to detect anomalies and protect such IoT devices from cyberattacks, different techniques have been proposed in the literature using diverse approaches going from the logic-based (knowledge bases and ontologies) ones to the statistical ones (Machine Learning-ML/Deep Learning-DL). In this paper, we focus on the later approaches (ML/DL) to identify, reproduce, evaluate, and compare different state-of-the-art machine learning algorithms for anomaly detection in IoT environments, along with the main datasets used in such research works. Once suitable ML models and datasets are identified, we investigated the potential for enhancing them by incorporating a feature selection algorithm. This aims to reduce the dataset's dimensionality while concurrently improving performance metrics such as accuracy, precision, recall, and F1-score.

Keywords: Anomaly detection · IoT networks · Security and Privacy · Machine/ Deep Learning · Model Reproducibility

1 Introduction

Internet of Things (IoT) is a domain mentioned for the first time in 1999 by Kevin Ashton, when he produced the idea of a radio-frequency identification chip (RFID) to track products in the supply chain [20]. IoT networks consist of many devices capable of data collection, storage, processing, and communication [11]. IoT has evolved significantly since then, thanks to the proliferation of connected objects, and the harvesting and processing of large volumes of data with Big Data and AI techniques. Nevertheless, the infrastructures, applications, and services associated with IoT devices introduced several threats and vulnerabilities as emerging protocols and workflows have exponentially increased attack surfaces [20]. Securing IoT devices is challenging for both academics and industries

L. Barolli (Ed.): AINA 2024, LNDECT 204, pp. 193–206, 2024.
https://doi.org/10.1007/978-3-031-57942-4_20

because of the heterogeneity of IoT environment. Besides, conventional security controls are not suitable for all the IoT devices, in particular for the most constrained ones. Since the distributed IoT networks are outside of the coverage of security perimeters, the existing solutions relying on the cloud also suffer from centralization and high delay. Furthermore, IoT device vendors commonly overlook security requirements in their design processes due to their rush-to-market proclivity. Moreover, the lack of security standards has contributed to the complexity of securing IoT devices. These challenges and the nature of IoT applications call for a monitoring system such as anomaly detection at device, edge and network levels beyond the organizational boundary [20].

The research community has already developed solid statistical and machine learning methods to detect anomalies inside IoT data, with real-time analysis, and prediction of unusual behaviors in IoT environments [13]. In this context, an anomaly in IoT data can be defined as a data point or a subset of data points that deviate from the normal patterns. These anomalies can be classified into 3 types, namely (i) **Point anomaly**, when it refers to a precise point/instant; (ii) **Contextual anomaly**, when an outlier is compared to other points according to a context and a time window; and (iii) **Collective anomaly**, when some points are individually normal, but as a group, they are suspicious.

This paper places particular emphasis on ML/DL approaches for anomaly detection in IoT data. First, we conducted a comprehensive review of recent works addressing anomaly detection in IoT environments. This analysis aimed to identify the main models along with their corresponding training datasets. We noticed that most of the research efforts in this area have been dedicated to point anomaly detection. Second, from the identified approaches and datasets, we selected 5 publicly available datasets and reproduced the identified models to assess their performance. Then, we selected the model demonstrating the highest performance, and further enhanced it through the implementation of a feature selection approach. The objective is to reduce the dimensionality of the dataset and the training time while increasing the performances. We showed that Random Forest-based model trained on TON-IoT dataset with only the 9 best features has a better Accuracy, Precision, Recall and F1-score than the model trained on the original dataset (with 45 features).

The rest of the paper is organized as follows. Section 2 analyzes the ML approaches developed in the literature for anomaly detection in IoT environments. Section 3 describes the existing datasets and trained ML/DL models for anomaly detection in IoT data. In Sect. 4 details our experiments with the selected datasets to reproduce and improve the ML/DL models for anomaly detection. Section 5 concludes the paper with some comments and future work.

2 Approaches for Anomaly Detection in IoT Data

We focused in this section on the recent papers published in the literature between 2019 and 2023 (summarized in Subsect. 2.1). We split them into 2 distinct types, namely: papers where a survey was conducted and papers where the authors compared amongst ML/DL models (summarized in Subsect. 2.2).

2.1 Surveys in Anomaly Detection in IoT

The authors in [10] focused on DL approaches such as Multilayer Perceptron (MLP) and Graph Neural Network (GNN) coupled with encoders or Recurrent Neural Networks (RNN). They also listed the recent available datasets that are used to train the models to detect attacks on IoT environments. In [3], the authors surveyed the existing literature on ML/DL methods used to detect cybersecurity attacks in IoT environments. The survey was conducted using the PRISMA method, wherein eighty studies from 2016 to 2021 were carefully selected and evaluated, along with the datasets available for IoT systems.

In [11,27], the authors provided an in-depth overview of the existing works in developing anomaly detection solutions using ML/DL techniques for protecting IoT systems. They focused on algorithms and datasets available in the field, but no model comparison has been performed. Similarly, in [5], the authors carried out a large overview of ML and DL techniques developed up to 2021, and the nature of data for IoT systems, identified types of anomalies, datasets, and the evaluation metrics to measure the performances of the trained models, as well. In [13], the authors studied anomaly detection in IoT environments, where huge amount of data can be collected, processed, and analysed to reduce risks, detect and prevent malicious activities, and avoid involuntary downtime. The study covered the period between 2000 and 2018 and fields of smart environments, transports, health, smart objects, and industrial systems. The literature review of the above articles is summarized in Table 1.

Table 1. Summary of surveys of ML/DL methods for anomaly detection in IoT.

Paper	Models	Dataset(s)
[11]	RF (Random Forest), DL, RL, LSTM, CNN (Convolutional Neural Network), GNN, Multiple, AE-ANN, AE-SNN, Ensemble, AE, Subspace, Self-Learning, TCN, AE-LSTM, DBN, DNN	N-BaIoT, CICIDS 2017, AWID, UNSW-NB15, NSL-KDD, Kyoto, KDD CUP 1999
[5]	C_LOF, AutoCloud, TEDA Clustering, BDLM & RBPF, HTM, ANN (Artificial Neural Networks), MDADM, Multi-kernel, xStream, Regression Model, SVM, CEDAS, MuDi-Stream, Extreme Learning Machines, AMAD, LSTM, Auto-encoder, DNN, Evolving spiking NN, ISTL, e-SREBOM	Space imager data stream, KDD29, Cover type, Spam-SMS, Spam-URL, KDDCup99, NAB, UCI, D1, DS2, DS3, Yahoo Webscope, HTTP, SMTP, SMTP+HTTP, COVERTYPE, SHUTTLE, Weather, Web traffic, Avocado, Temperature, and UCSD Pedestrian datasets
[10]	GDN, Gumbel softmax sampling strategy, OmniAnomaly, MLP, LR (Linear Regression), SVM, DT, RF, SS-TCVN, auto-encoder, LSTM, Gelenbe Network, Gaussian, HOT-SAX, GTA, SS-VTCN, CNN	Swat, WaDI, MSL, SMAP, DS2OS, BaIoT
[13]	SVM, PCA, Kernel nonLinear Regression, CNN, RNN, DT, Auto-encoders	no dataset mentioned
[27]	CNN, RED, DNN, Hybrid anomaly detection, clusters, IRESE unsupervised	No dataset
[3]	Naïve Bayes, BayesNet, DT, RF, SMV, SVR, KNN, FPT, Fuzzy C-Means Algorithm, XGBoost, LR, K-Means clustering, CDL, RNN, CNN, Deep Auto-encoders, DNN, DBN, MLPNN	BoT-IoT, AWID, MQTT regular traffic packets, KDD99, Vx-Heaven, Kaggle and Ransomware, NSL-KDD, ICS cyberattack dataset, IoT Traffic, UNSW-NB15, CICIDS2017, ISCX, UGR16

2.2 Comparing ML/DL Models for Anomaly Detection in IoT

The authors discussed in [17], the way IoTs collect data about their surrounding environments, and performed a comparative study of various ML/DL approaches for attack and anomaly detection, and concluded that Random Forest gives the best performances in terms of accuracy and precision. In [9], the authors trained and compared 10 models on the TON-IoT dataset: Bidirectional Gated Recurrent Unit Recurrent Neural Network (B-GRU-RNN), Bidirectional Long Short-Term Memory Recurrent Neural Network (B-LSTM-RNN), Random Forest (RF), Gradient Boosted Trees (GBT), K Nearest Neighbours (KNN), Deep Neural Network (DNN), eXtreme Gradient Boosting (XGB), MLP, Support Vector Machine (SVM), Naive Bayes (NB). They compared the efficiency of each model using 4 metrics (Accuracy, Precision, Recall, F1-score), and they identified B-GRU-RNN as the best model. In [24], the authors have trained a Two-tier Classification (TDTC) model combined with a Two-layer Dimension reduction, and compared it with Two-tier classification [23], NB, RF, SVM, and Decision Tree (DT) on the NSL-KDD dataset [31]. They compared the efficiency of each model using the Detection rate and showed that the best model still to be TDTC. The model is not only capable of detecting attacks but to distinguish the type of attacks as well. In [22], the authors introduced a ML-based approach for modeling IoT service behaviors by observing their communication patterns. The training process was performed on distributed nodes within multiple IoT sites, and the resulting models are combined together to produce a global model among different IoT sites. The authors showed that the combined model has a better anomaly detection rate than the local models.

In [29], the authors introduced an outlier detection procedure using the K-means algorithm coupled with Big Data techniques, to make the process scalable. The model was trained on Guildford's facility dataset, proposed within the framework of the European Smart Santander Project. In [18], the authors have considered 5 ML algorithms: Logistic Regression (LR), SVM, DT, Random Forest, ANN, to train models on DS2OS dataset[1] and to evaluate their performances by using 5 metrics (accuracy, precision, recall, F1 score, and area under the Receiver Operating Characteristic Curve). The authors showed that Random Forest-based ML model outperforms the other models on the used dataset, but they pointed out the need for a new robust algorithm for anomaly detection. In [26], the authors trained several ML/DL models for anomaly detection in IoT environments on NB-IoT EDGE DEVICE dataset. The comparison among the models has been carried out based on 3 metrics: Precision, Recall, F1-score. The auto-encoders found to be a better choice than ML for anomaly detection, when the detection is on the edge. In [33], the authors proposed a Convolution Neural Network (CNN) to detect and classify anomalies in IoT Networks using dimensionality 1D, 2D and 3D. They used transfer learning to do binary classification and multi-class classification and they trained their models on a combined dataset made of BoT-IoT, IoT Network Intrusion, MQTT-IoT-IDS2020

[1] https://www.kaggle.com/datasets/francoisxa/ds2ostraffictraces.

[19], and IoT-23 [15]. They compared each model with 4 metrics (Accuracy, Precision, Recall, F1-score), and they concluded that CNND1 performs better than CNN2D and CNN3D. In [12], the authors showed that DL models are better at handling the small variants due to their high-level feature extraction capabilities. The authors trained a DL model and a shallow model on NSL-KDD[2] dataset. They compared both models based on 6 metrics: Detection Rate (DR), False Alarm Rate (FAR), Accuracy, Precision, Recall, F1-score. The results showed that the DL model outperforms the shallow one for detecting distributed attacks.

In [1], the authors used a simulated IoT network to show that feature selection can help increase the accuracy of DDoS attack detection in IoT network traffic. The authors considered a variety of ML algorithms: KNN, LSVM, DT, RF, NN, and trained them on simulated data. The comparison among the models was based on the regular metrics of Accuracy, Precision, Recall, F1-score. They concluded that RF-based model outperforms the other models in both classification of the legitimate activities and DDOS attacks. In [32], the authors developed an intelligent intrusion-detection system tailored to IoT environments using a DL algorithm to detect malicious traffic inside such environments. They evaluated the models using both real-network traffic traces, and simulated data. They designed a DL model (DL-Sim) and compared it with existing IDS (Intrusion Detection Systems) solutions (IWC) using 3 metrics (Precision, Recall, F1-score). DL-SIM model outperforms the existing solutions. In [34], the authors designed a DL model to detect anomalies in Multivariate Time Series Data, which is resistant to noise. The model is MSCRED (Multi Scale Convolutional Recurrent Encoder Decoder) that correlates the inter-sensor data and uses an attention based Convolutional Long-Short Term Memory (ConvLSTM) network to detect patterns. The performance of the trained model is good but the training data used are synthetic. In [6], the authors proposed a clustering method to detect anomalies in Big Data. It is an improved optimization approach where a weight is assigned to each data point. The approach has been compared with K-means algorithm applied on Australian credit approval dataset [25], Heart dataset [30] and NSL-KDD [4]. The comparison has been performed by using 6 metrics: Purity, Mirkin, F-measure, Variation of Information (VI), Partition Coefficient (PC), V-measure. They showed that the clustering method detects anomalous values more accurately than K-means. The literature review of the above articles is summarized in Table 2.

[2] https://www.kaggle.com/datasets/hassan06/nslkdd?select=kddtest.

Table 2. State of the art summary of ML methods for anomaly detection in IoT.

Paper	Models	Best model	Dataset(s)
[9]	B-GRU-RNN, B-LSTM-RNN, RF, GBT, KNN, DNN, XGB, MLP, SVM, NB	B-GRU-RNN (Acc: 98.62%, P: 99.68%, R: 98.20%, F1: 98.93%)	TON-IoT
[24]	TDTC, Two-tier classification, Naïves Bayes, RF, SVM, DT	TDTC (Acc: 84.86%)	NSL-KDD
[22]	Mdeling IoT communicative behavior by observing traffic	Proposed model	2 datasets: Australian credit approval and Heart datasets
[29]	Outlier detection algorithm with Big Data processing	Outlier detection algorithm (AUC: 0.8967)	Guildford's facility (European Smart Santander Project)
[17]	LR, SVM, RF, Naïve Bayes, DT, CNN, MLP, GNB, RNN, GRU, LSTM, AdaB, KNN, DNN, XGBoost, ID3, QDA	RF	UCI ML, IoT-23, BoT-IoT, NSL-KDD, DS2OS, CICIDS-2017, UNSW-NB15, ICS Cyberattack, IoT Network Intrusion dataset, KDDCUP99
[18]	LR, SVM, DT, RF, ANN	RF (Acc: 99.4%, F1: 99%)	DS2OS traffic traces
[26]	ADM-EDGE, ADM-FOG, SVM, ABOD, KNN, PCA, HBOD	ADM-EDGE (P: 70.5%, R: 69%, F1: 60.7%)	NB-IoT EDGE DEVICE
[33]	CNN, CNN1D, CNN2D, CNN3D, C-LSTM-AE, C-CMU, FFN, SNN	CNN1D (Acc: 99.97%, P: 99.95%, R: 99.95%, F1: 99.95%)	BoT-IoT, IoT Network Intrusion, MQTT-IoT-IDS2020, IoT-23, IoT-DS-1/-2
[12]	Deep model, Shallow model	Deep model	NSL-KDD
[2]	GAAOD to approximate KNN	-	TAO, Stock, HPC
[34]	MSCRED	MSCRED	Synthetic Data, Power Plant Data
[32]	DNN, DL-Sim, IWC, DL-Testbed	DL-Sim (P: 96.88%, R: 98.02%, F1: 97.46%)	Synthetic data about a smart house. Around 60.000 data
[1]	KNN, LSVM, DT, RF, NN	RF (Acc: 99.9%, P-normal: 99.9%, P-attack: 99.9%, R-normal: 99.8%, R-attack: 99.9%, F1-normal: 99.8%, F1-attack: 99.9%)	synthetic data about IoT devices with normal activities and DoS attacks

3 Datasets and ML Approaches for Anomaly Detection

From the state of the art, we have identified the following ML/DL algorithms used for anomaly detection in IoT environments: Support Vector Machine (SVM), Random Forest (RF), K Nearest Neighbors (KNN), Basic gated Recurrent Unit- Recurrent Neural Network (B-GRU-RNN), Logistic Regression (LR), and Convolutional Neural Network 1 Dimension (CNN1D). The performances of their corresponding ML/DL models, trained on diverse datasets, in terms of Accuracy, Precision, Recall, F1-score are listed in Table 3. These results have been extracted from the state of the art, studied in the previous section.

We have also identified the publicly accessible datasets for training ML models. Their characteristics, in terms of dimensions, distribution and description, are detailed in Table 4.

4 Model Reproduction and Improvement

In this section, we detail the process of reproducing the results of the selected ML algorithms. We have carried out our training/testing processes on a computer with the following properties: Memory of 16 GB, Processor Intel(R) Core (TM) i5-7200U CPU @ 2.50 GHz 2.71 GHz, and Windows 10 Enterprise 22H2 Operating System. For the software, we used Python 3.11.3 on Visual Studio Code, with Pandas, Keras, Tensorflow and Scikit-learn, NumPy, Time, Datetime, IP address, ipynb, and os libraries.

Table 3. Ml/DL Model performances from the state of the art.

Paper	Model	Dataset	Accuracy	Precision	Recall	F1-score
[18]	SVM	DS2OS	98.2%	98%	98%	98%
[9]	SVM	TON-IoT	72.34%	82.91%	72.40%	77.30%
[14]	SVM	UR Fall Detection	98.39%	-	-	98.8%
[18]	RF	DS2OS	99.4%	99%	99%	99%
[9]	RF	TON-IoT	96.30%	96.36%	98.01%	97.18%
[28]	RF	UNSW-NB15	98.2%	98%	98%	98%
[9]	KNN	TON-IoT	95.79%	96.19%	97.38%	96.78%
[1]	KNN	Simulated DoS attacks	99.9%	99.8%	99.3%	99.5%
[14]	KNN	UR Fall Detection	98.79%	-	-	99.1%
[28]	KNN	UNSW-NB15	96%	96%	96%	96%
[9]	B-GRU-RNN	TON-IoT	98.62%	99.68%	98.20%	98.93%
[18]	LR	DS2OS	98.3%	98%	98%	98%
[33]	CNN1D	BoT-IoT (old TON-IoT)	99.97%	99.95%	99.95%	99.95%

Table 4. Dataset information summary.

Dataset	Dimensions	Data distribution	Description
NSL [3,6,11,12,17,24]	25192 × 43	13449 normal/11743 anomalies	Records of internet traffic seen by simple intrusion detection systems (IDS)
TON-IoT[a] [3,9,17,33]	461043 × 45	300000 normal/161043 anomalies	Heterogeneous data sources: IoT and IIoT (Industrial IoT) sensors, Operating systems logs (Windows 7 and 10, Ubuntu 14 and 18 TLS), and Network traffic
UNSW-NB15[b] [3,11,17]	82332 × 45	37000 normal/45332 anomalies	Hybrid of normal activities and synthetic attacks
DS2OS [10,17,18]	357952 × 13	347935 normal/10017 anomalies	Traces captured in the IoT environment DS2OS
IoT-23[c] [17,33]	8186879 × 23	497177 normal/7689702 anomalies	IoT Network traffic in Stratosphere Laboratory, AIC group, FEL, CTU (Czech Technical University)

[a]https://research.unsw.edu.au/projects/toniot-datasets
[b]https://research.unsw.edu.au/projects/unsw-nb15-dataset
[c]https://www.stratosphereips.org/datasets-iot23

4.1 Data Preprocessing

Because of dataset format, some features required a conversion step such as string values, which have been converted into integer values by using ASCII conversion, and True or False data were replaced by 1 and 0, respectively. Timestamp feature, which indicates the time when the data was collected, was broken down into 6 features: Year, Month, Day, Hour, Minute, and Second. Some other features, such as data identifier, have been discarded from the considered datasets since they are not relevant for anomaly detection. The size of datasets and the modifications performed on them are listed in Table 5.

Table 5. Dataset information summary before and after modifications.

Dataset & Dimension	New Dimensions	Modifications	Distribution
NSL-KDD 25192 × 43	25192 × 43	• 3 features converted	13449 normal/11743 anomalies
TON-IoT 461043 × 45	461043 × 43	• 20 features converted • Timestamp transformed • 8 features dropped	300000 normal/161043 anomalies
UNSW-NB1582332 × 45	82332 × 43	• 3 features converted • 2 features dropped	37000 normal/45332 anomalies
DS2OS357952 × 13	357952 × 18	• 10 features converted • Timestamp transformed • 1 feature dropped	347935 normal/10017 anomalies
IoT-238186879 × 23	8186879 × 27	• 11 features converted • Timestamp transformed • 2 features dropped	497177 normal/7689702 anomalies

4.2 Model Reproduction and Comparative Analysis

The comparison is performed by using 4 metrics: Accuracy, Precision, Recall, F1-score. The performances of each reproduced model are listed in Table 6 for both DL and ML approaches, where results in bold font represent the best trained models with regards the datasets used to train them. The results of the model comparison are relatively close to the results found in the respective papers. When comparing the models, we can see that Random Forest-based model outperforms the models when using TON-IoT dataset.

4.3 Random Forest-Based Model Improvement

Upon comparing our reproduced Random Forest-based model with the existing models based on Random Forest algorithm trained on TON-IoT dataset (as summarized in Table 7), it turns out that there is still room for improvement. To do so, one feasible way is to modify the dataset's dimensionality by working on its features with a feature selection approach.

Table 6. Reproduced Experimental results for ML/DL approaches.

Model	Dataset	Accuracy	Precision	Recall	F1-score
SVM	NSL-KDD	92.70%	91.99%	95.14%	93.54%
	TON-IoT	65.3060%	100%	65.3049%	79.0114%
	UNSW-NB15	55.8329%	2.3910%	66.6667%	4.6164%
	DS2OS	**97.1700%**	**100%**	**97.1700%**	**98.5647%**
	IoT-23	93.9236%	0.0010%	100%	0.0020%
Random Forest	NSL-KDD	98.3774%	99.7415%	99.3745%	99.5577%
	TON-IoT	**99.9972%**	**100%**	**99.9983%**	**99.9992%**
	UNSW-NB15	92.7240%	97.8987%	97.0749%	97.4851%
	DS2OS	99.9720%	99.9986%	100%	99.9993%
	IoT-23	99.9967%	99.9990%	99.9960%	99.9975%
KNN	NSL-KDD	98.7299%	98.4525%	99.1834%	98.8166%
	TON-IoT	**99.8449%**	**99.8583%**	**99.9033%**	**99.8808%**
	UNSW-NB15	82.9720%	84.1616%	79.3322%	81.6756%
	DS2OS	99.2541%	99.6132%	99.6189%	99.6161%
	IoT-23	93.8829%	41.5585%	49.0994%	45.0154%
Linear Regression	NSL-KDD	97.0034%	96.8000%	97.4310%	97.1145%
	TON-IoT	67.0292%	98.9666%	66.5971%	79.6176%
	UNSW-NB15	55.6628%	2.2349%	66.5323%	4.3245%
	DS2OS	**97.5178%**	**99.8231%**	**97.6727%**	**98.7362%**
	IoT-23	94.0720%	2.6400%	100%	5.1442%
B-GRU-RNN	NSL-KDD	99.4840%	99.5228%	99.5228%	99.5228%
	TON-IoT	99.8666%	99.9315%	99.8630%	99.8972%
	UNSW-NB15	95.7916%	95.0571%	95.7057%	95.3803%
	DS2OS	**99.9972%**	**99.9971%**	**100%**	**99.9986%**
	IoT-23	99.9155%	98.6767%	99.9350%	99.3019%
CNN1D	NSL-KDD	98.0353%	97.4141%	98.9122%	98.1575%
	TON-IoT	65.0132%	100%	65.0090%	78.7945%
	UNSW-NB15	44.6955%	100%	44.6955%	61.7787%
	DS2OS	**98.4300%**	**100%**	**98.4108%**	**99.1990%**
	IoT-23	6.0389%	100%	6.0389%	11.3900%

Table 7. Efficiency of multiple RF models using TON-IoT.

Dataset	Accuracy	Precision	Recall	F1-score
Our results	99.9972%	100%	99.9983%	99.9992%
IoT and IIoT Networks [9]	96.30%	96.36%	98.01%	97.18%
Network TON-IoT datasets [21]	99.9998%	n/a	n/a	n/a
TON-IoT Telemetry Dataset[7]	85%	87%	85%	85%
TON-IoT: Intrusion Data Sets [8]	98.075%	n/a	n/a	97.264%

Feature Selection: It is possible to improve the overall efficiency of the model by selecting only the best features of the dataset to train the model. To do so, we have used *SelectKBest* from Sci-kit-learn library. Once the features are ranked from the most important to the least important one, we progressively train the model in different cycles, by adding in each cycle one more feature in the ranked order. In each cycle, we compute the accuracy of the model. If the current accuracy is less than or equal the one of the previous cycle, we discard the current model and we return the previous one. Otherwise, we add the next best feature (if it remains) in the ranked order, and we proceed with a new cycle of training. This technique is based on the algorithm used in [16].

Obtained Results: Table 8 summarizes the obtained results when we removed 31, 33 and 34 features respectively. With Random Forest-based model, when using TON-IoT dataset along with features selection, the accuracy, precision, recall and F1-score increase. It seems that many features from TON-IoT dataset are not relevant for anomaly detection.

Table 8. Results of features selection applied to RF model with TON-IoT dataset.

Modifications	Accuracy	Precision	Recall	textbfF1-score
No modification	99.9972%	100%	99.9983%	99.9992%
Removing 31 features (12 left)	99.9976%	100%	99.9983%	99.9992%
Removing 33 features (10 left)	99.9978%	100%	99.9983%	99.9992%
SelectKBest (9 features left)	**99.9985%**	**100%**	**100%**	**100%**

Furthermore, we have tried to reproduce the results for Random Forest with other datasets, and to train other ML algorithms on TON-IoT dataset.

In Table 9, we summarized the following experiment results:

- Random Forest (RF) and 2 datasets: IoT-23 and DO2OS. In both cases, we noticed that feature selection is not relevant since the quality of the obtained models decreased when we applied *SelectKBest*.

- TON-IoT with KNN: the accuracy, precision, recall and F1-score increase when we remove the irrelevant features from the dataset TON-IoT.
- DL approach with DS20S dataset: we have also applied the same process on B-GRU-RNN algorithm with DS2OS dataset and we have also obtained less conclusive results.

Table 9. Results of the features selection experiment on other models and dataset

	Accuracy	Precision	Recall	F1-score
Random Forest with IoT-23 dataset				
No modification	**99.9967%**	**99.9990%**	**99.9960%**	99.9975%
SelectKBest: 7 features left	98.2123%	98.6970%	99.8292%	99.2599%
Random Forest with DS2OS dataset				
No modification	**99.9720%**	**99.9986%**	**100%**	**99.9993%**
SelectKBest: 6 features left	93.3768%	99.9756%	99.8408%	99.9082%
KNN model with TON-IoT dataset				
No modification	99.8449%	99.8583%	99.9033%	99.8808%
SelectKBest: 6 features left	**99.9902%**	**99.9500%**	**99.9084%**	**99.9292%**
B-GRU-RNN model with DS2OS dataset				
No modification	**99.4371%**	**99.9770%**	**99.4472%**	**99.7114%**
SelectKBest: 6 features left	97.2664%	100%	97.2664%	98.6143%

5 Conclusion and Future Work

We studied in this paper the provision of ML/DL approaches in anomaly detection within IoT data. We tested different algorithms on diverse IoT datasets to identify the best model for anomaly detection in IoT, and we were able to improve the efficiency of the best model by using feature selection approach. However, we noticed that the performances of ML/DL models often depend on the use-case on hand and the quality of the dataset used, in terms of size and diversity of anomalies represented. Besides, data preprocessing of the data is also crucial for the training process. We need to go beyond the empirical approach followed in this paper to formally prepare the data to train a ML/DL model in a generic way. Furthermore, most of the research work have been dedicated to point anomaly detection, which let open a large perspective for studying group and context-based anomaly detection with ML/DL approaches.

References

1. Machine learning DDoS detection for consumer internet of things devices, pp. 29–35. Institute of Electrical and Electronics Engineers Inc. (2018). https://doi.org/10.1109/SPW.2018.00013
2. KNN-based approximate outlier detection algorithm over IoT streaming data. IEEE Access **8**, 42,749–42,759 (2020). https://doi.org/10.1109/ACCESS.2020.2977114
3. Abdullahi, M., et al.: Detecting cybersecurity attacks in internet of things using artificial intelligence methods: a systematic literature review. Electronics **11**(2) (2022). https://doi.org/10.3390/electronics11020198
4. Aggarwal, P., Sharma, S.K.: Analysis of KDD dataset attributes - class wise for intrusion detection, pp. 842–851. Elsevier (2015). https://doi.org/10.1016/j.procs.2015.07.490
5. Al-amri, R., Murugesan, R.K., Man, M., Abdulateef, A.F., Al-Sharafi, M.A., Alkahtani, A.A.: A review of machine learning and deep learning techniques for anomaly detection in IoT data. Appl. Sci. **11**(12) (2021). https://doi.org/10.3390/app11125320
6. Alguliyev, R.M., Aliguliyev, R.M., Imamverdiyev, Y.N., Sukhostat, L.V.: An anomaly detection based on optimization. Int. J. Intell. Syst. Appl. **9**, 87–96 (2017). https://doi.org/10.5815/ijisa.2017.12.08
7. Alsaedi, A., Moustafa, N., Tari, Z., Mahmood, A., Anwar, A.: Ton_iot telemetry dataset: a new generation dataset of IoT and IIoT for data-driven intrusion detection systems. IEEE Access **8**, 165,130–165,150 (2020). https://doi.org/10.1109/ACCESS.2020.3022862
8. Booij, T.M., Chiscop, I., Meeuwissen, E., Moustafa, N., Hartog, F.T.H.D.: Ton_iot: the role of heterogeneity and the need for standardization of features and attack types in IoT network intrusion data sets. IEEE Internet Things J. **9**(1), 485–496 (2022). https://doi.org/10.1109/JIOT.2021.3085194
9. Da Silva Oliveira, G.A., et al.: A stacked ensemble classifier for an intrusion detection system in the edge of IoT and IIoT networks. In: 2022 IEEE Latin-American Conference on Communications (LATINCOM), pp. 1–6 (2022). https://doi.org/10.1109/LATINCOM56090.2022.10000559
10. DeMedeiros, K., Hendawi, A., Alvarez, M.: A survey of AI-based anomaly detection in IoT and sensor networks. Sensors **23**(3) (2023). https://doi.org/10.3390/s23031352
11. Diro, A., Chilamkurti, N., Nguyen, V.D., Heyne, W.: A comprehensive study of anomaly detection schemes in IoT networks using machine learning algorithms. Sensors **21**(24) (2021). https://doi.org/10.3390/s21248320
12. Diro, A.A., Chilamkurti, N.: Distributed attack detection scheme using deep learning approach for internet of things. Futur. Gener. Comput. Syst. **82**, 761–768 (2018). https://doi.org/10.1016/j.future.2017.08.043
13. Fahim, M., Sillitti, A.: Anomaly detection, analysis and prediction techniques in IoT environment: a systematic literature review. IEEE Access **7**, 81,664-81,681 (2019). https://doi.org/10.1109/ACCESS.2019.2921912
14. Galvao, Y.M., Albuquerque, V.A., Fernandes, B.J.T., Valenca, M.J.S.: Anomaly detection in smart houses: monitoring elderly daily behavior for fall detecting, pp. 1–6. IEEE (2017). https://doi.org/10.1109/LA-CCI.2017.8285701
15. Garcia, S., Parmisano, A., Erquiaga, M.J. (2020). https://www.stratosphereips.org/datasets-iot23

16. Haidar, N., Tamani, N., Nienaber, F., Wesseling, M.T., Bouju, A., Ghamri-Doudane, Y.: Data collection period and sensor selection method for smart building occupancy prediction. In: 2019 IEEE 89th Vehicular Technology Conference (VTC2019-Spring), pp. 1–6 (2019)https://doi.org/10.1109/VTCSpring.2019.8746447
17. Haji, S.H., Ameen, S.Y.: Attack and anomaly detection in IoT networks using machine learning techniques: a review. Asian J. Res. Comput. Sci. **9**, 30–46 (2021)
18. Hasan, M., Islam, M.M., Zarif, M.I.I., Hashem, M.: Attack and anomaly detection in IoT sensors in IoT sites using machine learning approaches (2019). https://doi.org/10.1016/j.iot.2019.10
19. Hindy, H., Tachtatzis, C., Atkinson, R., Bayne, E., Bellekens, X. (2020). https://ieee-dataport.org/open-access/mqtt-iot-ids2020-mqtt-internet-things-intrusion-detection-dataset
20. Merchant, N.: IoT technologies explained: History, examples, risks & future. https://www.visionofhumanity.org/what-is-the-internet-of-things/
21. Moustafa, N.: A new distributed architecture for evaluating AI-based security systems at the edge: Network ton_iot datasets. Sustain. Cities Soc. **72**, 102,994 (2021). https://doi.org/10.1016/j.scs.2021.102994
22. Pahl, M.O., Aubet, F.X.: All eyes on you: Distributed multi-dimensional IoT microservice anomaly detection. In: 2018 14th International Conference on Network and Service Management (CNSM), pp. 72–80 (2018)
23. Pajouh, H.H., Dastghaibyfard, G., Hashemi, S.: Two-tier network anomaly detection model: a machine learning approach. J. Intell. Inf. Syst. **48**(1), 61–74 (2017). https://doi.org/10.1007/s10844-015-0388-x
24. Pajouh, H.H., Javidan, R., Khayami, R., Dehghantanha, A., Choo, K.K.R.: A two-layer dimension reduction and two-tier classification model for anomaly-based intrusion detection in IoT backbone networks. IEEE Trans. Emerg. Top. Comput. **7**(2), 314–323 (2019). https://doi.org/10.1109/TETC.2016.2633228
25. Quinlan, R.: Statlog (Australian credit approval). (https://doi.org/10.24432/c59012)
26. Savic, M., et al.: Deep learning anomaly detection for cellular IoT with applications in smart logistics. IEEE Access **9**, 59,406–59,419 (2021). https://doi.org/10.1109/ACCESS.2021.3072916
27. Sharma, B., Sharma, L., Lal, C.: Anomaly detection techniques using deep learning in IoT: a survey. In: 2019 International Conference on Computational Intelligence and Knowledge Economy (ICCIKE), pp. 146–149 (2019). https://doi.org/10.1109/ICCIKE47802.2019.9004362
28. Shaver, A., Liu, Z., Thapa, N., Roy, K., Gokaraju, B., Yuan, X.: Anomaly based intrusion detection for IoT with machine learning. Institute of Electrical and Electronics Engineers Inc. (2020). https://doi.org/10.1109/AIPR50011.2020.9425199
29. Souza, A.M., Amazonas, J.R.: An outlier detect algorithm using big data processing and internet of things architecture, pp. 1010–1015. Elsevier B.V. (2015)https://doi.org/10.1016/j.procs.2015.05.095
30. Steinbrunn, A., Pfisterer, W., Detrano, M., Janosi, R.: Heart disease (1988). https://doi.org/10.24432/c52p4x
31. Tavallaee, M., Bagheri, E., Lu, W., Ghorbani, A.A.: A detailed analysis of the KDD cup 99 data set. In: 2009 IEEE Symposium on Computational Intelligence for Security and Defense Applications, pp. 1–6 (2009). https://doi.org/10.1109/CISDA.2009.5356528

32. Thamilarasu, G., Chawla, S.: Towards deep-learning-driven intrusion detection for the internet of things. Sensors (Switzerland) **19** (2019). https://doi.org/10.3390/s19091977
33. Ullah, I., Mahmoud, Q.H.: Design and development of a deep learning-based model for anomaly detection in IoT networks. IEEE Access **9**, 103906–103926 (2021). https://doi.org/10.1109/ACCESS.2021.3094024
34. Zhang, C., et al.: A deep neural network for unsupervised anomaly detection and diagnosis in multivariate time series data (2018)

A Survey and Planning Advice of Blockchain-Based Used Car Bidding Platform

Chin-Ling Chen[1,2], Yong-Yuan Deng[2(✉)], Yao-Yuan Hsu[2], Hsing-Chung Chen[3,4(✉)], Der-Chen Huang[5], and Ling-Chun Liu[5(✉)]

[1] School of Information Engineering, Changchun Sci-Tech University, Changchun 130600, China

[2] Department of Computer Science and Information Engineering, Chaoyang University of Technology, Taichung 41349, Taiwan
allendeng@cyut.edu.tw

[3] Department of Computer Science and Information Engineering, Asia University, Taichung 413305, Taiwan
cdma2000@asia.edu.tw

[4] Department of Medical Research, China Medical University Hospital, China Medical University, Taichung, Taiwan
shin8409@ms6.hinet.net

[5] Department of Computer Science and Engineering, National Chung-Hsing University, Taichung 402202, Taiwan
huangdc@nchu.edu.tw, d110056004@mail.nchu.edu.tw

Abstract. In recent years, due to the increasing gap between rich and poor around the world, more and more people prefer to buy second-hand cars instead of buying new ones. As a result, the production and demand for new cars have decreased and the demand for second-hand cars has increased significantly. However, traditional second-hand car transactions have problems such as information asymmetry, opacity, and lack of trust. Many researchers have previously proposed related research on second-hand car transactions to solve the above problems. Because of the research of these scholars, we propose planning suggestions for a second-hand car bidding platform. We suggest that the Hyperledger Fabric blockchain architecture can be used to ensure the authenticity of vehicle information through a third-party impartial organization. The entire bidding process is transparently recorded on the blockchain, and decentralized IPFS storage can be used to reduce the burden of data on the blockchain. In addition, we recommend joining an arbitration institution to resolve any disputes that may arise after the transaction is completed. We believe that such a system plan will serve as an excellent guide for subsequent research and will certainly help improve the current second-hand car trading environment.

1 Introduction

In recent years, due to changes in the global economic situation, shortages in the supply of new cars and delivery delays have become a common phenomenon, which has made the used car market even more attractive. Many consumers are turning to buying used

L. Barolli (Ed.): AINA 2024, LNDECT 204, pp. 207–216, 2024.
https://doi.org/10.1007/978-3-031-57942-4_21

cars as a viable alternative. Second-hand cars are not only more attractive in terms of price but are also a response to the shortage of new cars. According to statistics from the Society of Motor Manufacturers and Traders (SMMT), the British second-hand car market showed amazing growth in 2021, reaching 7,530,956 transactions, a year-on-year increase of 11.5% [1]. According to the automotive data provider (AAA Data), the French second-hand car transaction volume in 2021 was 3.9% higher than in 2019. On the contrary, new car registrations showed a significant decline, falling by 25.1% [2]. According to a report by the Spanish Manufacturers and Dealers Association (Aecoc), new car registrations in Spain fell sharply by 31.7% in 2021. Compared to this, despite the sharp decline in new car registrations, second-hand car purchases decreased by only 5.2% [3].

As the transaction volume in the second-hand car market increases sharply, this change has caused many problems and challenges. The traditional second-hand car market has problems such as unclear product authenticity, opaque transactions, and fraud. Consumers have difficulty confirming the actual condition of their vehicles and are susceptible to fraud by dishonest sellers. In this case, the transaction process becomes risky and may lead to unsatisfactory transactions and financial losses. There is a lack of transparency in the transaction process in the used car market. It is often difficult for consumers to obtain complete information on used cars, including important information such as vehicle history, maintenance records, and accident reports. This makes it difficult for consumers to make informed car purchasing decisions and also creates an unequal information environment that allows some dishonest sellers to benefit. Finally, information asymmetry is an important motivation in the used car market. Sellers typically have more vehicle information, while buyers are in a position to have less information. This asymmetry may lead to unequal trading conditions and affect the efficiency and fairness of the market.

Blockchain is a technical system that is jointly maintained by multiple parties, uses cryptography to ensure transmission and access security, can achieve consistent data storage, cannot be tampered with, and cannot be denied. At present, blockchain and the Internet of Things (IoT) have been widely applied in various fields, such as video surveillance concerning privacy preservation [4], transparent and anti-counterfeiting supply of COVID-19 vaccine vials [5], and smart building architecture with emerging technology [6]. The hash algorithm is used to ensure that the medical record data on the chain cannot be tampered with, and that patient identity authentication can be achieved through the combination of electronic authentication technology. The decentralized database technology realizes the safe storage and traceability of data. It can effectively solve the problems of data sharing, data authenticity and security, privacy protection, and real-time monitoring of used car transactions [7–12]. We have compared some current blockchain-based second-hand car transaction research in Table 1.

However, even with a blockchain architecture applied, the cloud can still be subject to a 51% trust attack (Sybil attack), and thus can only be considered a semi-trusted role. In addition, the structure of the standard blockchain, each role in the chain, as long as the registration is successful, will enjoy permanent data access rights, which is less flexible for access control. Therefore, we suggest that the Hyperledger Fabric blockchain architecture can be used to ensure the authenticity of vehicle information through a third-party impartial organization. The entire bidding process is transparently recorded on the

Table 1. Current blockchain-based second-hand car transaction research.

Author	Years	Goal	Advantage	Shortcoming
Yoo and Ahn [7]	2021	Construct a second-hand car trading platform based on the blockchain	Using Ethereum combined with IPFS	Built on a public chain with insufficient management capabilities
El-Switi and Qatawneh [8]	2021	Using blockchain to improve integrity issues in used car transactions	Detailed background research	Lack of system planning advice
Kumar et al. [9]	2021	Construct a second-hand car bidding platform based on the blockchain	Using Ethereum combined with IPFS	Built on a public chain with insufficient management capabilities
Yu et al. [10]	2021	Using blockchain to improve integrity issues in used car transactions	Detailed background research	Lack of system planning advice
Shen et al. [11]	2022	Using blockchain to improve integrity issues in used car transactions	Using Hyperledger Fabric to trace vehicle history	Lack of system design including transaction process
Butera et al. [12]	2022	Using blockchain to improve integrity issues in used car transactions	Using NFT to trace vehicle history	Lack of system design including transaction process

blockchain, and decentralized IPFS storage can be used to reduce the burden of data on the blockchain. In addition, we recommend joining an arbitration institution to resolve any disputes that may arise after the transaction is completed.

2 Preliminary

2.1 Chaincode

A chaincode is a computer program or transactional agreement designed to automatically execute, control, or record legally relevant events and actions according to the terms of the contract or agreement. The goal of chaincodes is to reduce the need for trusted intermediaries, arbitration and enforcement costs, and fraud losses, and reduce malicious and accidental anomalies [13, 14].

Chaincodes are simply programs stored on the blockchain that run when predetermined conditions are met. They are often used to automate the execution of an agreement so that all participants can immediately determine the outcome without the involvement of any intermediaries or waste of time. They can also automate workflows and trigger the next action when conditions are met.

2.2 ECDSA

The Elliptic Curve Digital Signature Algorithm (ECDSA) was proposed by Vanstone [15] in 1992. It provides a variant of the Digital Signature Algorithm (DSA) using Elliptic Curve Cryptography (ECC). The Digital Signature Algorithm is a federal information processing standard for digital signatures, based on the mathematical concepts of modular exponentiation and discrete logarithmic problems.

ECC is a public-key cryptography method based on elliptic curve algebraic structures over finite fields. Compared to non-EC encryption, ECC allows the use of smaller keys to provide equivalent security. The properties of ECC allow ECDSA to require significantly smaller key sizes and the same level of security, providing faster computation and less storage space. The security of the digital signature algorithm is based on the discrete logarithm problem [16].

2.3 Hyperledger Fabric

Hyperledger Fabric [17] is blockchain infrastructure. It was originally contributed by IBM and Digital Asset to the Hyperledger project. Provides a modular architecture that groups nodes in the architecture, the execution of chaincodes known in the Fabric project as "chaincode", and configurable consensus and membership services [18]. The Fabric network includes peer nodes that execute chaincode contracts and access ledger data, approve transactions, and invoke APIs. The ordering nodes are responsible for ensuring the consistency of the blockchain and transmitting endorsed transactions to peers on the network, and the MSP service is primarily used as a certificate authority to manage X.509 certificates to verify membership and roles [19].

2.4 InterPlanetary File System

The InterPlanetary File System (IPFS) is a peer-to-peer decentralized file system that attempts to connect to the same file system for all computing devices. Hyperledger Fabric does not directly use IPFS as its default data storage solution. Fabric has its ledger mechanism and world-state storage, typically storing data within its distributed ledger. However, in certain cases, people might consider integrating IPFS with Hyperledger Fabric to meet specific needs, such as storing large files or multimedia data. In such scenarios, IPFS can act as a decentralized file system to store this data, while Hyperledger Fabric serves as the blockchain platform for managing transactions and metadata (like data pointers or hashes).

3 Proposed Scheme

3.1 System Architecture

We recommend a complete second-hand car bidding system that needs to include the following roles.

Bidder: Customers can select the car they want to buy on the second-hand car bidding platform and bid with other buyers. Before bidding, they need to pay the bidding platform the bidding deposit set by the platform.

Original Owner: Set the selling price and bid deposit of the vehicle to be sold, transfer the vehicle to be sold to a third-party impartial agency for verification, and sign a contract with the highest bidder after the auction.

New Owner: After the bidding ends, the highest bidder signs a vehicle transaction contract with the original owner. After the original owner confirms receipt of the final payment from the highest bidder, the highest bidder becomes the new car owner.

Used Car Auction Platform: Responsible for the vehicle bidding process. The original car owner and the bidder must register an account on the bidding platform. Before bidding, the bidder will pay the deposit to the designated bank. This platform will display vehicle information, bidding records, bidding status, and results that have been verified by a third-party impartial organization.

Bank: Before bidding, the bidder deposits the deposit paid into the account of a specific bank according to the bank designated by the bidding platform. After the bidding ends and the highest bidder signs a vehicle transaction contract with the original owner, the bank is responsible for transaction settlement and ensuring that the balance paid by the bidder is safely transferred to the original car owner.

Arbitration Institution: After the transaction is completed, if a dispute occurs during the warranty period, the new car owner can file a complaint with the arbitration institution, which will investigate and provide a fair solution.

Blockchain Center: Responsible for the registration of participants. The Blockchain Center performs functions such as data upload, node management, and data verification.

Certificate Authority: After the participants register, they issue digital certificates and public and private keys.

InterPlanetary File System (IPFS): The blockchain center can upload participant information to IPFS, such as vehicle exterior photos, vehicle inspection reports, contracts signed by buyers and sellers, etc.

Third Trust Party: Responsible for verifying the authenticity of the vehicle's identity, origin, maintenance, and accident records, ensuring that the vehicle's information and conditions are consistent with those provided by the owner. After the verification is completed, based on the vehicle inspection report, the relevant personnel will negotiate with the original owner of the car for a reasonable bidding price and upload the inspection results report to IPFS.

3.2 System Communication Phases

Registration Phase: Relevant participants in the system need to register an account on the used car bidding platform. The certificate issuing authority will issue the corresponding digital certificate, public key, and private key to the participant.

Vehicle Transfer and Information Verification Phase: The original owner will hand over the vehicle, including the vehicle's basic information, vehicle identification number, maintenance records, accident history, and other information to a third-party impartial agency for vehicle information verification. The impartial third-party agency will upload the vehicle inspection report to the used car bidding platform and IPFS.

Bidding Phase: Bidders can find the vehicle they want to buy on the second-hand car bidding platform and place a bid, but before bidding, they need to pay the bidding deposit set by the original car owner to the bank designated by the bidding platform. During the bidding period, the bidder needs to sign his/her bidding information and upload it to the blockchain. After the bidding ends, the used car bidding platform will announce the bidding results and upload them to the blockchain after signing.

Transaction Contract Signing Phase: After the bidding ends, the original car owner will sign a contract with the highest bidder and the contract will be verified by the bidding platform. The highest bidder will deliver the remaining balance to the bank and the bank will pay the original owner of the vehicle. After the original owner confirms the final payment, the highest bidder becomes the new owner.

Dispute Arbitration Phase: If a dispute occurs during the warranty period after the transaction is completed, the new owner can file a complaint with the arbitration institution.

4 Features and Security Analysis

We recommend using the following items to perform system characteristics and security analysis [20–32]:

Decentralization for Information Sharing: In the blockchain network architecture, decentralized bookkeeping and storage are adopted. All nodes have the same rights and obligations. Therefore, any error or shutdown of any node will not affect the operation of the entire blockchain network. Decentralization is a phenomenon or structure that can only occur in a system with many users or many nodes, each of which can connect and influence other nodes. Everyone is the center and everyone can connect and influence other nodes. This flat, open-source, and equally large phenomenon or structure is called decentralization. At the same time, decentralization is one of the typical characteristics of the blockchain. It uses distributed storage and computing power, and the rights and obligations of all nodes of the network nodes are the same. The data in the system is essentially maintained by the entire network nodes, so the blockchain no longer relies on the central processing node to realize the decentralized storage, recording, and updating of data, and each blockchain follows a unified rule, which is based on a cryptographic algorithm rather than a credit certificate, and the data update process requires user approval. This establishes that the blockchain does not require the endorsement of intermediaries or trusted institutions.

Openness and Privacy: By using the blockchain mechanism based on ECDSA (Elliptic Curve Digital Signature Algorithm), combined with the encryption mechanism of the public key system, in addition to the encrypted privacy information of the transaction subject, the signature data information stored in the blockchain is open and transparent. All members of the same alliance can verify the accuracy of the message and reduce the asymmetry of the data, which is the openness of the blockchain system. Furthermore, the plaintext of the message is also encrypted by the public key, and only the members of the alliance with the corresponding private key can decrypt the message settings to achieve privacy protection.

Verifiable: In the proposed scheme, the sender's message will be signed and the receiver can verify the authenticity of the signature. The message sender will sign with his/her private key, and the message receiver will verify with the message sender's public key. Sharing: Initially, the dealer holds a secret as input, and each player holds an independent random input. The sharing phase may consist of several rounds. At each round, each player can privately send messages to other players and can also send a message. Each message sent or broadcast by a player is determined by its input, its random input, and messages received from other players in previous rounds. Reconstruction: In this phase, each player provides its entire view from the sharing phase, and a reconstruction function is applied, which is taken as the protocol's output.

Traceable: Due to the equality of information, blockchain technology guarantees that all participants have equal rights to know and choose. Anything exchanged between two roles is auditable. All transaction information is accessible to all participants in the alliance. Due to the decentralized storage verification feature, all change records are synchronized on the blockchain. In the event of disputes in the future, there is data to prove that the rights and interests of any two roles are protected.

Integrity and Unforgery: To ensure the integrity of the message in the communication process, the scheme adopts a signature mechanism and a timestamp to ensure that the message is not tampered with. For data, ensure the correctness of the data during transmission or storage, and unauthorized changes, substitutions, and forgery can be detected. For the system, it is necessary to ensure that the identity of any user cannot be impersonated, and that the system code cannot be changed to pass the test successfully. This can be achieved through the use of one-way hash functions or digital signature techniques.

Non-repudiation: Once the data are verified and added to the blockchain, it is stored forever, and the blockchain's inherent time stamping feature records the time of creation. Modifying the content of messages that have been uploaded to the blockchain center requires control of more than 51% of the consortium members, which will be difficult to achieve. The transmission of data can prevent the sender from denying that he/she has sent a certain message, and it can also prevent the receiver from denying that he/she has received or knows the message. For the system, the nonrepudiation mechanism will automatically leave relevant evidence of access activities to facilitate subsequent security audits and management.

Resist Sybil Attack: A Sybil attack is a type of attack on a computer network service in which an attacker subverts the service's reputation system by creating many pseudonymous identities and using them to gain a disproportionately large influence. It is named after the subject of the Sybil book, a case study of a woman diagnosed with dissociative identity disorder. An attacker can intercept messages in transit and send illegitimate messages to the user. The proposed scheme uses the ECDSA system to secure messages. As a result, attackers cannot properly intercept important information when conducting Sybil attacks.

5 Conclusions

Until now, there are still many problems in the traditional second-hand car market. Among them, the authenticity and transparency of information during the bidding process are very important issues. After the bidding is over, the security of the transaction is even more critical. Currently, many scholars have researched second-hand car transactions, but we believe there is still room for improvement. On the basis of the research of these scholars, we propose planning suggestions for a second-hand car bidding platform. We suggest that the Hyperledger Fabric blockchain architecture can be used to ensure the authenticity of vehicle information through a third-party impartial organization. The entire bidding process is transparently recorded on the blockchain, and decentralized IPFS storage can be used to reduce the burden of data on the blockchain. In addition, we recommend joining an arbitration institution to resolve any disputes that may arise after the transaction is completed. We believe that such a system plan will serve as an excellent guide for subsequent research and will certainly help improve the current second-hand car trading environment.

Funding. This work was supported in part by the National Science and Technology Council of Taiwan, R.O.C., under contract NSTC 112-2410-H-324-001-MY2, and contract MOST 111-2218-E-002-037. This work was supported by the Chelpis Quantum Tech Co., Ltd., Taiwan, under the Grant number of Asia University: I112IB120. This work was supported by the National Science and Technology Council (NSTC), Taiwan, under NSTC Grant numbers: 111-2218-E-468-001-MBK and 110-2218-E-468-001-MBK.

References

1. Used car market up despite volatile year. https://www.smmt.co.uk/2022/02/used-car-market-up-despite-volatile-year/
2. Europe's used-car markets on track to exceed pre-pandemic level in 2022. https://autovista24.autovistagroup.com/news/analysis-europe-used-car-market-exceed-pre-pandemic-levels/
3. Passenger car registrations: -2.4% in 2021; -22.8% in December. https://www.acea.auto/pc-registrations/passenger-car-registrations-2-4-in-2021-22-8-in-december/
4. Dave, M., Rastogi, V., Miglani, M., Saharan, P., Goyal, N.: Smart Fog-Based Video Surveillance with Privacy Preservation based on Blockchain, Wireless Personal Communications. Springer (2021). https://doi.org/10.1007/s11277-021-09426-8
5. Chauhan, H., et al.: Blockchain enabled transparent and anti-counterfeiting supply of COVID-19 Vaccine Vials, Vaccines, MDPI, 9(11), Article ID: 1239, October 2021. Doi:https://doi.org/10.3390/vaccines9111239
6. Kumar, A., Sharma, S., Goyal, N., Singh, A., Cheng, X., Singh, P.: Secure and energy-efficient smart building architecture with emerging technology IoT. Comput. Commun. **176**, 207–217 (2021). https://doi.org/10.1016/j.comcom.2021.06.003

7. Yoo, S.G., Ahn, B.: A study for efficiency improvement of used car trading based on a public blockchain. J. Supercomput. (2021). https://doi.org/10.1007/s11227-021-03681-z
8. El-Switi, S., Qatawneh, M.: Application of blockchain technology in used vehicle market: a review. In: 2021 International Conference on Information Technology (ICIT). IEEE (2021). https://doi.org/10.1109/ICIT52682.2021.9491670
9. Kumar, M.S.N., Akshatha, G.C., Bangre, M.D., Dhanush, M., Abhishek, C.: Decentralized used cars Bidding application using Ethereum. In: 2021 IEEE International Conference on Computation System and Information Technology for Sustainable Solutions (CSITSS). IEEE (2021). https://doi.org/10.1109/CSITSS54238.2021.9682892
10. Yu, Y., Yao, C., Zhang, Y., Jiang, R.: Second-hand car trading framework based on blockchain in cloud service environment. In: 2021 2nd Asia Conference on Computers and Communications (ACCC), IEEE (2021). https://doi.org/10.1109/ACCC54619.2021.00026
11. Shen, C.W., Koziel, A.M., Wen, C.: Application of hyperledger blockchain to reduce information asymmetries in the used car market. In: 14th Asian Conference, ACIIDS 2022, pp. 495–508. Springer (2022)
12. Butera, A., Gatteschi, V., Prattico, F.G., Novaro, D., Vianello, D.: Blockchain and NFTs-based Trades of Second-hand Vehicles. IEEE Access **10**. (2022). https://doi.org/10.1109/ACCESS.2023.3284676
13. Szabo, N.: Smart contracts: building blocks for digital markets, EXTROPY: The Journal of Transhumanist Thought **18**(2), 16 (1996)
14. Szabo, N.: The idea of smart contracts (1997). http://www.fon.hum.uva.nl/rob/Courses/InformationInSpeech/CDROM/Literature/LOTwinterschool2006/szabo.best.vwh.net/smart_contracts_idea.html
15. Vanstone, S.: Responses to NIST's proposal. Commun. ACM **35**, 50–52 (1992)
16. Johnson, D., Menezes, A., Vanstone, S.: The Elliptic Curve Digital Signature Algorithm (ECDSA). Int. J. Inf. Secur. **1**, 36–63 (2001). https://doi.org/10.1007/s102070100002
17. Hyperledger Fabric Docs. Available online: https://hyperledgerfabric.readthedocs.io/en/release-2.2
18. Foschini, L., Gavagna, A., Martuscelli, G., Montanari, R.: Hyperledger fabric blockchain: chaincode performance analysis. In: ICC 2020 - 2020 IEEE International Conference on Communications (ICC), 2020, pp. 1–6 (2020). https://doi.org/10.1109/ICC40277.2020.9149080
19. Uddin, M.: Blockchain Medledger: Hyperledger fabric enabled drug traceability system for counterfeit drugs in pharmaceutical industry. Int. J. Pharmaceutics **597**, Article 120235 (2021)
20. Roy, S., Das, A.K., Chatterjee, S., Kumar, N., Chattopadhyay, S., Rodrigues, J.J.: Provably secure fine-grained data access control over multiple cloud servers in mobile cloud computing based healthcare applications. IEEE Trans. Industr. Inf. **15**(1), 457–468 (2018)
21. Wazid, M., Das, A.K., Kumari, S., Li, X., Wu, F.: Provably secure biometric-based user authentication and key agreement scheme in cloud computing. Secur. Commun. Networks **9**(17), 4103–4119 (2016)
22. Sureshkumar, V., Amin, R., Vijaykumar, V.R., Sekar, S.R.: Robust secure communication protocol for smart healthcare system with FPGA implementation. Futur. Gener. Comput. Syst. **100**, 938–951 (2019)
23. Roy, S., Chatterjee, S., Das, A.K., Chattopadhyay, S., Kumari, S., Jo, M.: Chaotic map-based anonymous user authentication scheme with user biometrics and fuzzy extractor for crowdsourcing Internet of Things. IEEE Internet Things J. **5**(4), 2884–2895 (2017)
24. Banerjee, S., et al.: A provably secure and lightweight anonymous user authenticated session key exchange scheme for the Internet of Things deployment. IEEE Internet Things J. **6**(5), 8739–8752 (2019)
25. Shuai, M., Yu, N., Wang, H., Xiong, L.: Anonymous authentication scheme for smart home environment with provable security. Comput. Secur. **86**, 132–146 (2019)

26. Abbas, A., Khan, S.: A review on the state-of-the-art privacy preserving approaches in e-health clouds. IEEE J. Biomed. Health Inform. **18**(4), 1431–1441 (2014)
27. Yang, J., Li, J., Niu, Y.: A hybrid solution for privacy preserving medical data sharing in the cloud environment. Futur. Gener. Comput. Syst. **43–44**, 74–86 (2015)
28. Soni, P., Pal, A.K., Islam, S.H.: An improved three-factor authentication scheme for patient monitoring using WSN in remote health-care system. Comput. Methods Programs Biomed., 182, Article 105054 (2019)
29. Masdari, M., Ahmadzadeh, S.: A survey and taxonomy of the authentication schemes in telecare medicine information systems. J. Netw. Comput. Appl. **87**, 1–19 (2017)
30. Amin, R., Islam, S.H., Biswas, G.P., Khan, M.K., Kumar, N.: A robust and anonymous patient monitoring system using wireless medical sensor networks. Futur. Gener. Comput. Syst. **80**, 483–495 (2018)
31. Chen, L., Lee, W.K., Chang, C.C., Choo, K.K.R., Zhang, N.: Blockchain based searchable encryption for electronic health record sharing. Futur. Gener. Comput. Syst. **95**, 420–429 (2019)
32. Tanwar, S., Parekh, K., Evans, R.: Blockchain-based electronic healthcare record system for healthcare 4.0 applications. J. Inf. Secur. Appl., 50, Article 102407 (2020)

Misuse Detection and Response for Orchestrated Microservices Based Software

Mohamed Aly Amin, Adnan Harun Dogan, Elif Sena Kuru, Yigit Sever, and Pelin Angin(✉)

Middle East Technical University, Ankara, Turkey
{mohamed.amin,elif.kuru}@metu.edu.tr,
{adhd,yigit,pangin}@ceng.metu.edu.tr

Abstract. In the evolving landscape of cloud computing, containerized microservices have emerged as a dominant architecture, presenting unique security challenges. This paper introduces a novel security framework, harnessing the power of machine learning, to enhance the detection and response capabilities against misuse in Kubernetes-based microservices environments. Central to our approach is the Dynamic Topology Adjustment (DTA) operator, seamlessly integrated with Kube-OVN's advanced networking features, enabling proactive and dynamic adaptation of the network topology in response to real-time security threats. We implement an AI-driven misuse detection model based on the SGDOneClassSVM algorithm, tailored to analyze network flows within these complex systems. Our framework not only addresses immediate security concerns but also sets a foundation for adaptive, intelligent security management in cloud-based microservices. Experimental results, derived from a specially curated dataset targeting container-specific vulnerabilities, demonstrate the efficacy of our approach in detecting a range of security threats with high accuracy, showcasing its potential as a robust solution for container security in cloud environments.

1 Introduction

The advent of cloud computing has profoundly transformed software architecture, particularly with the shift towards microservices-based designs. These architectures offer unprecedented scalability and isolation, especially in containerized deployments, but also introduce significant security challenges within orchestrated environments such as Kubernetes. The inherent dynamic and distributed nature of these systems demands sophisticated and adaptable security measures.

In response to these challenges, this paper presents a novel security framework designed for Kubernetes environments, leveraging advanced machine learning (ML) to achieve real-time misuse detection and response. At the heart of our approach is the Dynamic Topology Adjustment (DTA) operator, seamlessly integrated with Kube-OVN, facilitating dynamic network topology adjustments

L. Barolli (Ed.): AINA 2024, LNDECT 204, pp. 217–226, 2024.
https://doi.org/10.1007/978-3-031-57942-4_22

in reaction to emerging security threats. Our solution employs an AI model grounded in the SGDOneClassSVM algorithm, specifically tailored to address the complexities of microservices network flows.

A pivotal element of our research is the development and utilization of a comprehensive dataset, curated to assess the effectiveness of the proposed framework. This dataset features an array of simulated attacks, targeting prevalent vulnerabilities within containerized settings, thus offering insights into the dataset's depth and its applicability to real-world scenarios.

2 Related Work

Abbas et al. [1] introduced PACED, designed to identify container-escape attacks by focusing on cross-namespace flows, often indicative of such attacks. The paper proposes "privileged flows" to detect attacks in Docker and Kubernetes environments. This rule is tested against contemporary Common Vulnerabilities and Exposures (CVEs), providing a practical and verifiable solution. While the system demonstrates high accuracy in detecting container escape attacks, its scope is confined to this particular type of threat. Chen et al. [3] proposed Informer, a two-phase ML framework, to detect irregular traffic for Remote Procedure Calls (RPCs) in microservices. Their framework first extracts RPC chain patterns by clustering. Then, it builds a graph model for each RPC chain pattern, which simultaneously learns spatial and temporal features to predict future RPC traffic. This framework accurately captured anomalies within dynamic and complex microservice production scenarios. Sever et al. [6] developed and deployed a microservice-based application onto a Kubernetes cluster, and executed ten different attack scenarios, each showcasing different security vulnerabilities. Cui et al. [4] proposed an unsupervised introspection approach for containerized applications. The proposed LSTM-based framework includes a scaling factor normalization module and a tunable threshold module, enabling the framework to classify unseen behaviors within an unsupervised manner accurately. One-Class Support Vector Machine (OC-SVM) was used by Zhang et al. [9] to identify anomalies in containerized applications. They evaluated the detection capability of the proposed system against modern attacks like adversarial attacks with a customized attack dataset. According to their experimental findings, the OC-SVM algorithm can detect contemporary attacks with a satisfactory false positive rate (FPR) of 0.02 for brute force attacks and 0.12 for adversarial ML attacks. Sever et al. [7] presented reproducible steps towards generating a dataset of benign and malicious traffic generated through interactions with real-world vulnerable container images discovered using the CVE database, which includes data on system calls and network flow data. By employing both network and host-based monitoring techniques, they built a ML-based intrusion detection system (IDS) for containers. Their results show that basing the IDS on the network layer performs better than the host-based IDS for the investigated vulnerabilities, which underscores the importance of network monitoring as a more effective approach for enhancing the security of containerized environments. Tien et al. [8] proposed *KubAnomaly*, a system that offers security monitoring capabilities for anomaly detection on the Kubernetes orchestration platform. Their experimental results showed that its accuracy is nearly 96% for detecting anomalies in Docker containers. Aktolga et al. [2] conducted a comprehensive review of the container security literature, which illustrates that the integration of ML into security systems has the potential to enhance misuse detection, anomaly detection, and inter-container security within container clusters in orchestrated microservices-based software systems.

3 Proposed System Architecture

Our system introduces the Dynamic Topology Adjustment (DTA) operator, integrated with Kube-OVN, to dynamically manage Kubernetes-clustered microservices' network topology in response to security threats. Utilizing SGDOne ClassSVM for ML-based misuse detection, it proactively safeguards microservices by leveraging Kube-OVN's enhanced networking capabilities (Fig. 1).

3.1 Kubernetes Cluster

3.1.1 Networking and Virtualization

- **Kubernetes Cluster**: Orchestrates diverse virtualizations optimized for applications, including those requiring different virtualization technologies.
- **Kube-OVN**: Enhances Kubernetes with advanced networking features like overlay-underlay sub-net intercommunication and internal load balancing.
- **Open Virtual Switch (OVS)**: Enables network automation within the Kubernetes environment.

3.1.2 ML Integration and Cluster Coordination

- **DTA Operator**: Dynamically re-configures network topology in response to ML-informed security threats.
- **SGDOneClassSVM Model**: Analyzes security threats in real time for informed decision-making.
- **Cluster Controller Manager**: Continuously updates the DTA Operator with the latest cluster state, informing its action plans.

3.1.3 SGDOneClassSVM ML Model: Threat Detection and System Interoperability

- **Core ML-based Detection**: Forms the backbone of the intrusion detection system and is capable of on-cluster training.
- **Data Processing with Prometheus**: Enriches network flow data for compatibility and enhanced threat analysis. It minimizes and pre-processes the data that is being fed to the ML model to achieve high performance.
- **API Endpoints**: Facilitates interaction with both on-cluster and off-cluster security systems through the Kubernetes API server.

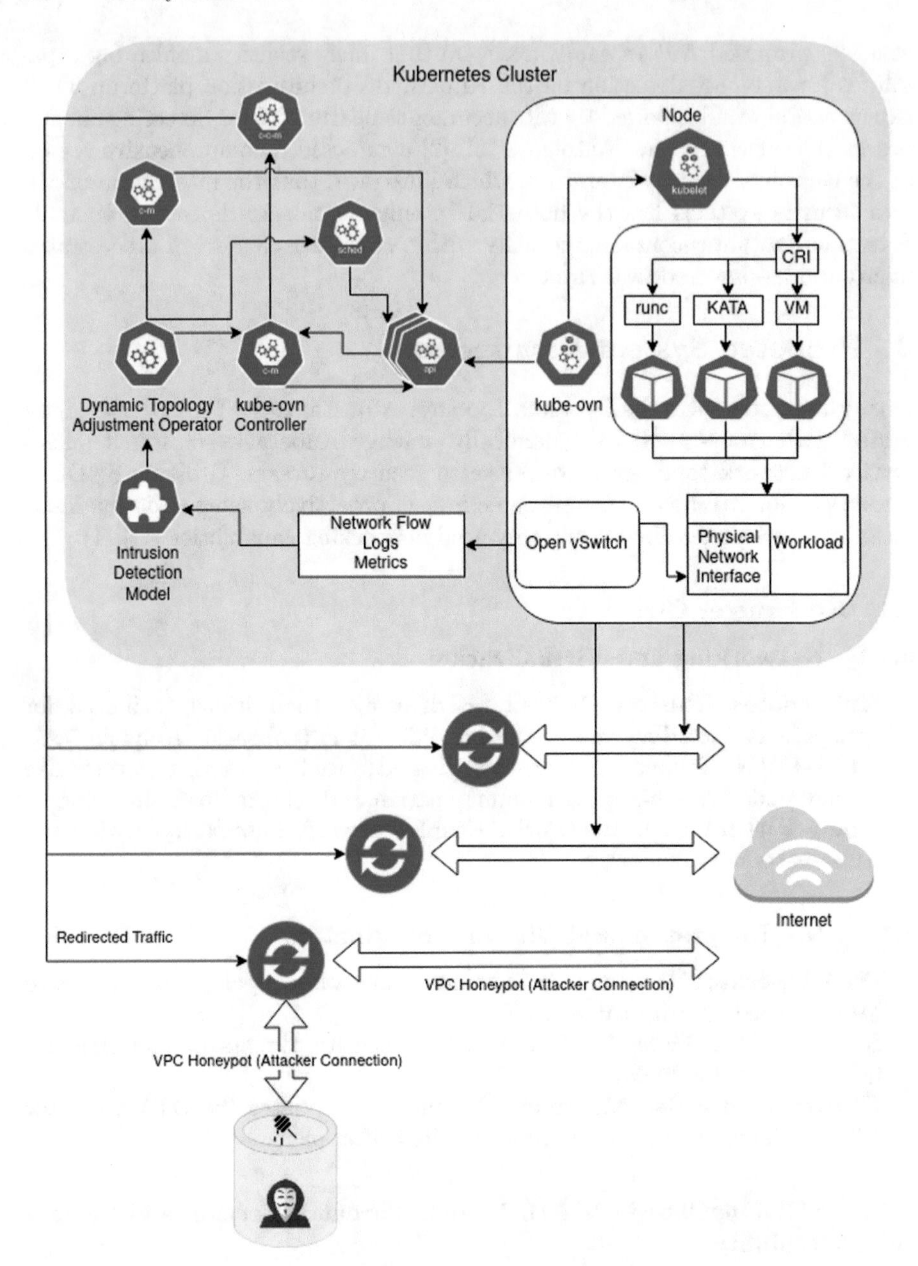

Fig. 1. Dynamic Topology Adjustment (DTA) in the Proposed System Architecture

3.2 Automated Incident Response and DTA Controller Operations

Our system ensures robust security through a dual approach: automated incident response mechanisms and strategic network management via the DTA Controller. These components work in tandem to maintain the security and efficiency of the micro-services environment.

- **Network Topology Changes and Configuration Management**: The DTA Operator and Controller collaboratively manage network configurations, dynamically adjusting the topology to redirect attacker traffic to isolated subnets or remote clusters, thus protecting the primary cluster. This is achieved through real-time analysis and strategic planning, ensuring optimal resource allocation and network settings.
- **Incident Response and Pod Management**: In response to threats, infected pods are quickly rescheduled for removal, replaced by secure pods, with the DTA Controller coordinating these actions alongside the Kubernetes Scheduler for seamless deployment and security enforcement.
- **Strategic Coordination**: The DTA Controller ensures continuous alignment of security strategies and operational responses, liaising with the Cluster Controller Manager to adapt to emerging threats efficiently.

This streamlined approach ensures proactive threat mitigation and the resilience of Kubernetes-clustered microservices.

4 Evaluation

In evaluating our security framework, we relied on a meticulously constructed dataset, designed to reflect the diverse range of threats encountered in Kubernetes environments. This dataset includes implementations of several attack scenarios, each simulating different security vulnerabilities in containerized systems.

4.1 Container Vulnerabilities and Attack Simulations

In this work, we used the Social Network and Media Service of DeathStarBench [5]. Deploying the said microservices-based applications on our cluster, we used the provided load testing suite as the benign traffic and deployed attacks against the cluster and the application as malicious traffic. Our dataset features a range of simulated attacks, including:

- **CVE-2019-5736**: A container escape vulnerability targeting the runc tool, allowing attackers to gain root access on the host machine.
- **CVE-2021-30465**: A container escape vulnerability exploiting a race condition in the runc tool during volume mounting.
- **CVE-2021-25741**: A Kubernetes-specific vulnerability in the kubelet component, enabling container escape through a race condition during host file path mounting.
- **CVE-2022-23648**: Discovered in the containerd runtime, this vulnerability allows for arbitrary host file access.

4.2 Experiments

Dataset Preprocessing

The dataset consists of two subsets: benign and malicious rows based on the'Label' feature. A downsampling technique was applied to obtain a subset of both classes since the dataset's proportion of malicious and benign subsets is not balanced. Normalization was performed on the selected columns using the `StandardScaler` from sci-kit-learn, ensuring that features have zero mean and unit variance. Additionally, Principal Component Analysis with 20 components was applied to reduce the dimensionality of the data while retaining relevant information.

Model Selection

We applied model selection by considering a diverse set of ML models across different categories. Within the Ensemble Learning Models category, we explored the efficacy of Random Forest, AdaBoost, XGBoost, and GradBoost. In the Tree-based Models group, Decision Tree, Random Forest, AdaBoost, and XGBoost were evaluated for their ability to construct decision trees and enhance accuracy through ensemble methods. In Linear Models, Logistic Regression, Perceptron, and SGD were considered. Support Vector Machines (SVM) was explored through SVM itself, specialized OneClassSVM for one-class classification, and SGDOneClassSVM utilizing Stochastic Gradient Descent for efficient handling of large datasets and online learning scenarios. Additionally, K-nearest neighbor (KNN), a non-parametric lazy learning algorithm, was also assessed. Finally, the Naive Bayes model GaussianNB, specifically designed for tasks with Gaussian-distributed features and assuming conditional independence, was included in the model selection process. This comprehensive approach aims to identify suitable models for effectively detecting intrusions.

Hyperparameter Tuning.

Grid search is a hyperparameter tuning technique that involves defining a grid of hyperparameter values and searching through the grid to find the best combination for an ML model.

Tree-Based Models: For the `DecisionTree` model, grid search is performed on the criterion parameter with values 'gini' and 'entropy'. The max depth of the tree is explored with values 'None', 10, 20, 30, 40, and 50. The min samples split and min samples leaf parameters are tuned with values 2, 5, and 10, and 1, 2, and 4.

Tree-Based Ensemble-Learning Models: Grid search for the `RandomForest` model involves exploring the number of estimators (trees) with values 50, 100, and 200. The criterion parameter is tested with'gini' and'entropy'. The max depth of the trees is considered with values'None', 10, 20, 30, 40, and 50. The min samples split and min samples leaf parameters are varied similarly. Grid search for the `AdaBoost` model involves tuning the number of estimators with values 50, 100, and 200. The learning rate, controlling the contribution of each weak learner, is explored with values 0.01, 0.1, and 0.2. Grid search for the `XGBoost` model includes tuning the learning rate with values 0.01, 0.1, and 0.2.

The number of estimators is explored with values 50, 100, and 200. The max depth of the trees is tested with values 3, 4, and 5. Similarly, for the `GradBoost` model, grid search is performed with different combinations of hyperparameter values. The number of estimators (trees in the ensemble) is explored with values 100, 200, and 300. The learning rate, controlling the contribution of each tree, is varied with values 0.01, 0.1, and 0.2. The max depth of the trees is considered with values 3, 4, and 5. The min samples split and min samples leaf parameters, determining the conditions for tree node splitting and leaf creation, respectively, are explored with values 2, 5, and 10, and 1, 2, and 4, respectively. The subsample parameter, specifying the fraction of samples used for fitting the individual base learners, is tested with values 0.8, 0.9, and 1.0.

Linear Models: For the `LogisticRegression` model, a grid search is conducted over the hyperparameters. The regularization parameter C is explored with values 0.001, 0.01, 0.1, 1, 10, 100, and 1000. The penalty parameter, specifying the type of regularization ('None', 'l1', 'l2','elasticnet'), is tested. The solver parameter, defining the optimization algorithm, is explored with multiple options. The maximum number of iterations for the solver is set to 50000, and the dual parameter is explored with values True and False. The `Perceptron` model undergoes grid search with the penalty parameter ('none', 'l2', 'l1', 'elasticnet') and the alpha parameter, controlling the regularization strength, tested with values 0.0001, 0.001, 0.01, and 0.1. Grid search for the `SGD` model includes tuning the loss function ('hinge', 'log', 'modified_huber', 'perceptron'), the penalty parameter ('l2', 'l1', 'elasticnet'), and the alpha parameter with values 0.0001, 0.001, 0.01, and 0.1.

Support Vector Machines: For the `SVM` model, grid search is conducted with the C parameter, controlling the regularization strength, tested with values 0.1, 1, 10, and 100. The kernel parameter is explored with 'linear', 'rbf', and 'poly', and the gamma parameter is tested with 'scale' and 'auto'. For the `OneClassSVM` model, a grid search is performed with different values for the hyperparameters. The kernel hyperparameter is explored using the values 'linear', 'rbf', 'poly', and 'sigmoid'. The nu parameter, controlling the upper bound on the fraction of training errors and a lower bound of the fraction of support vectors, is varied with values 0.01, 0.1, 0.2, and 0.5. The gamma parameter, responsible for the kernel coefficient, takes on values 'scale', 'auto', 0.1, and 1.0. The degree parameter, applicable for the 'poly' kernel, is explored with values 2, 3, and 5. Finally, the coef0 parameter, relevant for the'poly' and 'sigmoid' kernels, is varied with values 0.0, 0.1, and 0.5.

Neighbor-Based Models: The `KNN` model undergoes grid search with the number of neighbors (n_neighbors) tested with values 3, 5, 7, and 9. The weights parameter, specifying the weight function used in prediction ('uniform' or 'distance'), is explored, and the p parameter, determining the power parameter for the Minkowski distance, is tested with values 1 and 2.

Naive Bayes Models: For the `GaussianNB` model, grid search is performed on the var_smoothing parameter with values 1e−09, 1e−08, 1e−07, and 1e−06.

5 Experiment Results

Table 1 lists the best performances of the different ML models. As seen in the table, while most models achieve high precision, their recall performances are suboptimal, except for the OneClassSVM model. The Decision Tree model, with a best score of 0.17, demonstrated superior performance with hyperparameters criterion = 'entropy', max depth = 40, min samples leaf = 2, and min samples split = 5. Achieving a best score of 0.06, the Random Forest model performed optimally with hyperparameters criterion = 'entropy', max depth = 30, min samples leaf = 1, min samples split = 2, and number of estimators = 50. The `AdaBoost` model achieved a best score of 0.11, with optimal hyperparameters learning rate = 0.2 and number of estimators = 200, demonstrating improved performance and ensemble learning effectiveness. With a best score of 0.11, the `XGBoost` model excelled using hyperparameters learning rate = 0.1, max depth = 4, and number of estimators = 200. The `GradBoost` model excelled with a best score of 0.13, utilizing optimal hyperparameters; learning rate = 0.2, max depth = 5, min samples leaf = 4, min samples split = 10, number of estimators = 300, and subsample = 0.8. With a best score of 0.08, the Logistic Regression model excelled using hyperparameters C = 100, dual = False, max iter = 50000, penalty = 'l1', and solver = 'liblinear'. The `Perceptron` model achieved a best score of 0.13, with optimal hyperparameters alpha = 0.001 and penalty = 'elasticnet'. The `SGD` model, achieving a best score of 0.14, demonstrated this performance with hyperparameters alpha = 0.0001, loss = 'perceptron', and penalty = 'l1'. `SVM` achieved a best score of 0.10, and the optimal hyperparameters for the SVM model are C = 100, 'gamma' = 'auto', and kernel = 'poly'. The `OneClassSVM` model obtained an outstanding best score of 0.98, with optimal hyperparameters coef0 = 0.5, degree = 2, gamma = 1.0, kernel = 'sigmoid', and nu = 0.01, indicating its suitability for one-class classification tasks. The `KNN` model attained a best score of 0.12 with optimal hyperparameters number of neighbours = 3, p = 1, and weights = 'distance'. With the best score of 0.32, the Gaussian Naive Bayes model performed optimally using hyperparameter variance smoothing = 1e−09.

Table 1. The best performances of different ML models

ML Classifier Model	Acc.	F1	Prec.	Recall	TP	FP	Normal data	Attack data	Test data
OneClassSVM	**98.11%**	**0.99**	1.0	**0.98**	**88.3**	**1.7**	2000	10	90
GaussianNB	31.67%	0.42	1.0	0.32	28.5	61.5	2000	10	90
DecisionTreeClassifier	16.56%	0.28	1.0	0.17	14.9	75.1	2000	10	90
SGDClassifier	14.11%	0.24	1.0	0.14	12.7	77.3	2000	10	90
GradBoostingClassifier	13.0%	0.23	1.0	0.13	11.7	78.3	2000	10	90
Perceptron	12.67%	0.20	0.7	0.13	11.4	78.6	2000	10	90
KNeighborsClassifier	11.67%	0.21	1.0	0.12	10.5	79.5	2000	10	90
AdaBoostClassifier	11.11%	0.19	1.0	0.11	10.0	80.0	2000	10	90
XGBClassifier	10.78%	0.19	1.0	0.11	9.7	0.0	2000	10	90
SVC	9.67%	0.17	1.0	0.10	8.7	81.3	2000	10	90
LogisticRegression	8.11%	0.15	1.0	0.08	7.3	82.7	2000	10	90
RandomForestClassifier	5.78%	0.11	1.0	0.06	5.2	84.8	2000	10	90

6 Conclusion

In this paper we introduced a security framework for Kubernetes-based microservices, leveraging the Dynamic Topology Adjustment (DTA) operator and advanced ML techniques. Our approach, integrating with Kube-OVN, demonstrates a significant advancement in dynamically adapting network topologies to counteract real-time security threats in containerized environments. The employment of the SGDOneClassSVM model represents a notable stride in enhancing the efficacy of threat detection and response mechanisms. The diversity and complexity of the attacks included in the created dataset underscore its practical relevance, offering a realistic benchmark for evaluating the framework's performance. The results derived from this dataset confirm the robustness of our solution in accurately detecting and responding to various security threats, showcasing its potential as a tool in the domain of cloud-based container security.

Future work will explore the integration of additional ML algorithms, the expansion of the dataset to include emerging threats, and the application of our framework in various cloud-based architectures.

Acknowledgement. This research has been supported by the TÜBİTAK 3501 Career Development Program under grant number 120E537 and the TÜBA GEBİP Program. The entire responsibility of the publication belongs to the owners of the research.

References

1. Abbas, M., et al.: PACED: provenance-based automated container escape detection. In: 2022 IEEE International Conference on Cloud Engineering (IC2E), pp. 261–272 (2022). https://doi.org/10.1109/IC2E55432.2022.00035
2. Aktolga, I.T., Kuru, E.S., Sever, Y., Angin, P.: AI-driven container security approaches for 5G and beyond: a survey. ITU J-FET **4**(2), 364–382 (2023). https://doi.org/10.52953/ZRCK3746
3. Chen, J., Huang, H., Chen, H.: Informer: irregular traffic detection for containerized microservices RPC in the real world. High-Confidence Comput. **2**(2), 100,050 (2022). https://doi.org/10.1016/j.hcc.2022.100050
4. Cui, P., Umphress, D.: Towards unsupervised introspection of containerized application. In: 2020 the 10th International Conference on Communication and Network Security, ICCNS 2020, New York, NY, USA, pp. 42–51. Association for Computing Machinery (2021). https://doi.org/10.1145/3442520.3442530
5. Gan, Y., et al.: An open-source benchmark suite for microservices and their hardware-software implications for cloud & edge systems. In: Proceedings of the Twenty-Fourth International Conference on Architectural Support for Programming Languages and Operating Systems, pp. 3–18. ACM, Providence RI USA (2019). https://doi.org/10.1145/3297858.3304013
6. Sever, Y., Dogan, A.H.: A Kubernetes dataset for misuse detection. ITU J. Futur. Evolving Technol. **4**(2), 383–388 (2023). https://doi.org/10.52953/FPLR8631
7. Sever, Y., et al.: An empirical analysis of ids approaches in container security. In: 2022 International Workshop on Secure and Reliable Microservices and Containers (SRMC), pp. 18–26 (2022). https://doi.org/10.1109/SRMC57347.2022.00007

8. Tien, C.W., Huang, T.Y., Tien, C.W., Huang, T.C., Kuo, S.Y.: KubAnomaly: anomaly detection for the Docker orchestration platform with neural network approaches. Eng. Rep. **1**(5), e12,080 (2019). https://doi.org/10.1002/eng2.12080
9. Zhang, L., Cushing, R., de Laat, C., Grosso, P.: A real-time intrusion detection system based on OC-SVM for containerized applications. In: 2021 IEEE 24th International Conference on Computational Science and Engineering (CSE), Shenyang, China, pp. 138–145. IEEE (2021). https://doi.org/10.1109/CSE53436.2021.00029

Quantum Microservices: Transforming Software Architecture with Quantum Computing

Suleiman Karim Eddin[1,2], Hadi Salloum[1,2(✉)], Mohamad Nour Shahin[1,2], Badee Salloum[3,4], Manuel Mazzara[2], and Mohammad Reza Bahrami[2,5]

[1] QDeep, Innopolis, Russia
h.salloum@qdeep.net
[2] Innopolis University, Innopolis, Russia
[3] Syrian Virtual University, Damascus, Syria
[4] SCASE, Tartous, Syria
[5] Samarkand International University of Technology, Samarkand, Uzbekistan

Abstract. This paper conducts an exhaustive exploration of the evolutionary journey of microservices within the domain of software architecture. It meticulously traces the historical trajectory, current status, and potential future pathways of microservices in software design. Additionally, this study introduces a pioneering concept known as Quantum Microservices. Quantum Microservices represent a novel approach aimed at augmenting software design by leveraging concepts from quantum computing. In this paper, we will delve into defining their architecture, core features, challenges, and future prospects. This study envisions their pivotal role in reshaping the landscape of software development by offering enhanced efficiency and innovation opportunities.

Keywords: Microservices · Software Architecture · Distributed Systems · Quantum Computing

1 Introduction

The landscape of software architecture has undergone a transformative evolution, responding to the ever-increasing complexities and demands of modern applications. One of the pivotal strides in this journey is marked by the advent of microservices, an architectural paradigm that revolutionizes the way software systems are designed, developed, and deployed.

Microservices represent a departure from the traditional monolithic approach, advocating for the decomposition of applications into smaller, autonomous services. These services, operating as cohesive units, communicate via well-defined interfaces, fostering modularity, scalability, and agility in software development. The trajectory of microservices can be traced through the annals of

L. Barolli (Ed.): AINA 2024, LNDECT 204, pp. 227–237, 2024.
https://doi.org/10.1007/978-3-031-57942-4_23

software architecture, stemming from the foundational principles of modularization and evolving through Service-Oriented Computing (SOC) to its current iteration as microservices [17,23,24].

This exploration delves into the historical evolution of microservices, elucidating their foundational principles and the context in which they emerged. It delves into their present state, showcasing the dynamic landscape and the challenges and opportunities inherent in their implementation. Moreover, it casts a visionary gaze towards the future, exploring the convergence of microservices with quantum computing - an emerging frontier that holds promises of unparalleled computational capabilities and innovative applications.

The emergence of Quantum Microservices signifies a paradigmatic fusion, where the computational supremacy of quantum computing intersects with the service-oriented architecture [9]. This synergy opens vistas for groundbreaking advancements, offering specialized quantum functionalities encapsulated within modular and autonomous services.

This paper delves into the evolution and future prospects of microservices, particularly exploring their role in the realm of quantum computing. It presents the architecture, defines Quantum Microservices, outlines their core characteristics, operations, challenges, and anticipates their future evolution and prospects within the quantum computing domain.

2 Microservices and Their Architecture

Microservices represent a software architectural style characterized by the decomposition of applications into small, independent services that communicate via messages. These services operate as cohesive, self-contained processes, offering specific functionalities and interacting through well-defined interfaces. An exemplar of this architecture involves a service dedicated to performing arithmetic operations, named Calculator, which solely offers mathematical computations upon receiving messages. However, it refrains from encompassing extraneous functionalities, such as graph plotting or display features [1,12].

The hallmark of a microservice architecture lies in its distributed nature, where every module within the application constitutes a microservice. This architectural style emphasizes the composition and coordination of these individual components via message passing. For instance, to create a service that plots a function graph, one can introduce a new microservice, the Plotter, which orchestrates the Calculator to compute the graph's shape and engages the Displayer to render and exhibit the result back to the user. This orchestration, depicted through a workflow, showcases the collaborative interaction among the microservices involved in fulfilling the user's request.

In the realm of microservice architectures, the focus is on building discrete functionalities as separate microservices. This approach allows developers to concentrate on implementing and testing specific, coherent functionalities within individual microservices. Moreover, higher-level microservices oversee the coordination of functionalities among other microservices, demonstrating a scalable and modular system design [10].

These services operate independently and communicate with each other through well-defined interfaces, typically via message passing protocols. This approach offers several advantages [3,21]:

- **Decomposition and Independence:** Microservices focus on breaking down applications into smaller, manageable services, allowing for easier development, deployment, and maintenance. Each service operates autonomously, enabling independent scaling, updates, and enhancements without impacting the entire system.
- **Modularity and Scalability:** By dividing the application into smaller components, microservices promote modularity, making it simpler to understand and modify individual services. Additionally, they facilitate scalability, where specific services experiencing higher demand can be scaled independently without affecting the entire system.
- **Fault Isolation and Resilience:** Isolating functionalities into separate services minimizes the impact of failures. If one microservice encounters an issue, it's less likely to disrupt the entire application, promoting overall system resilience.
- **Technology Diversity:** Microservices permit the use of different programming languages, frameworks, and technologies for each service, allowing developers to choose the most suitable tools for specific functionalities.

The architecture of a microservice-based system involves the following key components and principles:

- **Service Components:** Each microservice encapsulates a specific business capability and exposes a well-defined interface for interaction with other services. These services can be independently developed, deployed, and maintained.
- **Message Passing and Communication:** Services communicate with each other using lightweight protocols like message queues, or event-driven mechanisms. This communication allows for asynchronous interactions, fostering flexibility and resilience.
- **Independent Data Management:** Microservices often manage their data using dedicated databases or data storage systems. This decentralized data management reduces dependencies and improves scalability.
- **Orchestration and Coordination:** Higher-level microservices, known as orchestrators, coordinate the interactions between various microservices to fulfill complex user requests. Orchestration services manage the workflow and choreography of communication among multiple microservices.

3 Evolution of Microservices

The evolution of microservices traces its roots through the annals of software architecture. Software architecture has undergone significant evolution since its inception in the 1960s when concerns regarding large-scale software development surfaced. Over the years, software architecture progressed from being perceived as distinct from implementation to an integrated component essential for software construction [6,16].

Notable works in the field, such as Perry and Wolf's foundational definition of software architecture in 1992 [2], laid the groundwork for extensive research and practical applications in both academic and industrial domains. The emergence of object orientation in the 1980s and its subsequent influence in the 1990s, notably exemplified by design patterns like Model-View-Controller (MVC) [11], contributed substantially to the software architecture domain.

Service-Oriented Computing (SOC), an offshoot of object-oriented and component-based computing, introduced the paradigm of services offering functionalities through message passing. SOC emphasized interface decoupling from implementation and the creation of workflow languages to manage complex service actions. This paradigm shift facilitated the development of formal models for understanding and verifying service interactions, leading to the classification of business modeling approaches and benefiting from foundational process models.

The initial service-oriented architectures (SOA) imposed complex requirements on services, which were later streamlined in the evolution toward microservices. Microservices, representing the second iteration of SOA and SOC, advocate for simplicity by focusing on the development of small, self-contained services that implement singular functionalities effectively. This paradigm shift, while rooted in the componentization concept akin to object-oriented programming, emphasizes independent deployment and message passing, necessitating specialized tools and design patterns to support its implementation and development.

Through this historical lens, the journey of microservices unfolds as an evolutionary stride in software architecture, leveraging the principles of modularization, scalability, and simplicity to address the challenges of modern distributed applications.

4 Present State of Microservices

The microservice architecture represents a relatively recent paradigm in software design and development, gaining prominence in recent years due to its potential to address modern software challenges. Rooted in the concepts of Service-Oriented Architecture (SOA), the term "microservices" emerged in 2011 during an architectural workshop, initially described as a means to encapsulate software architecture patterns. This approach, also previously recognized as fine-grained SOA, has evolved into a new trend emphasizing the creation of highly maintainable and scalable software systems [20].

Microservices functionally decompose complex systems into independent services, aiming for loose coupling and high cohesion. They operate as independent entities during both development and deployment phases, fostering modularity and maintaining focus on providing singular business capabilities. This architectural approach introduces several distinct characteristics setting it apart from its predecessors:

- **Size:** Microservices maintain comparably smaller sizes compared to traditional services, adhering to the principle that larger services should be divided into smaller, more manageable ones. This granularity enhances service maintainability and extendability.
- **Bounded Context:** Related functionalities are grouped within specific business capabilities, each implemented as an individual service.
- **Independence:** Every microservice operates autonomously, communicating solely through their published interfaces and maintaining operational independence from other services.

The current state of microservices is defined by its key system characteristics:

- **Flexibility:** Microservices enable systems to adapt and evolve within the dynamic business landscape, ensuring continual support for necessary modifications to remain competitive.
- **Modularity:** Systems are structured into isolated components, with each component contributing to overall system behavior rather than providing complete functionality.
- **Evolution:** Systems maintain their maintainability while continuously evolving and incorporating new features.

The landscape of microservices, despite gaining popularity, still lacks complete consensus on its defining characteristics. Works by M. Fowler, J. Lewis, S. Newman, L. Krause, among others, have established foundational concepts and best practices. Notably, studies have detailed the design and implementation aspects of systems utilizing microservices, showcasing diverse applications and implementations. For instance, research by Rahman and Gao demonstrates the application of behavior-driven development (BDD) [22] to alleviate the maintenance burden and promote acceptance testing in microservice architectures. The present state of microservices represents a relatively young and evolving architectural paradigm, continuously expanding its principles, practices, and real-world applications to address contemporary software complexities.

5 Future of Microservices

The exploration of microservices is in its nascent stages, and envisioning future directions is pivotal in advancing this paradigm. While microservices offer tremendous advantages, their pervasive distribution, which fosters loosely coupled systems, also poses significant challenges in programming distributed systems compared to monolithic architectures [18].

5.1 Dependability

Programming with microservices introduces various pitfalls, particularly in building dependable systems. An essential aspect in ensuring dependability is the specification of interfaces. However, the diversity of technologies used in microservices creates challenges in specifying contracts for service composition. Presently, informal documentation mostly describes service usage, leading to potential ambiguities and errors in client development [3].

Efforts to address this issue involve formal specification tools for defining technology-agnostic interfaces. Tools like Jolie, Apache Thrift, and Google's Protocol Buffers attempt to bridge the gap by enabling language-specific interfaces or verifying well-typedness of messages. Challenges remain in adapting these tools to architectures like REST and applying static type-checking to dynamic languages like JavaScript, commonly used in microservices.

5.2 Behavioural Specifications and Choreographies

While APIs define interfaces, ensuring compatibility during execution sessions remains a challenge. Behavioural types, exemplified by session types, attempt to describe and verify service behaviors, yet widespread adoption faces limitations due to restrictive nature, especially in dealing with nondeterminism in protocols.

Behavioural interfaces and choreographies, derived from W3C's global behavior definitions, offer promise in formalizing communications envisioned during software design [15]. Choreographic Programming, leveraging choreographies for automatic compliant implementations, signifies a correctness-by-construction approach, guaranteeing crucial properties like deadlock freedom.

5.3 Moving Fast with Solid Foundations

Aligning behavioural types and choreographies with established logical models holds promise. The Curry-Howard correspondence linking session types to -calculus and its subsequent developments offer insights into ensuring correct microservice systems by leveraging logical models and formal methods. Empirical investigations in microservice programming will likely delineate precise research directions to address practical scenarios [17].

5.4 Trust and Security

The microservices paradigm poses intricate trust and security challenges [19], expanding the attack surface area due to network exposure and heightened network complexity. The diverse nature of microservices' interactions necessitates robust security measures to address potential vulnerabilities arising from the increased attack surface and intricate network activity. The future of microservices relies on addressing challenges related to dependability, behavioural specifications, leveraging established formal methods, and enhancing security measures to ensure a robust and secure ecosystem for distributed systems.

6 Quantum Microservices: Exploring the Future Frontier

The advent of quantum computing heralds a paradigm shift in computational capabilities, promising unprecedented processing power that could revolutionize various industries. As quantum computing matures, its integration with microservices introduces a novel frontier with profound implications for the technological landscape. Quantum computing operates on quantum bits or qubits, leveraging quantum phenomena like superposition and entanglement to perform computations exponentially faster than classical computers. This immense computational power offers an array of possibilities, ranging from complex simulations, cryptography, optimization problems, to drug discovery and artificial intelligence [7,8].

Quantum microservices signify a groundbreaking paradigm in software architecture that integrates quantum computing capabilities into the service-oriented approach, presenting a tailored solution to the intricacies of quantum software development. Let's delve deeper into the components and implications of quantum microservices:

6.1 Defining Quantum Microservices

Quantum microservices encapsulate specialized quantum functionalities [14], serving as modular, independent, and focused service units. These units are designed to handle specific quantum computing tasks or algorithms, enabling granular control over intricate quantum operations.

The architecture of Quantum Microservices, as shown in Fig. 1, embodies a layered structure that encompasses distinct components:

1. **Quantum Algorithms Integration Layer:** This layer enables the seamless incorporation of quantum algorithms into microservices functionalities, empowering the system to execute quantum operations as part of its computational tasks.
2. **Quantum Data Handling Module:** Specifically designed to manage and process quantum data structures leveraging quantum properties like superposition and entanglement, this module ensures efficient storage, retrieval, and manipulation of quantum information within the microservices framework.
3. **Quantum Computing Infrastructure Integration:** This component integrates with quantum computing hardware and infrastructure, allowing microservices to access quantum processors, annealers, or simulators, thereby facilitating the execution of quantum operations within the architecture.
4. **Microservices Framework:** At the core of the architecture lies the microservices framework, embodying the principles of modularity, scalability, and flexibility. It provides the foundation for integrating quantum capabilities while maintaining the advantages of a distributed computing environment.

6.2 Core Characteristics and Operation

- **Modularity and Specialization:** Quantum microservices focus on executing precise quantum tasks, such as implementing algorithms (e.g., Shor's algorithm for factorization) or simulating quantum systems. Each microservice concentrates on a distinct quantum operation, fostering efficiency and dedicated functionality.

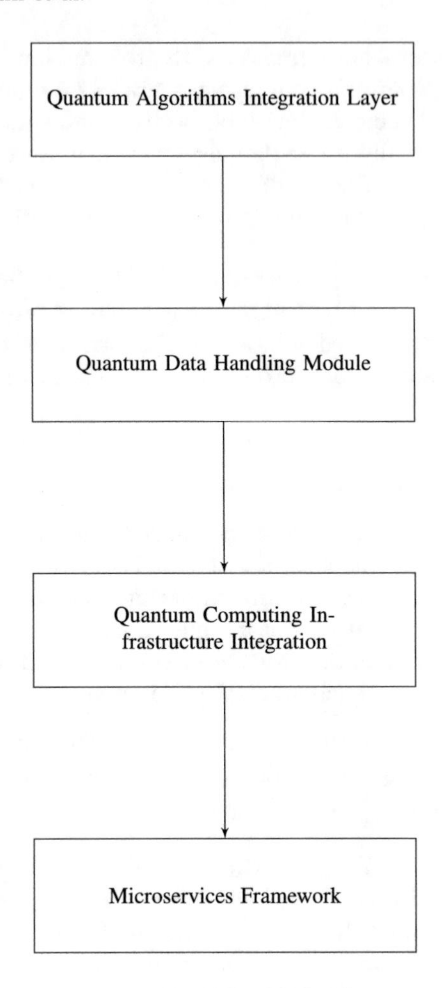

Fig. 1. Architecture of Quantum Microservices

- **Independence and Scalability:** Similar to traditional microservices, quantum microservices operate autonomously, facilitating independent deployment, updates, and scalability. This autonomy ensures that changes or modifications in one service don't impact the functioning of others.
- **API-Driven Architecture:** Quantum microservices expose well-defined APIs that act as gateways for external systems to interact with and utilize their quantum capabilities. These APIs abstract the complexity of quantum operations, enabling seamless integration into various applications [5,13].
- **Integration and DevOps Compatibility:** The alignment of quantum microservices with DevOps methodologies streamlines their development, testing, and deployment. Leveraging Continuous Integration (CI) and Continuous Deployment (CD) tools ensures efficient delivery and reliability, mirroring practices in traditional software development.

- **Utilizing Quantum Development Tools:** Quantum microservices leverage dedicated quantum development frameworks like Qiskit, Cirq, or similar Quantum Software Development Kits (SDKs). These platforms provide the necessary tools, libraries, and simulators to construct quantum algorithms or circuits, forming the foundation of these microservices' functionality.
- **Diverse Applications:** The potential applications of quantum microservices span across diverse domains, from optimizing complex systems and cryptography to material science, finance, and beyond. Their ability to offer specialized quantum functionalities ignites innovation in various industries.

6.3 Challenges and Future Evolution

The evolution of quantum microservices faces notable challenges, including the absence of standardized communication protocols, API specifications, and language-agnostic code generation tools specifically tailored for quantum services. Researchers are actively exploring these realms, aiming to define communication frameworks and implement quantum microservices for real-world applications.

Deployment challenges persist, with current platforms like Amazon Braket providing limited control over quantum service execution platforms. Efforts are underway to gain finer control over execution environments and adapt API Gateways for efficient handling of quantum services [4,5,13,14].

6.4 Future Prospects of Quantum Microservices

As of the present state, Quantum Microservices represent an area of ongoing research and development at the intersection of quantum computing and microservices architecture. While theoretical models and experimental frameworks have been proposed and explored, widespread practical implementation and deployment are still in their infancy.

As quantum computing continues to advance, the evolution of quantum microservices presents opportunities for standardization, improved integration, and harnessing the full potential of quantum technology within service-oriented frameworks. Addressing existing challenges will lead to a refined and mature ecosystem for quantum microservices, fostering innovation and advancement across industries.

7 Conclusion

This paper serves as a comprehensive exploration of the evolutionary trajectory of microservices, paralleled by an in-depth investigation into the realm of quantum microservices. Through a meticulous analysis, it has elucidated the fundamental definitions, core characteristics, operational modalities, architectural frameworks, and the spectrum of challenges and opportunities inherent in this nascent field. It is crucial to acknowledge the infancy of quantum software engineering, where various facets are just beginning to capture the curiosity and attention of researchers. This signifies the need for sustained inquiry and rigorous research efforts to fully harness the transformative capabilities of quantum computing within the expansive landscape of software engineering. As the field continues to mature, fostering a deeper understanding of quantum principles in software development will undoubtedly shape the future of technological innovation.

References

1. Carneiro, C., Schmelmer, T.: Microservices from Day One. Apress, Berkeley, CA (2016). https://doi.org/10.1007/978-1-4842-1937-9
2. Gacek, C., Abd-Allah, A., Clark, B., Boehm, B.: On the definition of software system architecture. In: Proceedings of the First International Workshop on Architectures for Software Systems, Seattle, WA, pp. 85–94 (1995)
3. Wolff, E.: Microservices: Flexible Software Architecture. Published by Addison-Wesley Professional (2016)
4. Moguel, E., Garcia-Alonso, J., Haghparast, M., Murillo, J.M.: Quantum Microservices Development and Deployment. arXiv preprint arXiv:2309.11926 (2023)
5. Moguel, E., Rojo, J., Valencia, D., Berrocal, J., Garcia-Alonso, J., Murillo, J.M.: Quantum service-oriented computing: current landscape and challenges. Softw. Q. J. **30**(4), 983–1002 (2022)
6. De Giacomo, G., Lenzerini, M., Leotta, F., Mecella, M.: From component-based architectures to microservices: a 25-years-long journey in designing and realizing service-based systems. In: Aiello, M., Bouguettaya, A., Tamburri, D.A., van den Heuvel, W.-J. (eds.) Next-Gen Digital Services. A Retrospective and Roadmap for Service Computing of the Future. LNCS, vol. 12521, pp. 3–15. Springer, Cham (2021). https://doi.org/10.1007/978-3-030-73203-5_1
7. Salloum, H., et al.: Integration of machine learning with quantum annealing. In: International Conference on Advanced Information Networking and Applications. Springer (2024)
8. Salloum, H., et al.: Quantum advancements in securing networking infrastructures. In: International Conference on Advanced Information Networking and Applications. Springer (2024)
9. Kumara, I., Van Den Heuvel, W.-J., Tamburri, D.A.: QSOC: quantum service-oriented computing. In: Barzen, J. (ed.) SummerSOC 2021. CCIS, vol. 1429, pp. 52–63. Springer, Cham (2021). https://doi.org/10.1007/978-3-030-87568-8_3
10. Thönes, J.: Microservices. IEEE Softw. **32**(1), 116 (2015)
11. Deacon, J.: Model-view-controller (MVC) architecture, vol. 28 (2009). http://www.jdl.co.uk/briefings/MVC.pdf. Cited 10 Mar 2006
12. Bogner, J., Fritzsch, J., Wagner, S., Zimmermann, A.: Microservices in industry: insights into technologies, characteristics, and software quality. In: IEEE International Conference on Software Architecture Companion (ICSA-C), pp. 187–195. IEEE (2019)
13. Garcia-Alonso, J., Rojo, J., Valencia, D., Moguel, E., Berrocal, J., Murillo, J.M.: Quantum software as a service through a quantum API gateway. IEEE Internet Comput. **26**(1), 34–41 (2021)

14. Rojo, J., Valencia, D., Berrocal, J., Moguel, E., Garcia-Alonso, J., Rodriguez, J.M.: Trials and tribulations of developing hybrid quantum-classical microservices systems. arXiv preprint arXiv:2105.04421 (2021)
15. Baresi, L., Garriga, M.: Microservices: the evolution and extinction of web services? In: Microservices, pp. 3–28. Springer, Cham (2020). https://doi.org/10.1007/978-3-030-31646-4_1
16. Bass, L. Clements, P., Kazman, R.: Software Architecture in Practice. Addison-Wesley Professional (2003)
17. Mazzara, M., Dragoni, N., Bucchiarone, A., Giaretta, A., Larsen, S.T., Dustdar, S.: Microservices: migration of a mission critical system. IEEE Trans. Serv. Comput. **14**(5), 1464–1477 (2018)
18. Dragoni, N., et al.: Microservices: yesterday, today, and tomorrow. In: Present and Ulterior Software Engineering, pp. 195–216. Springer, Cham (2017). https://doi.org/10.1007/978-3-319-67425-4_12
19. Mateus-Coelho, N., Cruz-Cunha, M., Ferreira, L.G.: Security in microservices architectures. Procedia Comput. Sci. **181**, 1225–1236 (2021)
20. Zimmermann, O.: Microservices tenets: agile approach to service development and deployment. Comput. Sci. Res. Dev. **32**, 301–310 (2017)
21. Li, S., et al.: Understanding and addressing quality attributes of microservices architecture: a systematic literature review. Inf. Softw. Technol. **131**, 106449 (2021)
22. Rahman, M., Gao, J.: A reusable automated acceptance testing architecture for microservices in behavior-driven development. In: 2015 IEEE Symposium on Service-Oriented System Engineering, pp. 321–325. IEEE (2015)
23. Vernon, V., Jaskula, T.: Strategic Monoliths and Microservices: Driving Innovation Using Purposeful Architecture. Published by Addison-Wesley Professional (2021)
24. Abgaz, Y.: Decomposition of monolith applications into microservices architectures: a systematic review. IEEE Trans. Softw. Eng. (2023)

Handover Management Scheme in Macrocell-Femtocell Networks

Maroua Ben Gharbia[1,2](✉) and Ridha Bouallegue[1]

[1] LR11TIC03, Innov'Com Research Laboratory, University of Carthage, High School of Communication of Tunis, Tunis, Tunisia
maroua.bengharbia@fsb.ucar.tn

[2] University of Tunis El Manar, National Engineering School of Tunis, Tunis, Tunisia

Abstract. The deployment of Femtocell is considered as one of the most important solution for increasing the capacity and the coverage of the wireless networks. Femtocells are short-range, low-cost and low-power access points designed especially for Home and small organization. As the number of user equipment is increasing every day so the existing networks are unable to provide a high data rates. So Femtocells play a key role to offload the data traffic from Macrocell. The mobility in the presence of Femtocells is one of the most challenging issues owing to the dense network layout, the short cell radii and the potentially unplanned deployment. In this paper, we present a new Handover strategy in Femtocell. Unlike the traditional algorithms, the potential reduction of the service capacity due to failures are considered as well as various performances parameters such as total time transmission and throughput. To evaluate the performance of our Handover management scheme, the simulation results are compared with traditional algorithms. The simulation results show that the number of Handover is considerably reduced and the system performance is improved.

Keywords: Handover · Femtocell · Management Scheme · Parameters · Macrocell

1 Introduction

Nowadays, mobile communication systems are becoming more complex due to the emergence of several technologies and the demand to improve indoor coverage and high data rates [1, 2]. Femtocell is considered as a key promising technology to extend mobile network coverage and to enhance the system capacity. It is a short-range, low-cost and low- power cellular access point, known as Home Base Station or Home Node B [3, 4]. The coverage ranges of Femtocells are in the tens of meters. They are initially designed for residential use to get better indoor communication and data coverage, improving the Macrocell reliability and increasing the peak-bit rate in low coverage areas.

The integration of Femtocell in the coverage area of Macrocell can increase the utilization of wireless capacity which is not covered by Macrocell. So, the deployment of Femtocell allows for better quality communication services and data transmission.

L. Barolli (Ed.): AINA 2024, LNDECT 204, pp. 238–248, 2024.
https://doi.org/10.1007/978-3-031-57942-4_24

Femtocell offers the following benefits. The coverage and capacity are improvement because transmit receive distance is short. These translate into improved reception and higher capacity. The improved coverage enables user equipment to work at the peak of their capabilities so the data rates and call quality are highest. Femtocell allows operators to offload a significant amount of traffic away from Macrocell so the reliability of Macrocell is improved. Femtocell minimizes the capital and operational expenditure by reducing the additional time for installation and operation cost.

Using the specific knowledge of the user equipment location Femtocell can offer benefits such as control of devices around the home. Femtocells are designed to be simple deployment: simple to install, to configure and to operate [5]. Figure 1 shows a Macrocell-Femtocell network.

There are many technologies that have deployed Femtocells. One of the first these technologies are Long Term Evolution (LTE). LTE is a 4G network standard introduced by 3GPP. It can enhance capacity and data rates and reduce delays [6]. The availability of Femtocells in particular area increases the technological challenges in Handover procedure. Another challenge is the mitigation the unnecessary Handover since large number of Femtocells can trigger the very frequent Handover even before the current initiated Handover procedure is completed. Handover, in the presence of Femtocells, is one of the most challenging issues because of the short cell radius, the dense network layout and the unplanned deployment.

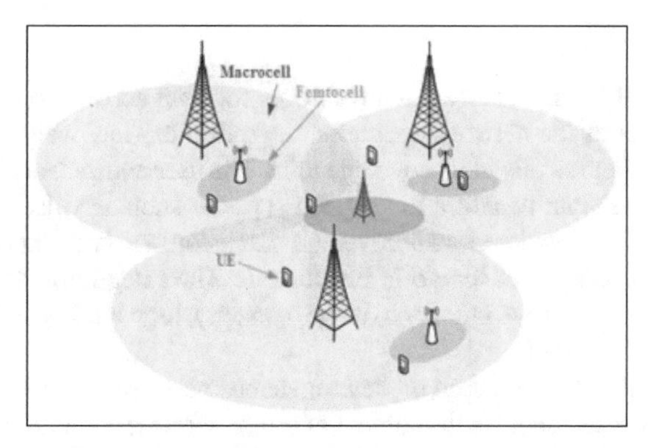

Fig. 1. Macrocell-Femtocell network

On the other hand, Handover is complicated to give for user equipment (UE) access to Femtocells. In the access control management, there are three Femtocell access modes. The first is open access mode which any user can access the Femtocell and can provide its services but it's not secure. Network operators prefer open access mode this mode extends the coverage of UE. It improves the overall network throughput. Although, the signaling and the number of Handover are increased causing certain security issues. The second mode is closed access mode, if you are not the closed subscriber group members (CSG), you can't access the Femtocell. This mode facilitates high quality coverage and higher rate multimedia services to indoor UE but its resources are limited. Finally, hybrid access mode is the combination of the open and closed access mode. It is easily vulnerable to security challenges. To choose the appropriate access mode is very important [7–9].

There are many security problems regarding access control to block unauthorized and authorized users equipments. The knowledge of the owner its neighbors may use its services and can lead potential network intrusions and hacking. So, the system capacity can be reduced. Several researches have defined techniques used to resolve the problem of Handover in Femtocell. They used the parameters to make the Handover decision as Reference Signal Received Power (RSRP), speed of user equipment, distance, bandwidth, energy and interference. The majority of existing Handover algorithms uses the RSRP as a primary parameter of decision Handover to compare the RSRP of serving and target cells, with the aim to lower the Handover and minimize the ping-pong effect [10].

In this paper a new Handover management scheme in Macrocell-Femtocell LTE networks is proposed. The main advantage of our proposed scheme is that many parameters of Handover decision are considered. One of the most challenging tasks in communication networks is to maintain seamless mobility and service continuity during a mobility of user equipment. The rest of this paper is organized as follows: Sect. 2 reviews some related works on Handover in Macrocell-Femtocell networks. The Handover procedure is presented in Sect. 3. Section 4 illustrated our proposed Handover algorithm. Then, the performance of simulation is analyzed in Sect. 5. Finally, Sect. 6 concludes the paper.

2 Related Work

The deployment of Femtocell present advantages for both network operators and users equipments. As far as the network operators, the traffic in Macrocell is reduced duo to the integration of Femtocells in its coverage area. The user equipments in the proximity of the coverage area can benefit a high speed services such as voice, multimedia and video. Femtocell is considered as a coverage extension for both indoor and outdoor coverage when Macrocell coverage is insufficient. The integration of Femtocell into cellular network can improve indoor coverage, capacity, high quality and high data rates to discharge the Macrocell.

In [11] the authors are focused on heterogeneous networks. Many approaches have been proposed by researchers in the context of future wireless communication networks. A Handover algorithm is proposed and based on multiple conditions that need to be satisfied for the execution of the Handover and select a suitable Femtocell. The role of Femtocells in the heterogeneous is presented in [12].

The interference mitigation techniques in two-tier networks are discussed. The performance of LTE network in a Macro Femtocell scenario is evaluated in [13]. The authors have described the mobility management and presented the Handover strategy based on hysteresis margin and time to trigger. To improve the throughput of Heterogeneous network, the authors in [14] proposed an optimal placement of Femtocell model, a dynamic bandwidth allocation scheme, a dynamic power mechanism and an enhanced priority scheduling. The random position of Femtocell comes with challenges of network performance and cross-tier interference.

In [15], the authors presented a positioning algorithm for deployment of Femtocell within the context of the indoor environment. In the paper [16], the Handover strategy between Macrocell and Femtocell is proposed to reduce the unnecessary Handover. The strategy is mainly based on calculating equivalent received signal strength along with dynamic margin. The existing Handover decision algorithms for Macrocell-Femtocell networks is presented and is classified based on the primary Handover decision criterion used. The authors presented a comparative summary of the principal decision parameters and the existing Handover decision algorithms.

The CoMP-based Handover scheme is proposed in order to minimize unnecessary Handover decision [17]. A combination of RSRP and RSRQ facilitate a decision. The random position of Femtocells comes with challenges of network performance and cross-tier interference.

To minimize the unnecessary Handover in two-tier heterogeneous networks with dense deployment of Femtocells, the authors in [18] proposed multicriteria Handover strategy based on different velocity thresholds and number of users equipment in Femtocell. To eliminate the redundant Handover in Femtocell, the combination of received signal strength indicator and mobile speed of the current cell and neighboring cells is presented in [19]. The authors proposed a novel approach to reduce the number of Handover. In the same way, the Handover strategy is based on calculating the equivalent received signal strength with the dynamic margin for performing Handover [20]. The interference in Femtocell due to neighboring Femtocell and Macrocell can be reduced by Handover. The Handover algorithm is proposed for uplink co-channel interference mitigation and based on time to stay and signal to interference plus noise ration thresholds.

3 Handover Procedure

The network systems including the Femtocell network implement a Handover procedure to support the user mobility. Handover is a process of transferring UE from one cell to another while a session is in progress. The mobility in the presence of Femtocell is one of the most challenging issues, owing to the short cell radii, the potentially unplanned deployment and the dense network.

The UE moves randomly according to its movements. In Femtocell network, there are three types of Handover scenarios, i.e., Hand-In, Hand-Out and Inter-Femtocell Handover shown in Fig. 2.

In the Hand-In scenario, the UEs are moving from the coverage area of Macrocell to the coverage area of Femtocell [18–20]. It is the most complex type of scenario compared to other Handover scenarios as there are many Femtocells in the coverage of Macrocell. During this scenario, the UE should be selecting the appropriate Femtocell. The second scenario is the Hand-Out. The UE moves from the coverage area of Femtocell to the coverage area of Macrocell. This scenario does not pose many problems as compared to Hand-in scenario. Although the target cell is Macrocell, there is no option to select the target cell since there only the Macrocell. When the RSSI from Macrocell is stronger than the RSSI of Femtocell, the UE will be connected directly. The third scenario is the Inter-Femtocell. It corresponds to the scenario of Handover from one Femtocell to another Femtocell. In this scenario all Femtocells are assumed to be placed at the same location and served by the same service provider.

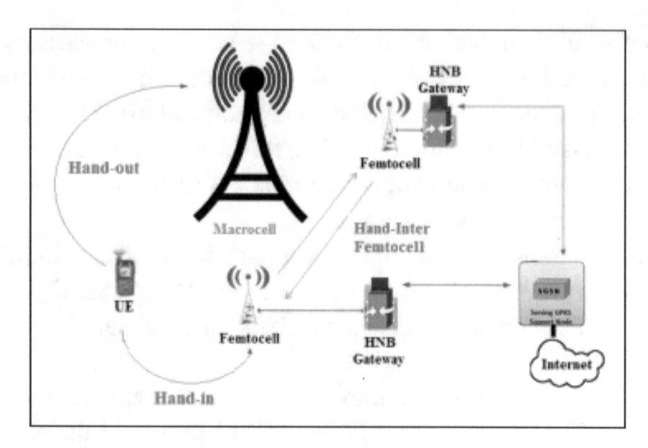

Fig. 2. Handover Scenario

4 Proposed Strategy

The problem of designing a Handover management scheme in Macrocell-Femtocell network is formally described as follows. Hand-In scenario is discussed. Our proposed scheme can be shown in the flowchart in Fig. 3. When the UE is initially connected to Macrocell and moves through many Femtocells in the coverage of Macrocell. If the UE enters in the coverage of Femtocell and detects the signal of this Femtocell, two cases are discussed.

First case, the UE moves to the Femtocell according to following conditions. The RSRP of Femtocell is greater than the RSRP of the Macrocell with a certain margin, the UE moves with a medium speed and the transmission time of UE in the coverage of Femtocell is less than the transmission time of UE in the coverage of Macrocell and the dwell time of the UE in the coverage of Femtocell.

Second case, the UE stays connect to the Macrocell when the conditions mentioned previously are not checked. The Handover procedure is started to make a decision and to choose the appropriate Femtocell. This decision is based on estimated values of T_{Dwell} and T_E.

The system model of Macrocell-Femtocell network is illustrated in Fig. 2. A large number of Femtocells are deployed in the coverage area of Macrocell.

Definition 1: Dwell Time (T_{Dwell}) is the time period from the UE starting to communicate with the Macrocell/ Femtocell to the UE stopping to communicate with the Macrocell/Femtocell, where the UE is in the coverage of the Macrocell/Femtocell.

Definition 2: Transmission Time between the UE and the Femtocell, where is equal to the remainder data of the UE divided by the average transmission rate between the UE and the Femtocell within dwell time.

Let $SNR_{UE\text{-}F}$ is the signal power between the UE and Femtocell. Let $SNR_{UE\text{-}M}$ is the signal power between the UE and Macrocell. Let $SNR_{M\text{-}F}$ is the signal power between the Macrocell and Femtocell.

The Dwell Time (T_{Dwell}) is given as follows.

1. If a UE enters the coverage of Femtocell at time t_i, there are two values of signal power can be received at Femtocell, $SNR_{UE\text{-}F}$ is obtained by Femtocell and $SNR_{UE\text{-}M}$ is obtained from Macrocell. The distance between UE and Femtocell $D_{UE\text{-}F}$, the distance between UE and Macrocell $D_{UE\text{-}M}$, and the distance between Macrocell and Femtocell $D_{M\text{-}F}$ can be obtained.
2. If the UE moves with a fixed speed, we obtain three distances values $D'_{UE\text{-}F}$, $D'_{UE\text{-}M}$, and $D'_{M\text{-}F}$ can at time t_{i+1}.

   An angle θ_i between $D_{M\text{-}F}$ and $D_{UE\text{-}F}$ at time t_i and an angle θ_{i+1} between $D'_{M\text{-}F}$ and $D'_{UE\text{-}F}$ at time t_{i+1} can be calculated.

   An angle θ_c is between $D_{UE\text{-}F}$ and $D'_{UE\text{-}F}$. An angle θ between $D_{UE\text{-}F}$ and the moving direction of the UE is determined.

   Let the maximum signal power of the UE at time t_{i+1} be denoted as SNR_{MAX}. The effect of the path loss is considered.

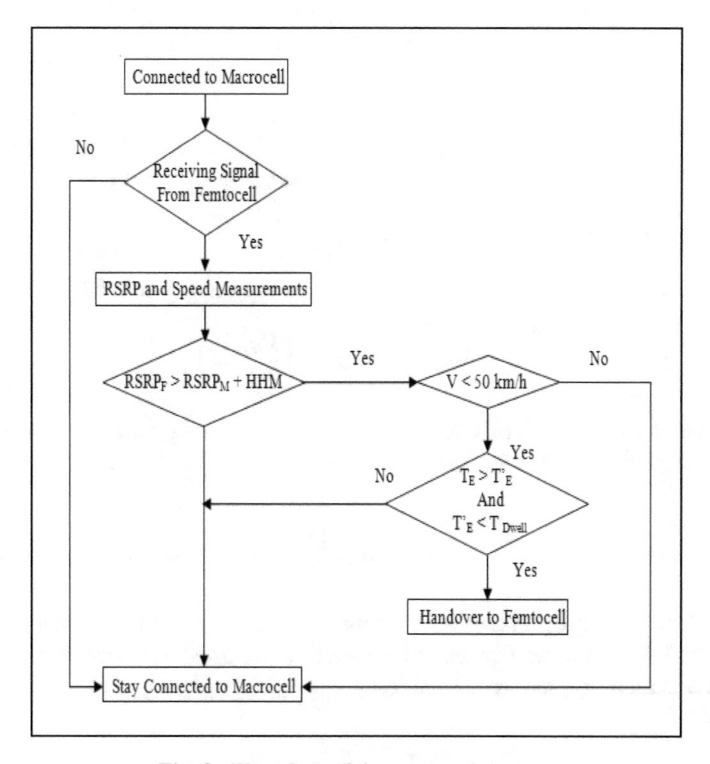

Fig. 3. Flowchart of the proposed strategy

The path loss model [7] is adopted:

$$PL(d) = 38.46 + 20\log_{10} + 0.7d_{2D} \tag{1}$$

where d is the distance between the UE and Femtocell, and $0.7d_{2D}$ is the penetration loss due to internal walls. The signal power becomes

$$SNR_{PL} = SNR_{UE} - PL(R\sin\theta) \tag{2}$$

where SNR_{UE} is the SNR of UE at time t_{i+1}. The effect of the fast fading and shadowing are considered.
The maximum received signal strength can be represented as

$$SNR_{MAX} = \frac{1}{\sqrt{2\psi_{dB}^2}} e^{\left(-\frac{(SNR_{PL}-\mu^2)}{2\sigma_{dB}^2}\right)} \tag{3}$$

where σ_{dB} is the standard deviation of shadowing and μ is the average power.
The T_{Dwell} is calculated as follows. Let T denoted the time of UE entering the coverage of Femtocell to leaving the coverage of the Femtocell.

$$T = 2 \times \frac{SNR_{MAX}}{\alpha} \tag{4}$$

where $\alpha = \frac{\int_R^r SNR_{MAX}}{\Delta t}$ is the average variation of SNR at Δt, where R and r are the distances between UE and Femtocell at time t_i and t_{i+1} and $\Delta t = t_{i+1} - t_i$.

The Dwell time is equal to $T_{Dwell} = T - \Delta t = T - (t_{i+1} - t_i)$.

Transmission Time is calculated as follows.

1. SNR is obtained from the signals received from the serving cell, where

$$SNR_{dB} = 10\log_{10}\left(\frac{P_{signal}}{P_{Noise}}\right) \tag{5}$$

2. Let D denotes the data size, R denotes the transmission rate, C denotes the maximum channel capacity and B is the bandwidth $C = B \times \log_2(1 + SNR_{dB})$ the transmission time is

$$T = \frac{D}{R} \tag{6}$$

3. Let P_u denotes as the probability of resources unoccupied by the UE within the time period of T. To predict the unoccupied resources, we use the Poisson distribution, the formal equation of Poisson distribution is

$$P(k,T) = \frac{(\lambda T)^k}{k!} e^{-(\lambda T)} \tag{7}$$

where k is the number of events, T is the period time and λ is the proportion of average event happened. Let k set as zero and T set as transmission time T. So, the probability of resources unoccupied by the UE within the time period of T is

$$P = \Pi_{n=1}^{m} P_n(0, T) = e^{-\Sigma_{n=1}^{m}\lambda_n T} \tag{8}$$

Par consequently, the transmission time in current cell is

$$T_E = \frac{D}{P_u \times R} + (1 - P_u) \times T_C \tag{9}$$

where T_C denotes the layer switch time and (1-P_u) is the probability of resources can be occupied within the time period T.

4. The transmission time in new cell is

$$T'_E = \frac{D}{P'_u \times R'} + \left(1 - P'_u\right) \times T_C + T_H \tag{10}$$

where T_H denotes the Handover time, T_C denotes the layer switch time and (1-P'_u) is the probability of resources can be occupied within the time period T' in the new cell.

5 Simulation Results

Our paper presents a new Handover management scheme in Macrocell-Femtocell networks. To evaluate our Handover management scheme, the Handover decision [18] and the traditional algorithm are mainly implemented using the MATLAB Simulator. It is an open source simulator in which a scenario based on single Macrocell with multiple Femtocells. The UE are randomly placed in coverage area of Macrocell near Femtocells.

Table 1. Simulation Parameters

Parameter	Values
Radius of Macrocell	1000 m
Radius of Femtocell	50 m
Number of Macrocell	1
Number of Femtocell	10
Number of UE	5–50
Data size	5–50 Mb
Simulation time	100–1000 s

The simulation parameters are given in Table 1. The performance parameters to be observed are as follows:

The performance parameters are:

- Total Transmission Time (TTT) is the time interval of the data transmission between a UE and current Femtocell/Macrocell or new Macrocell/Femtocell. It is estimated from the first packet transmitted from current Femtocell/Macrocell until the final packet received by UE through the current Femtocell/Macrocell or new Femtocell/Macrocell.
- Throughput (TP) is the total number of data packets which can be transmitted and received between a UE and the Femtocell/Macrocell.

1. **Total Transmission Time (TTT)**

The simulation results of the TTT under various data sizes and number of UE are shown in Fig. 4 (a) and (b). Figure 4 (a) shows the simulation results of TTT under various data sizes, ranging from 5 to 50 Mb, when the number of UE is fixed at 10. We observed that the TTT increases as the data sizes increases. The curve of TTT of proposed algorithm is lower than the TTT of cross layer algorithm and the TTT of traditional algorithm. This is because that our Handover management scheme uses multiple parameters of Handover decision to choose an appropriate Femtocell. Figure 4 (b) illustrated the impact of TTT with various numbers of UEs, ranging from 5 to 50, when the data size is fixed at 25 Mb. In general, the TTT increases as the number of UEs increases. The TTT of our proposed strategy is lower than the TTT of cross algorithm and the TTT of traditional algorithm.

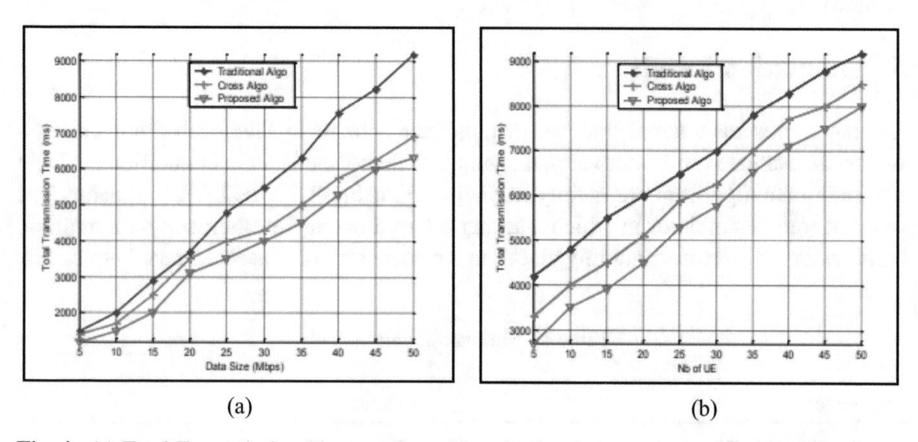

Fig. 4. (a) Total Transmission Time vs. Data Size, (b) Total Transmission Time vs. Number of UE

2. **Throughput (TP)**

The simulation results of the TP under various data size and number of UEs are shown in Fig. 5 (a) and (b). Figure 5 (a) shows the TP variation under various data size, ranging from 5 to 50Mb, when the number of UEs is fixed at 10. The TP drops as the data size increases. Figure 5 illustrates that the curve of the TP of our proposed strategy is higher than the TP of cross layer algorithm and the TP of traditional algorithm. This is because that our proposed strategy uses multiple parameters of Handover decision.

Figure 5 (b) illustrates the simulation results of TP under various numbers of UEs, ranging from 5 to 50, when the data size is fixed at 25Mb. The TP drops as the number of UEs increases. The curve of TP of cross layer algorithm and traditional algorithm are lower than the TP of proposed strategy because the cross layer algorithm and the traditional algorithm use few parameters of Handover decision.

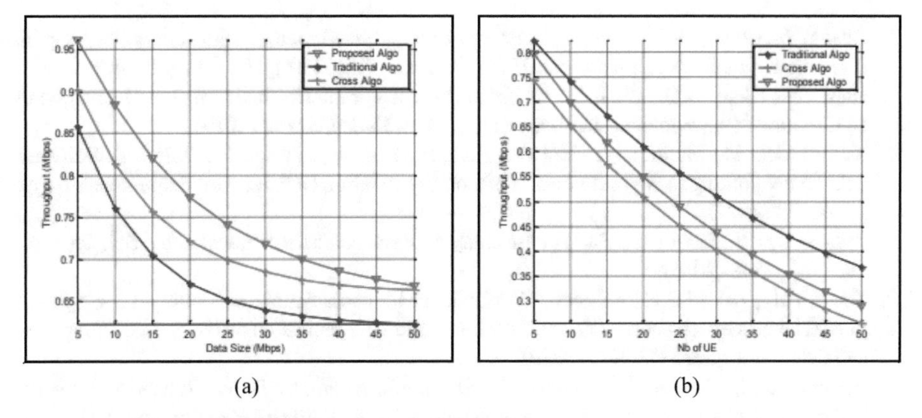

(a) (b)

Fig. 5. (a) Throughput vs. Data Size, (b) Throughput vs. Number of UE

6 Conclusion

The integration of Femtocells into Macrocell network dictates architectural and procedural enhancements that go beyond the standard cellular networks. Femtocell can provide a higher quality network access in indoor coverage and discharge the Macrocell. As the Femtocell coverage area is very small, there are some unnecessary Handover occurs in Macrocell to Femtocell. Therefore Handover management plays a significant role in cellular networks. Aiming to execute the Handover procedure, in this paper we have developed a new Handover management scheme in Macrocell-Femtocell networks. A new strategy is implemented and based on several Handover decision parameters that lead to increase overall network performance. In particular, our strategy aims to minimize total transmission time and throughput. It is clearly shown by the results that the proposed strategy can significantly offer more throughput result and less total transmission time compared with the cross layer Handover and traditional algorithm.

An interesting suggestion for future work could be the import of more parameters of the proposed strategy such as will be done to further develop the power consumption and the packet loss ratio.

References

1. Sapwatcharasakun, C., Kunarak, S., Prasertwong, K.: Green energy of handover management in seamless networks. In: 2023 International Conference On Cyber Management and Engineering (CyMaEn) (2023)
2. Haghrah, A., Abdollahi, M.P., Azarhava, H., Niya, J.M.: A survey on the handover management in 5g-NR cellular networks: aspects, approaches and challenges. Wirel. Commun. Netw. **2023**, 52 (2023https://doi.org/10.1186/s13638-023-02261-4
3. Raheem, R., Lasebae, A., Raheem, A.: Novel group handover mechanism for cooperative and coordinated mobile femtocells technology in railway environment. ELSEVIER **15**, 1–10 (2022)
4. Maiwada, U.D., Danyaro, K.U., Sarlan, A.B.: An improved mobility state detection mechanism for Femtocells in LTE network. In: 2022 International Conference on Decision Aid Sciences and Applications (DASA) (2022)

5. Chang, W., Cheng, W.Y., Meng, Z.T.: Energy-efficient sleep strategy with variant sleep depths for open-access femtocell networks. IEEE Commun. Lett. **23**(4), 708–711 (2019)
6. Habib, A., Khan, W.U., Ullah, T.: An efficient cross-tier interference mitigation technique in LTE femtocell environment. IEEE Commun. Mag. **18**(49), 37–43 (2019)
7. Karimzadeh, M., Valtulina, L., Van Den Berg, H., Pras, A., Liebsch, M., Taled, T.: Software defined networking to support IP address mobility in future LTE network. In: Wireless Days, pp. 46–53 (2017)
8. Padmapriya, S., Tamilarasi, M.: A case study on femtocell access modes. Eng. Sci. Technol. **19**, 1534–1542 (2016)
9. Yusof, A.L., Ali, M.S.N., Ya'acob, N.: Handover implementation for femtocell deployment in LTE heterogeneous networks. In: 2019 International Symposium on Networks, Computers and Communications (ISNCC) (2019)
10. Alhammadi, A., Roslee, M., Alias, M.Y., Shayea, I., Alraih, S.: Dynamic handover control parameters for LTE-A/5G mobile communications. In: 2018 Advances in Wireless and Optical Communications (RTUWO), pp. 39–44 (2018)
11. Ahmed, A.U., Islam, M.T., Ismail, M.: A review on femtocell and its diverse interference mitigation techniques in heterogeneous network. Wirel. Pers. Commun. **78**, 85–106 (2014)
12. Krasniqi, B., Rexha, B..: Analysis of macro-femto cellular performance in LTE under various transmission power and scheduling schemes. J. Commun. **13**(3), 119–123 (2018)
13. Ghosh, S., Sathya, V., Ramamurthy, A., Tamma, B.R.: A novel resource allocation and power control mechanism for hybrid access femtocells. Comput. Commun. **109**, 53–75 (2017)
14. Akinlabi, O.A., Joseph, K.: Positioning algorithm for deployment of femtocell network in mobile network. World Congr. Eng. **1**, 1–5 (2018)
15. Deswal, S., Singhrova, A.: A vertical handover algorithm in integrated macrocell femtocell networks. Int. J. Electr. Comput. Eng. (IJECE) **7**(1), 299–308 (2017)
16. Ahmed, R., Kouvatsos, D.: An efficient CoMP-based handover scheme for evolving wireless networks. Electron. Notes Theor. Comput. Sci. **340**, 85–99 (2018)
17. Deswal, S., Singhrova, A.: Quality of service provisioning using multicriteria Handover strategy in overlaid Network. In: 2020 Malysian Jounal of Computer Science, pp.1–6 (2020)
18. Sapwatcharasakun, S., Kunarak, S., Prasertwong, K., Abdel Hamid, G.A., Abdelaziz, A.: Green energy of handover management in seamless networks. In: 2023 International Conference on Cyber Management and Engineering (CyMaEn), pp.1–6 (2023)
19. Chen, Y.S., Wu, C.Y.: A green handover protocol in two-tier OFDMA Macrocell-femtocell networks. Math. Comput. Model. **57**, 2814–2831 (2013)
20. Rasheed, M., Ajmal, S.: Interference and resource management strategy for handover in femtocells. In: Spring Wireless Networks, pp.1–14 (2019)

An α-Rotated Fourier Transform Used as OTFS Enhancement

Marwa Rjili[1(✉)], Abdelhakim Khlifi[2], Fatma Ben Salah[1], Belghacem Chibani[1], and Said Chniguir[3]

[1] MACS laboratory: Modeling Analysis and Control of Systems LR16ES22, National Engineering School of Gabes, University of Gabes, Gabes, Tunisia
marwa.rjili@isimg.tn, fatmabensalah89@gmail.com, abouahmed17@gmail.com

[2] Innov'COM Laboratory, Sup'COM, University of Cartahge, Tunis, Tunisia
abdelhakim.khlifi@gmail.com

[3] Mathematical Department, IPEIG, University of Gabes, Gabes, Tunisia
saidchneguir@yahoo.fr

Abstract. Orthogonal-time-frequency-space (OTFS) modulation has received significant attention in wireless communication research due to its exceptional ability to reliable mobile communication. Network links censure high-speed connection. OTFS represents a promising candidate for next-generation wireless communication systems. Operating in the delay-Doppler (DD) domain, it relies on channel invariance. Furthermore, the OTFS system can be implemented on top of an existing Orthogonal Frequency Division Multiplexing (OFDM) system, thereby reducing installation expenses. This advantage is especially noteworthy, given that the Fractional Fourier Transform (FrFT)-based OFDM scheme has demonstrated its superiority over traditional OFDM schemes. We utilize the FrFT-based OFDM system to enhance the performance of the OTFS system. The Weighted Fractional Fourier Transform (WFrFT), as a variant of FT, is a generalized Fourier transform this induces the discrete Fourier transform (DFT) as a special case. In this article, we have chosen to apply FrFT-based OFDM to assess the performance of the OTFS system. Specifically, we have employed the Weighted Fractional Fourier Transform (WFrFT) as a variant of FrFT. We evaluate the usage of an OTFS system that incorporates the inherent WFrFT- based OFDM processing. We obtained various optimal alpha values for SNR values ranging from 0 to 9 dB. Additionally, we have demonstrated that the OTFS system based on WFrFT maintains a lower error rate compared to the conventional OTFS system for SNR ranging from 0 to 9 dB.

1 Introduction

The interaction between emitter and a specific receiver, depends on the channel character. Consequently, it becomes imperative to meticulously select the appropriate waveforms to overcome potential constraints. Addressing this challenge, numerous novel waveforms have been proposed to cater to the multifaceted

L. Barolli (Ed.): AINA 2024, LNDECT 204, pp. 249–260, 2024.
https://doi.org/10.1007/978-3-031-57942-4_25

requirements and scenarios of next mobile network generation that extend past the fifth generation [1].

A variety of waveforms have been proposed as an enhanced version of orthogonal frequency division multiplexing (OFDM) using various techniques including, generalized frequency division multiplexing (GFDM)[2,3], Filtered OFDM (F-OFDM), and Asymmetric OFDM (A-OFDM) [4,5]. Recently, a novel waveform known as OTFS has emerged, demonstrating an effective capacity to handle time-varying channels with pronounced Doppler effects [6,7]. OFDM modulation creates a channel that is equivalent to the Time-Frequency domain (TF), while OTFS modulation creates a channel that is equivalent to the Delay Doppler domain [8].

In a recent study [9], an integrated waveform framework was introduced, wherein the Weighted Fractional Fourier Transform (WFrFT) serves as a precoding technique in the OTFS system [1]. In summary, the Fractional Fourier Transform (FrFT) emerges as a potent signal processing tool that enhances system robustness, also finding application in OFDM systems as a replacement for the conventional Fourier Transform (FT) to enhance performance [9]. The DFrFT OFDM system exhibits a lower symbol error rate compared to conventional OFDM [10,11]. This holds true even in frequency-selective channels, effectively mitigating Inter-Symbol Interference (ISI) [12,13]. However, these limitations can potentially be overcomemed by adopting the newly proposed OTFS operator with the use DFrFT [9].

This paper presents the evaluation analysis of the OTFS system combined with WFrFT- OFDM. We will evaluate the design of an OTFS system based on WFrFT-OFDM and determine the optimal fractional orders for various SNR values, ranging from 0 to 9 dB with a 3 dB increment, based on Monte Carlo simulations.

2 Fractional Fourier Transform

The time-frequency plan has orthogonal axes that are formed by frequency and time. In comparison, the conventional Fourier transform can be considered as a transformation that redefines a signal in the time domain with respect to frequency; this transformation can be represented as a rotation up to $\pi/2$ radians in the trigonometric direction within the time-frequency plan.

On the other hand, the Fractional Fourier Transform is an extension of the classic Fourier Transform, it can be rotated to an angle in the time-frequency plane as shown in Fig. 1. The Fractional Fourier Transform (FrFT) presents the original signal function in time for each unique value of α. As a result, the FrFT can be considered as a one-parameter family of transformation. When $\alpha = 0$, the Fractional Fourier Transform is used as an identity transformation which essentially restores the original signal in the time domain conversely, when $\alpha = \pi/2$, it gives the conventional Fourier Transform, the FrFT of a signal is mathematically defined as [14]:

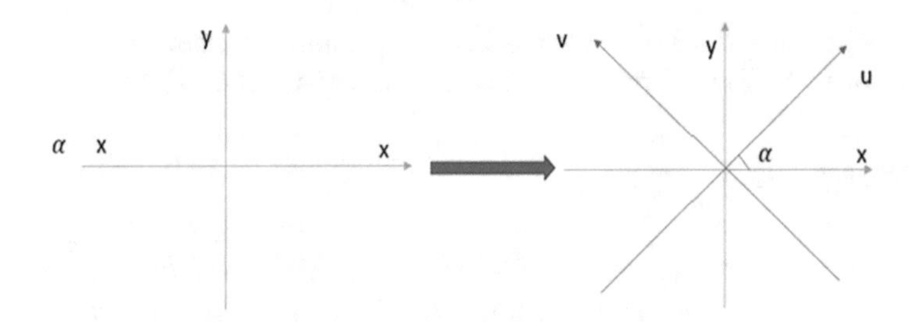

Fig. 1. (Ox, Oy) plane rotation

$$X_\alpha = \int_{-N/2}^{N/2} x(t) k_\alpha(t, v)\, dt \tag{1}$$

where $\alpha = a\pi/2$ is the angle of FrFT, a is the real number defined as FrFT order and $k_\alpha(t, v)$ is defined as transformation Kernel [14]:

$$\begin{cases} \delta(t+v) \quad if \quad \alpha+\pi \quad is \quad a \quad multiple \quad of \quad 2\pi \\ \delta(t-v) \quad if \quad \alpha \quad is \quad a \quad multiple \quad of \quad 2\pi \\ \sqrt{\frac{1-j\cot\alpha}{2\pi}} \exp(j\frac{(t^2+v^2)}{2}\cot(\alpha) - jvt\cos(\alpha)) \quad if \quad \alpha \quad is \quad not \quad a \quad multiple \quad of \quad 2\pi) \end{cases} \tag{2}$$

Then, the inverse Fractional Fourier Transform can be expressed as [14]:

$$x(t) = \frac{1}{2\pi} \int_{-N/2}^{N/2} X_\alpha(v) k_{-\alpha}(t, v)\, dv \tag{3}$$

2.1 The Weighted Fourier Fractional Transform

A weighted Fractional Fourier Transform consists of Fourier Transforms of four different integer orders with different weighting factors which is defined 4-WFrFT. This approach allows for the development of a new method to defining the FrFT [19], you can express it's definition as:

$$F^\alpha[f(t)] = \sum_{k=0}^{3} B_k^\alpha f_k(t) \tag{4}$$

with

$$B_k^\alpha = \cos(\frac{(\alpha-k)\pi}{4})\cos(\frac{2(\alpha-k)\pi}{4})\exp(\frac{-3(\alpha-k)\pi}{4}). \tag{5}$$

You can write:

$$\begin{aligned} f_0(t) &= f(t) \\ f_1(t) &= F[f_0(t)] \\ f_2(t) &= F[f_1(t)] \\ f_3(t) &= F[f_2(t)] \end{aligned} \tag{6}$$

where F represents the Fourier transform. Equation (7) presents an alternative definition of four-WFrFT primarily utilized in signal processing [19,20].

$$\begin{aligned} F^{\alpha}[f(t)]) &= \sum_{k=0}^{3} B_k^{\alpha} f_k(t) \\ &= B_0^{\alpha}.f_0(t) + B_1^{\alpha}.f_1(t) + B_2^{\alpha}.f_2(t) + B_3^{\alpha}.f_3(t) \\ &= B_0^{\alpha}.I.f(t) + B_3^{\alpha}.F.f(t) + B_3^{\alpha}.F^2.f(t) + B_3^{\alpha}.F^3.f(t) \\ &= (B_0^{\alpha}.I + B_1^{\alpha}.F + B_2^{\alpha}.F^2 + B_3^{\alpha}.F^3) f(t) \end{aligned} \tag{7}$$

In this context, I represents the identity matrix. while $I = F^4$ denotes the use of the Fourier Transform in both vector and matrix forms. This allows us to write the equation as follows:

$$F^{\alpha}[f(t)] = \begin{bmatrix} I, F_1, F_2, F_3 \end{bmatrix} \begin{bmatrix} B_0^{\alpha} \\ B_1^{\alpha} \\ B_2^{\alpha} \\ B_3^{\alpha} \end{bmatrix} f(t) \tag{8}$$

To maintain the consistency of dimension, the correct form of B_k^{α} in Eq. (8) must be $B_l^{\alpha}.I$. To make things easier; it is also known as B_k^{α} $(k = 0, 1, 2, 3)$, the weightings coefficients B_k^{α} can be defined as [19]:

$$\begin{bmatrix} C_0^{\alpha} \\ C_1^{\alpha} \\ C_2^{\alpha} \\ C_3^{\alpha} \end{bmatrix} = \begin{bmatrix} 1 & 1 & 1 & 1 \\ 1 & i & -1 & -i \\ 1 & -1 & 1 & -1 \\ 1 & -i & -1 & i \end{bmatrix} \begin{bmatrix} B_0^{\alpha} \\ B_1^{\alpha} \\ B_2^{\alpha} \\ B_3^{\alpha} \end{bmatrix} \tag{9}$$

By using this transformation, the coupled equations are completely separated into a new set of equation that have considerably simpler forms:

$$\begin{aligned} C_0^{(\alpha+\beta)} &= C_0^{\alpha} C_0^{\beta}, C_1^{(\alpha+\beta)} = C_1^{\alpha} C_1^{\beta} \\ C_2^{(\alpha+\beta)} &= C_2^{\alpha} C_2^{\beta}, C_3^{(\alpha+\beta)} = C_3^{\alpha} C_3^{\beta} \end{aligned} \tag{10}$$

where $C_n^{\alpha} = \exp(2\pi n\alpha/2)$, $\alpha = 0, 1, 2, 3$. Solutions for the original set of coefficients are obtained by inverse transformation of Eq. (10) as follows:

$$B_0^{\alpha} = \exp(3i\pi\alpha/4)\cos(\pi\alpha/2)\cos(\pi\alpha/4), B_1^{\alpha} = B_1^{(\alpha-1)}, B_2^{\alpha} = B_0^{(\alpha-2)}, B_3^{\alpha} = B_3^{(\alpha-3)} \tag{11}$$

As a result, the fractional-order Fourier transform can be expressed in the following manner:

$$\begin{aligned} F^{\alpha}[f(t)]) &= B_0^{\alpha} f_0(t) + B_1^{\alpha} f_1(t) + B_2^{\alpha} f_2(t) + B_3^{\alpha} f_3(t) \\ &= \sum_{n=0}^{3} \exp(\frac{i3\pi(\alpha-n)}{4}) \cos(\frac{\pi(\alpha-n)}{2}) \cos(\frac{\pi(\alpha-n)}{4}) f_n(t) \end{aligned} \tag{12}$$

3 WFrFt-OFDM Based OTFS System

3.1 OTFS Transmitter

At the transmitter, in the first step, the information symbols X reside in the Delay Doppler domain (DD) of size $M \times N$, where M represents the number of delay intervals and N represents the doppler intervals respectively. As demonstrated in Fig. 2, the time-frequency domain OTFS symbols are realized based on X via the Inverse Symplectic Fast Fourier Transform (ISFFT), the transmitter signal is given as [16]:

$$X_{TF} = \frac{1}{MN} \sum_{k=0}^{N-1} \sum_{l=0}^{M-1} X_{DD} e^{(j2\pi[\frac{nk}{N} - \frac{ml}{M}])} \tag{13}$$

where the variable X_{TF} represents values in the TF domain, while X_{DD} represents values in the DD domain. Additionally, the index m takes integer values from 0 to $M-1$, and the index n takes integers from 0 to $N-1$. A more concise representation of the Inverse Symplyctic Fast Fourier Transform (ISFFT) can be obtained using the discrete Fourier Transform (DFT) matrices $T_N \in C^{N\times M}$ and $T_M \in C^{N\times M}$, let $X_{DD} \in C^{N\times M}$ contain the symbols $x[k,l]$ of the delay doppler domain, and $X_{TF} \in C^{N\times N}$ contains the symbols $X[m,n]$ of the time frequency domain. Then, the Eq. (14) can be written in matrix form as [9]:

$$X_{TF} = T_N^H X_{DD} T_M \tag{14}$$

Then, Time Frequency domain signal X is processed with an IWFrFT transform of order α [16] to generate the signal as shown in the following Fig. 2 [9]: which is referred as WFrFT-OFDM modulation.

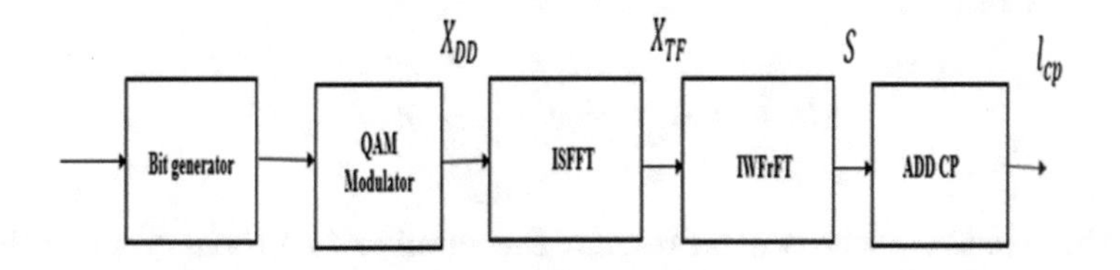

Fig. 2. OTFS transmitter

$$S = X_{TF} T_{-\alpha,M} \tag{15}$$

where X_{TF} represents the time-frequency signal matrix and $T_{-\alpha,M}$ represents the DFT matrix of order α. Using (14) in (15), the equation for S can be reduced to:

$$S = T_N^H X_{DD} T_M T_{-\alpha,M} \tag{16}$$

The signal matrix in the delay-doppler domain, denoted as X_{DD}, can be transformed into an equivalent vector signal x, within the same domain. This vector, represented as x, has a size of $MN \times 1$ and is obtained by vectorizing X. This means that each element in X, at position (k, l), corresponds to the $(kM + l)^{th}$ indexed element in the resulting x vector. As a result, the time-domain (TD) signal can be represented as a vector as it follows:

$$s = vec\{T_M T_{-\alpha,M} X_{DD} T_N^H\} = (T_N^H \bigotimes T_{1-\alpha,M}) \tag{17}$$

where s is located within the TD domain signal vector of size $MN \times 1$. The notation $\bigotimes$ denote the Kronecker product operation and $T_{(1-\alpha,M)}$ represents $(1-\alpha)^{th}$ order WFrFT matrix of size M. Then the signal is transformed to passband signal and a cyclic prefix(cp) with length l_{cp} is added.

3.2 DD Domain Channel

The signal $s(t)$ is conveyed through a time-varying channel possessing a complex baseband channel impulse response denoted as $h(\tau, v)$. This impulse response characterizes how the channel responds to an impulse signal with a delay of τ and a Doppler shift of v, as described in [17]. The received signal, $R(t)$ can be expressed as follows [18]:

$$R(t) = \int\int h(\tau, v) k_\alpha s(t, v) e^{2j\pi(t-\tau)}\, d\tau dv \tag{18}$$

Equation (18) signifies a continuous Heisenberg transform that is parameterized by $s(t)$, as described in [17]. Since there are generally only a limited number of reflectors within the channel, each associated with specific delays and Doppler shifts, we require only a minimal number of parameters to characterize the channel within the delay-Doppler domain. The representation of the channel $h(\tau, v)$ is presented as [18]:

$$h(t - v) = \sum_{j=1}^{q} h_i \delta(t - \tau_j)\delta(t - v_j) \tag{19}$$

In this context, where q represents the number of propagation paths, h_j, τ_j, and v_j respectively stand for the path gain, delay, and Doppler shift (or frequency) associated with the $j-th$ path. Additionally, $\delta(.)$ represents the Dirac delta function. We label the delay and Doppler taps for the $i-th$ path as follows:

$$\tau_l = \frac{l_j}{M\Delta f} \quad and \quad v_l = \frac{k_j}{NT} \tag{20}$$

where l_j and k_j represents the delay and doppler taps, respectively corresponding to the lth path can take numbers that are either integers or fractions. At the receiver, the received signal S is converted to baseband. The cyclic prefix of the baseband signal is then removed to obtain a vector/signal of the form:

$$R = \{R(n)\}_{n=0}^{MN-1} \tag{21}$$

The relationship between the entries of $R(n)$, the transmitted data symbols $S(n)$ and the additive white Gaussian noise $w(n)$ in Time Delay can be described by the following equation [16].

$$R(n) = \sum_{j=1}^{q} h_j e^{j2\pi(k_j(n-l_j))/MN)} S([n-l_j]_{MN} + w(n) \tag{22}$$

The notation [.]n represents the modulo-n operation. Equation (22) can also be expressed as a generalized linear system equation involving a channel matrix and transmitted and received vectors, as illustrated below [9]:

$$R = Hs + w \tag{23}$$

Let R, s, w denote the vectors that respectively represent the received signal, and Additive white Gaussian Noise (AWGN) samples of size $MN \times 1$. On the order, $H \in C^{MN \times MN}$ refers to a complex matrix.

3.3 OTFS Receiver

After having prior knowledge that relevant information has been obtained, appropriate treatments were performed in the transmitting unit, which then transferred them to a desired receiving destination. A The receiver, an orthogonal time frequency space (OTFS) demodulator initially, the signal is received, and the cyclic prefix is removed, as shown in the following Fig. 3 [9]. Subsequently, the resulting discrete signal, as defined in Eq. (15), is converted into a 2-D signal represented in matrix form as $R = vec^{-1}(R)$. Following this, the Weighted Fractional Fourier Transform (WFrFT) is applied to this matrix to compensate for the Inverse Weighted Fractional Fourier Transform (IWFrFT), resulting in the generation of the Time-Frequency (TF) domain matrix denoted as Y, as depicted below:

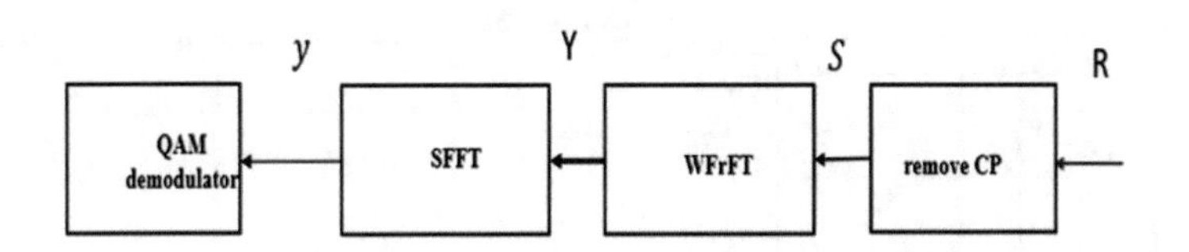

Fig. 3. OTFS receiver

$$Y = RT_{\alpha,M} \tag{24}$$

The SFFT is used to transform the symbols Y obtained in the time frequency domain into the delay doppler domain after the Wigner transform [19]:

$$y = \frac{1}{MN} \sum_{n=0}^{N-1} \sum_{m=0}^{M-1} Y e^{-j2\pi[nk/N - ml/M]} \tag{25}$$

The matrix operation equivalent to the (SFFT) can be represented as follows:

$$Y = T_M^H Y T_N \tag{26}$$

where we substitute Eq. (24) for Eq. (26), we get:

$$Y = T_M^H F_{\alpha,M} R T_N \tag{27}$$

After $T_M^H = T_{-1,M}$, Eq. (27), similar to Eq. (23) can be inscribed in vector form:

$$y = vecT_{-1,M} T_{\alpha,M} R T_N = (T_N \bigotimes T_{\alpha-1,M})R \tag{28}$$

The received signal vector y is located within the Delay Doppler domain, and R is derived from Eq. (24). $T_{\alpha-1,M}$ represents the inverse of the WFrFT (Weighted Fractional Fourier Transform) matrix with a fractional parameter $\alpha - 1$.

4 Simulation Results

In this section, we present an assessment of the performance of OTFS (Orthogonal Time Frequency Space) systems utilizing the Weighted Fractional Fourier Transform, focusing on their bit-error-rate (BER). The outcomes are obtained through simulations conducted on an OTFS data frame with dimensions 16×32, where $N = 16$ delay bins, each further divided into $M = 32$ Doppler bins. These bins are then grouped together to form an OTFS grid comprising MN cells, each containing a single symbol. This data frame is subsequently employed in Monte Carlo simulations. We consider the vehicular A channel for the simulations and employ QAM modulation to generate graphical representations of error performance.

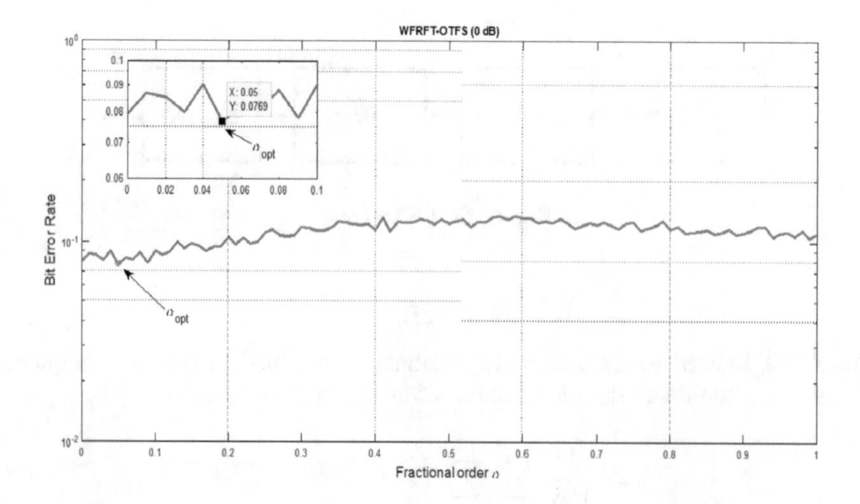

Fig. 4. WFrFT-OTFS for SNR = 0 dB

Table 1. Various Optimal Fractional Orders

SNR(dB)	0	3	6	9
BER_{min}	75.10^{-3}	72.10^{-3}	52.10^{-3}	33.10^{-3}

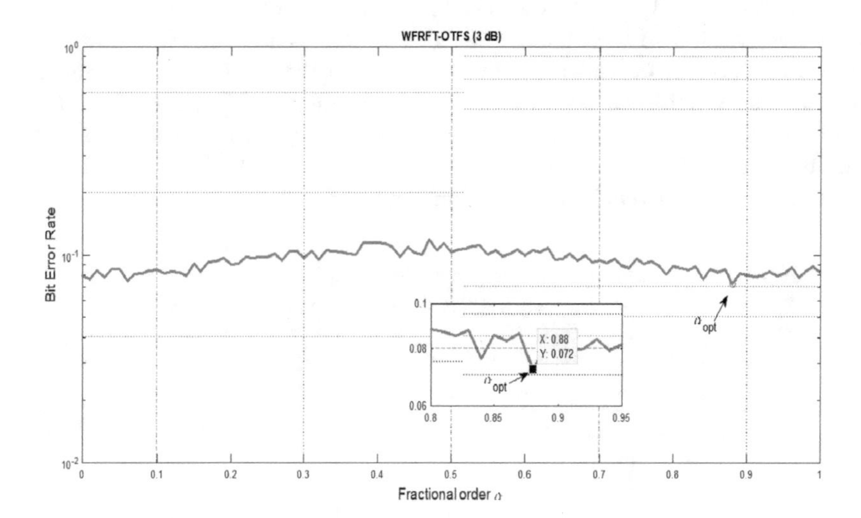

Fig. 5. WFrFT-OTFS for SNR = 3 dB

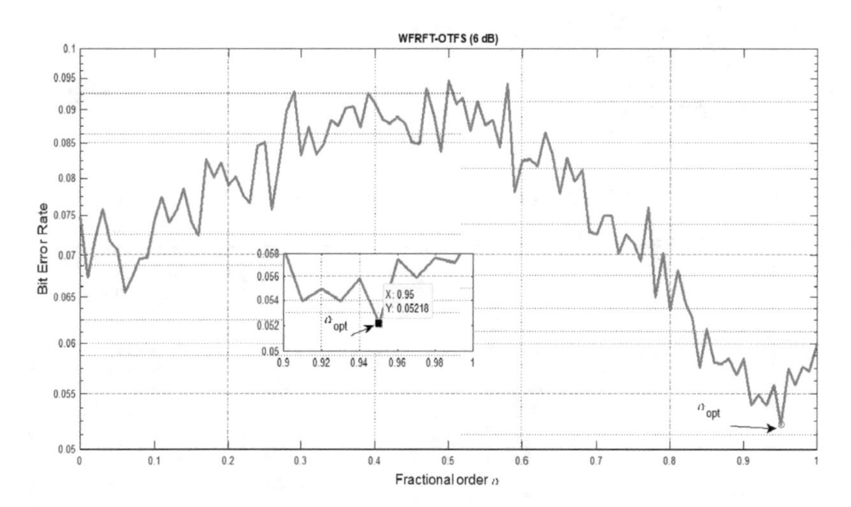

Fig. 6. WFrFT-OTFS for SNR = 6 dB

To determine the optimal fractional value α_{opt} that yields the lowest BER rates under various scenarios, we conduct 1000 Monte Carlo simulations, varying the fractional orders in the range of [0.1], with a step size of 0.01. The results reveal different optimal fractional order values across a range of SNR (Signal-to-Noise

Ratio) values, spanning from 0 to 9 dB, with 3 dB increments and a minimum Bit Error Rate (BER) for each SNR value. These optimal values are summarized in Table 1 below. In Fig. 4 can be seen that we have achieved an optimal alpha (α_{opt}) of 0.05 at a SNR of 0 dB, which is equivalent to a minimum Binary Error Rate (BER) of 75.10^{-3}. The results of Fig. 5 indicate that we have obtained an optimal alpha of 0.88 with a SNR of 3 dB, which is equal to a minimum Binary Error Rate (BER) of 72.10^{-3}. While Fig. 6 illustrates that an optimal alpha value of 0.95 is achieved with an SNR of 6 dB and a BER value of 52.10^{-3}. In Fig. 7, we show that the optimal alpha value is 0.95 with an SNR of 9 dB and a BER value of 33.10^{-3}. While Fig. 8 shows that the WFrFT-OTFS systems outperforms the conventional OTFS one ($\alpha = 1$).

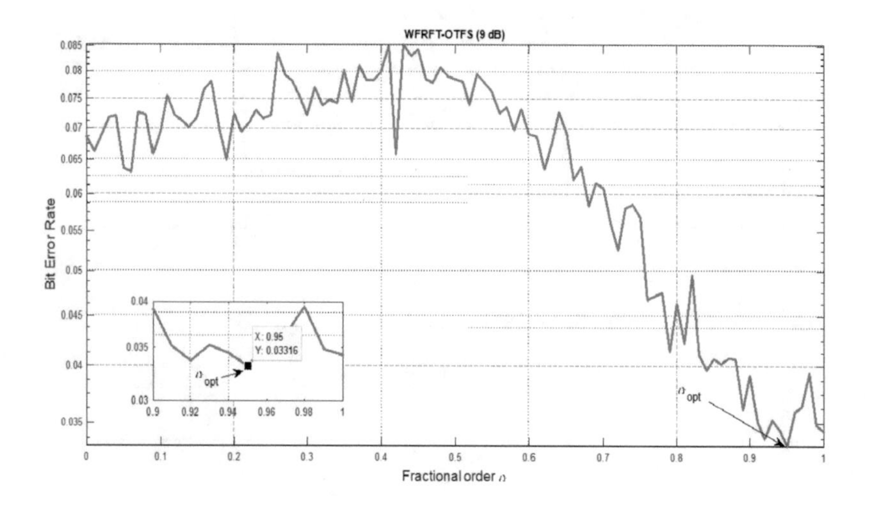

Fig. 7. WFrFT-OTFS for SNR = 9 dB

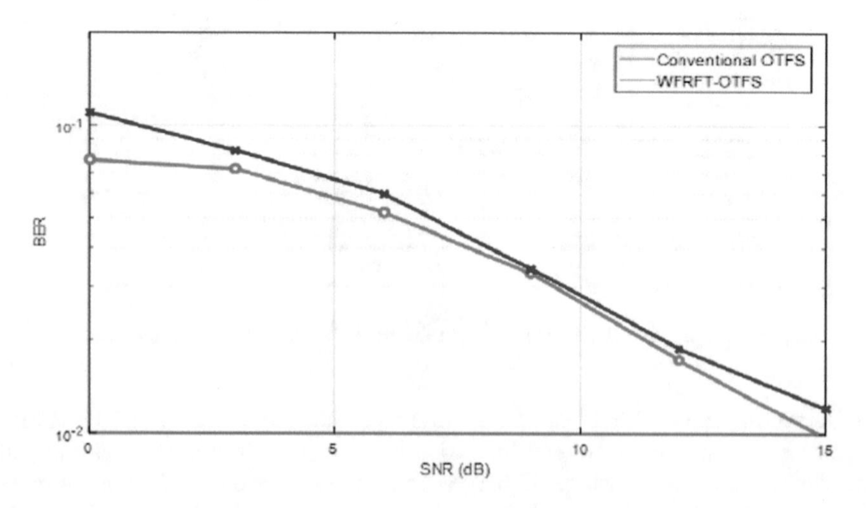

Fig. 8. WFRFT-OTFS performance in terms of BER

5 Conclusion

This article presents a general OTFS system designed on the basis of the WFRFT. Since WFRFT is considered as a generalized Fourier transform with alpha equal to 1, inducing the conventional Fourier transform, we have also demonstrated the relationship between the input and output signals of the OTFS system as well as the Delay-Doppler channel. We obtained different optimal alpha values for different SNR values, ranging from 0 to 9 dB with a step size of 3 dB. Furthermore, we concluded that the WFRFT-based OTFS system maintains a lower error rate than the conventional OTFS system.

References

1. Wang, Z., et al.: BER analysis of integrated WFRFT-OTFS waveform framework over static multipath channels. IEEE Commun. Lett. **25**(3), 754–758 (2020)
2. Fettweis, G., et al.: GFDM-generalized frequency division multiplexing. In: VTC Spring 2009-IEEE 69th Vehicular Technology Conference. IEEE (2009)
3. Darghouthi, A., et al.: Link performance analysis for GFDM wireless systems. In: IEEE 21st international Conference on Sciences and Techniques of Automatic Control and Computer Engineering (STA) (2022)
4. Abdoli, J., et al.: Filtered OFDM: a new waveform for future wireless systems. In: 2015 IEEE 16th International Workshop on Signal Processing Advances in Wireless Communications (SPAWC). IEEE (2015)
5. Zhang, J., et al.: Asymmetric OFDM systems based on layered FFT structure. IEEE Signal Process. Lett. **14**(11), 812–815 (2007)
6. Murali, K.R., et al.: On OTFS modulation for high-Doppler fading channels. In: Information Theory and Applications Workshop (ITA). IEEE (2018)
7. Hadani, R., et al.: Orthogonal time frequency space (OTFS) modulation for millimeter-wave communications systems. In: IEEE MTT-S International Microwave Symposium (IMS) (2017)
8. Raviteja, P., et al.: Practical pulse-shaping waveforms for reduced-cyclic-prefix OTFS. IEEE Trans. Veh. Technol. **68**(1), 957–961 (2018)
9. Mallaiah, R., Mani, V.V.: A novel OTFS system based on DFrFT-OFDM. IEEE Wirel. Commun. Lett. **11**(6), 1156–1160 (2022)
10. Martone, M.: A multicarrier system based on the fractional Fourier transform for time-frequency-selective channels. IEEE Trans. Commun. **49**(6), 1011–20 (2001)
11. Arya, S., et al.: ICI analysis for FRFT-OFDM systems in doubly selective fading channels. In: IEEE International Conference on Signal Processing, Computing and Control (ISPCC) (2013)
12. Rawat, S., et al.: An overview: OFDM technology an emerging trend in wireless communication. IJIRCCE **5**, 3947 (2017)
13. Mattera, D., et al.: Comparing the performance of OFDM and FBMC multicarrier systems in doubly-dispersive wireless channels. Signal Process. **179**, 107818 (2021)
14. Rezgui, C.: Analyse performance of fractional Fourier transform on timing and carrier frequency offsets estimation. Int. J. Wirel. Mob. Netw. (IJWMN) **8**(2), 1–10 (2016)
15. Wang, X., et al.: Analysis of weighted fractional Fourier transform based hybrid carrier signal characteristics. J. Shanghai Jiaotong Univ. **25**, 27–36 (2020)

16. Hadani, R., et al.: Orthogonal time frequency space modulation. In: Proceedings of IEEE Wireless Communications and Networking Conference, pp. 1–6 (2017)
17. Jakes, W.C. Jr.: Microwave Mobile Communications. Wiley, New York (1974)
18. Raviteja, P., et al.: Interference cancellation and iterative detection for orthogonal time frequency space modulation. IEEE Trans. Wirel. Commun. **17**(10), 6501–6515 (2018)
19. Shih, C.-C.: Fractionalization of Fourier transform. Optics Commun. **118**(5–6), 495–498 (1995)
20. Mei, L., et al.: Research on the application of 4-weighted fractional Fourier transform in communication system. Sci. China Inf. Sci. **53**, 1251–1260 (2010)

Monitoring a Rice Field with a Low-Cost Wireless Sensor Network

Chaima Bejaoui[1(✉)] and Nasreddine Hajlaoui[2,3]

[1] LETI Lab, National School of Engineering, University of Sfax, Sfax, Tunisia
bejaouichaima@gmail.com
[2] Hatem Bettaher IResCoMath Research Lab, University of Gabes, Gabes, Tunisia
nasreddine.hajlaoui@fsg.rnu.tn
[3] Applied College, Qassim University, Buraydah, Saudi Arabia

Abstract. Precision agriculture enables significant productivity increases, as well as enable to anticipate or reduce risks. In this paper, we propose to monitor a rice field with a low-cost architecture. The architecture is based on a low-cost wireless sensor network, composed of static sensors in the field and mobile sensors carried by farmers working in the field. The data from the static sensor is first uploaded to the mobile sensors and then uploaded to the sink (located in the village). Mobile sensors also exchange data and can help to establish permanent connectivity between the sink and farmers in the field. We discuss several aspects of this architecture and present several related challenges in terms of sensor node hardware, network protocols, and data management. Simulation results indicate that our proposed architecture outperforms the random architecture proposed in the literature in terms of lifetime and the forwarded packets per CH.

Keywords: WSN · WMSN · precision agriculture

1 Introduction

Precision agriculture is a recent concept to improve agriculture productivity based on frequent localized observations and appropriate responses. Using precision agriculture principles significantly improves crop efficiency and reduces financial costs while increasing production. Thus, precision agriculture requires three components: a monitoring component, a data analysis component, and a response notification component. Several existing works on precision agriculture are based on expensive solutions, rely on farmers to perform specific actions (such as gathering data), or use distant monitoring (through satellite imaging, or aerial drone monitoring for instance). We believe that it is possible to benefit from advances in wireless sensor networks (WSNs) to provide a cheap, automated, and localized solution. In this paper, we propose a low-cost architecture for precision agriculture applied to rice fields. The monitoring component is based on a low-cost WSN, composed of three types of nodes: (i) static sensor

L. Barolli (Ed.): AINA 2024, LNDECT 204, pp. 261–272, 2024.
https://doi.org/10.1007/978-3-031-57942-4_26

nodes that monitor the rice field, (ii) mobile sensor nodes that collect the data, collaborate to improve the confidence in the data (and reduce uncertainties due to measurement errors), and carry the data to the sink, (iii) a sink that is used as a gateway to a database on the Internet. The data analysis component is localized in the cloud, and is responsible for archiving the data, estimating the expected productivity based on localized meteorological data and market trends, and generating warnings based on possible risks. The response notification component takes as input these warnings, and disseminates them from the sink to the mobile sensor nodes in a multi-hop manner. It includes a simple yet effective graphical user interface to help farmers understand the possible risks and the actions to perform. The remainder of this paper is organized as follows. Section 2 describes the related work on WSNs for monitoring applications and for precision agriculture. Section 3 presents the architecture we propose, with the three types of nodes: static sensor nodes, mobile sensor nodes, and static sink node. Section 4 describes the relevant research issues and challenges, in terms of sensor node hardware, network protocols, and data management. Section 5 evaluates our architecture. Section 6 concludes our paper.

2 Related Works

In the following, we describe related works. We start by showing how WSNs can be used for various environmental monitoring applications. Then, we focus on precision agriculture with WSNs. Finally, we describe the special case of wireless multimedia sensor networks (WMSNs), that is WSNs using multimedia data, which are of special interest for monitoring rice fields.

2.1 Wireless Sensor Networks for Environmental Monitoring Applications

WSNs are composed of cheap battery-powered devices that are able to sense their environment and to communicate in a wireless manner. With these communications capabilities, they are able to transmit data from nodes to nodes to a sink, which is generally connected to the Internet. The low-cost and energetic autonomy of WSNs make them good candidates for environmental monitoring applications, and several such applications have emerged in recent years. The monitoring of volcanic activity with WSNs [1,2] helps learning how volcanoes behave, and enables early warning systems. They are less expensive than seismometers, provide data more frequently than satellite imaging, and can be deployed directly on the volcano (rather than in a safe but distance place). The proposed WSNs usually contain the following three types of nodes.

- Sensor nodes measure geophysical parameters (such as temperature, vibrations, volume of gas emissions, etc.) to infer the activity of the volcano.
- Relay nodes are used to transfer the data generated by the sensor nodes to a distant sink.

- The sink node collects the data, archives it, and makes it available for the scientists monitoring the volcano.

The monitoring of forest with WSNs allows the early detection of forest fires in regions with low population density [3]. Indeed, in these regions, traditional monitoring is infrequent, and is often partial (especially when the forests cover a large area). Sensor nodes monitoring temperature are disseminated in the forest (usually in several spots), and communicate to a sink having a GPRS or GSM connection (if available) or a satellite connection.

WSNs for environmental monitoring applications often share similar constraints.

- Autonomy: As WSN nodes are battery-powered and are expected to operate for years without human intervention, the nodes have to save as much energy as possible. Since the radio module consumes a significant part of the energy of a node, it is important for nodes to communicate as little as possible.
- Robustness: As WSN nodes are deployed outdoor, they are subject to a wide range of weather conditions, and the hardware has to be designed to resist the impact of the environment. Radio communication is also severely hindered in outdoor deployment, especially when nodes are not deployed in line of sights.
- Scalability: As the monitored zones are usually large areas, numerous nodes are needed. Thus, the network protocols have to be scalable in order to be able to deal with all these nodes.

2.2 Precision Agriculture with Wireless Sensor Networks

Precision agriculture is a special case of environmental monitoring where WSNs are particularly suitable. In [4], the authors proposed a real deployment of a WSN based crop monitoring in order to realize precision agriculture. End-users can tailor the node operation to a large variety of setups, which allow farmers to collect data on temperature, humidity, or soil moisture, from locations previously inaccessible on a micro-measurement scale. The WSN system they proposed has a low-power consumption and is able to deliver the data collected directly to the farmer's mobile via the GSM technology, and to automatically actuate the water sprinklers during the period of water scarcity. The architecture implemented in the sensor nodes builds a wireless network based on the XMesh protocol. A sink acts as a gateway between the nodes and the database connection on the Internet. In [5], the authors proposed a new WSN architecture with low-cost sensor nodes for soil moisture in order to optimize the irrigation processes. The sensors and actuators collected sensed data from different places of soil to measure the level of water. The authors used three types of nodes to monitor and exchange data to enable or disable the irrigation system. The irrigation system is scheduled periodically, once a sensor registers an alarm, the system collects the data on the designated area. If the number of sensors exceeds 5 then the ditch gate is activated in that area for drip irrigation. Otherwise, the system goes into idle mode. In [6,7], the authors proposed a monitoring system for a rice field in the

Kuttanad region of India. Electro-mechanical sensors measure the water level and the resulting data is then transmitted through a WSN. The communication is achieved using a cluster architecture on a mesh topology. Each cluster of nodes has a cluster head which receives data from its members nodes and aggregates them. Then, the cluster head analyzes the aggregated value with respect to a set of predefined thresholds. If all the data is within the expected values, no data is sent to the user and the cluster head assumes that the situation is normal. Otherwise, the cluster head switches to the alert state and sends alarm to the sink. In turn, the sink informs the user. Simultaneously, the sink requests all sensor nodes to increase the frequency of their measurements. Once the sink detects that the situation is back to normal, it requests all sensor nodes to go back to the normal state, where they perform measurement at a low frequency, and are able to save energy.

2.3 Precision Agriculture with Wireless Multimedia Sensor Networks

Precision agriculture can greatly benefit from WMSNs. Recall that WMSNs are WSNs that sense, process and transmit multimedia data such as audio or image. They have been made possible by recent advances in WSNs [8]. Examples of precision agriculture with WMSNs include ultrasonic audio sensors to detect wild rodent in fields [9], or automatic detection of diseases on leaves with camera sensors, as shown on Fig. 1. In [10], the authors presented a comprehensive survey on the detection of diseases in plants based on images. They describe several image processing techniques that use various criteria: measurement of plant lengths, measurement of leaves lengths, detection of the shape of leaves, detection of colors in leaves, etc. They show that it is possible to detect diseases using image processing algorithms, which makes the detection automatic. They also show that some techniques can even detect diseases that farmers cannot detect with the naked eye. In [11], the authors proposed a robust multi-support and modular platform using multimedia sensors that are camera CDD with low

Fig. 1. Camera sensors can be used to detect some plant diseases, such as narrow brown spots caused by the fungus Sphaerulina oryzina on rice leaves.

cost to detect plant diseases and insects. The platform called MiLive allows the processing of images and signals. Multimedia sensors use batteries as an energy source, which presents a challenge with large data processing, which is why Milive uses cooperative processing with the server to save energy. In [12], the authors proposed a system that can quickly detect plants affected by fungus or parasites in agricultural fields, based on images. The system uses cameras that record videos in all directions of a field. To do this, the field is decomposed into clusters, and each cluster contains scalar sensors, video sensors, image sensors and a sink. Again, the sink is in charge of collecting the data from the sensor nodes uploading them to a database, and notifying farmers in case of an alert (through her mobile phone).

3 Description of Our Architecture

Existing architectures for precision agriculture based on WSNs and WMSNs generally possess one or many of the following drawbacks.

- They assume that the farmers have access to mobile phones with GPRS or GSM capabilities. However, in some countries, there are still many rural areas that are not yet covered. Moreover, the cost of the mobile phone subscription might be too expensive for some farmers.
- They assume that the sensor nodes form a connected network, that is all sensor nodes are in the range of others. When monitoring large areas, this is difficult to achieve without deploying many sensor nodes or even relay nodes, which increases the overall cost of the architecture.
- They propose architectures focusing either on the data acquisition or on the networking aspects, but rarely on both.
- There are few existing WMSNs tailored to the need for rice monitoring.

In the following, we describe our proposed architecture, which addresses all these drawbacks. Then, we explain the types of sensors that can be used, while focusing on camera sensors.

3.1 General Overview

Our architecture is based on the following design choices:

- Limited infrastructure: Our architecture relies on a single Internet access at the sink. Otherwise, there is no need for existing infrastructure. Nodes communicate only in wireless ad-hoc mode (thus, there is no need for GPRS or GSM access, nor other wireless infrastructures).
- No permanent connectivity: We do not assume that our network is fully connected. Indeed, we believe that it is more realistic to consider that sensor nodes are distant from each other, and cannot communicate directly. To allow communications, we rely on mobile nodes, carried by farmers as they work in their fields.

- Rice monitoring: Our architecture and protocols are designed specifically for rice monitoring. Note that some features can be used for other types of precision agriculture as well.
- Additional services: In addition to the monitoring of rice fields, we are able to provide farmers with additional services, such as Internet connectivity in fields (when the density of farmers is large enough, and with limited bandwidth).

Figure 2 presents our architecture as an example. The figure depicts a sink on the left (represented by a double circle) and a distant field on the right (represented by a dashed box). The field is equipped with four static sensors (denoted by s_1, s_2, s_3, and s_4) that are disseminated in the field. These sensors monitor various parameters such as temperature or water level. They are not able to communicate with each other and store their data locally when they are isolated (see s_2 in the figure). Farmers carry devices (such as smartphones without phone service subscriptions) that act as mobile sensors (denoted by m_1, m_2, m_3, m_4 and m_5). When a farmer works in her field, the mobile sensor she carries attempts to communicate with neighbor nodes. When a mobile sensor is in range of a static sensor (such as m_1 and s_1 on the figure), the mobile sensor can upload the data from the static sensor, and the static sensor can purge its memory. When the mobile sensor is in range of the sink (such as m_3 on the figure), the mobile sensor can download the data from its memory to the database located on the Internet. The mobile sensor can also receive notifications and warnings for other mobile sensors. When the mobile sensor is in range of another mobile sensor (such as m_4 and m_5), the two sensors can exchange data, notifications or warnings. Notice that our architecture does not require that the network is connected at all time. Depending on the mobility and number of mobile sensors, it is possible to establish clusters of interconnected nodes (on the figure, there are three small groups: s_1 and m_1, m_3 and the sink, and m_4

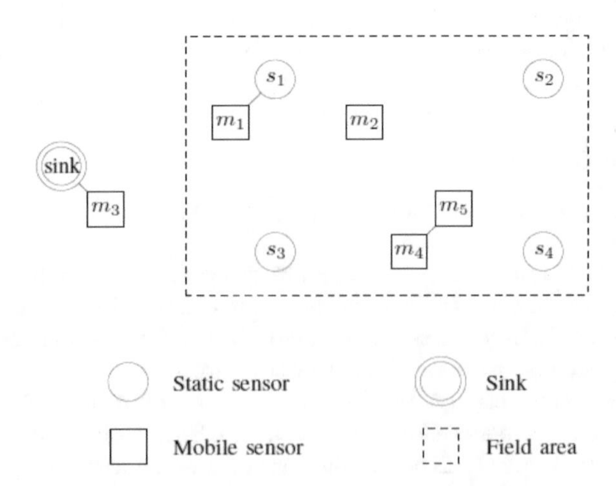

Fig. 2. Proposed architecture for our WSN.

and m_5, but in practice such groups could be larger). If the sink is in a cluster, all the nodes of the cluster have access to the Internet via the sink. Thus, if the number of mobile sensors is large, it is possible for a cluster to be formed with the sink and nodes on the path from the sink to the field, thus providing Internet connectivity to farmers in the field.

3.2 Types of Sensors

Our architecture supports a large variety of sensors to collect various data. We classify the collected data into three categories, and we give the example of a few sensors for each category.

- Meteorological data are used to obtain microclimate information. Indeed, localized data is necessary for the accurate prediction of productivity and risks. Meteorological data can be captured by temperature sensors, wind force and direction sensors, humidity sensors, light sensors, etc.
- Water data is crucial for the growth of rice. Water data can be captured by water level sensors, water flow sensors, water conductivity sensors, water pH sensors, and oxidation redox potential sensors.
- Rice data is used to obtain data about rice growth. Such sensors are all camera sensors but can be used to compute the height of leaves, the color of the leaves, the shape of leaves, etc. Note that the color of leaves is often an indication of disease in rice, such as the narrow brown spot disease caused by a fungus (see Fig. 1) or the yellow mottle virus causing chlorosis of the leaves and making them turn yellow (see the zoom of the leaf on Fig. 3). Other symptoms might require a large field of view (FoV), such as altered panicle exertion or stunting symptoms (causing loss of productivity).

In the following, we focus on the camera sensors. The main challenge is to process images with devices having limited computational capabilities, as well as limited memory. Several types of cameras exist:

- CMOS cameras [13] have the ability to provide high image quality due to the high quality of their lenses (focal length of 2.10 mm, with a working focal length from 100 mm, and 52° of field of vision). They are controlled by a low-power consumption module.
- Cyclops cameras [14] can be interfaced with several existing sensor nodes, such as Mica2 or MicaZ. They are designed for lightweight imaging and include image-processing processing embedded circuits and programmable memory. They can send data using a low-power consumption mode.
- Charged-coupled devices (CCD) cameras [15] are suitable for high-precision data acquisition, which can be used for color detection or shape recognition.
- Small webcams can also be used for our purpose. They often have a very low resolution, but they are inexpensive and do not consume a lot of energy. They also natively produce small images that can be processed by sensor nodes.

Fig. 3. A camera installed in the field can monitor various rice parameters, such as average height of crops or crop density. A good quality camera can also detect symptoms such as the yellow mottle virus (see zoom) by comparing the amount of green and yellow colors in the image.

4 Issues and Challenges

Our proposed architecture poses several research challenges that have to be addressed. We classify these challenges into hardware challenges, communication protocol challenges, data management challenges.

4.1 Hardware Challenges

Radio communications in rice fields are difficult for the following reasons.

- Communications in harsh conditions. As the WSN is deployed outdoors, it suffers from harsh communication conditions. Indeed, radio propagation is greatly impacted by the presence of vegetation (which depends on the rice height) and weather conditions (e.g., in foggy or rainy weather). The radio frequency used by the node has to be carefully selected in order to ensure a good communication range with low power consumption, for various weather conditions. It is likely that frequencies lower than 2.4 GHz are more suitable, but they yield longer antenna sizes and reduce the available bandwidth.
- Communications without a line of sight. The static sensor is not always in line of sight of the mobile sensors. Indeed, this depends on their deployment and on the height of the crop. The nodes have to be able to communicate in all conditions.

The tropicalization of nodes. Nodes have to be robust against weather conditions, as they are deployed outdoors for a long duration. They have to be waterproof, and able to support temperatures from 0 °C to 40 °C, for example.

4.2 Communication Challenges

Medium Access Control (MAC) protocols have the following constraints.

- Energy-efficiency. The MAC protocol is in charge of managing the radio module of the nodes (both static and mobile), which is usually the most energy-consuming component of a node. Thus, it is very important that the MAC protocol used in our architecture is highly energy efficient so that the overall network lifetime is high.
- Neighbor discovery. The process of discovering neighbors is used whenever a mobile entity becomes in the range of another entity (either a static sensor, another mobile sensor, or the sink). The shorter the delay to discover the neighbor, the more time nodes can use to communicate and exchange data.

Network protocols have to provide the following services.

- Management of mobile nodes. Many nodes in our architecture are mobile. Thus, the network protocol has to be able to maintain routing tables that lead to these nodes as they move, with a limited overhead in terms of the control message.
- Efficient flooding. When the data analysis model generates a warning for farmers, it is important that farmers receive this warning as fast as possible, even if they do not come back in the range of the sink. Thus, the data has been flooded in the network of mobile sensors, in order to reach as many sensor nodes as possible in a short time.
- Delay-tolerant capabilities. The network protocol has to deal with a network that is often disconnected. As it is not possible to predict the mobility pattern of nodes, the network protocol has to communicate in an opportunistic manner, that is, contacting nodes without knowing whether these neighbors will move close to the sink or not. Aiming to reduce the delays in such networks is often achieved by tailoring the flooding algorithm with information on the sensitivity of each data.
- Scalability. The network protocol has to be able to deal with many nodes, as we expect our architecture to span a few square kilometers, with tens of fields and hundreds of mobile sensors.

4.3 Data Management Challenges

Multimedia data are difficult to handle by WSNs because of their size, mostly.

- Compression of images. The large size of high-resolution images require nodes to use compression algorithms for the image to fit into their limited memory. Depending on the processing, lossless compression algorithms have to be used, or lossy algorithms (having higher compression ratio) that do not hide the important features of the image.
- Fragmentation of images into frames. Low-power protocols often limit the frame size to about one hundred bytes (this is the case for the physical layer

of IEEE 802.15.4 for instance). As high-resolution images tend to have a large size, it is necessary to fragment the image into several frames and to ensure that the images can be reconstructed at the reception (for instance, lost fragments have to be retransmitted).

A data analysis model has to be computed from the data collected by the sensors. The role of this model is to predict rice productivity based on microclimate data and to generate notifications and warnings for the farmers.

- Accurate model from incomplete data. The model has to provide accurate results, from incomplete data. In addition, the model has to be auto-adaptive, as it depends on localized data from which no information is known a priori.
- Distributed implementation of a partial model. A partial model can be implemented into static sensors and mobile sensors. For instance, if a mobile sensor receives data indicating that the water level is low and that the temperature is high, it can directly generate a warning for the farmer, rather than having to wait for a communication with a sink and the reception of the notification from the complete model implemented in the cloud.
- Improvement of data through collaboration. We can benefit from having several sensors (both static and mobile) in the same area to improve the confidence in the data. For instance, if a camera sensor identifies some brown spots on leaves, it might be interesting to confirm whether these brown spots are caused by a disease, or if it comes from something else, e.g. dirt. Neighboring nodes can be requested to perform additional measures (such as taking pictures from a different angle or a close-by location) in order to confirm the measurement.

5 Simulation Results

In this section, we present both the simulation results and their analyses. We compare the performance of our architecture and the random architecture applied in the protocol cited in [16]. We used the NS2 simulator. We deploy 10, 20 and 40 nodes in an area of 100 m × 100 m. We use the MAC layer of 802.11. In our simulation, we evaluate the percentage of network lifetime based on the first node die, and the forwarded packet per Cluster Head (CH). Figure 4 shows the average forwarded packet per CH vs the number of nodes. Our proposed architecture presents less number of packets; which ensures that our nodes are less congested than the nodes in random architecture. Figure 5 present the network lifetime. We can notice that our architecture prolongs the network lifetime compared to random architecture. Our stability period is longer due to the well-studied architecture.

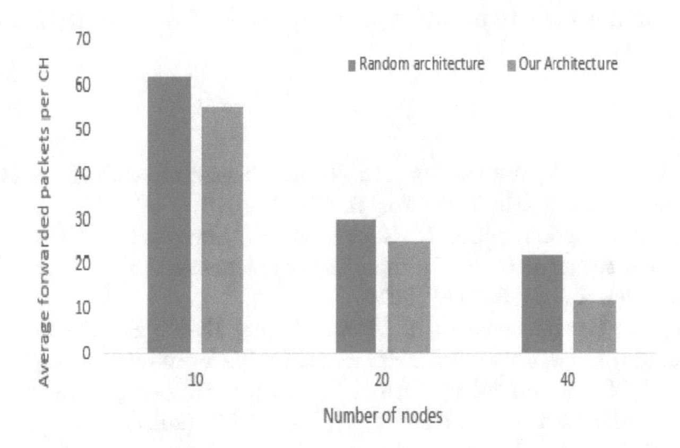

Fig. 4. Average forwarded packets per CH.

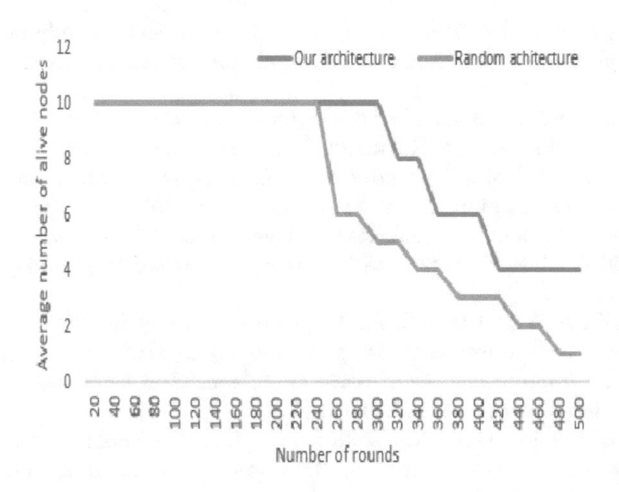

Fig. 5. Network lifetime.

6 Conclusion

WMSNs can be used to achieve precision agriculture for rice field monitoring. In this paper, we proposed a low-cost architecture which does not rely on an existing infrastructure, which does not assume that the WMSN is connected, and which provides additional services to farmers. Then, we discuss the research challenges closely related to this architecture by focusing on three research fields: node hardware design, communication protocols design, and data management. Simulation results showed that our architecture reduces the congestion in CH and extends network lifetime compared to random architecture. The perspective of this work includes the experimental study of our low-cost architecture.

References

1. Hulu, E., Riyanto, B., Widyantoro, S.: Wireless sensor networks for volcano activity monitoring: a survey. Sci. J. Inform. **2**, 53–62 (2016)
2. Lopes Pereira, R., Trindade, J., Goncalves, F., Suresh, L., Barbosa, D., Vazao, T.: A wireless sensor network for monitoring volcano-seismic signals. Nat. Hazards Earth Syst. Sci. **14**, 3123–3142 (2014)
3. Dampage, U., Bandaranayake, L., Wanasinghe, R., et al.: Forest fire detection system using wireless sensor networks and machine learning. Sci. Rep. **12**, 46 (2022)
4. Sakthipriya, N.: An effective method for crop monitoring using wireless sensor network. Middle-East J. Sci. Res. **20**(9), 1127–2232 (2014)
5. Lloret, J., Sendra, S., Garcia, L., Jimenez, J.M.: A wireless sensor network deployment for soil moisture monitoring in precision agriculture. Sensors **21**(21), 7243 (2021)
6. Simon, S., Paulose Jacob, K.: Wireless sensor networks for paddy field crop monitoring application in Kuttanad. Int. J. Mod. Eng. Res. (IJMER) **2**(4), 2017–2020 (2012)
7. Simon, S., Paulose Jacob, K.: Development and deployment of wireless sensor networks in paddy fields of Kuttanad. Int. J. Eng. Innov. Technol. (IJEIT) (2012)
8. Akyildiz, I.F., Melodia, T., Chowdhury, K.R.: A survey on wireless multimedia sensor networks. Comput. Netw. **51**(4), 921–960 (2007)
9. Schwedhelm, P.: Real-time analysis of ultrasonic rodent vocalizations on wearable sensor nodes. Master's thesis, Eberhard Karls Universitat, Tubingen, Germany (2010)
10. Wang, J., Chen, Y., Chanet, J.-P.: An integrated survey in plant disease detection for precision agriculture using image processing and wireless multimedia sensor network. In: International Conference on Advanced in Computer, Electrical and Electronic Engineering (ICACEEE) (2014)
11. Shi, H., Hou, K.M., Diao, X., Xing, L., Li, J.-J., De Vaulx, C.: A wireless multimedia sensor network platform for environmental event detection dedicated to precision agriculture (2018)
12. Tummala, H., Goli, K.M.: Wireless sensors & video networks (WSVN) for indication of fungus affected plants in an agricultural field. Int. J. Eng. Res. Appl. (IJERA) **2**(3), 1388–1390 (2012)
13. Moorhead, T.W.J., Binnie, T.D.: Smart CMOS camera for machine vision applications. In: Image Processing and Its Applications, no. 465 (1999)
14. Rahimi, M., et al.: Cyclops: in situ image sensing and interpretation in wireless sensor networks. In: SenSys 2005 (2005)
15. Aslan, Y.E., Korpeoglu, I., Ulusoy, O.: A framework for use of wireless sensor networks in forest fire detection and monitoring. Comput. Environ. Urban Syst. **36**(6), 614–625 (2012)
16. Béjaoui, C., Guitton, A., Kachouri, A.: Efficient data for monitoring rice field based on WMSN. In: ACS/IEEE International Conference on Computer Systems and Applications (AICCSA), Hammamet, Tunisia, pp. 1–4, October 2017

Machine Learning Based-RSSI Estimation Module in OMNET++ for Indoor Wireless Sensor Networks

Ghofrane Fersi[1(✉)], Mohamed Khalil Baazaoui[2], Rawdha Haddad[1], and Faouzi Derbel[2]

[1] Research Laboratory of Development and Control of Distributed Applications (ReDCAD), National School of Engineers of Sfax, University of Sfax, BP 1173-3038, Sfax, Tunisia
ghofrane.fersi@redcad.org
[2] Smart Diagnostic and Online Monitoring, Leipzig University of Applied Sciences, Leipzig, Germany

Abstract. Network simulation is a crucial step in studying the behavior and the performance of a given network before its real deployment. The more the simulator mimics the real world, the more its results are constructive and reliable. Received Signal Strength Indicator (RSSI) is one of the most important metrics in Wireless Sensor Networks (WSNs). However, such a metric is very sensitive especially indoors to many factors such as temperature (T) and relative humidity (RH) that have spatial and temporal variations, in addition to its known sensitivity to the distance between the sender and the receiver. The existing simulators are not able to generate realistic RSSI values which hurts the accuracy of all the simulated protocols using the RSSI metric such as routing and localisation protocols. Having the ability to estimate RSSI accurately, improves the overall simulation performance and makes it much more closer to reality. For this reason, we have proposed in this paper as a first step, a novel machine learning-based system that considers the distance between the sender and the receiver as well as the temperature and the relative humidity to estimate the RSSI accordingly. The experimental results have shown that our proposed system has improved drastically the accuracy of the RSSI estimation and made it extremely close to the real values. Then, we have included our proposed system as a new module in the OMNET++ network simulator. The simulation results have shown that our added module has improved drastically the simulations' veracity by offering more realistic RSSI measurements.

Keywords: Simulation · RSSI · WSN · ANN

1 Introduction

The Internet of Things (IoT) is invading our lives and is revolutionizing the way we interact with everyday objects, connect devices, and gather and analyze

L. Barolli (Ed.): AINA 2024, LNDECT 204, pp. 273–285, 2024.
https://doi.org/10.1007/978-3-031-57942-4_27

data. As we witness the transformative impact of IoT on various aspects of our daily routines, it is crucial to delve into the core component of IoT, which is the Wireless Sensor Networks (WSN) [1]. WSN are made up of tiny, autonomous, resource-constrained wireless sensors used to collect and transmit data from their environment to ensure efficient monitoring, control, and decision-making across numerous applications.

To unveil the full capabilities and the various characteristics of WSN, the role of simulation tools becomes paramount [2]. In fact, it is of high importance to leverage WSN simulations before venturing into the intricate deployment and management of physical sensor networks. These simulators offer researchers and engineers the opportunity to meticulously analyze, refine, and optimize network designs in a risk-free virtual environment. This not only allows us to explore, test, and evaluate complex networks but also ensures a finely tuned, high-performing, and resilient WSN. Simulation reduces as well the cost and time of implementation [3]. Besides, the simulator provides researchers with a general view of WSN behavior for future use. Especially when using large-scale WSNs, it is crucial to test and simulate the network before implementing the network.

One of the most important challenges for the simulators is that they should offer a simulation environment that mimics the real environment. They should hence behave realistically by using reliable models. However, such a requirement is not always easy to ensure. There are in fact in reality numerous complex phenomenons that can not be reproduced easily.

For example, in indoor WSN, the Received Signal Strength Indicator (RSSI) fluctuates even at fixed distances and in the same environment having the same physical obstacles. This problem has been extensively studied in [4]. It has been proved in that paper that the temperature and relative humidity variations have an impact on the RSSI. Numerous network simulators do not take into consideration external factors and hence are unable to reproduce the RSSI phenomenon as in reality. This presents a negative point for all the protocols relying on RSSI such as routing and WSN-based localization protocols [5,6].

Effectively, in most of the WSN-based localization systems, the location of an object is specified according to the RSSI. The more the RSSI is specified accurately, the better the location estimation is. Since the simulator is unable to reproduce a similar RSSI behavior as in reality, all the localization processes will be disturbed and the performance evaluation of these protocols become far from reality.

Similarly, the RSSI is among the essential factors that specify the perfect next node in the routing process to improve the transmission quality. Having a simulated RSSI value that is far from reality, reduces significantly the reliability of the simulator and may lead to wrong node selection which in turn leads to erroneous protocol performance that is far from reality which raises many problems when this protocol is implemented in a real WSN. It is hence pivotal to find a new method that represents the RSSI in the simulator more pragmatically.

In this paper, we propose a machine learning-based RSSI estimation system that offers accurate RSSI estimation by taking into account the distance, temperature, and relative humidity. Since the performance evaluation of our proposed system has proved that it offers RSSI readings that are very close to the real ones, the integration of such a system in a network simulator like Omnet++ [7] would certainly improve the simulator's credibility by offering realistic RSSI readings that are aware of the environmental factors. We have hence developed a novel RSSI module that integrates our RSSI estimation system on OMNET++. The simulation results have shown that thanks to our new RSSI module, the simulator becomes more realistic and the RSSI fluctuations are similar to the real ones.

The outline of this paper is as follows. Section 2 presents our related work. Our proposed machine learning (ML) RSSI estimation system is presented in Sect. 3. Section 4 presents the experimental results of our proposed system. The integration of our ML-based module in OMNET++ is described in Sect. 5. Section 6 presents the simulation results of our RSSI module. Finally, Sect. 7 concludes the paper.

2 Related Work

This section is divided into two parts. In the first part, we outline the researches that have studied the impact of temperature and relative humidity on RSSI. In the second part, we present a study of the WSN simulators.

2.1 Impact of Environmental Factors on RSSI in Wireless Sensor Networks

RSSI is the power level of the signal received at the device's antenna. Usually, the RSSI is defined by the power of the transmitter, the distance between the transmitter and receiver, and the radio environment fields that can interfere with the RF wave signal beneath the receiver's antenna as depicted in the Eq. 1.

$$RSSI|_{dBm} = RSSI_0|_{dBm} - 10|_{dB}\, n \times log\left(\frac{d}{d_0}\right) + X_\sigma|_{dB} \tag{1}$$

where $RSSI_0$ is the RSSI when the reference distance is d_0; n is the path loss index in a specific environment; X_σ, is a distribution with σ-level standard deviation.

The signal strength of radio waves propagating between sensor nodes in WSN can be affected by atmospheric conditions, even in line-of-sight circumstances. These atmospheric conditions provoke a variation in the power of the propagated waves causing its fluctuations due to the nature of radio-frequency signals. A lot of researches have been carried out to study such a phenomenon.

The study [4] has proved the impact of temperature and humidity variation on the RSSI in the indoor environment. The experiment is conducted in a 9 $\times$ 9 m^2 laboratory room where a CC1101 868 MHz radio module is used. The experiment was carried out to demonstrate that even a slight variation in temperature and humidity has an impact on the RSSI where the correlation is more significant for higher distances between the anchors and the senders.

In [3], authors have proved the existence of a correlation between the temperature, humidity, and RSSI in larger intervals using the CC1101 868 MHz radio chip. The experiment was carried out in a climatic chamber and different levels of sending power have been used. The results of the experiment are modeled in later steps using the linear regression in the OMNeT++ discrete event simulator. No superposition of the two factors nor an estimation of the model was studied in this paper.

An experimental study on the effects of meteorological conditions such as solar radiation, humidity, temperature, and rain on LoRa modules in an outdoor tropical environment has been demonstrated in [8]. A Regular RSSI pattern is observed. The RSSI values show a positive correlation with the temperature during the day but degrade in the evenings. However, the results of the experiment show no correlation with humidity.

The contribution [9] investigated the effects of ambient temperature and humidity on the RSSI of an Atmel ZigBit 2.4 GHz radio module in an outdoor WSN. Experimental results showed that changes in weather conditions affected the received signal strength. Temperature affects the signal strength in general, whereas higher relative humidity might have had some effect, especially below 0 °C.

In [10] a study has been carried out to show the impact of temperature on the RSSI in an oil refinery with CC2420 radio chips. Experimental results have shown that temperature has a significant effect on signal strength and link quality.

2.2 Wireless Sensor Networks Simulators

NS-3 [11] and OMNeT++ [7] in [12] are considered the largest and most popular simulation tools in the research community, particularly the OMNeT++ simulator since its academic, and user community is quite large. It is also popular and the researcher can find many additional modules or libraries for specific needs.

The paper in [13] presented a chronological overview of available Internet of Things (IoT) and WSN simulation tools. It categorized and analyzed the academic papers published in the IoT and WSN simulation tools research area by highlighting simulation tools, research types, research scope, and research performance metrics. It has demonstrated that OMNeT++ and NS-2 [14] are the most top cited WSN simulators. In the modular discrete-event simulator, the OMNeT++ shows better performance than NS-2 due to its large frameworks such as INET which integrates more realistic models of a WSN, besides the large libraries that have been included which makes OMNeT++ one of the largest academic open-source communities in the latest years. Furthermore, the simulator is conceived based on a hierarchical structure of the models making it one of the best software that can integrate new models for simulating real-world scenarios. Based on this comparison, we have chosen to simulate WSN networks using the Omnet++ simulator.

3 Machine Learning-Based RSSI Estimation System

In this section, we present our proposed machine learning system to accurately estimate RSSI. We first start by describing our system design.

3.1 Experimental Setup

We present in this subsection a brief description of our experimental setup that has conducted us to acquire the measurements data-set used in our system training. Then, we describe our proposed model. The experiments were carried out in an indoor environment, inside a building of $12 \times 10\,\mathrm{m}^2$. TelosB motes are used with integrated temperature, light, and humidity sensors. These sensors function according to the IEEE 802.15.4 standard and operate within the 2.4 GHz frequency ISM band, utilizing a bandwidth of 2 MHz. There are two sensors and a base station connected to a PC in our proposed system. These sensors broadcast beacons every 5 s, in their scheduled time frames, with a transmission power fixed to 0 dBm. For each received frame from the sender to the base station, RSSI, temperature, and humidity for the receiver have been extracted. The locations of sensors have been modified during the experiments to vary the distance between the sender and the receiver. This distance has been measured manually and added to the dataset. Our dataset includes 26141 data measurements.

3.2 Large Indoor Variations for Temperature and Relative Humidity

Based on our measurements, we observe that for the receiver relative humidity, measured values vary between 54.8% and 80.9% as depicted in Fig. 1. The measured values for the receiver's temperature vary from 21.8 °C to 26.7 °C as depicted in Fig. 2. These variations are due to the use of cooling and heating systems.

3.3 Proposed Artificial Neural Network (ANN)-Based System

We have developed our new RSSI estimation system using the Artificial Neural Network (ANN) since ANN is among the machine learning technologies that can approximate any arbitrary function (that is not necessarily linear) to any degree of accuracy using the multi-layer architecture. Based on the findings in our previous work [4] showing that the slight modification of temperature and/or

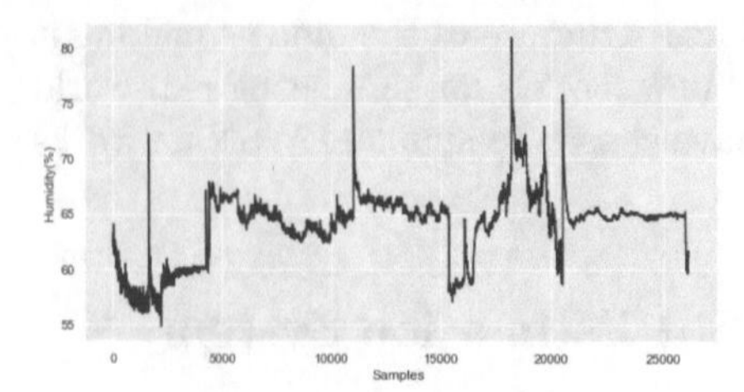

Fig. 1. Humidity variation during experiments

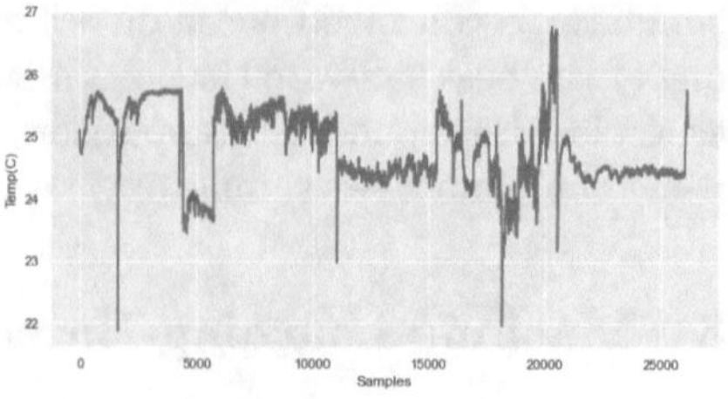

Fig. 2. Temperature variation during experiments

relative humidity affects RSSI, we have set temperature, humidity as well as distance as inputs of our ANN model and the estimated RSSI as the output as depicted in the Fig. 3.

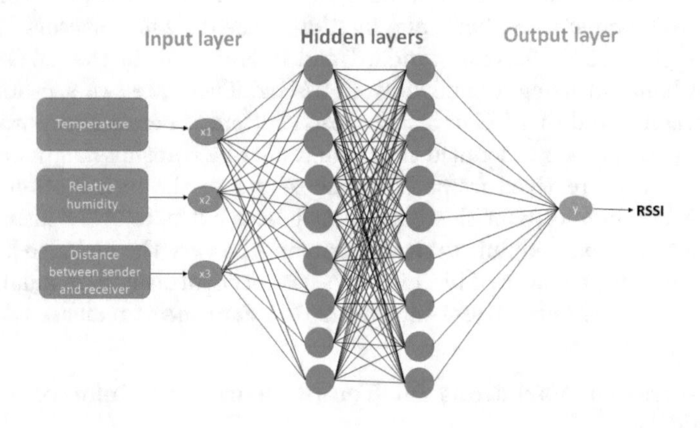

Fig. 3. Proposed ANN system to estimate RSSI in indoor systems

We have split our data so that 70% of the data are used for the model training and the remaining 30% of data are used for the model evaluation. Both training and testing subsets were selected randomly.

4 Experimental Results of Our Machine Learning-Based System

To evaluate the performance of our proposed ANN system, we have conducted experiments with our ANN system and with the Multiple Linear Regression (MLR) systems, and we compared the obtained results. We have the same input for both systems: temperature, RH, and distance. The output is the estimated RSSI.

Table 1. ANN performance through the different configurations

Number	Parameters				Evaluation metrics		
	AF	Number hidden layers	N° epochs	batch size	R^2	MAE	MSE
1	Linear	1	100	20	0.79	3.30	18.34
		1	200	10	0.53	5.15	41.04
		1	200	20	0.85	2.75	12.50
		2	200	20	0.82	2.64	15.36
2	ReLu	1	100	20	0.79	3.15	17.88
		1	200	10	0.79	3.09	18.12
		1	200	20	0.84	2.59	13.80
		2	**200**	**20**	**0.88**	**2.14**	**10.50**
		3	200	20	0.87	2.35	11.24

4.1 ANN Configuration

Since there is no predefined optimal ANN configuration, we have varied the ANN parameters to specify the optimal configuration. The evaluation results are depicted in Table 1. We notice from the Table 1 that the most suitable parameters for our ANN systems are as follows: The Rectified Linear Unit (Relu) as activation function, 200 epochs, 2 hidden layers, and batch size of 20. Effectively, using these parameters has led to having the highest coefficient of determination R^2 (0.88) which represents smaller differences between the observed data and the fitted values. Similarly, the computed MSE and MAE are very low reflecting a low average distance between the observed data values and the estimated data values.

Figure 4 represents the training and validation accuracy using the parameters that have led to the most accurate results marked in bold in Table 1: ReLu activation function, 2 hidden layers, 200 epochs, batch size = 20. Figure 5 depicts the loss training results using the previously mentioned parameters. These two figures show the high accuracy of our proposed ANN system and its extremely low loss.

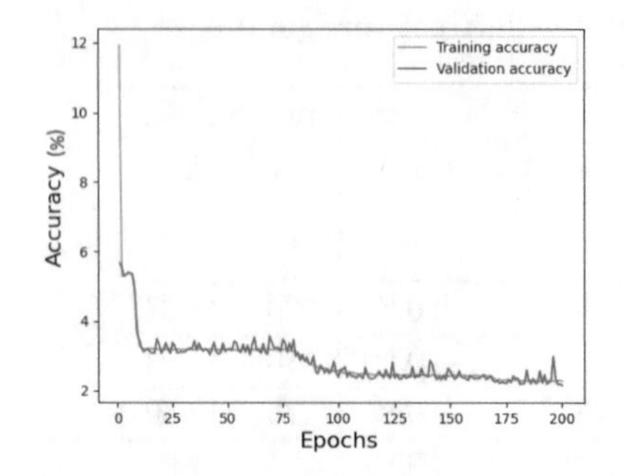

Fig. 4. Accuracy training results for 200 epochs using two hidden layers based on ReLu activation function

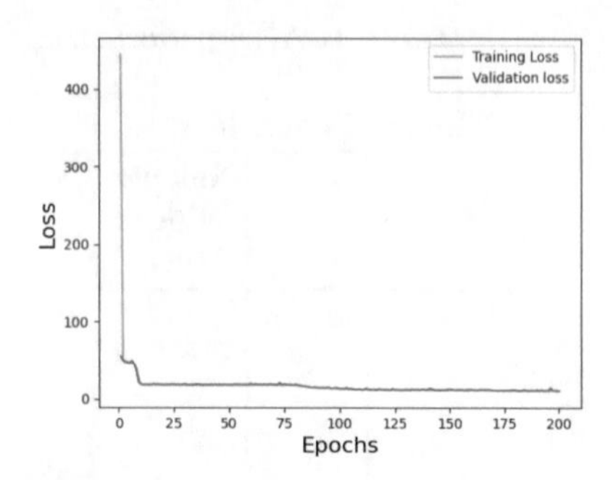

Fig. 5. Loss training results for 200 epochs using two hidden layers based on ReLu activation function

4.2 Performance Evaluation

To evaluate the performance of our proposed ANN system, we have compared its estimation accuracy as a first step with the MLR system. An extract from the RSSI estimation results using our proposed MLR and ANN is illustrated in Tables 2 and 3. We notice from these tables that the estimated RSSI values are much closer to the real RSSI in our ANN model than in the MLR model which further proves that the relationship between the RSSI and environmental factors is not linear.

As a second step, we have compared our ANN system to the MLR and to the ANN system proposed in [15] that has as inputs the distance between the sender and the receiver and the distance between the sensors and the walls and does not take as inputs the temperature and the RH as we have done in our system. Table 4 illustrates the performance results of MLR, our proposed ANN, and the ANN system proposed in [15]. We notice that our proposed ANN system outperforms MLR and offers less error and more accurate estimation. This proves that the relationship between temperature/humidity and RSSI is not linear. Our proposed ANN outperforms the ANN system proposed in [15] because our ANN system takes into account the temperature and relative humidity in additionally to the distance between the sender and the receiver which proves the positive impact of adding these parameters on the accuracy of the ANN estimation system.

Table 2. An extract from estimated values for the testing set using MLR model

N°	Inputs			Estimated RSSI (dBm)	Real RSSI (dBm)
	T°	HR	d		
1	24.41	66.43	9	−74.4	−70.0
2	24.49	64.8	4.5	−64.4	−66.0
3	25.28	65.3	6.5	−74.69	−75.0
4	25.18	65.23	6.5	−73.8	−76.0
5	24.54	59.54	2	−50.70	−51.0

Table 3. An extract from estimated values using our ANN system for the testing set

N°	Inputs			Estimated RSSI (dBm)	Real RSSI (dBm)
	T°	HR	d		
1	24.47	64.52	4.5	−65.23	−66.0
2	24.29	65.23	4.5	−66.50	−66.0
3	24.88	64.12	11	−70.45	−70.0
4	24.39	66.62	9	−68.67	−68.0
5	24.41	66.43	9	−68.66	−70.0

5 Modeling of the ANN-Based RSSI Estimation in OMNeT++ Discrete Event Simulator

In Sect. 4, we have proved that our proposed ANN-based estimation system offers very accurate RSSI estimation. Hence, including this system in the simulator would improve drastically its performance and make its behavior more realistic.

Modeling the RSSI using path-loss equations including the shadowing effects in the Eq. 1 becomes a traditional method. It cannot fulfill the same performance as using machine learning estimation methods with high accuracy and a low margin of error. In the previous section, we have proved the accuracy of our proposed ANN system in the RSSI estimation.

In this section, we will include our proposed ANN system into OMNET++ as a new RSSI module. It uses Keras open-source software library APIs to estimate the RSSI values according to distance, temperature, and humidity employing the imported JSON file results.

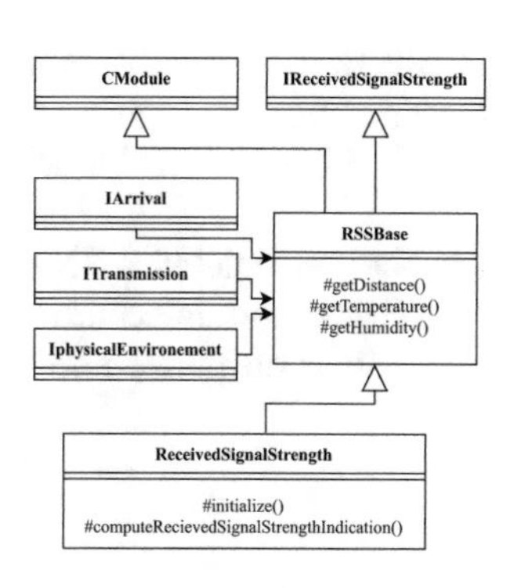

Fig. 6. RSSI Module presentation in OMNeT++ discrete event simulator

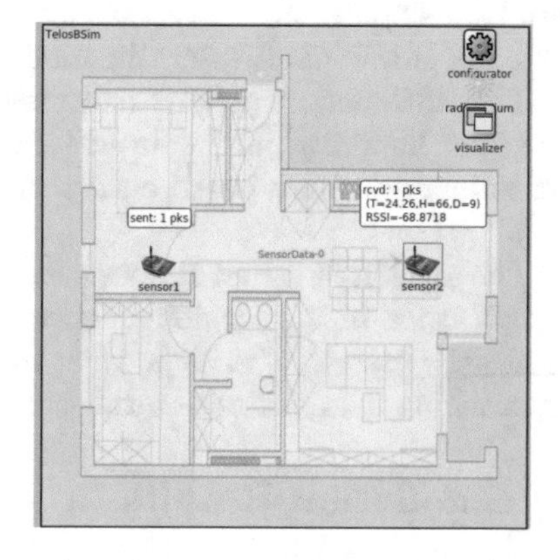

Fig. 7. TelosB Simulation under OMNeT++ discrete event simulator

5.1 Integration of RSSI Module

The work starts with the integration of the RSSI module in OMNeT++. The RSSI module is implemented below the physical layer of the narrow-band receiver composite module. This module is inherited from the RSSBase which is a child class of the base classCModule. The RSSBase module calculates the distance between the transmitter and the receiver modules (IArrival and ITransmission modules in our case) and gets the current temperature and humidity from the IphysicalEnvironement to pass them into our main RSSI module. Therefore, the parameters of the RSSI module are:

- Distance: is computed using the position of both the receiver and sender. The sender's location is obtained from the ITransmission module, and the receiver position is obtained from the IArrival module. A particular API built in the RssBase file calculates the distance between the nodes using the Euclidean method.
- Temperature: is specified using the INI file of the simulation. It is recorded in the IphysicalEnvironement module and converted into the RssBase file.
- Humidity: is assigned by the user employing the INI file of the simulation. It is registered in the IphysicalEnvironement module and transformed into the RssBase file.

The Fig. 6 presents the construction of the modules in the OMNeT++ Simulator.

5.2 Exporting Our ANN-Based Estimation Module in OMNET++

After developing the new RSSI estimation system using ANN, a model was created based on the output results. This model is exported to the OMNeT++ simulator as a JSON file. The ReceivedSignalStrength module uses Keras open-source software library for estimating the output RSSI results and exploits the output results from that JSON file. The built-in architecture in the simulator uses Keras open-source software library on top of the TensorFlow layer to estimate the RSSI values using the model estimate API.

The RSS module gets its parameters from the RSSBase (distance, temperature, and humidity) and converts them into a standard vector, then computes the estimated values based on the previously created JSON file that stores the data.

The estimated values are basically the same as in the constructed ANN model using Python in the previous section to enhance the RSSI module accuracy and reduce the margin of error leading to boosting the simulation performance and approaching it to the real-world scenarios.

6 Simulation Results

To prove the validity of our work, we have conducted simulations where two sensors are peer-to-peer communicating as shown in Fig. 7. The sending interval is configured every 10 s. The frame length is 10 Bytes. The RSSI module gets the vector of data composed of (distance, temperature, humidity) as we have specified in Subsect. 5.1, and estimates the current RSSI using the developed ANN model. Figure 8 shows the implemented module results for 24.26 °C, 66.0 RH, 9.0 m vector.

The RSSI outcome is −68.87 dBm which is the same result as in the designed ANN model. This proves the success of the integration of our estimation module in Omnet++.

Table 5 presents a comparison between the estimated RSSI using our ANN-based module, the RSSI module proposed in [3], and the real RSSI values. We can notice that the simulated RSSI values in our ANN-based module are the nearest values to the real values in all the distances and temperature and humidity values. This proves that the integration of our ANN-based estimation system into OMNET++ has improved the RSSI behavior making it more realistic.

To verify the impact of the integration of the RSSI estimation module on the overall simulation performance, we have measured the time required to load the RSSI module in the simulator. As depicted in Fig. 9, the time required demonstrates the running time doesn't exceed a few milliseconds. This proves that our model doesn't affect the simulation time thanks to its lightness and simplicity.

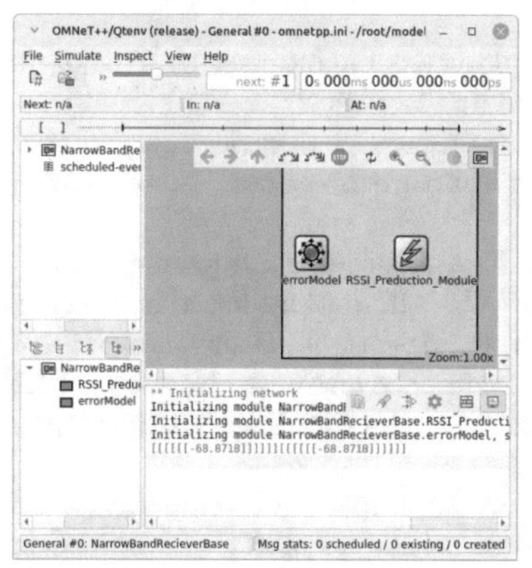

Fig. 8. RSSI Module Estimation

```
Loading json ... done. elapsed time: 0.000221 s
Building model ... done. elapsed time: 0.002326 s
Running test 1 of 1 ... done. elapsed time: 0.000062 s
Loading, constructing, testing of fdeep_model.json took 0.002732 s overall.
Loading json ... done. elapsed time: 0.000186 s
Building model ... done. elapsed time: 0.000706 s
Running test 1 of 1 ... done. elapsed time: 0.000047 s
Loading, constructing, testing of fdeep_model.json took 0.001033 s overall.
```

Fig. 9. Overall required time for RSSI module estimation

Table 4. Performance evaluation results

Metrics	MLR	ANN [15]	Proposed ANN
Adjusted R^2	0.564	Not Applied	0.88
MSE	37.6842	44	10.5
MAE	5.028	5.39	2.14

Table 5. Comparison of RSSI estimation accuracy between the new ANN module and the linear regression approach mentioned in [3]

Temperature (°C)	Humidity (%)	Distance (m)	Our estimated RSSI (dBm) in OMNeT++	Simulated RSSI (dBm) [3]	Actual RSSI (dBm)
24.26	66	9	−68.87	−91.39	−69
23.76	66.68	7	−70.82	−89.29	−71
24.62	64.59	4.5	−66.38	−85.12	−66

7 Conclusion

In this paper, we have proposed an ANN system that estimates accurately the RSSI in indoor environments. The inputs of our proposed ANN system are the distance between the sender and the receiver, the temperature and the relative humidity. The experimental results have shown that our proposed ANN system outperforms MLR since the relationship between the environmental factors and the RSSI is not linear. It also outperforms other ANN systems since the latter does not consider the variation of environmental factors. The estimation results given by our ANN system are very accurate and confirm the importance of adding temperature and RH as inputs.

Furthermore, we built an innovative RSSI module with outstanding performance to emulate real-world scenarios by integrating our ANN estimation system into the simulation. This opens the horizons for us and researchers to experiment with the behavior of WSNs more accurately and realistically with a lower price and reduced time of implementation.

References

1. Sharma, S., Verma, V.K.: An integrated exploration on Internet of Things and wireless sensor networks. Wireless Pers. Commun. **124**(3), 2735–2770 (2022)
2. Mansour, M., et al.: Internet of Things: a comprehensive overview on protocols, architectures, technologies, simulation tools, and future directions. Energies **16**(8), 3465 (2023)
3. Baazaoui, M.K., Ketata, I., Fersi, G., Fakhfakh, A., Derbel, F.: Implementation of RSSI module in Omnet++ for investigation of WSN simulations based on real environmental conditions. In: Prasad, R.V., Pesch, D., Ansari, N., Benavente-Peces, C. (eds.) Proceedings of the 11th International Conference on Sensor Networks, SENSORNETS 2022, 7–8 February 2022, pp. 281–287. SCITEPRESS (2022). https://doi.org/10.5220/0011012600003118
4. Guidara, A., Fersi, G., Derbel, F., Jemaa, M.B.: Impacts of temperature and humidity variations on RSSI in indoor wireless sensor networks. Procedia Comput. Sci. **126**, 1072–1081 (2018). Knowledge-Based and Intelligent Information and Engineering Systems: Proceedings of the 22nd International Conference, KES-2018, Belgrade, Serbia. https://www.sciencedirect.com/science/article/pii/S187705091831322X
5. Guidara, A., Derbel, F., Fersi, G., Bdiri, S., Jemaa, M.B.: Energy-efficient on-demand indoor localization platform based on wireless sensor networks using low power wake up receiver. Ad Hoc Netw. **93**, 101902 (2019)
6. Guidara, A., Fersi, G., Jemaa, M.B., Derbel, F.: A new deep learning-based distance and position estimation model for range-based indoor localization systems. Ad Hoc Netw. **114**, 102445 (2021)
7. Omnet++. https://omnetpp.org/. Accessed Dec 2023
8. Elijah, O., et al.: Effect of weather condition on LoRa IoT communication technology in a tropical region: Malaysia. IEEE Access **9**, 72835–72843 (2021)
9. Luomala, J., Hakala, I.: Effects of temperature and humidity on radio signal strength in outdoor wireless sensor networks. In: 2015 Federated Conference on Computer Science and Information Systems (FedCSIS), pp. 1247–1255 (2015)

10. Boano, C.A., Tsiftes, N., Voigt, T., Brown, J., Roedig, U.: The impact of temperature on outdoor industrial sensornet applications. IEEE Trans. Ind. Inf. **6**(3), 451–459 (2010)
11. Ns3. https://www.nsnam.org/. Accessed Dec 2023
12. Owczarek, P., Zwierzykowski, P.: Review of simulators for wireless mesh networks. J. Telecommun. Inf. Technol. (2014)
13. Idris, S., Karunathilake, T., Förster, A.: Survey and comparative study of LoRa-enabled simulators for Internet of Things and wireless sensor networks. Sensors **22**(15), 5546 (2022). https://www.mdpi.com/1424-8220/22/15/5546
14. Ns2. https://www.isi.edu/nsnam/ns/. Accessed Dec 2023
15. Raj, N.: Indoor RSSI prediction using machine learning for wireless networks. In: 2021 International Conference on COMmunication Systems NETworkS (COMSNETS), pp. 372–374 (2021)

Path Planning in UAV-Assisted Wireless Networks: A Comprehensive Survey and Open Research Issues

Henda Hnaien[1(✉)], Ahmed Aboud[2], Haifa Touati[2], and Hichem Snoussi[3]

[1] Hatem Bettahar IReSCoMath Research Lab, Gabes National Engineering School, University of Gabes, Gabes, Tunisia
hnaienhenda@gmail.com

[2] Hatem Bettahar IReSCoMath Research Lab, University of Gabes, Gabes, Tunisia
ahmedaboud@isimg.tn, haifa.touati@univgb.tn

[3] University of Technology of Troyes, Troyes, France
hichem.snoussi@utt.fr

Abstract. UAV-assisted wireless networks are becoming more and more used and are invading many fields thanks to their performance and efficiency. However, there are still some challenges that need to be addressed before this technology can be widely adopted. One of the main issues is UAV path planning. This task is challenging due to various factors such as data collection, energy consumption, limited battery life, and dynamic changes in the environment. Efficient path planning algorithms are crucial to ensuring safe and efficient UAV operations, minimizing collision risks, and maximizing mission success. To give a complete and clear view of recent papers dealing with this crucial trajectory tracing problem, this survey aims to present a collection of work carried out in this line of research, and for ease of convenience, we have classified existing solutions according to the optimization method selected: heuristic, genetic, machine learning and game theory. Our analysis and qualitative comparison of the current literature on UAV path planning, unveil open research challenges in this field. These challenges serve as a roadmap for future research efforts in the deployment of UAV-assisted wireless networks and steer the exploration of innovative solutions.

Keywords: UAVs · path planning · heuristic algorithm · genetic algorithm · machine learning · game theory

1 Introduction

An Unmanned Aerial Vehicle (UAV) is an aircraft that is operated without a pilot on board. It is controlled either remotely by a human operator or autonomously by onboard computers [1]. In recent years and with the massive growth of wireless networks, using UAV in wireless communication has become increasingly popular. A UAV-assisted wireless network allows for efficient and reliable communication in areas with limited or no existing network infrastructure. By deploying UAVs equipped with wireless transmitters, this technology can provide connectivity to remote locations, disaster-stricken areas, or events with high network demand [2]. Additionally, UAV-assisted

L. Barolli (Ed.): AINA 2024, LNDECT 204, pp. 286–297, 2024.
https://doi.org/10.1007/978-3-031-57942-4_28

wireless networks have the potential to revolutionize industries such as agriculture, surveillance, and emergency response by enabling real-time data collection and analysis. UAVs occupy several roles in wireless networks, such as an aggregate node to gather or disseminate in- formation to Sensor Nodes (SNs) or a relay station to establish wireless connections with remote users who are not directly connected to a base station. Additionally, UAVs can enhance network coverage by acting as mobile base stations, extending the reach of wireless signals to areas with limited infrastructure [3]. Despite the benefits of using UAVs in wireless networks, there are a number of energy and data-related challenges to overcome. All of these issues about saving energy, minimizing time, maximizing data collection, reducing data loss, increasing coverage, and escaping threats revolve around an essential pillar of this type of network, which is UAV path planning. UAV path planning is the process of determining the optimal route for a UAV to navigate from its starting point to its destination while avoiding obstacles and adhering to certain constraints. This involves considering factors such as the UAV's speed, fuel efficiency, weather conditions, and any airspace regulations that may be in place. Additionally, path planning algorithms can also take into account mission-specific objectives, such as maximizing coverage or minimizing risk.

Given the major importance of designing and optimizing UAV trajectories during a mission, numerous research initiatives have been undertaken to develop effective trajectory planning algorithms and techniques. Some of the key factors considered in trajectory planning include energy efficiency, collision avoidance, and mission completion time. This article provides a comprehensive survey of proposed solutions for modeling and optimizing the path planning problem. To enhance clarity in understanding these solutions, we have chosen to categorize them based on the type of algorithm employed-whether heuristic, genetic, machine learning, or game theory. Furthermore, our goal is to offer a qualitative evaluation of the studied solutions, comparing them across various criteria to provide a more insightful analysis.

The remainder of the paper is organized as follows: Sect. 2 is devoted to the classification, presentation, and analysis of solutions considered to solve the UAV path planning problem under different constraints. Section 3 presents a qualitative evaluation of the solutions studied. Section 4 identifies open research issues that have not been fully explored in existing literature. Finally, Sect. 5 concludes the paper and emphasizes future perspectives.

2 Literature Review of Path Planning Algorithms in UAV-Assisted Wireless Networks

In UAV-assisted wireless network, path planning plays a crucial role in optimizing the performance and efficiency of the system. It involves determining the most optimal route for the UAV to follow while considering various factors such as terrain, obstacles, communication range, and energy consumption. By carefully planning the path, the UAV can effectively assist in providing reliable and seamless network coverage, ensuring efficient data transmission and connectivity for users on the ground. For further clarification, we have classified path planning solutions into four categories, as shown in Fig. 1.: (i) Heuristic-based solutions, (ii) Genetic-based solutions, (iii) Machine learning-based solutions, and (iv) Game theory-based solutions.

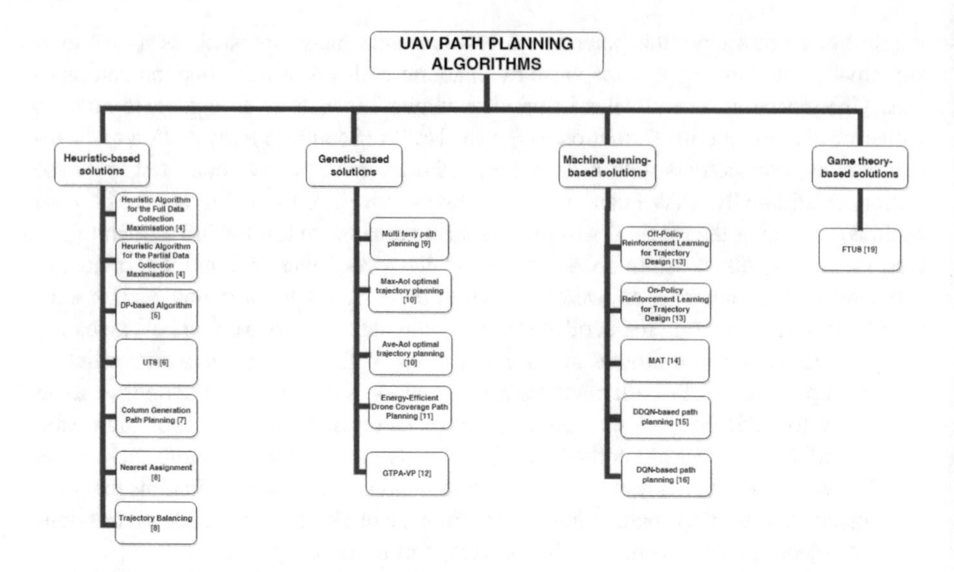

Fig. 1. Classification of path planning solutions for UAV-assisted wireless networks

2.1 Heuristic-Based Solutions

Heuristic algorithms are problem-solving techniques that prioritize finding a solution quickly, even if it may not be the most optimal one. They can be particularly useful in situations where finding an exact solution is computationally expensive or time-consuming. Additionally, they can provide approximate solutions that are good enough for practical purposes. That's why this type of algorithm has been repeatedly used in UAV-assisted wireless network for path planning, task allocation, resource allocation, and deployment problems.

Li et al. [4] design a data collection framework that allows the UAV to gather data from several IoT devices within the UAV's coverage range, using the Orthogonal Frequency Division Multiple Access (OFDMA) technique. Using a single UAV, they aim to collect the maximum amount of data while minimizing energy consumption. In such network, some of the IoT devices are considered as aggregate SNs to store their own and neighbors' sensing data. The authors divide the overflight region statically into square zones and assume that their centers are the UAV's hovering positions. However, this may be unworkable in real-life scenarios. They also assumed that, sensors have the same data transmission rate but in reality it can vary from one sensor to another even if they have the same amount of data to transmit.

Hu et al. [5] consider a wireless network where a UAV launches from a data center, flies to SNs according to a predetermined flight trajectory, and supplies them with energy. SNs then upload their data to the UAV one at a time, using the energy they have collected. The authors formulate an optimization problem to reduce the average Age of Information (AoI) of the data collected from all SNs. Therefore, three factors are studied to be minimized, namely the time required for energy harvesting, the time spent collecting data, and finally the trajectory followed by the UAV. They also note that although the proposed algorithm outperforms both the Greedy algorithm [5] and the Random algorithm [5], it consumes more time and memory when the number of sensors is increased.

In [6], Wang et al. focus on large-scale sensor networks, and point out the drawbacks of using UAVs in these networks. Indeed, because of its limited coverage radius and communications techniques, the UAV can only serve a relatively restricted number of users which could increase data loss. To overcome this problem, user scheduling is the obvious alternative. Therefore, authors suggest a joint user scheduling and trajectory planning data collection strategy. According to this strategy, the UAV performs actions in a time-division manner, it means that at each time slot, the UAV will schedule several SNs as communication objects in the following time slot and choose the coming waypoint to gather data from the involved SNs. We note that this solution outperforms the benchmark algorithms in terms of data loss rate, however, up to 20% of data can be lost in certain scenarios.

All of the previously listed solutions focus on scenarios with a single UAV, in the following paragraph, we will elaborate on solutions that consider Multi UAV-assisted Wireless Network.

In [7], Garraffa et al. consider UAVs as mobile data collection sinks in a large dispersed WSN. They aim to reduce UAVs flight times while guaranteeing energy autonomy, balanced path lengths, and collision avoidance. They have also implemented a tracking feature that allows a remote control system to follow the paths taken by UAVs by communicating their positions. To improve the quality of service of tracking message transmission and to minimize packet loss, they have integrated into their UAV trajectory calculation model a threshold on the minimum average packet delivery rate that must be respected. For this purpose, they adopted the heuristic column generation approach to generate optimal UAV routes by computing the length and average received packet rate for each path. Applied in realistic scenarios with a variable number of static and mobile nodes, the proposed algorithm presents a reasonable complexity and ensures fairness between the lengths of the different paths taken by the UAVs. However, the increase in the number of mobile nodes leads to a decrease in the received packet rate.

Authors of [8] focus on path planning for multi-UAV-assisted wireless networks in a post-disaster scenario where UAVs serve as cellular base stations. They formulate the problem as a Multi-Depot Vehicle Routing Problem (MDVRP), where each UAV is deployed at a temporary air station and there is a set of service points in the area, and they look at how to generate trajectories while improving coverage performance and energy efficiency. They design two heuristic algorithms: Nearest Assignment (NA) and Trajectory Balancing (TB), to maximize the number of service points that UAVs can serve while also ensuring that every UAV can reach its air station before the battery runs out. The first solution assigns each service point to the nearest air station. However, the second one iteratively updates the number of service points assigned to each UAV in order to balance trajectories using Linear Programming (LP). In terms of coverage ratio, the TB algorithm outperforms the NA algorithm, but its computational complexity is higher.

2.2 Genetic-Based Solutions

Genetic algorithms are a type of optimization algorithm inspired by the process of natural selection. They are used to solve complex problems by mimicking the principles of genetics and evolution. These algorithms have been successfully applied to optimize the performance of UAV-assisted wireless communication networks by finding the most efficient path-planning strategies for UAVs and improving energy efficiency and reliability for optimal coverage and connectivity.

Authors of [9] consider the case study of multi message ferry networks where UAVs transport messages between remote wireless nodes by acting as message ferries. To reduce the average weighted delay of message delivery in the network, they propose a new genetic-based path planning solution that builds clusters of nodes, assigns UAVs to clusters, and determines the schedule for UAVs to visit the nodes inside the clusters based on traffic flows between nodes and messages load in nodes. Authors model a path as a chromosome where a gene is constituted of a node and the ferry which is assigned to this node. To measure a solution's quality, they suggest a new fitness function that computes the average delay of all messages generated in the network for a solution. The algorithm converges smoothly and generates a path with the shortest message delay but not the shortest travel distance. Moreover, a major gap in this work is that the physical ferry collision is ignored.

Liu et al. [10] consider the case study of a WSN where a UAV departs from the data center to gather data from ground SNs, flies along a predetermined trajectory, and afterward returns to the data center to perform data analysis. They highlight the importance of the AoI metric, which is the amount of time that has passed between the time that information is sensed and the time that it is delivered to the data center, in the calculation and selection of the optimal path. To solve this problem, they have designed a model based on a genetic algorithm that takes into account two different metrics: the maximum AoI and average AoI of SNs. Indeed, they seek to determine two age-optimal paths for UAV-assisted data collection: one to reduce the age of the first sensing data among the SNs. The other to reduce the average age of the sensing data. Results demonstrate that the suggested algorithms can find near optimal trajectories and help to keep the sensed data fresh, especially for large-scale WSN. However, there are some drawbacks: inaccurate results could be produced because sensing time is neglected during the analysis. Additionally, energy consumption has not been taken into account in the proposed model.

To reduce the amount of energy required for a UAV to complete a task, Shivgan et al. [11] model the UAV path planning problem as a traveling salesman problem and suggest using a genetic algorithm to solve it. As drone's energy consumption depends on its speed, they develop an energy consumption model that takes into account the various flight phases including acceleration, deceleration, hovering, and turning. Population and fitness value are two crucial terms used to describe the genetic algorithm. In this specific problem, all potential drone traveling routes are considered as population and the optimized solution is the fitness value calculated based on energy consumption along a path. This genetic algorithm has the advantage that energy savings become more noticeable as the number of waypoints increases, nevertheless it needs to be enhanced to reduce its computation complexity.

Genetic algorithms have also been exploited to solve UAV path planning issues in Multi-UAV-Assisted Mobile Edge Computing (MEC) Systems. Asim et al. [12] consider a network where several UAVs are deployed to serve Internet of Things devices and propose an adaptive path planning solution with a variable number of UAVs and unknown UAV hovering positions. The formulated model aims to deploy a variable number of stop points, to associate UAVs with these points, and to schedule UAV travel in order to minimize the hovering and flying energy of the system. According to this solution, the population is formed by the location of the stop points, and the evaluation is done by the fitness function, which is calculated based on the energy consumed. The suggested algorithm efficiently predicts the optimal number of UAVs and ensures better performance in terms of energy consumption.

2.3 Machine Learning-Based Solutions

Machine learning has invaded UAV-assisted wireless networks and given multiple solutions for several problems, such as path planning problems. It offers the ability to optimize routes based on real-time data. By analyzing various factors such as signal strength, network congestion, and terrain conditions, machine learning algorithms can determine the most efficient paths for UAVs to navigate.

Zhang et al. [13] integrate Reinforcement Learning (RL) algorithms to optimize UAV trajectory based on transmission data rate, energy consumption and coverage fairness. Authors take into account infrastructure real-world conditions, i.e. ability to adapt to the temporal and dynamic conditions of communication networks. They designed a model for harvesting energy from two feeds, solar power and charging stations, to avoid battery depletion. Next, they develop a Q-learning based solution for regular scenarios and a SARSA based solution for disaster rescue scenarios and remote monitoring. The two solutions prevent energy outages and outperform greedy and random algorithms in terms of data throughput and power consumption.

Wang et al. [14] optimize each UAV's trajectory in order to jointly maximize geographical fairness among converted User Equipments (UEs), fairness of UE-load of each UAV, and overall energy consumption of UEs. In a MEC context, they conceive a multi-agent Deep RL based solution, where each UAV is considered as an agent. Each agent explores its environment and use the gathered information to determine the best path to take in carrying out its assigned task. Authors note that their solution reaches a high level of fairness coverage index and fairness UAV's UE-load index.

Similarly to [14], Peng et al. [15] couple MEC and RL concepts to optimize UAV's trajectory. They propose an online path planning solution based on double deep Q-learning network. They model the UE mobility using Gauss-Markov random movement model and plot the trajectory of a single UAV that will attempt to accomplish the computing tasks transferred by several devices. Furthermore, when modeling this system, the authors took into account two major factors, namely the amount of data migrated from IoT devices to the UAV, and the amount of energy consumed by the latter, either for flying or for task processing. Without forgetting to mention that in the scenario considered, they have adopted a static partition of the UAV's overflight zone, which may not be possible in a real environment.

In the context of WSN, Guo et al. [16] introduce UAV-assisted cooperative computing, offering the SN close-range computing services and enabling them to save energy and increase the WSN lifetime. They built a model wherein nodes, UAV, and sink nodes can process computation tasks. Moreover, they propose a deep Q network-based path planning algorithm to dynamically alter the trajectory in response to the environment's variations in order to maximize energy efficiency and attain energy balance. In the proposed model, the UAV is considered as an agent and the Markov's Decision Process is used to model its action decision problem. Here, the UAV needs to revise instantly its action strategy based on node's computing capacity, tasks' number of bits, tasks' deadline, and flight distance.

2.4 Game Theory-Based Solutions

Game theory has been widely used to solve UAV path planning problems, especially in the military domain, and more specifically on the battlefield [17, 18]. However, in the context of wireless networks, this concept is rarely used. Authors of [19], have used game theory to model UAV path planning in wireless networks. They introduced a new term: Value Added Internet of Things Services (VAIoTS), to represent a set of tasks that can be assigned to a UAV in parallel with its original mission.

Moreover, they have defined a framework in which a central system orchestrator is used to allocate tasks and organize UAV flights. They have developed three solutions based on two conflicting objectives: Minimise UAVs' energy consumption and reduce UAVs' operation time. The first solution schedules a set of tasks using a minimum number of UAVs in order to optimize the total energy spent. The second one optimizes the UAVs time operation a mission and also finds the threshold that should not be exceeded. In addition, a third solution is proposed, based on Kalai-Smorodinsky Bargaining model, to achieve a trade-off between energy consumption and system operation time. One of the drawbacks noted from the simulation results is that, by increasing the number of tasks, the third algorithm takes more than 4 min to run.

2.5 Discussion

As previously mentioned, there are important factors, such as data rate, energy consumption, flight time, and others, that have a significant impact on the modeling of path planning problems in the context of UAV-assisted wireless networks. Despite the great effort made in several studies to model this type of problem and the different methodologies adopted such as heuristic algorithms, genetics, machine learning, game theory, etc., we have noted various shortcomings as shown in Table 1. These drawbacks vary from one study to another in terms of the number of UAVs used, the number of constraints to be met and the factors involved in optimization. Among the disadvantages identified are the complexity of the proposed model, network overload, data loss, inequitable coverage, and services offered to users equipments. Moreover, there are suggested solutions [4, 5, 7, 13, 16, 19] that are not adaptable to real-life scenarios or large-scale networks or have a high execution time. In addition, there are solutions [9, 10, 14] that have chosen to gain complexity by ignoring crucial factors such as the UAV's computational capacity,

Table 1. Literature review of UAV- assisted wireless network path planning

Method	Solution	Algorithm	Year	UAV		Communication Technology	Evaluation		Advantages	Limits
				Multi	Single		Tools	Metrics		
Heuristic	Heuristic Algorithm for the Full Data Collection Maximisation [4]	NA	2020		✓	5G	Experimental Evaluation	Collected data volume	Minimizes energy consumption	Static partition of the overflight region
	Heuristic Algorithm for the Partial Data Collection Maximisation [4]									
	DP-Based Algorithm [5]	DynamicProgramming	2020		✓	NA	NA	AoI	Reduces the average AoI	Not adapted to large-scale wireless networks
	UTS [6]	NA	2021		✓	NA	NA	Data loss rate	Decreases data loss	Up to 20% of data can be lost in certain scenarios
	Column Generation Path Planning [7]	Column Generation	2018	✓		wifi	OMNeT++	Path Length Received packet Rate	Reasonable complexity Balanced path lengths	Overloading the network with tracking packets Significant loss of data packets when mobile nodes are increased
	Nearest Assignment [8]	NA	2019	✓		Cellular/Satellite	NA	Coverage Rate	Reasonable complexity	More energy consumed
	Trajectory Balancing [8]	NA							Improves coverage performance	Complexity
Genetic	Multi ferry path planning [9]	GA	2018	✓		wifi	Python	Average delay Travel distance	Converges smoothly Generates a path with the shortest message delay	Ignore path length Ignore UAV collision
	Max-AoI optimal trajectory planning [10]	GA	2018		✓	NA	NA	AoI	Keep the sensed data fresh	Sensing time is neglected Ignore energy consumption
	Ave-AoI optimal trajectory planning [10]	GA								
	Energy-Efficient Drone Coverage Path Planning [11]	GA	2020		✓		Matlab	Energy consumption	Improves energy saving	Complexity
	GTPA-VP [12]	GA	2021	✓		5G		Energy consumption	Predicts the optimal number of UAVs Reduces energy consumption	Complexity
Machine learning	Off-PolicyReinforcement Learning for Trajectory Design [13]	Q-Learning	2021		✓	LTE	Python	Energy consumption Energy outage ratio Throughput	Improve data throughput Prevent energy outages	Coverage fairness declines as the number of IoT terminals rises
	On-Policy Reinforcement Learning for Trajectory Design [13]	SARSA								
	MAT [14]	MADDPG	2020	✓		Cellular Network	PythonTensorflow	Energy consumption	Enhances fairness coverage Improves fairness UAV's UE-load	Don't take into account each UAV's maximum computing resource constraint
	DDQN-based path planning [15]	DDQN	2021		✓	Cellular Network	NA	Energy consumption Offloaded Data bits	Reduces UAV energy consumption	There is no service equity between user equipments
	DQN-based path planning [16]	DQN	2023		✓	NA	Python	Energy consumption Offloaded Data bits	Enhances energy efficiency	Even in a small-scale network, execution time is important.
Game theory	FTUS [19]	KSBS	2019	✓		5G	PythonGurobi	Energy consumption UAVs' operation time	Achieves a trade-off between UAVs' energy consumption and UAVs' operation time	More tasks lead to rising time complexity.

energy consumption, data processing time, and collision risk in the case of multi-UAV networks.

Table 2. Comparison of path planning solution in UAV-assisted wireless networks

Algorithm/Ref.	Complexity	Scalability	Data-Driven	Path Planning
Heuristic Algorithm for the Full Data Collection Maximisation [4]	low	yes	yes	static
Heuristic Algorithm for the Partial Data Collection Maximisation [4]	low	yes	yes	static
DP-Based Algorithm [5]	high	no	no	static
UTS [6]	low	yes	yes	static
Column Generation Path Planning [7]	low	no	yes	static
Nearest Assignment [8]	low	yes	no	static
Trajectory Balancing [8]	high	yes	no	static
Multi ferry path planning [9]	low	yes	yes	dynamic
Max-AoI optimal trajectory planning [10]	low	yes	no	static
Ave-AoI optimal trajectory planning [10]	low	yes	no	static
Energy-Efficient Drone Coverage Path Planning [11]	high	yes	no	static
GTPA-VP [12]	high	yes	no	dynamic
Off-Policy Reinforcement Learning for Trajectory Design [13]	high	yes	no	dynamic
On-Policy Reinforcement Learning for Trajectory Design [13]	high	yes	no	dynamic
MAT [14]	high	no	no	dynamic
DDQN-based path planning [15]	high	no	yes	dynamic
DQN-based path planning [16]	high	no	yes	dynamic
FTUS [19]	low	no	no	dynamic

3 Qualitative Evaluation

To qualitatively evaluate the solutions already mentioned, we analyse, in Table 2, each UAV path planning solution according to four criteria:

- *Complexity:* refers to the execution time of a solution and the resources used to achieve its path planning objective.
- *Scalability:* indicates whether the network size can be increased by adding a number of user devices or sensors without affecting the solution's performance.
- *Data-Driven:* a solution is said to be data-driven if it gives major importance to the amount of data disseminated and its effectiveness is assessed from the point of view of data loss.
- *Path Planning:* indicates whether the proposed solution adopts a static trajectory planning that takes no account of changes in the environment under study, or ensures dynamic adaptation of the UAV's path.

As shown in Table 2, only one of the studied solutions meets all four criteria: low complexity, scalability, data-driven and dynamic path planning. Furthermore, only three other solutions guaranteed low complexity, scalability, data-driven modeling, but in return they adopted static path planning.

Moreover, according to Fig. 2, we notice that on the one hand, we have more than 60% of scalable solutions, and on the other, half of the solutions are of high complexity.

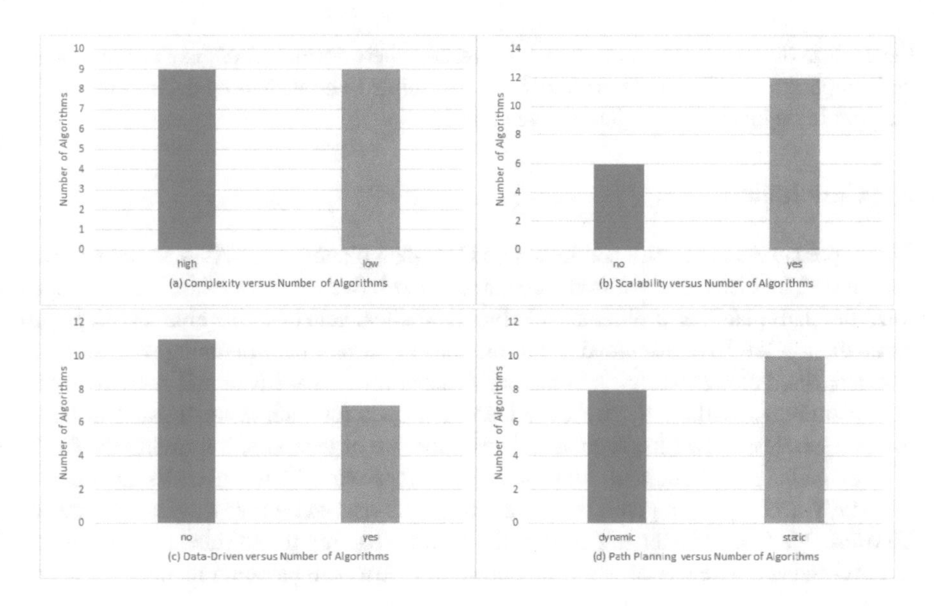

Fig. 2. Analysis of the performance of UAV path planning solutions based on Complexity, Scalability, Data-Driven and Path Planning criteria.

What's more, 55% of the papers studied proposed a static path planning solution, so we can deduce that having a scalable and dynamic solution implies high complexity. Now, if we talk about data-driven solutions, we notice that 61% of these solutions are not data-driven, i.e. the majority of solutions have sought to improve energy consumption, reduce UAV flight time and path length, whereas data disseminated in networks is one of the major criteria marking network performance and efficiency, and we need to involve it in UAV trajectory planning solutions as a modeling constraint.

4 Open Research Issues

Even though UAV-assisted wireless networks have been the focus of numerous studies, we notice according to Sect. 3 that the majority of the articles we have already studied have not given due weight to the data collection factor. Despite the fact that this factor plays an important role in path tracking issues, it has been largely overlooked or underestimated. In this context, we can speak about data-oriented path planning, and instead of having a static path planning, which is the solution adopted in most papers, we can vote for a dynamic path planning according to the volume of data to be processed in each zone. In this way, the path taken by the UAV will be adapted in real time according to the amount of data available in each area.

In brief, there are still a number of intriguing issues that need to be resolved. Firstly, maximize data collection by developing a path-planning model that aims to reduce the rate of data loss. Secondly, to give more importance to the crucial term AoI, which measures the freshness of collected information. Finally, study the planning problem in the context of multi-UAV cooperative networks, where several UAVs can communicate and cooperate together to determine the set of points to be visited by each of them and the path to be taken to complete a given mission.

5 Conclusion

This paper surveys the work conducted on UAV path planning in UAV-assisted wireless networks. The discussed solutions are categorized based on the methods used to optimize the path planning problem, including heuristics, genetics, machine learning, and game theory. We have discussed the limitations of current path-planning solutions and compare them along four axes: complexity, scalability, data-driven and static/dynamic path planning. Additionally, we have identified open research issues related to UAV-assisted wireless networks, such as reducing the rate of data loss, improving the freshness of gathered information, and managing cooperation between UAVs in order to efficiently accomplish a given task in a multi-UAV-assisted wireless network context. To overcome these challenges, further research is required to optimize data collection in UAV-assisted wireless networks and develop robust path planning models.

References

1. Touati, H., Chriki, A., Snoussi, H., Kamoun, F.: Cognitive radio and dynamic TDMA for efficient UAVs swarm communications. Comput. Netw. **196**, 108264 (2021)
2. Chriki, A., Touati, H., Snoussi, H., Kamoun, F.: Centralized cognitive radio based frequency allocation for UAVs communication. In: IWCMC 2019, pp. 1674–1679 (2019)
3. Chriki, A., Touati, H., Snoussi, H., Kamoun, F.: FANET: communication, mobility models and security issues. Comput. Netw. **163**, 106877 (2019)
4. Li, Y., et al.: Data collection maximization in IoT-sensor networks via an energy-constrained UAV. IEEE Trans. Mob. Comput. **22**(1), 159–174 (2021)
5. Hu, H., Xiong, K., Qu, G., Ni, Q., Fan, P., Letaief, K.B.: AOI-minimal trajectory planning and data collection in UAV-assisted wireless powered IoT networks. IEEE Internet Things J. **8**(2), 1211–1223 (2020)
6. Wang, X., Liu, X., Cheng, C., Deng, L., Chen, X., Xiao, F.: A joint user scheduling and trajectory planning data collection strategy for the UAV-assisted WSN. IEEE Commun. Lett. **25**(7), 2333–2337 (2021)
7. Garraffa, M., Bekhti, M., Létocart, L., Achir, N., Boussetta, K.: Drones path planning for WSN data gathering: a column generation heuristic approach. In: WCNC, pp. 1–6 (2018)
8. Lin, Y., Wang, T., Wang, S.: Trajectory planning for multi-UAV assisted wireless networks in post-disaster scenario. In: GLOBECOM 2019, pp. 1–6 (2019)
9. Harounabadi, M., Bocksberger, M., Mitschele-Thiel, A.: Evolutionary path planning for multiple UAVs in message ferry networks applying genetic algorithm. In: PIMRC 2018, pp. 1–7 (2018)
10. Liu, J., Wang, X., Bai, B., Dai, H.: Age-optimal trajectory planning for UAV-assisted data collection. In: INFOCOM WKSHPS 2018, pp. 553–558 (2018)
11. Shivgan, R., Dong, Z.: Energy-efficient drone coverage path planning using genetic algorithm. In: HPSR 2020, pp. 1–6 (2020)
12. Asim, M., Khan, W.U., Belhaouari, S.B., Hassan, S.: A novel genetic trajectory planning algorithm with variable population size for multi-UAV-assisted mobile edge computing system. IEEE Access **9**, 125569–125579 (2021)

13. Zhang, L., Çelik, A., Dang, S., Shihada, B.: Energy-efficient trajectory optimization for UAV-assisted IoT networks. IEEE Trans. Mob. Comput. **21**(12), 4323–4337 (2022)
14. Wang, L., Wang, K., Pan, C., Xu, W., Aslam, N., Hanzo, L.: Multi-agent deep reinforcement learning-based trajectory planning for multi-UAV assisted mobile edge computing. IEEE Trans. Cogn. Commun. Netw. **7**(1), 73–84 (2020)
15. Peng, Y., Liu, Y., Zhang, H.: Deep reinforcement learning based path planning for UAV-assisted edge computing networks. In: WCNC 2021, pp. 1–6 (2021)
16. Guo, Z., Chen, H., Li, S.: Deep reinforcement learning-based UAV path planning for energy-efficient multitier cooperative computing in wireless sensor networks. J. Sens. **2023**, 1–13 (2023)
17. Moolchandani, D., Prathap, G., Afanasyev, I., Kumar, A., Mazzara, M., Sarangi, S.R.: Game theory-based parameter-tuning for path planning of UAVs. In: VLSID 2021, pp. 187–192 (2021)
18. Van Nguyen, L., Phung, M.D., Ha, Q.: Game theory-based optimal cooperative path planning for multiple UAVs. IEEE Access **10**, 108034–108045 (2022)
19. Motlagh, N.H., Bagaa, M., Taleb, T.: Energy and delay aware task assignment mechanism for UAV-based IoT platform. IEEE Internet Things J. **6**(4), 6523–6536 (2019)

Design of Low Sidelobe Microstrip Antenna Array with Non-uniform Excitation Amplitude Based on Dolph-Tschebyscheff Synthesis at 28 GHz

Emna Jebabli[1(✉)], Mohamed Hayouni[1,2], and Fethi Choubani[1]

[1] Innov'COM Research Laboratory, Higher School of Communications of Tunis (Sup'Com), Université de Carthage, Ariana, Tunisia
{emna.jebabli,fethi.choubani}@supcom.tn, mohamed.hayouni@supcom.rnu.tn
[2] Higher Institute of Computer Sciences of Kef, Université de Jendouba, Kef, Tunisia

Abstract. The sidelobes in antenna array radiation patterns affect its performance and cause interferences with other communication devices, which degrades the wireless communication. Reduction of the sidelobes level is then inevitable. A variety of methods, such as the Taylor distribution and Dolph-Chebyshev approach, can be employed to lower the sidelobe level (SLL). In this paper, a 1×6 series fed linear microstrip antenna array based on Chebyshev distribution is designed to reduce the SLL below 26 dB in the E-plane at 28 GHz. Simulation results show a SLL reduction of 12.2 dB (from -7.9 dB to -20.1 dB), a bandwidth of 1.59% and a gain of 14 dB at 28 GHz. This antenna has then the potential to be used in mm-wave 5G applications.

1 Introduction

Antenna arrays with low sidelobe are needed for some precision wireless technology applications such as satellite communication, radar, and traffic monitoring systems [1]. However, the SLL of the conventional uniform antenna array is about -13.2 dB, which is not able to fulfill the demands of modern communication [2]. In general, the SLL in a uniform power distribution antenna array is increased by the addition of radiation elements, resulting in waste energy [3]. Fortunately, in an antenna array, it is possible to control the array excitation to produce extremely low sidelobe patterns [2]. Thus, several methods exist to control the SLL level, including Dolph-Tschebyscheff or Chebyshev synthesis and Taylor line source synthesis [4].

The Dolph-Chebyshev array is the simplest way to obtain a beam pattern with a specified low SLL [5]. This method provides a compromise between two important parameters: beamwidth and minimum SLL [6]. However, the SLL is controlled by using a particular weighting for each element in the array that is specified by the Chebyshev polynomial [7]. In other words, the width of the antenna array elements is reduced or tapered as function of the coefficients of the Chebyshev polynomial [7]. Hence, the

L. Barolli (Ed.): AINA 2024, LNDECT 204, pp. 298–306, 2024.
https://doi.org/10.1007/978-3-031-57942-4_29

array factor of the correspondent antenna array must be expressed as a polynomial with coefficients that correspond to the magnitudes of the current elements [7]. Then, this array factor is compared to a Chebyshev polynomial with the same degree [7]. After that, a resolution of a linear system of equations is done, and the excitation coefficient are found. Each independent variable represent an excitation coefficient of one element [7]. Generally, the elements of a symmetrical array have the same excitation coefficient.

The aim of Taylor line source synthesis is to produce an antenna array that emits a specific radiation pattern with a low SLL and a controlled beam shape [4,6]. The Taylor synthesis method is based on the idea of approximating a desired radiation pattern by using a series of line sources, each one with a specific magnitude and phase [4,6]. However, a linear array is composed of equally spaced elements, the weights assigned to each element determine the magnitude and phase of the line source in question [4,6]. An application of Taylor distribution has been presented in [3]. The antenna corporate-feed lines are designed by using Taylor coefficient, which results in an unequal power divider. The SLL has been then minimized by 18 dB [3].

The design of Dolph-Chebyshev array ensures a same SLL, while Taylor creates a pattern where the inner sidelobes are constant and the others remaining decrease monotonously [6]. For some applications, such as radar, the Taylor distribution is considered a best choice because it results in reduced sidelobes when the number of elements increases [6].

Regardless of the Chebychev approach and Taylor line source synthesis, there are other methods that can be applied to minimize the SLL such as the Bayless Synthesis.

The design of a 1×6 linear antenna array fed on series based on Doplh-Chebyshev distribution will be presented in this paper.

In Sect. 2, a theoretical design of a Chebyshev antenna array by applying it to a linear antenna of six elements, will be presented. In Sect. 3, a comparison between the two proposed antennas will be carried out. Finally, in Sect. 4, the main results are summarized in the conclusion.

2 Design of Antennas Arrays

The used electromagnetic simulation tool is CST STUDIO. The substrate material used to model the antenna arrays is Rogers RT Duroid 5880 with a thickness and a permittivity of 0.254 mm and 2.2 respectively. The conductive material is the copper with a thickness of 17 μm.

2.1 Design of Rectangular Antenna Array

The 1×6 rectangular series fed antenna array is shown in Fig. 1.

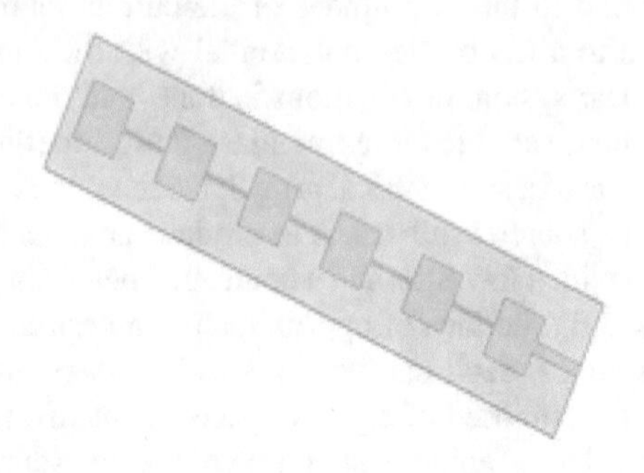

Fig. 1. 1×6 series fed rectangular antenna array.

Each rectangular patch constituting the 1×6 antenna array has a width W and a length L of 4.24 mm and 3.45 mm, respectively, calculated at 28 GHz by using Eqs. (1) and (2) [8]:

$$W = \frac{v_0}{2f_r}\sqrt{\frac{2}{\varepsilon_r + 1}} \tag{1}$$

$$L = \frac{v_0}{2f_r\sqrt{\varepsilon_{reff}}} - 2\Delta L \tag{2}$$

where:

v_0 is the speed of light in free space
f_r is the resonate frequency
ε_{reff} can be found by using the following Eq. (3)

$$\varepsilon_{reff} = \frac{\varepsilon_r + 1}{2} + \frac{\varepsilon_r - 1}{2}[1 + 12\frac{h}{w}]^{\frac{-1}{2}} \tag{3}$$

The ΔL, is calculated by the following Eq. (4):

$$\frac{\Delta L}{h} = 0.412\frac{(\varepsilon_{reff} + 0.3)(\frac{w}{h} + 0.264)}{(\varepsilon_{reff} - 0.258)(\frac{w}{h} + 0.8)} \tag{4}$$

2.2 Design of Chebychev Array Antenna

To design a Chebyshev array, the first step is to find the Array Factor (AF) as given by the following equations [4,6] and [9]:

$$AF = \begin{cases} \sum_{n=1}^{M} a_n cos[(2n-1)u], & \text{N=2M even} \\ \sum_{n=1}^{M+1} a_n cos[2(n-1)u], & \text{N=2M+1 odd} \end{cases} \tag{5}$$

With :
N is the number of the antenna array elements.

a_n is the excitation coefficient of the array elements and

$$u = \frac{\pi d}{\lambda} \cos\theta \tag{6}$$

with d is the separation distance between the antenna array elements and λ is the wavelength.

In the second step, in the AF that has been founded in the first step, we have to replace each cos(mu) with (m = 0, 1, 2, 3, etc.) by its corresponding series expansion as given by the following equations [4,6] and [9]:

By using Euler's formula and $sin^2 u = 1 - cos^2 u$ we obtain, for:

$$m = 0, \ \ cos(mu) = 1; \tag{7}$$

$$m = 1, \ \ cos(mu) = \cos u; \tag{8}$$

$$m = 2, \ \ cos(mu) = cos(2u) = 2cos^2 u - 1; \tag{9}$$

$$m = 3, \ \ cos(mu) = cos(3u) = 4cos^3 u - 3\cos u; \tag{10}$$

$$m = 4, \ \ cos(mu) = cos(4u) = 8cos^4 u - 8cos^2 u + 1; \tag{11}$$

$$m = 5, \ \ cos(mu) = cos(5u) = 16cos^5 u - 20cos^3 u + 5\cos u; \tag{12}$$

Then, we must substitute

$$\cos u \ \ by \ \ \frac{z}{z_0} \tag{13}$$

To determine the value of the point z_0, we must use the following equations [4,6] and [9]:

$$R_0 = \cosh(m \cosh^{-1}(z_0)), \quad here \ \ m = N - 1 \tag{14}$$

So,

$$z_0 = \cosh(\frac{\cosh^{-1}(R_0)}{m}) \tag{15}$$

where R_0 is the desired SLL to be attended. To be used in the equation above, it needs to be transformed into a linear form.

$$R_0 = 10^{R_0(dB)/20} \tag{16}$$

In the third step, we have to make a comparison between the polynome found in the second step and a Chebyshev polynome T_m(z) of the same order. Then, we calculate the excitation coefficients a_n [4,6] and [9].

The Chebyshev polynomials are determine by:

$$T_m(z) = \begin{cases} (-1)^m \cosh(mcosh^{-1}|z|), & z < -1 \\ cos(mcos^{-1}(z)), & -1 < z < 1 \\ cosh(mcosh^{-1}(z)), & z > 1 \end{cases} \tag{17}$$

The first five chebyshev polynomes are:

$$T_0(z) = 1; \tag{18}$$

$$T_1(z) = z; \tag{19}$$

$$T_2(z) = 2z^2 - 1; \tag{20}$$

$$T_3(z) = 4z^3 - 3z; \tag{21}$$

$$T_4(z) = 8z^4 - 8z^2 + 1; \tag{22}$$

$$T_5(z) = 16z^5 - 20z^3 + 5z; \tag{23}$$

The higher order of Chebyshev can be found using the following equation:

$$T_m(z) = 2zT_{m-1}(z) - T_{m-2}(z) \tag{24}$$

Finally, we have to write the a_n excitation coefficient that has been found in the third step in a normalized form which is done by using the excitation coefficient of the elements at the edge or at the center [4,6] and [9].

In our case, we have used an antenna array with six elements (N = 6) with an SLL ratio of R_0 = 26 dB. So, based on the above Eqs. (5)–(24), we have calculated the coefficients.

The normalized coefficients, from the central patch to the edge, are respectively [0.37, 0.72, 1, 1, 0.72, 0.37]. So, the new widths of the antennas are [1.5688, 3.0528, 4.24, 4.24, 3.0528, 1.5688]. The proposed 1 × 6 Chebyshev antenna array is shown in Fig. 2.

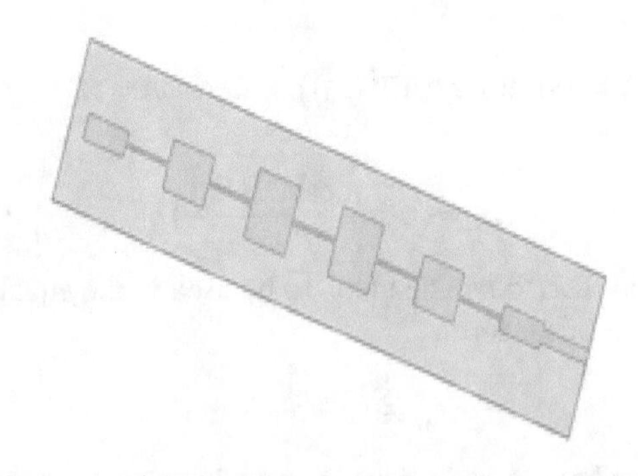

Fig. 2. 1 × 6 Chebyshev antenna array series fed.

3 Simulation Results

Figure 3 depicts the S_{11}. As a result, for the Chebyshev antenna array, a level of −22.7 dB and a bandwidth of 1.59% are obtained at 28,07 GHz, whereas a level of −44.8 dB and a bandwidth of 1.4% is attended at 28,7 GHz for the rectangular antenna array.

Figure 4 plots a comparison of the two suggested antennas with respect to SLL. It is evident that the Chebyshev array has reduced the SLL in the E-plane by an order of 12.2 dB (from −7.9 dB to −20.1 dB). Nevertheless, the rectangular array's gain is 14.2 dB, whereas the Chebyshev array's gain is only slightly reduced by 0.2 dB, yielding a gain of 14 dB. Hence, Fig. 5 shows the radiation pattern of the two proposed antennas in 3D at 28 GHz. However, the surface current distribution of the Chebyshev antenna array at 28 GHz is shown in Fig. 6.

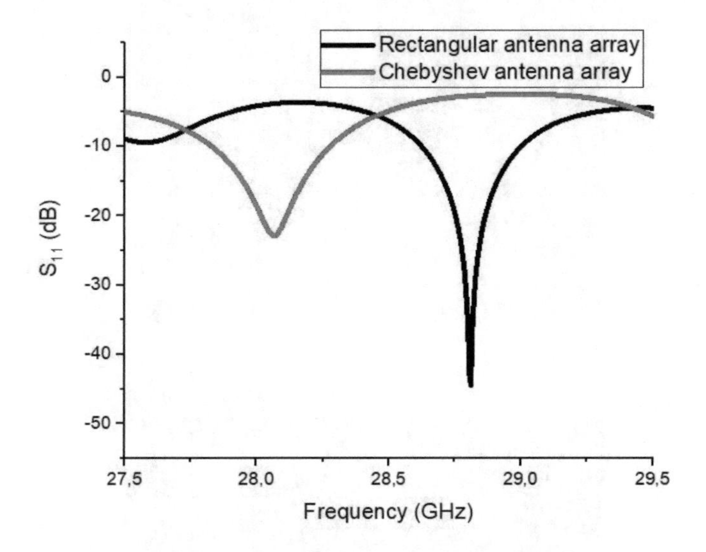

Fig. 3. Simulated S-parameter for rectangular and Chebyshev antenna array.

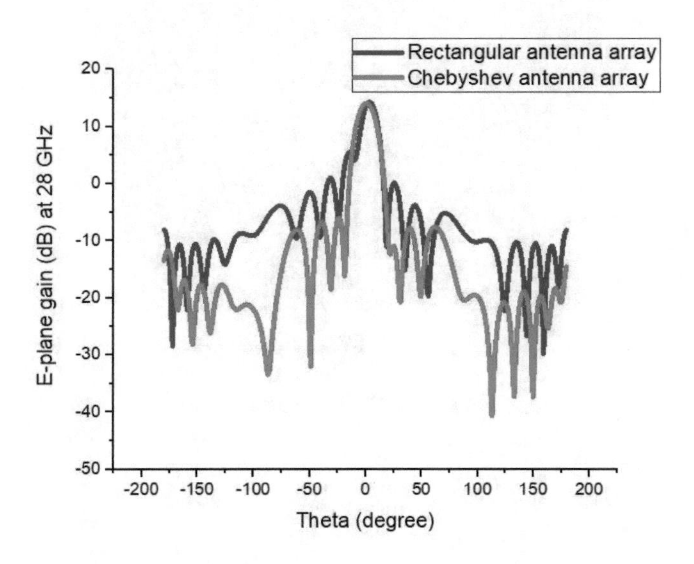

Fig. 4. 2D Radiation pattern.

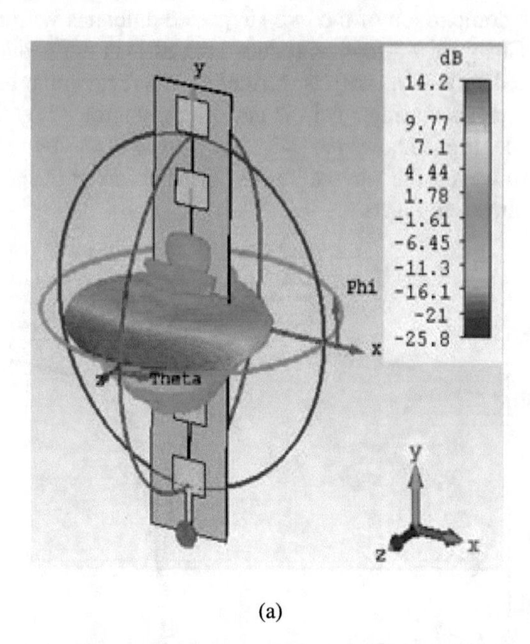

(a)

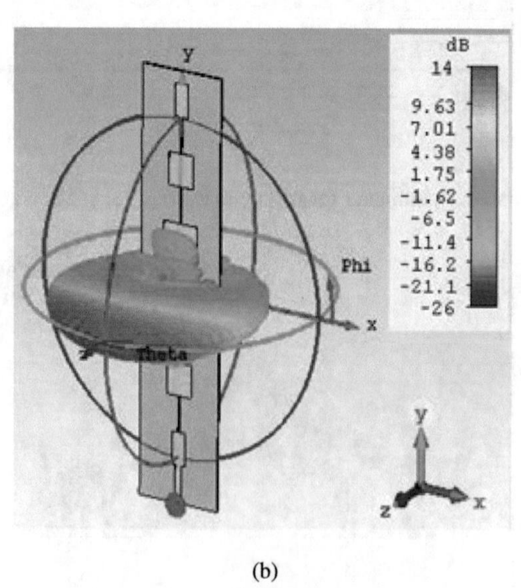

(b)

Fig. 5. 3D Radiation pattern.

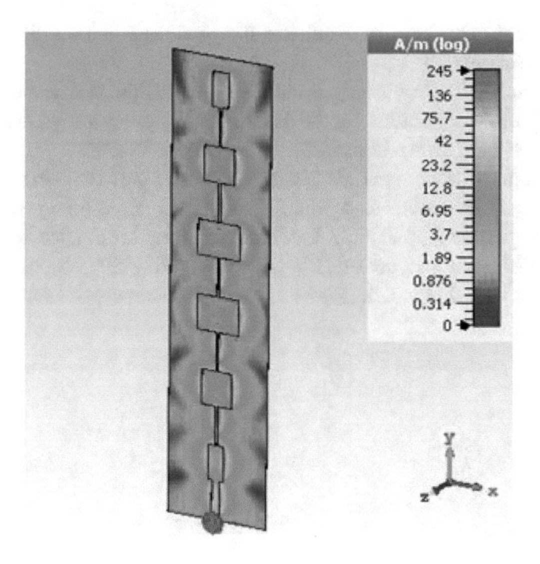

Fig. 6. Chebyshev antenna array surface current distribution at 28 GHz.

4 Conclusion

A theoretical design of a rectangular and Chebyshev microstrip antenna arrays has been presented and a comparison between the two in terms of SLL has been carried out. The main drawback of a uniform distribution antenna array is the presence of sidelobes. Hence, to minimize the SLL, we have designed an antenna array based on non-uniform distribution through the Chebyshev method. Indeed, we have firstly presented the design of a 1×6 rectangular antenna array. Then, the design of 1×6 Chebyshev antenna array at 28 GHz. Thus, the SLL has been minimized by 12.2 dB in the E-plane. A bandwidth of 1.59% and a gain of 14 dB have been obtained at 28 GHz. This antenna array can be used for some 5G precious applications at 28 GHz.

References

1. Cao, J., Liu, Y., Wang, Y., Han, R.: Design of a new microstrip antenna array with high gain and low side-lobe. In: 2018 International Conference on Microwave and Millimeter Wave Technology (ICMMT), Chengdu, China, pp. 1–3 (2018). https://doi.org/10.1109/ICMMT.2018.8563673
2. Design of low sidelobe series microstrip array antenna with non-uniform spacing and excitation amplitude
3. Saif, Z., Shahid, I., Arif, M.: Implementation of 1×48 stripline feed network for 30 dB first sidelobe level using Taylor aperture distribution. In: International Bhurban Conference on Applied Sciences and Technology (2017). https://doi.org/10.1109/ibcast.2017.7868138
4. Mailloux, R.J.: Phased Array Antenna Handbook, 2nd edn. Artech House, Inc. (2005)

5. A Modified Dolph-Chebyshev Approach for the Synthesis of Low Sidelobe Beampatterns With Adjustable Beamwidth
6. Balanis, C.A.: Antenna Theory, Analysis and Design, 3rd edn. Wiley, New York (2016)
7. Safaai-Jazi, A.: A new formulation for the design of Chebyshev arrays. IEEE Trans. Antennas Propag. **42**(3), 439–443 (1994). https://doi.org/10.1109/8.280736
8. Jebabli, E., Hayouni, M., Choubani, F.: Phased millimeter-wave antenna array for 5G handled devices. In: Zbitou, J., Hefnawi, M., Aytouna, F., El Oualkadi, A. (eds.) Handbook of Research on Emerging Designs and Applications for Microwave and Millimeter Wave Circuits, pp. 97–118. IGI Global (2023). https://doi.org/10.4018/978-1-6684-5955-3.ch005
9. Pozar, D.M.: Microwave Engineering, 4th edn. Wiley, University of Massachusetts at Amherst (2012)

Analyzing Security and Privacy Risks in Android Video Game Applications

Ratiros Phaenthong and Sudsanguan Ngamsuriyaroj(✉)

Faculty of Information and Communication Technology, Mahidol University, Salaya, Thailand
ratiros.pha@student.mahidol.ac.th, sudsanguan.nga@mahidol.ac.th

Abstract. In today's world, where smartphones are nearly universal and their user base is growing rapidly, the security and privacy of users have become paramount concerns. This paper investigates the risks related to security and privacy in video game applications, which have seen a surge in popularity. Our extensive study involves the detailed analysis of most popular 400 free Android games from the Google Play Store in the Thailand region which spread across action, role-playing, simulation, and strategy genres. We conducted an in-depth static and dynamic analysis focusing on aspects such as malware detection, permission requests, third-party tracking, and security of server connections. Our work uncovers the existence of potential malware, incorporation of third-party trackers in apps, discrepancies in permission requests, inconsistencies between third-party tracker identification between static and dynamic analysis, and possibly malicious connections. The insights from this study are intended to assist game developers and end-users to be aware of the security and privacy standards of their games.

1 Introduction

The widespread use of smartphones, with a projected rise to 6,162 million users by 2028 [1], emphasizes the pivotal role these devices play in our daily life. Among the myriad applications, video games stand out as one of the most popular choices, with approximately 490,267 available on Google Play [2].

However, the surge in smartphone usage, particularly in gaming, raises concerns about user security and privacy, especially for vulnerable groups like children. Issues such as personal information collection and data tracking have become seriously urgent [3, 4]. In addition, despite varying levels of technical knowledge, adult users consistently overlook privacy concerns when choosing apps [5].

This paper aims to address security and privacy concerns through a comprehensive analysis of the most popular 400 free Android game applications, we found from static analysis that there are malware risks, heavy usage of third-party trackers, identification of Android and custom permissions. Our dynamic analysis unveiled inconsistencies of third-party tracker identification between static and dynamic analysis, and possibly malicious connections. By providing insights into potential security and privacy risks, we seek to raise awareness among game developers and end-users regarding the security and privacy standards of their games, as well as safeguard their personal information and digital well-being.

L. Barolli (Ed.): AINA 2024, LNDECT 204, pp. 307–319, 2024.
https://doi.org/10.1007/978-3-031-57942-4_30

2 Background

2.1 Android Structure Overview

The Android Operating System is an open-source platform based on Linux, and designed for smartphones and various devices [6]. Its components include System Applications, Java API Framework, Native C and C++ Libraries, Android Runtime, Hardware Abstraction Layer (HAL), and Linux Kernel [7].

2.2 Mobile Application Analysis

Mobile application analysis consists of two primary methods: static analysis, which involves examining components without execution, and dynamic analysis, which examines applications during execution [8]. To facilitate these analyses, a range of tools is available. However, in this research context, we utilized the game genre criteria proposed by Apperley [9] which categorized into four genres: Action, Role-playing, Simulation, and Strategy for dataset selection, although not universally agreed upon [10, 11], and Google Play Unofficial API [12, 13] for file downloading. For static analysis, we employed Apktool [14], smali/baksmali [15], the tracker list from Exodus Privacy [16, 17], and VirusTotal [18, 19], for which we selected malicious criteria proposed by Arp et al. [20] and malware classification based on Avast [21] which consists of Ransomware, Spyware, Worm, Adware, Trojan, and Botnet. Finally, we used Burp Suite [22] for dynamic analysis.

3 Related Work

In recent years, several studies have delved into the privacy and security implications of Android applications, uncovering many vulnerabilities. Ikram *et al.* [23] analyzed 283 VPN Android apps, exposing concerns like malware presence and sensitive permission requests. Similarly, Ikram and Kaafar [24] focused on 97 ad-blocking apps, highlighting issues such as third-party tracking. Gamba *et al.* [25] explored risks in 82,501 pre-installed apps, identifying threats like third-party tracking and malware. Notably, 612 potentially dangerous pre-installed apps were discovered. Papageorgou *et al.* [26] investigated 20 health apps, revealing issues like invalid privacy policies. Monogios *et al.* [27] exclusively used dynamic analysis for 5 GPS navigation apps, identifying concerns like invalid privacy policies and third-party tracking. Building on those studies, our work proposes a comprehensive analysis of 400 free Game Applications on the Android OS, spanning various genres. We aim to identify components posing risks to users' privacy and security, by using both static and dynamic methods. Our key focuses include malware presence, sensitive permission requests, third-party tracking, and insecure server connections. Our investigation seeks to enhance understanding and contribute insights into specific risks users face in the gaming domain. The summary of studies is shown in Table 1.

Table 1. Summary of the studies

Study / Findings	Ikram *et al.* [23]	Ikram and Kaafar [24]	Gamba *et al.* [25]	Papageorgou *et al.* [26]	Monogios *et al.* [27]	This Study
	283 VPN Apps	97 Ad-blocking Apps	82,501 Pre-installed Software	20 Health Apps	5 GPS Navigation Apps	400 Game Apps
App Mechanism		x				
Dev Ecosystem			x			
Insecure Coding			x	x		
Insecure server connections				x		x
Insecure Tunneling Protocol	x					
Invalid Privacy Policy				x	x	
Regulations Compliance				x		
Malware Presence	x	x				x
Personal Information Access				x		
Sensitive Permission Requests	x	x	x	x	x	x
Third-party Tracking	x	x			x	x
Traffic Interception / Leaks / Manipulation	x					

4 Proposed Work

4.1 System Overview and Data Collection

The system overview of our work, depicted in Fig. 1, emphasizes two main processes: data collection and analysis, incorporating both static and dynamic analysis methods. Static analysis involves examining elements such as the presence of malware, permission requests, and third-party tracking. In contrast, dynamic analysis concentrates on analyzing insecure server connections and domain name access. For data collection, we employed Google Play Unofficial APIs [12, 13] to compile and download a list of the top 100 free Android games across four genres in the Thailand region directly from Google Play.

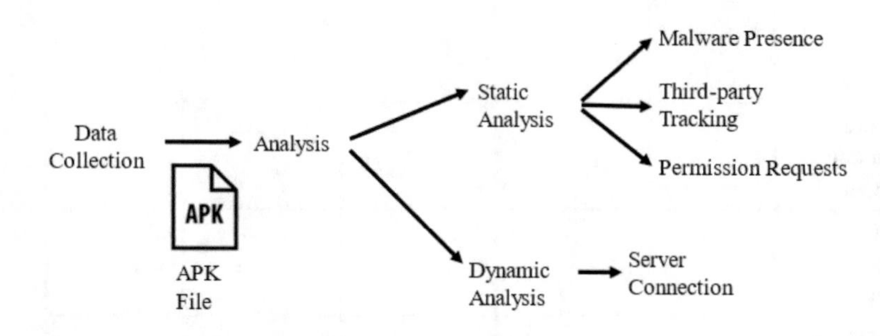

Fig. 1. The System Overview

4.2 Static Analysis

During the static analysis phase, we analyzed 400 downloaded APK files for malware presence using the VirusTotal API [18, 19]. To determine whether a file is malicious, we adopted the approach suggested by Arp et al. [20], which requires inputs from more than one vendor ($N \geq 2$) to ensure accuracy and minimize false positives. In addition, we identified third-party trackers embedded in these files by using smali/baksmali [15] and cross-referenced them with the tracker list from Exodus Privacy [16], following the guidelines provided by oF2pks [17]. We also employed Apktool [14] to extract permission requests from the "AndroidManifest.xml" file of each game, as well as compare these requests with the Android's official documentation to comprehend their potential implications.

4.3 Dynamic Analysis

The dynamic analysis involved examining the server connections of the games. Using Burp Suite [22] as an HTTP Proxy and a rooted OnePlus 2 phone, we intercepted and inspected HTTP and HTTPS traffic from the games. This process enabled us to analyze communication patterns, especially those related to domain name requests, by playing selected games from each genre for approximately 10 min. The details of this server connection analysis are depicted in Fig. 2.

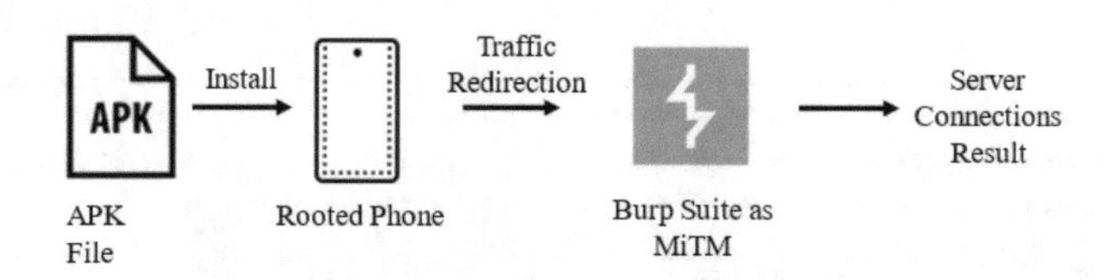

Fig. 2. The Server Connections Analysis Process

5 Experiments and Results

5.1 Malware Presence

Our analysis, focusing on the security and privacy threats posed by malware in mobile games, revealed significant findings. We used VirusTotal to identify potential malware within our dataset. We classified games as potential malware if flagged by two or more antivirus vendors, aligning with the criteria proposed by Arp *et al.* [20]. Our analysis spanned four game genres and identified four games as potential malware, as detailed in Table 2. Each identified malware falls into distinct categories as per Avast's malware taxonomy [21], including Adware, Spyware, and Trojan. Notably, all these games remain accessible on the Google Play Store, posing a significant risk to users, especially children and non-technical individuals.

Table 2. Summarized Data on Detected Malware

Package Name	Version Code	Genre	Detected by Vendors	Type of Malware
com.mgc.RopeHero.ViceTown	650	Action	2	Spyware and Trojan (Ikarus-Trojan:AndroidOS/Banker.M)
vng.games.revelation.mobile	18	Role-playing	2	Trojan (Ikarus-Trojan-Dropper.AndroidOS.Agent)
com.craftsman.go	31	Simulation	12	Adware and Trojan (K7GW-Adware, Dr.Web-Android.Packed.57083)
com.viouk.free.gamecrft.big.ops	2344	Strategy	13	Adware and Trojan (K7GW-Adware, Dr.Web-Android.Packed.57083)

5.2 Third-Party Tracking

Our analysis of third-party trackers in mobile games aims to identify and categorize the libraries developers used, often unbeknownst to players. The findings, as shown in Fig. 2, reveal "com.applovin (AppLovin (MAX and SparkLabs))" as a prevalent tracker across various categories like Advertisement, Analytics, Identification, and Profiling. Significantly, role-playing games frequently employ "Facebook-related trackers (com.facebook.login and com.facebook.share)" for identification and analytics, enhancing user engagement and providing business insights to developers (Fig. 3).

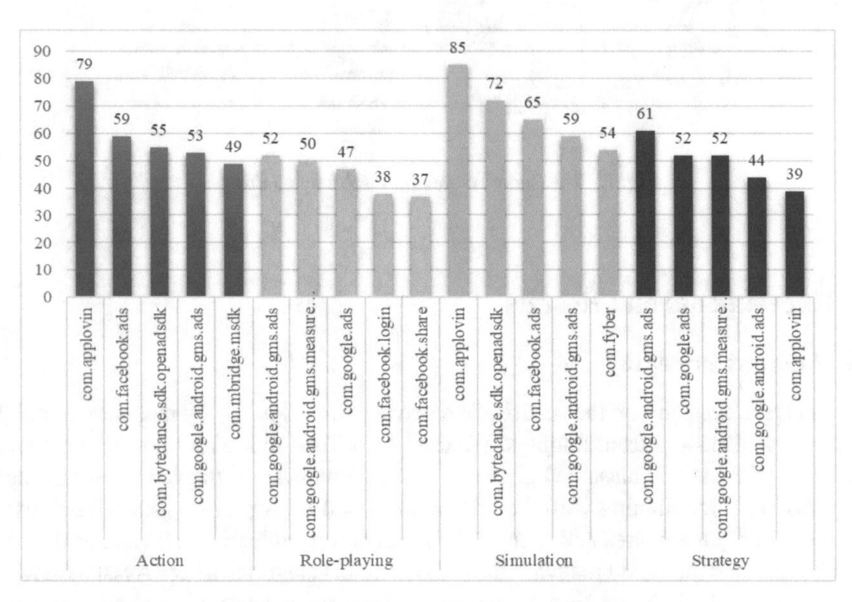

Fig. 3. Most Popular Trackers by Genre

Figure 4 categorizes identified trackers into seven types, including Advertisement, Analytics, Crash Reporting, Identification, Location, Profiling, and Unidentified. Advertisement trackers emerged as the most common across all genres, highlighting the emphasis on monetization strategies.

A striking observation, illustrated in Fig. 4, is the contrast between games with and without third-party trackers. Some games, especially those listed in Table 3, including potential malwares like "com.craftsman.go" and "com.viouk.free.gamecrft.big.ops", lack documented trackers. The absence of trackers could indicate either a genuine lack of third-party libraries, unrecognized trackers by Exodus Privacy, or malwares operating without trackers.

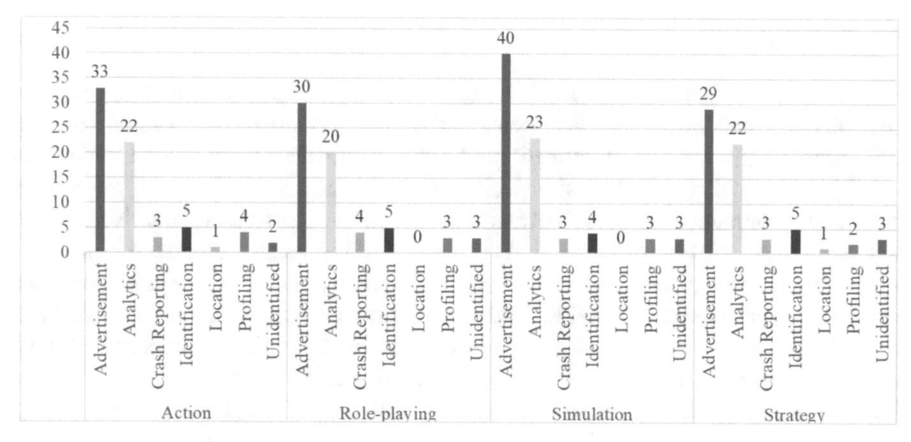

Fig. 4. Distribution of Third-party Trackers by Type Across Game Genres

The prevalent use of trackers in free mobile games underscores the reliance on user data for targeted advertising and monetization, though this practice may raise concerns regarding user privacy and security.

Table 3. List of Games Possibly Without Trackers

Genre	Package Name	Version Code	Comment
Action	com.innersloth.spacemafia	2874	
	snake.game.ular.cacing.permainan.worm	14101	
	com.netease.idv.googleplay	90	
Role-playing	com.pearlabyss.blackdesertm.gl	30689	
	com.rsg.myheroesen	130	
Simulation	com.craftsman.go	31	Potential Malware
	com.farmadventure.global	873	
	diy.makeup.artist.jogo.maquiagem.maquillaje.makyaj.oyunu	11101	
	game.latto.latto.tek.tek.game.bola	1090	
Strategy	com.MegaAwaken	30	
	com.netease.g104na.gb	156	
	com.percent.aos.randomdicego	11300	
	com.percent.royaldice	7100400	
	com.viouk.free.gamecrft.big.ops	2344	Potential Malware
	tower.defense.ant.conquer.games	106	

5.3 Permission Requests

Our analysis of permissions in the dataset revealed significant findings regarding privacy and security. We categorized the permissions declared in the 400 games into Android-defined and custom-defined permissions, identifying 209 in total (119 Android-specific and 90 custom). Figure 5 highlights the most common Android permissions, with android.permission.INTERNET and android.permission.ACCESS_NETWORK_STATE being consistently predominant across all genres. However, other frequently declared permissions like android.permission.WAKE_LOCK and android.permission.ACCESS_WIFI_STATE varied by genre. The results indicated over-declaration of permissions in certain games, as detailed in Table 4, which increased a potential attack surface, privacy, and security risk to users [28], even though this concern was raised in the Android and research community years ago [29].

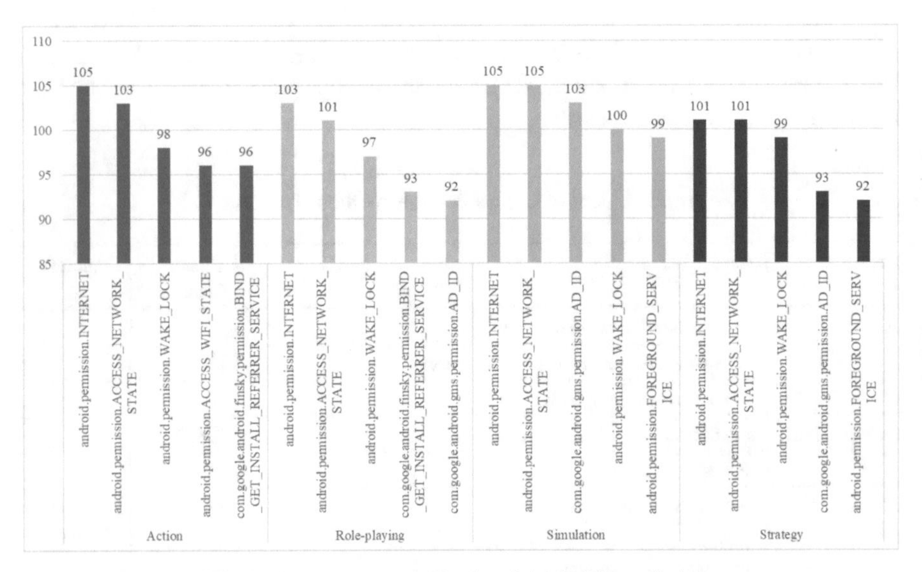

Fig. 5. Most Frequently Declared Android Permissions

Furthermore, custom permissions were categorized into nine types as illustrated in Fig. 6, which presented unique challenges due to their non-standardized nature and sparse documentation. The most frequently discovered custom permission across all genres was "com.applovin.array.apphub.permission.BIND_APPHUB_SERVICE" as depicted in Fig. 7. These custom permissions could pose risks to user privacy and security, as suggested by certain scenarios proposed by Li *et al.* [30].

Table 4. Games Exhibiting Over-Declaration of Permissions

Genre	Package Name	Version Code
Action	com.mobile.legends	17698401
	com.tencent.ig	17325
	jpark.AOS5	188
Role-playing	com.archosaur.madtale.global.gp	24
	com.bandainamcoent.saoifww	168
	com.pazugames.avatarworld	24
Simulation	com.hapno.wdss	2
	diy.makeup.artist.jogo.maquiagem.maquillaje.makyaj.oyunu	11101
	diy.makeup.master.artist.stylist	1392
Strategy	tower.defense.ant.conquer.games	106

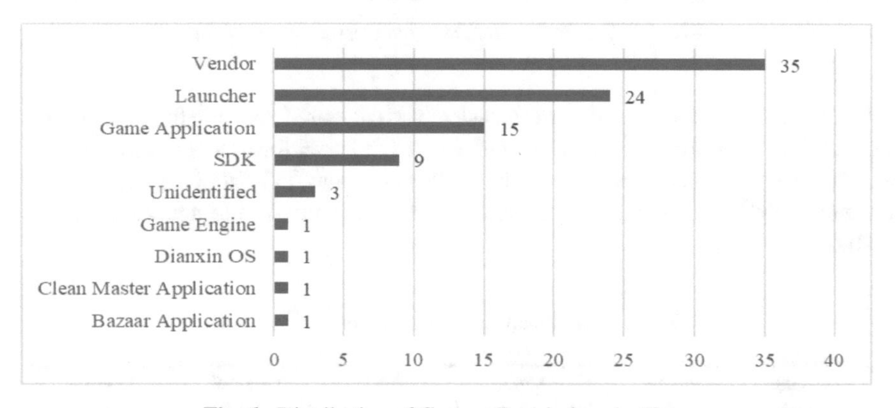

Fig. 6. Distribution of Custom Permissions by Type

5.4 Dynamic Analysis

Our dynamic analysis of HTTP and HTTPS traffic in mobile games aims to uncover potential vulnerabilities and privacy risks in game-server communications. We selected four games from all genres and analyzed network traffic for selected games using Burp Suite, focusing on domain name requests. An important observation emerged from the comparison between static and dynamic analyses. While the static analysis detected no third-party trackers, the dynamic analysis revealed the presence of tracker-related domain names, including applovin.com. This finding revealed differences in third-party trackers identification between the two methodologies.

The dynamic analysis also identified counterfeit Minecraft games, namely "com.craftsman.go" and "com.viouk.free.gamecrft.big.ops," which contain a malicious connection flagged by VirusTotal. However, it is important to note that this domain was flagged by only one vendor, which does not meet the criteria established by Arp et al.

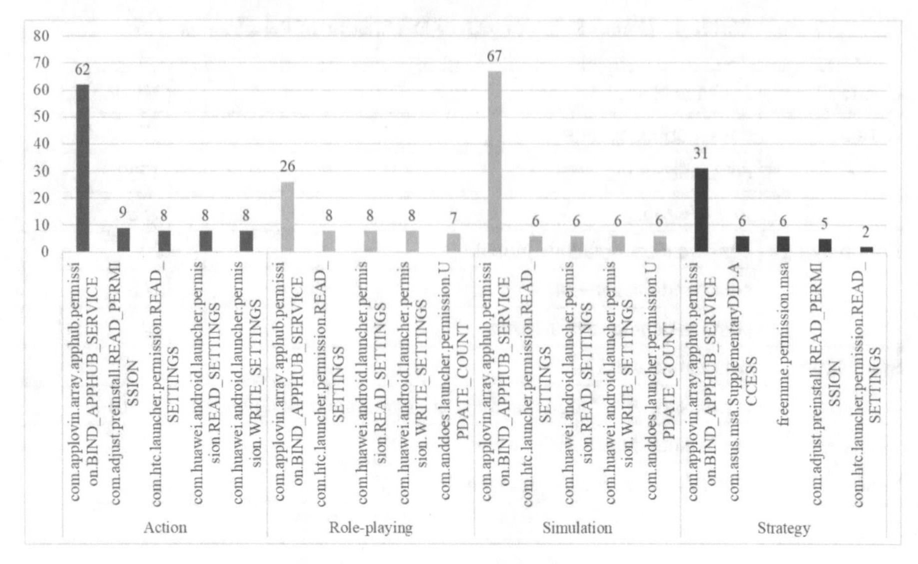

Fig. 7. Most Frequently Declared Custom Permissions

[25]. Nevertheless, these results are consistent with potential malware findings discussed in Sect. 5.1. Therefore, this analysis underscores the need for users to exercise caution and discernment when interacting with mobile game apps, in light of the identified risks and vulnerabilities in their network communications. Our main findings are summarized in Table 5.

Table 5. Summarized Data on Dynamic Analysis

Package Name	Version Code	Trackers Identification using Static Analysis	Trackers Identification using Dynamic Analysis	Potential Malware from #5.1	Potential Malicious Connection
com.mgc.RopeHero.ViceTown	650	☒	☑	☒	☒
com.nexon.bluearchive	203922	☑	☑	☑	☑
com.craftsman.go	31	☒	☑	☒	☒
com.viouk.free.gamecrft.big.ops	2344	☑	☑	☑	☑

6 Conclusions

Our static analysis on 400 mobile games revealed four games that are potential malware and widespread use of third-party trackers, primarily for advertising and analytics. We identified 115 Android and 95 custom permissions in these games. Our dynamic analysis unveiled inconsistencies in third-party tracker identification compared to static analysis, demonstrating varied developer approaches in integrating tracker functionalities. The findings emphasize the importance of user security and privacy in the mobile gaming ecosystem.

Our research, while insightful, has limitations. The dataset, limited to 400 free Android games from four genres, may not fully represent the mobile gaming market. The static analysis, especially custom permission analysis, required more in-depth analysis. The dynamic analysis, constrained in duration and scope, might not have captured all security analysis aspects. Moreover, our focus was solely on Android games, potentially limiting the applicability of our findings to other platforms. Future research could include further dynamic analysis of the identified malware, examining privacy policies of third-party SDKs, more detailed permission analysis, and developing guidelines for mobile game developers to enhance security and privacy practices.

Acknowledgments. Our work received partial support from the Faculty of Information and Communication Technology at Mahidol University. I am profoundly grateful to my advisor, Assoc. Prof. Sudsanguan Ngamsuriyaroj, for her invaluable guidance, and to the instructors in the Cybersecurity and Information Assurance program for their teachings. My heartfelt appreciation goes to my life partner, Ms. Haruetai Pratumchart, for her steadfast support, and to my cats, Hacker and Panther, for their comforting presence. I also thank my family for their love and encouragement, and my colleagues at the Royal Thai Armed Forces Cyber Security Center for their understanding and support which is integral to the completion of this research.

References

1. Degenhard, J.: Global: number of smartphone users 2013–2028, Statista. https://www.statista.com/forecasts/1143723/smartphone-users-in-the-world. Accessed 07 Jul 2023
2. Clement, J.: Google play: number of available games by quarter 2022, Statista (1997) 415–438. https://www.statista.com/statistics/780229/number-of-available-gaming-apps-in-the-google-play-store-quarter/. Accessed 07 Jul 2023
3. Liu, M., Wang, H., Guo, Y., Hong, J.: Identifying and analyzing the privacy of apps for kids. In: Proceedings of the 17th International Workshop on Mobile Computing Systems and Applications (2016). https://doi.org/10.1145/2873587.2873597
4. Sobel, K., et al.: It wasn't really about the Pokémon: parents' perspectives on a location-based mobile game. In: Proceedings of the 2017 CHI Conference on Human Factors in Computing Systems (2017). https://doi.org/10.1145/3025453.3025761
5. Barth, S., de Jong, M.D.T., Junger, M., Hartel, P.H., Roppelt, J.C.: Putting the privacy paradox to the test: online privacy and security behaviors among users with technical knowledge, privacy awareness, and financial resources. Telematics Inform. **41**, 55–69 (2019). https://doi.org/10.1016/j.tele.2019.03.003
6. Alphabet: Android open source project. https://source.android.com/. Accessed 07 Jul 2023

7. Alphabet: Platform architecture: android developers. Android Developers. https://developer.android.com/guide/platform. Accessed 07 Jul 2023
8. Schleier, S., Holguera, C., Mueller, B., Willemsen, J.: Mobile application security testing guide version v1.6.0. The OWASP Foundation (2023)
9. Apperley, T.H.: Genre and game studies: toward a critical approach to video game genres. Simul. Gaming **37**(1), 6–23 (2006). https://doi.org/10.1177/1046878105282278
10. Arsenault, D.: Video game genre, evolution and Innovation. Eludamos J. Comput. Game Culture **3**(2), 149–176 (2009). https://doi.org/10.7557/23.6003
11. Vargas-Iglesias, J.J.: Making sense of genre: the logic of video game genre organization. Games and Culture **15**(2), 158–178 (2018). https://doi.org/10.1177/1555412017751803
12. 89z: googleplay. https://github.com/89z/googleplay. Accessed 18 Sep 2023
13. JoMingyu: google-play-scraper. https://github.com/JoMingyu/google-play-scraper. Accessed 18 Sep 2023
14. iBotPeaches: Apktool. Apktool - A tool for reverse engineering 3rd party, closed, binary Android apps. https://ibotpeaches.github.io/Apktool/. Accessed 27 Sep 2023
15. JesusFreke, J.: Smali/Baksmali. https://github.com/JesusFreke/smali. Accessed 27 Sep 2023
16. Exodus Privacy: Exodus Privacy. https://exodus-privacy.eu.org/en/. Accessed 6 Jul 2023
17. oF2pks: 3xodusprivacy-toolbox. https://gitlab.com/oF2pks/3xodusprivacy-toolbox. Accessed 5 May 2023
18. VirusTotal: Virustotal, VirusTotal. https://www.virustotal.com/. Accessed 29 Sep 2023
19. Shearer, C.: get-vtfilereport. get-VTFileReport. https://github.com/cbshearer/get-VTFileReport. Accessed 26 Sep 2023
20. Arp, D., Spreitzenbarth, M., Hübner, M., Gascon, H., Rieck, K.: Drebin: effective and explainable detection of android malware in your pocket. In: Proceedings 2014 Network and Distributed System Security Symposium (2014). https://doi.org/10.14722/ndss.2014.23247
21. Belcic, I.: What is malware and how to protect against malware attacks? What is malware and how to protect against malware attacks? https://www.avast.com/c-malware. Accessed 07 Jul 2023
22. PortSwigger: Burp Suite documentation. PortSwigger. https://portswigger.net/burp/documentation. Accessed 24 Jul 2023
23. Ikram, M., Vallina-Rodriguez, N., Seneviratne, S., Kaafar, M.A., Paxson, V.: An analysis of the privacy and security risks of android VPN permission-enabled apps. In: Proceedings of the 2016 Internet Measurement Conference (2016). https://doi.org/10.1145/2987443.2987471
24. Ikram, M., Kaafar, M.A.: A first look at mobile ad-blocking apps. In: 2017 IEEE 16th International Symposium on Network Computing and Applications (NCA) (2017). https://doi.org/10.1109/nca.2017.8171376
25. Gamba, J., Rashed, M., Razaghpanah, A., Tapiador, J., Vallina-Rodriguez, N.: An analysis of pre-installed android software. In: 2020 IEEE Symposium on Security and Privacy (SP) (2020). https://doi.org/10.1109/sp40000.2020.00013
26. Papageorgiou, A., et al.: Security and privacy analysis of mobile health applications: the alarming state of practice. IEEE Access **6**, 9390–9403 (2018). https://doi.org/10.1109/access.2018.2799522
27. Monogios, S., Limniotis, K., Kolokotronis, N., Shiaeles, S.: A case study of intra-library privacy issues on android GPS navigation apps. In: Katsikas, S., Zorkadis, V. (eds.) E-Democracy – Safeguarding Democracy and Human Rights in the Digital Age: 8th International Conference, e-Democracy 2019, Athens, Greece, December 12-13, 2019, Proceedings, pp. 34–48. Springer International Publishing, Cham (2020). https://doi.org/10.1007/978-3-030-37545-4_3

28. Lardinois, S., Beckers, J.: How malicious applications abuse Android permissions. NVISO Labs. https://blog.nviso.eu/2021/09/01/how-malicious-applications-abuse-android-permissions/. Accessed 01 Feb 2024
29. Johnson, R., Wang, Z., Gagnon, C., Stavrou, A.: Analysis of android applications' permissions. In: 2012 IEEE Sixth International Conference on Software Security and Reliability Companion (2012). https://doi.org/10.1109/sere-c.2012.44
30. Li, R., Diao, W., Li, Z., Du, J., Guo, S.: Android custom permissions demystified: from privilege escalation to design shortcomings. In: 2021 IEEE Symposium on Security and Privacy (SP) (2021). https://doi.org/10.1109/sp40001.2021.00070

Detecting and Mitigating MitM Attack on IoT Devices Using SDN

Mohamed Ould-Elhassen Aoueileyine[1(✉)], Neder Karmous[1], Ridha Bouallegue[1], Neji Youssef[1], and Anis Yazidi[2]

[1] Innov'COM Laboratory, SUPCOM, University of Carthage, Tunis, Tunisia
Mohamed.ouldelhassen@supcom.tn

[2] OsloMet University, Pilestredet 52, N-0166 Oslo, Norway

Abstract. The Software-Defined Network (SDN) is an innovative network architecture designed to offer enhanced flexibility and operational simplicity in network management through a centralized controller. While these qualities empower SDN to effectively address evolving network demands, they also expose security vulnerability. Given its centralized structure, SDN becomes susceptible to cyber attacks, particularly those targeting internet of things (IoT) devices. These attacks aim to target IoT devices and can lead to congestion and disruption. In this study, we introduce an Intrusion Detection and Prevention System (IDPS) framework based on SDN to detect Man-in-the-Middle (MitM) Attacks by decodes network packets, extracting ARP headers with source and destination Internet Protocol (IP) and Media Access Control (MAC) addresses, monitors Address Resolution Protocol (ARP) packet counts during flood attacks and add flow table to block attackers if the count exceeds a threshold. The research covers simulation outcomes as well as the implementation of a practical SDN model for applying our methodology. The results highlight the model's ability to rapidly and accurately detect MitM attacks targeting IoT devices and mitigate it in real time.

1 Introduction

IoT devices frequently communicate with each other and with central servers. Attackers can intercept and manipulate this communication using MitM attacks [1]. By eavesdropping on data exchanges, attackers can steal sensitive information or inject malicious commands, potentially leading to device compromise or unauthorized actions. Figure 1 shows the MitM attacks between user and IoT device in SDN.

In contrast to traditional networks [2], SDN can easily and flexibly configure networks, making it easier to detect and respond to attacks [3–5]. In SDN architecture [6], the control plane and data plane both contribute to detecting and mitigating MitM attacks. Control Plane manages network policies and routes, allowing for dynamic adjustments to traffic flows, which can help identify unexpected changes in network behavior indicative of MitM attacks. Data Plane executes packet forwarding based on control plane instructions through the OpenFlow (OF) protocol [7]. Monitoring traffic anomalies and patterns in the data plane can reveal suspicious activities and potential MitM attack attempts and mitigate it. Our main objective in this paper is to implement an

L. Barolli (Ed.): AINA 2024, LNDECT 204, pp. 320–330, 2024.
https://doi.org/10.1007/978-3-031-57942-4_31

IDPS-based-SDN [8] framework to detect and mitigate MitM attacks with a guarantee of preventing from spying: when a legitimate user wants to communicate within the IoT device. The Ryu controller [9] as a control plane based on its network policies and application requirements, creates flow entries in the switches' flow tables [10] to ensure proper communication. The flow entries created by the Ryu controller for legitimate users have match fields that correspond to the characteristics of legitimate traffic. These match fields might include source and destination MAC addresses, IP addresses, port numbers, and other relevant attributes. The action fields would specify the appropriate forwarding actions, such as sending the traffic to the desired port. When an MitM attacker tries to intercept and alter the communication between two parties and since the Ryu controller is responsible for managing flow entries, it can implement security policies to detect and mitigate malicious activities, it creates specific flow entries to identify potential attack patterns. When an attacker's packet arrives at a switch, the flow entry matching process occurs. If the packet's characteristics match an entry designed to detect malicious behavior, the controller can take an action. This action might involve redirecting the suspicious traffic to a dedicated monitoring port for further analysis or even blocking the traffic entirely. Our paper is divided into three sections. In the first section, we provide an overview of related works that are relevant to our research. The second section delves into our methodology, covering the hardware and software components we utilized, our SDN topology, and the detection and mitigation algorithms we employed. The third section presents a comparative analysis of our results with those from related studies. Finally, we conclude our paper.

2 Background and Related Works

The author in [11] using IP-MAC address bindings to detect MitM attacks. SDN is emulated using Mininet and the MitM attack over this network is performed using arpspoof, which is the segment of the tool named Dsniff. For the evaluation of their proposed algorithm, various network parameters are compared and analyzed in both RYU and POX controllers [12], their proposed algorithm mitigates the MitM attack successfully by dropping the attacked packets using POX or RYU controller but POX controller takes more time to process the packets once an attack occurs but consumes less CPU (CENTRAL PROCESSING UNIT) usage than that of the RYU controller.

The limitations of this paper is the need to reduce the detected time and decrease the consumption of CPU usage for Ryu controller.

Author in [13] presents a secure SDN-IoT architecture that combines IoT data collection, IoT gateway, SDN components, and a cloud-based storage platform. Within this framework, the innovative Belief-Based Secure Correlation (BBSC) algorithm enhances connectivity and security. BBSC employs ciphertext-policy attribute-based encryption, aligning with both SDN controllers and OpenFlow switches. BBSC streamlines user-switch communication, orchestrating access authentication and preventing unauthorized network entry. Users' uniqueness and access compatibility are rigorously verified. This

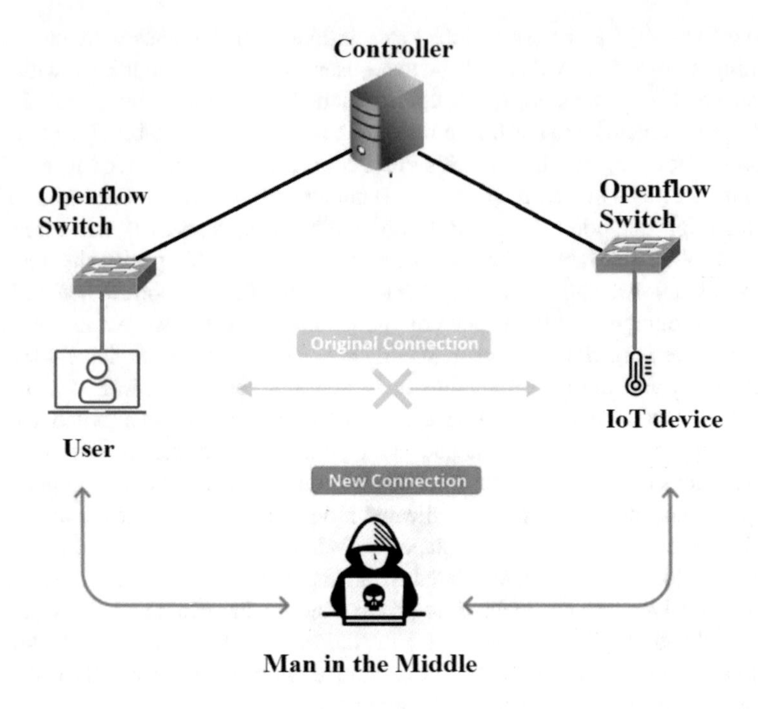

Fig. 1. MitM Attack in Software defined network.

SDN-enabled IoT architecture excels in adaptability and scalability, responding dynamically to shifting traffic patterns—a substantial improvement over traditional IoT networks. The cloud platform utilized for data storage is Thingsboard. The proposed framework can efficiently detect and mitigate KRACK framework. For 50 nodes the attack detection time and accuracy are 6.3ms and 86% respectively.

The limitations of this paper is the need for more accuracy and to minimize the detection time.

The authors in [14] use SDN controller to detect MitM based attacks that allow attackers to impersonate other devices proposes a more efficient approach by implementing security measures directly in the data plane. This approach results in faster response times, enhancing network reliability by automatically blocking attackers. The solution is designed to be flexible and adaptable to different network environments and attack scenarios. The authors use 4 steps: traffic monitor, traffic data extractor, traffic data analyzer, decision maker. The experiments conducted using KaliLinux, Mininet, and OpenVSwitch provide valuable insights and customization options for addressing ARP spoofing attacks [15]. Results shows the network with ten devices to the network with three devices, the processing time increases by 63.9%. Comparing the network with twenty devices to the network with ten devices, the processing time increases by 73.3%. And again, a non-linear increase is visible, showing that the larger the network becomes, the slower the solution becomes.

The limitations of this work are related to the environment being simulated and not being deployed on an actual SDN.

3 Implementation

3.1 Motivation

Compared to other related works, our approach is applicable to both small and large Software-Defined Networking environments. It focuses on real-time detection and mitigation of MitM attacks, which involves actively monitoring network traffic, analyzing anomalies, and promptly taking action to prevent or block suspicious activities. This comprehensive process includes packet inspection, flow analysis, and the use of various detection techniques, including ARP spoofing and ARP flood attacks.

3.2 Assumed Environment

For our research, we utilized a hardware environment consisting of Windows 10 equipped with 16 GB of RAM and powered by an Intel® Core™ i5-1235UL Processor. This processor, part of the Q3′22 release, offers 10 cores and operates at a clock speed of 4.40 GHz. This hardware setup played a crucial role in facilitating our SDN network to work and detect attacks effectively, thanks to its robust processing power and memory capacity. Additionally, the multi-core capabilities of the processor allowed for parallel processing, enhancing the overall efficiency of our detection and mitigation algorithms. The combination of this powerful hardware and our innovative approach resulted in the successful detection and mitigation of threats in real-time, contributing significantly to the success of our research efforts. We utilized as a software Mininet as a network emulator for creating virtual network environments, facilitating network protocol testing and research, Ryu Controller which is an open-source SDN controller framework in Python, enabling the development of custom SDN applications and interactions with OpenFlow-enabled devices, Python 3 which is the programming language used for developing Ryu controllers, Oracle Virtual Machine (Oracle VM VirtualBox): which is a virtualization software that allows us to run Ubuntu system, and enables us to create and manage SDN applications and network configurations, Mosquitto which is an open-source MQTT broker that enabling IoT devices to communicate securely and asynchronously. And Ettercap which is an open-source network security tool used for conducting MitM attacks on host networks.it allows to eavesdrop on network traffic and manipulate data being exchanged between devices on the same network.

3.3 Topology

Figure 2 illustrates our network topology, comprising a Ryu controller named C0, three switches named S1, S2, and S3, and four hosts denoted as h1 to h4. Among them, h2 and h3 are legitimate hosts, h1 acts as a MitM attacker, and h4 is an IoT device capable of publishing or subscribing to h2 and h3.

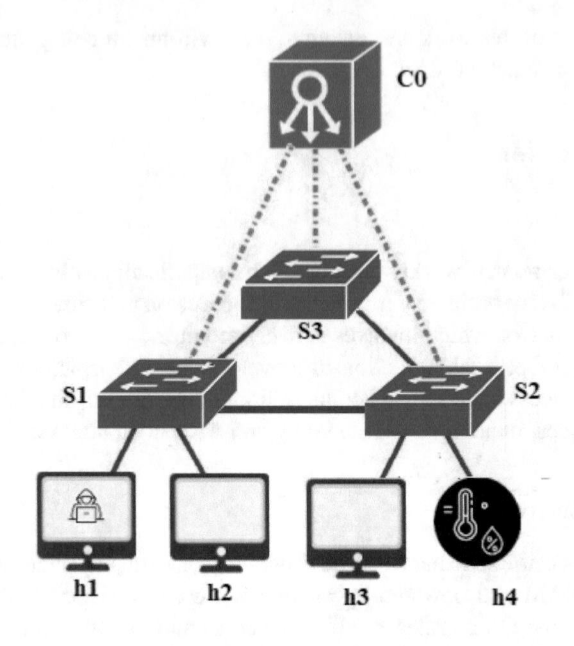

Fig. 2. SDN network topology.

The devices have already been configured based on the specifications provided in Table 1 below.

Table 1. SDN Topology Network details.

Device name	IP Address	Mac address	Related with Switch
h1	**10.0.0.1**	**00:00:00:00:00:01**	**S1**
h2	**10.0.0.2**	**00:00:00:00:00:02**	**S1**
h3	**10.0.0.3**	**00:00:00:00:00:03**	**S2**
h4	**10.0.0.4**	**00:00:00:00:00:04**	**S2**
C0	**127.0.0.1**	**nan**	**S1,S2 and S3**

4 Simulation and Results

In this section, we utilized virtual machines running the Ubuntu 20.04.6 LTS operating system. We installed our SDN network topology as described in Table 1 using Mininet to create the network infrastructure. Additionally, we employed the Ryu controller to handles packet control, utilizing the OF protocol to ensure efficient network management and control, all while maintaining a secure channel [16] for communication. When run this command: "mosquitto_sub -h 10.0.0.4 -t sensor/humidity -t sensor/temperature_celsius -v" on the host named h2 with the IP address 10.0.0.2, it will connect to the MQTT broker at IP address 10.0.0.4 and subscribe to the two specified topics related to sensor data. As messages are published to these topics on the MQTT broker, they will be displayed on the h2 terminal that is running the command, along with their respective topics. This allows to monitor and receive sensor data published to these MQTT topics in real-time, data is being exchanged between h4 and h2, as shown in the Fig. 3 below.

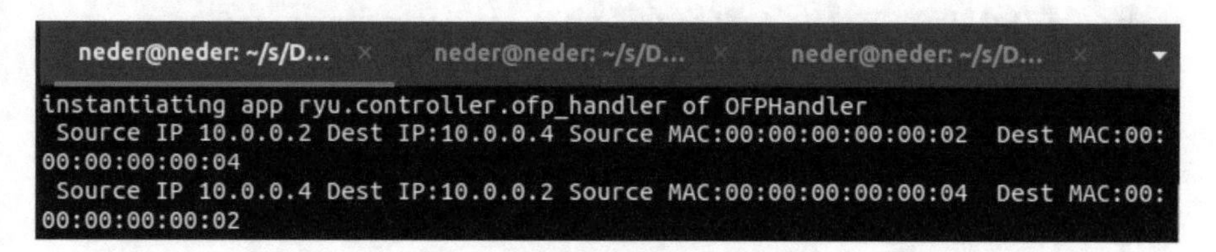

Fig. 3. Data exchanged from h2 and h4 using SDN.

We used host h1's functions as a MitM attacker attempting to intercept data being exchanged between h4 and h2 through the following ettercap command: "ettercap -T -i h1-eth0 -M ARP /10.0.0.2// /10.0.0.4//".

Algorithm 1 outlines the process for detecting MitM attacks by monitoring network traffic, specifically ARP packets. In the first scenario, known as ARP Flood attacks, the algorithm raises a flag named "MiD" when an excessive number of ARP requests occur on a port. If the port is found in the "PortCount" dictionary, it checks whether the count for that port surpasses a predefined threshold, set at a value of 20. If the count exceeds this threshold, the algorithm sets "MiD" to "True", indicating the presence of a MitM attack. The choice of the threshold value in the algorithm serves to impose a limit on the rate of ARP requests for a specific network port. This value is established through empirical testing and a comprehensive analysis of network behavior. An elevated rate of ARP requests exceeding the set threshold may suggest the existence of ARP flood or other anomalous network activities.

In the second scenario, referred to as ARP spoofing attacks, the algorithm verifies whether the source IP address of the ARP packet (arp_src_ip) mismatches with the source MAC address (arp_src_mac) of the host. If a disparity is detected between these addresses, it triggers the setting of "MiD" to "True," signifying the presence of a MitM attack.

Algorithm 2 presents the mitigation of MitM attacks based on the input information provided in mac and msg. It blocks all incoming network traffic on a specific port (in_port) by creating and sending an OF rule to the associated OF switch. This action is taken in response to a suspected MitM network attack to mitigate the potential threat by blocking traffic from the identified source.

Algorithm1: Detection of MitM attack.

```
1: Inputs: packetarp,msg
2: Outputs: MiD = {True, False}
3:host={h1,h2,h3,h4}
4: Threshold_ARP_Rate←20
5 :PortCount ← {}
6 :port ← msg.match['in_port']
7 :MiD← False
8: if port not in PortCount:
9:          PortCount.update({port:1})
10:        end if
11: else :
12:         if     PortCount[port]> Threshold_ARP_Rate
13:           MiD=True:
14:        end if
15:          else if pkt_arp.opcode== 1 :
16:            PortCount[port] += 1
17:          end else if
18:end else
19: if  host[arp_src_ip]) != arp_src_mac:
20:          MiD=True
21: end if
22: return MiD
```

Algorithm2: Mitigate MitM attack.

```
1: Inputs: mac,msg
2: Outputs: blocking attack
3: actions = []
4: in_port = msg.match['in_port']
5: datapath = msg.datapath
6:ofproto = datapath.ofproto
7 :parser = datapath.ofproto_parser
8 :inst =
[parser.OFPInstructionActions(ofproto.OFPIT_CLEAR_ACTIO
NS,[])]
9 :match = parser.OFPMatch(in_port = in_port)
10:mod = parser.OFPFlowMod(datapath = datapath, match =
match,idle_timeout = 60, hard_timeout = 60, priority =
20, instructions = inst)
11 :datapath.send_msg(mod)
```

Figure 4 illustrates that after deploying our SDN network and implementing our proposed approach, we were able to detect and mitigate MitM attacks in real-time. Our proposed approach effectively prevents h1, which was identified as a MitM attacker, from connecting with h4 while allowing h2 to connect with h4.

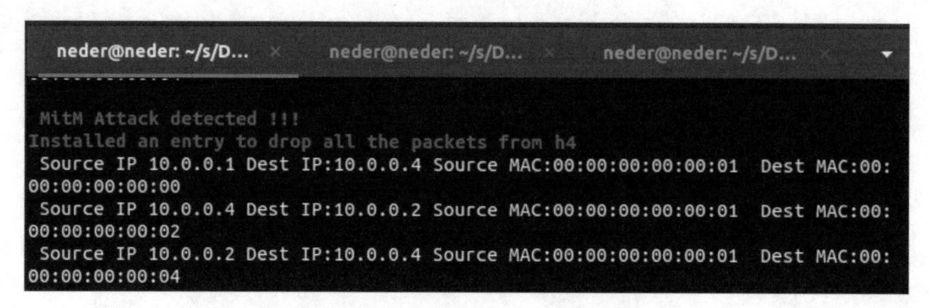

Fig. 4. Detect and mitigate MiTM attack in SDN.

5 Discussion

As shown in Figs. 5 and 6, during MitM attacks, the CPU usage averages around 38%. After our approach successfully detected and blocked the MitM attacks, and CPU usage decreased to an average of 6%.

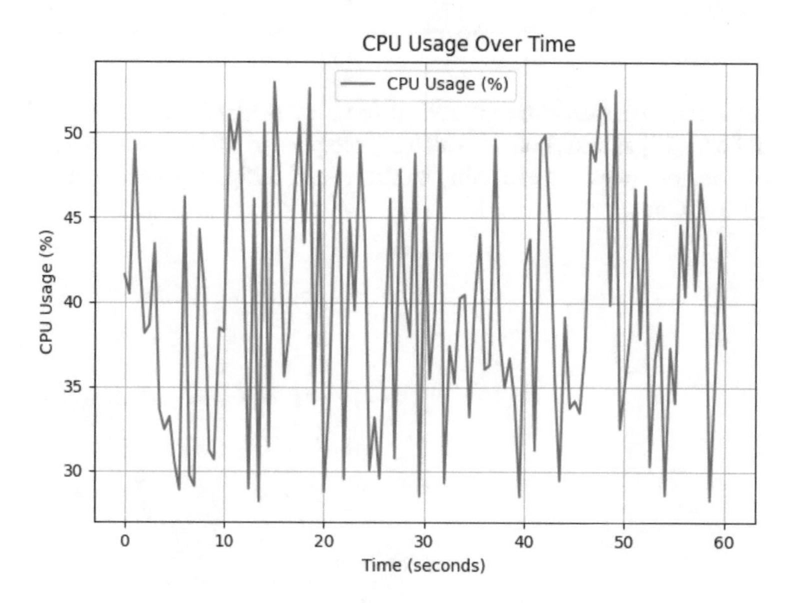

Fig. 5. CPU usage before mitigating MiTM attack.

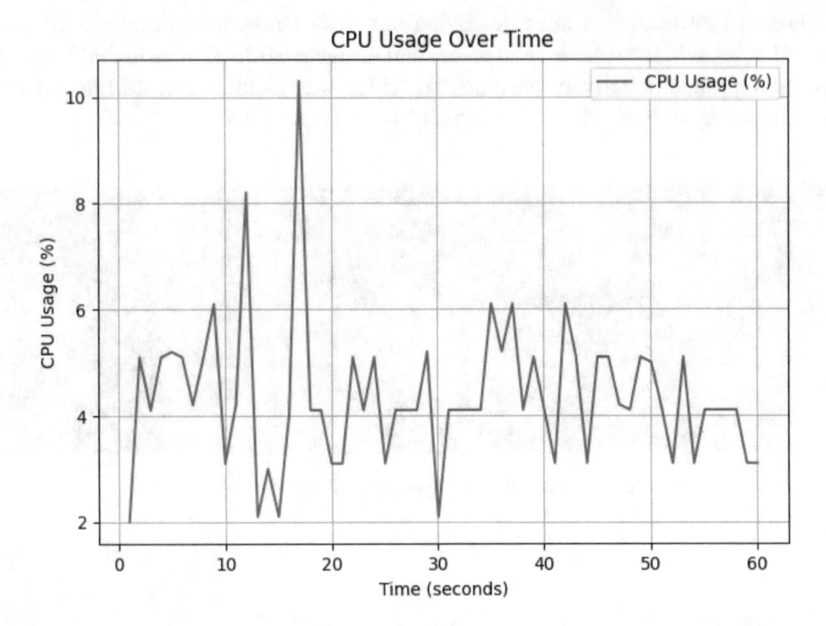

Fig. 6. CPU usage after mitigating MitM attack.

Table 2. Detection time for MitM attacks.

MitM attack name	ARP spoofing	ARP flood attacks
Detect and mitigate time (ms)	1	1.6

Table 2 presents the detection and mitigation times for identifying ARP spoofing [15] and ARP flood [17] attacks, both of which are subtypes of MitM attacks. Our approach effectively demonstrates the capability to detect and mitigate MitM attacks within a range of 1 to 1.6 ms.

Table 3. Comparison between our approach and related works.

Research	Used approach	Before Mitigation (%)	After Mitigation (%)	Time detection (ms)
[11]	Using IP-MAC address	71.4	47.4	3.6
[13]	Ciphertext-policy attribute-based encryption	–	–	6.3
[14]	- Trafic data extractor - Trafic data analyzer - Decision maker	–	–	1800
Our approach	- MAC address - ARP packet rate	38%	6%	1–1.6

As Table 3 illustrates, our proposed approach demonstrates greater effectiveness compared to other related works. It successfully detects and mitigates MitM attacks within a remarkable 1-ms timeframe, resulting in a substantial 32% reduction in CPU usage.

6 Conclusion and Future Works

This paper presents an IDPS designed to identify MitM attacks targeting IoT devices within an SDN environment. MitM attack detection typically relies on a Ryu controller, which continuously monitors and analyzes network traffic patterns and behaviors for any signs of suspicious activity. Our approach employs a mitigation technique that involves matching MAC and IP addresses while also calculating ARP packet rates and comparing them to predefined thresholds. Our results demonstrate the effectiveness of our approach, achieving flawless attack detection with an impressive 1-ms detection time.

The Future work in this area will involve exploring supervised machine learning methods and collecting MitM attack traffic dataset using the Ryu monitoring controller. These advancements are intended to improve the accuracy of MitM attack detection, reduce the false alarm rate, and shorten detection times. Furthermore, applying this method in a larger SDN network will help evaluate its suitability and effectiveness in environments with diverse traffic patterns, further enhancing the security of IoT devices within SDN networks.

References

1. Mallik, A.: Man-in-the-middle-attack: understanding in simple words. Cyberspace: Jurnal Pendidikan Teknologi Informasi **2**(2), 109–134 (2019)

2. Jaramillo, A.C., Alcivar, R., Pesantez, J., Ponguillo, R.: Cost effective test-bed for comparison of SDN network and traditional network. In 2018 IEEE 37th International Performance Computing and Communications Conference (IPCCC), pp. 1–2. IEEE (2018)
3. Sultana, N., Chilamkurti, N., Peng, W., Alhadad, R.: Survey on SDN based network intrusion detection system using machine learning approaches. Peer-to-Peer Networking Appl. **12**, 493–501 (2019)
4. Hande, Y., Muddana, A.: A survey on intrusion detection system for software defined networks (SDN). In: Research Anthology on Artificial Intelligence Applications in Security, pp. 467–489. IGI Global (2021)
5. Alhaj, A.N., Dutta, N.: Analysis of security attacks in SDN network: a comprehensive survey. Contemp. Issues Commun. Cloud Big Data Analytics Proc. CCB **2020**, 27–37 (2022)
6. Waseem, Q., Din, W.I.S.W., Aminuddin, A., Mohammed, M.H., Aziza, R.F.A.: Software-defined networking (SDN): a review. In: 2022 5th International Conference on Information and Communications Technology (ICOIACT), pp. 30–35. IEEE (2022)
7. Alsaeedi, M., Mohamad, M.M., Al-Roubaiey, A.A.: Toward adaptive and scalable OpenFlow-SDN flow control: a survey. IEEE Access **7**, 107346–107379 (2019)
8. Mazhar, N., Salleh, R., Zeeshan, M., Hameed, M.M., Khan, N.: R-IDPS: real time SDN based IDPS system for IoT security. In: 2021 IEEE 18th International Conference on Smart Communities: Improving Quality of Life Using ICT, IoT and AI (HONET), pp. 71–76. IEEE (2021)
9. Hardwaj, S., Panda, S.N.: Performance evaluation using RYU SDN controller in software-defined networking environment. Wireless Pers. Commun.Commun. **122**(1), 701–723 (2022)
10. Yu, M., Xie, T., He, T., McDaniel, P., Burke, Q.K.: Flow table security in SDN: adversarial reconnaissance and intelligent attacks. IEEE/ACM Trans. Networking **29**(6), 2793–2806 (2021)
11. Saritakumar, N., Anusuya, K.V., Balasaraswathi, B.: Detection and mitigation of MITM attack in software defined networks. In: Proceedings of the First International Conference on Combinatorial and Optimization, ICCAP 2021, December 7-8 2021, Chennai, India (2021)
12. Patel, R., Patel, P., Shah, P., Patel, B., Garg, D.: Software defined network (SDN) implementation with POX controller. In: 2022 3rd International Conference on Smart Electronics and Communication (ICOSEC), pp. 65–70. IEEE (2022)
13. Cherian, M.M., Varma, S.L.: Mitigation of DDOS and MiTM attacks using belief based secure correlation approach in SDN-based IoT networks. Int. J. Comput. Network Inf. Secur. **14**(1) (2022)
14. Buzura, S., Lehene, M., Iancu, B., Dadarlat, V.: An extendable software architecture for mitigating ARP spoofing-based attacks in SDN data plane layer. Electronics **11**(13), 1965 (2022)
15. Alina, A., Saraswat, S.: Understanding implementing and combating sniffing and ARP spoofing. In: 2021 4th International Conference on Recent Developments in Control, Automation & Power Engineering (RDCAPE), pp. 235–239. IEEE (2021)
16. Gulati, P., Kaur, G., Et Verma, G.N.: A review on secure channel establishment technique to increase security of IoT. Int. J. Comput. Sci. Mob. Comput.Comput. Sci. Mob. Comput. **8**(4), 01–06 (2019)
17. Du, J., et al.: Research on an approach of ARP flooding suppression in multi-controller SDN networks. In: 2021 IEEE International Conference on Parallel & Distributed Processing with Applications, Big Data & Cloud Computing, Sustainable Computing & Communications, Social Computing & Networking (ISPA/BDCloud/SocialCom/SustainCom), pp. 1159–1166. IEEE (2021)

Anonymous Credentials and Self-Sovereign Identity - An Initial Assessment

Katja Assaf(✉)

Hasso Plattner Institute, University of Potsdam, Potsdam, Germany
katja.assaf@hpi.de

Abstract. Privacy-enhancing authentication allows the use of a service requiring authentication while protecting privacy. Two main research lines try to achieve the trade-off between authentication and privacy: *Anonymous Credentials* and *Self-Sovereign Identity.* Anonymous Credentials are a cryptographic protocol closely related to group signatures and identity escrow schemes with several available implementations. Self-Sovereign Identity is a collection of primitives focusing on the users' control over their digital identity with a strong push towards practicability in the form of W3C Verifiable Credentials. Both concepts are rarely compared despite their similar settings and use cases. This work provides an analysis of the current intersection between the two research fields.

1 Introduction

With the anonymity offered by the internet, reliable identity management becomes necessary to provide services with restricted access. The focus on usability and the considerable amount of digital identities led to the development of federated identity management, with a few providers controlling most of our digital identities. Empowering individuals to control their digital identities and protect their privacy was envisioned by Chaum in 1985 [15]. The idea was developed further and became the origin of *Anonymous Credentials* (AC). Depending on who is asked, AC are understood as a cryptographic concept defined by their features, a set of protocols or the idea behind certain implementations. However, the common theme is that it provides a means for authentication, which is unlinkable between showings. Mostly independent from this line of work, Allen envisioned the next evolution in identity management and coined the term *Self-Sovereign Identity* (SSI) by formulating ten design principles in 2016 [3]. SSI is usually understood as a system implementing the design principles with varying degrees of compliance. Some practitioners consider SSI to be any implementation compliant with the W3C Verifiable Credential standard [50] or anything using W3C Decentralized Identifiers (DID).

Thus, both concepts aim to enhance users' control and privacy but lack a precise definition. The compatibility of *Anonymous Credentials* and *Self-Sovereign Identity* is still an open problem. Richter et al. [40] analysed the cryptographic

L. Barolli (Ed.): AINA 2024, LNDECT 204, pp. 331–343, 2024.
https://doi.org/10.1007/978-3-031-57942-4_32

requirements for verifiable credentials identifying Hyperledger AnonCreds as a suitable solution. Other researchers have identified a connection as well. Papers about AC mention SSI as a use case, and vice versa, AC are used as a building block, black box or concept in papers developing SSI solutions. Still, a structured approach could broaden awareness of the synergies and help researchers benefit from a wider range of results.

Our Contribution

We analysed the architecture (Sect. 2.1), projects and standards (Sect. 2.2) of *Anonymous Credentials* (AC) and *Self-Sovereign Identity* (SSI) trying to identify overlaps and differences. Further, we performed a systematic literature review to identify the research done within the intersection of both realms. Analysing the metadata, we identified where and when research on both topics combined in a single article was conducted (Sect. 3.1). Classifying the available literature into *AC-Research* and *SSI-Research* based on their keywords enables us to understand what one research field knows about the other. In Sect. 3.2 we answer the following research questions: How do SSI researchers define AC? How does it differ from AC researchers' understanding? And vice versa.

2 Measuring Two Worlds

Self-Sovereign Identity (SSI) and *Anonymous Credentials* (AC) aim at providing a privacy-preserving identity management system. Both ideas are found as concepts, protocols and implementations with limited overlap.

Historically, the concept of *Self-Sovereign Identity* developed from practitioners' view through whitepapers and blog posts [3,14,39] rather than traditional academic publications. The first academic publications, not concerned with a single implementation but the bigger picture, was a translation of an article by Der et al. [18] and Mühle et al.'s survey of the SSI's essential components [35]. Additional mapping and review studies from different perspectives have followed. Cucko and Turkanovic [17] performed a systematic mapping study of the field but made no distinction between "blockchain" and "self-sovereign" in their search term. The systematic review and mapping study by Schardong and Custódio [43] compares itself to many other overview papers and informs about the most influential works within the research area. Further, they identified the practical problems, formal definitions and cryptographic tools used in SSI-related works. *Anonymous Credentials* was not explicitly mentioned although the theoretical foundations [11,12,15] were cited. Sedlmeir et al. [45] provide a comprehensive history of the development of SSI from a socio-technical perspective. They reformulate the principles of SSI with Allen's Design Principles (DP) [3] as a starting point informed by a systematic literature review of available SSI solutions and revised through interviews with experts from academia and industry.

Following Sedlmeir et al. [45, p.11], *Self-Sovereign Identity* is a privacy-preserving (DP5) identity management (DP1) with the user in control (DP2) providing a secure and trustworthy solution (DP4, DP7, DP8), which is usable enough for all actors (DP3, DP6, DP9).

In *Anonymous Credential* research, there is the problem of synonyms and, at the same time, ambiguity of the terms *Anonymous Credential* and *(privacy-preserving) Attribute-Based Credential*. Some researchers call single-show credentials issued with blind signature protocols *Anonymous Credentials*, while others make a distinction and call them *Anonymous Tokens*. The term *(privacy-preserving) Attribute-Based Credential* was introduced within the ABC4Trust[1] project, an EU-funded project developing a privacy-protecting credential system from 2010 until 2015. However, *Attribute-Based Credential* is often used outside the community to describe any credential system with the option to encode attributes.

Although AC dating back to Chaum [15] is the far older research field, few overview works have been published. Kakvi et al. [31] categorise available academic papers and projects in zero-knowledge-based and self-blindable constructions. They also provide an overview of the available features extending their basic definition, such as *updatable* or *delegatable*. Connolly et al. [16] classify the available AC systems concerning their underlying signature scheme into (1) CL-signatures, (2) aggregatable signatures, (3) sanitisable signatures, (4) redactable signatures, and (5) structure-preserving signatures on equivalence classes. Following Fuchsbauer et al. [23], *Anonymous Credentials* are a multi-show credential system protecting the user's privacy, achieving unforgeability and unlinkability.

2.1 Architecture

Both SSI and AC are usually based on a generic credential system architecture as described by Brands [10]. An issuer issues a digital certificate, which the user can show to a verifier as illustrated in Fig. 1. Naming differs, as shown in Table 1, but the purpose of the respective roles stays the same. Depending on the focus and the use case, additional roles and features are introduced. However, they all comply with the original architecture (Fig. 1).

Many SSI systems refer to the W3C verifiable credential standard [50] as their basis [19,37,38,51,52]. The *verifiable data registry*, used as the central point of truth, is often implemented as a blockchain [21,52] emphasizing the close connection between SSI and blockchain. AC developed from the original idea of Chaum [15], who writes about a credential-issuing organisation and a credential-receiving organisation. Consequently, AC and SSI architecture are largely compatible since they originate from the same credential system architecture. Even if additional roles are added, the core architecture remains the same.

[1] https://abc4trust.eu/index.php.

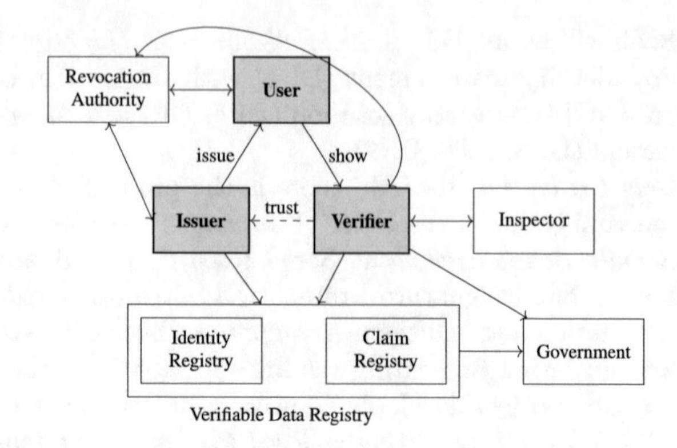

Fig. 1. Credential system architecture with additional roles

Table 1. Names for the credential system roles in the literature (extract)

Names for User	Names for Issuer	Names for Verifier
User [3,4,7,11,21,35]	*Issuer* [7,10,35,38]	*Verifier* [7,10,11,21,35,38]
Holder [38,50]	*Certificate Authority* [21]	*Relying Party* [7,14,35]
Subject [14]	*Group Manager* [4]	*Application Provider* [4]
Individual [15]	*Credential Issuing Organization* [15]	*Credential Receiving Organization* [15]
Prover [10]	*Organization* [11]	*Authentication Mechanism* [10]
Certificate Holder [10]		*Business* [38]

List of Additional Roles:
Inspector [7], *Revocation Authority* [7], *Identifier Registry* [35], *Claims Registry* [35], *Blockchain* [21], *Government* [38], *Verifiable Data Registry* [38,50]

2.2 Standards and Projects

The European Self-Sovereign Identity Framework Lab (eSSIF-Lab)[2], an EU-funded project running until 2022, aiming towards interoperable SSI solutions, put together a list of eight SSI standardisation activities. The relationship of these eight activities and other AC- or SSI-related projects is visualised in Fig. 2.

The Hyperledger Foundation comprises several open-source projects, such as Fabric, Aries, Indy, Ursa and AnonCreds. Hyperledger Indy was developed initially from Sovrin, an early project trying to find a solution for the inter-

[2] https://essif-lab.eu/.

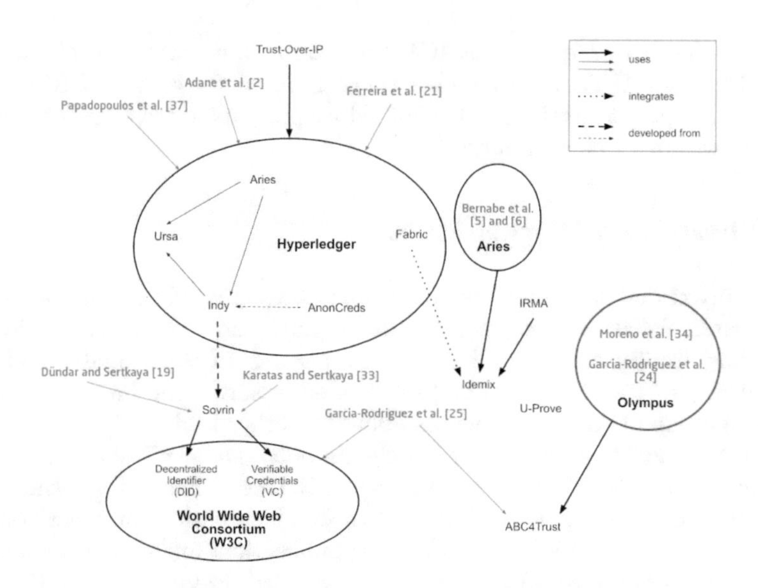

Fig. 2. SSI and AC related projects and their relationship

net's missing *identity layer* [49]. Sovrin was already building upon Decentralised Identifier (DID) and Verifiable Credentials (VC), called Verifiable Claims at that point, which were later standardised by the World Wide Web Consortium (W3C). The Sovrin network still exists and now builds upon Hyperledger Indy. Similarly, the relation between the Hyperledger projects is relatively close. The newest project, AnonCreds[3], developed from Indy, and Ursa, a shared cryptographic library, became a separate project in 2018 to avoid duplicating efforts within other projects such as Aries or Indy. The oldest project in the Hyperledger framework is Hyperledger Fabric dating back to the introduction of Hyperledger in 2016. In 2018, Fabric released v1.3[4] integrating IBM's Idemix into its framework. Idemix implements the AC system as initially defined by Camenisch and Lysanskaya [11] and provides the basis for other implementations of AC, such as IRMA and EU's Aries, not to be confused with Hyperledger Aries. The Aries (reliAble euRopean Identity EcoSystem) project[5] ran from fall 2016 until spring 2019. They build an SSI system integrating with Idemix for authentication [5]. IRMA[6], an abbreviation for *I reveal my attributes*, based on Idemix, is also active since 2016. However, IRMA has only had a significant impact on a national level within the Netherlands.

IBM Research not only developed Idemix but was also active in the ABC4Trust project[7], developing privacy-preserving attribute-based credentials

[3] https://www.hyperledger.org/use/anoncreds.
[4] https://hyperledger-fabric.readthedocs.io/en/release-1.3/whatsnew.html.
[5] https://www.aries-project.eu.
[6] https://privacybydesign.foundation/irma-explanation/.
[7] https://abc4trust.eu/index.php.

(p-ABC). Another party in the ABC4Trust Consortium was Microsoft Research, which developed U-Prove in parallel. Building upon the results of ABC4Trust, the EU Olympus project[8] uses distributed cryptographic techniques to develop a system without a central authority.

3 Where Two Worlds Meet

A literature review was conducted to evaluate the extent of the relation between the research fields *Self-Sovereign Identity* (SSI) and *Anonymous Credentials* (AC). The literature review allows for the analysis of the available metadata, a classification into SSI-related and AC-related research, and finally, finding out what one group of researchers writes about the other field.

In March 2023, a search for *("self-sovereign identity" OR "SSI") AND "anonymous credential"* was performed on Google Scholar. Everything older than 2015 were mishits with 'SSI' used as an abbreviation for other concepts. Master theses, project reports, non-English publications and papers not available through institutional or open access were excluded, leaving us with 47 results. During the next step, 16 papers were excluded, which had solely referenced sources containing '*self-sovereign*' or '*anonymous credential*' as part of their title and providing no further information. The remaining 31 sources provide some interpretation or relation to both SSI and AC within their text.

3.1 Quantitative Analysis

The earliest scientific work connecting the concepts of AC and SSI was published in 2019, and the number of publications increased from five in 2020 to eleven in 2022 (Table 2). However, researchers working on Sovrin must have been aware of both concepts, as AC is mentioned in the Sovrin whitepaper [39].

Visualising authors and institutions (blue) of the 31 sources (red) as well as the institutions' nationalities (green) in Fig. 3 shows that researchers worldwide have identified AC and SSI as related topics. Research is focused in Europe with an emphasis on the *Intelligent Systems and Telematics Research Group* in Spain and several unrelated projects in Germany. Researchers such as Sedlmeir and Ferreira da Silva facilitate exchange through their work with different co-authors. According to Sedlmeir et al. [45], the major part of SSI research is done in Europe and Northern America. However, only one source is associated with an institution from Northern America. Further, only a fraction of SSI research is concerned with *Anonymous Credentials* as Schardong and Custódio [43] already identified 82 sources providing significant conceptual or practical contributions towards SSI in 2022.

The 31 sources were categorised into AC or SSI research based on the available keywords. During the ABC4Trust project[9] *Anonymous Credentials* were

[8] https://olympus-project.eu/.
[9] https://abc4trust.eu/index.php.

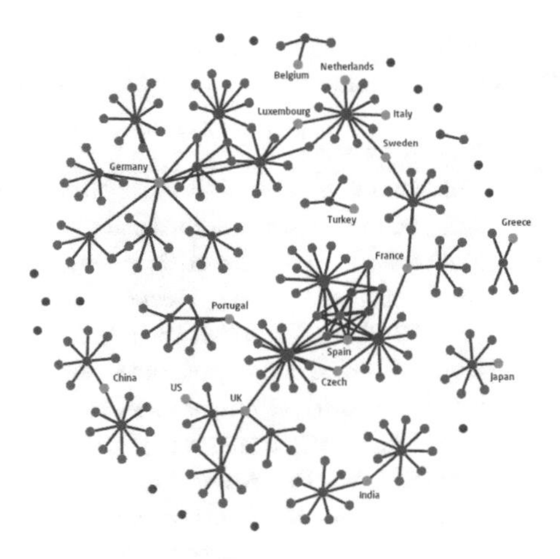

Fig. 3. Relation between Sources, red - sources, green - country, blue - author or institution. Black indicates the 16 sources excluded in the last step.

called *attribute-based credentials*. Thus, sources with this keyword or its abbreviation *ABC* were identified as AC-related. Similarly, SSI is sometimes called *decentralised identity* due to the absence of authority. Consequently, sources with the keyword *decentralised identifier* were sorted into SSI-related research. The result is shown in Table 2 with the keywords defining classification highlighted.

An investigation whether a relation between the classification and the citation of fundamental research in AC [11–13,15] or SSI [3,14,35,39] exists, remained inconclusive. Researchers seem to cite AC and SSI papers independent of their background.

3.2 Definitions - Understanding Each Other

From 31 sources, eight were classified as *AC-Research*, ten as *SSI-Research*, and two additional sources are in both categories. Eleven sources did not provide keywords or at least no relevant ones (Table 2). For related fields, it is expected that keywords appear together often, as is the case for '*SSI*' and '*verifiable credential*'. '*Verifiable credential*' is a keyword in five of the 11 sources tagged as SSI. Further, IEEE Xplore has 22 results tagged with the keywords '*SSI*' and '*verifiable credential*', while no results can be found for '*SSI*' and '*anonymous credential*'. The descriptions of *SSI* in sources classified as *AC-research* show the understanding the authors had of the concept and vice versa:

Cryptographers' view on SSI

Muth et al. [36] and Bosk et al. [9] base their understanding of SSI on the often referenced articles from Allen [3] and Mühle et al. [35]. Schanzenbach et al. [42] and Bobolz et al. [8] reference concrete projects such as Sovrin [39,49] to define the term. Sovrin itself states that they work toward an "Internet-like identity

Table 2. List of sources, their categorisation into AC- or SSI-research, and an extract of their (abbreviated) keywords, omitted keywords are indicated with '...'.

Source	Year	Keywords (extract)
Schanzenbach et al. [42]	2019	Zero-Knowledge, **Attribute-Based Credentials (ABC)**, ...
Bernabe et al. [5]	2020	Privacy, security, risk reduction, ...
Dündar and Sertkaya [19]	2020	**SSI**, indirect SSI control, privacy, guardianship, ...
Halpin [28]	2020	immunity passports, **Decentralized Identifier (DID)**, Verifiable Credentials (VC), W3C, security, privacy, ...
Karatas and Sertkaya [33]	2020	..., **SSI**, identity management, **AC**, VC, blockchain, ...
Moreno et al. [34]	2020	..., security, identity management, IoT, digital identities, ...
Abramson et al. [1]	2021	**SSI**, ..., VC, cryptography, verifier, trust, assurance, risk
Adane et al. [2]	2021	**SSI**, VC, Decentralized System, Security
Bobolz et al. [8]	2021	..., issuer-hiding, privacy-preserving, **AC**, authentication
Ferreira et al. [21]	2021	..., EV charging process, blockchain, IoT, mobile app
Garcia-Rodriguez et al. [24]	2021	..., Identity management, Multi-signatures, **ABC**, ...
Garcia-Rodriguez et al. [25]	2021	VC, p-**ABC**, privacy
Gross et al. [27]	2021	Anonymity, CBDC, Compliance, Design Science, Digital Identity, Digital Wallet, ..., **SSI**, Zero-Knowledge Proof
Papadopoulos et al. [37]	2021	trust, machine learning, federated learning, **DID**, VC
Pauwels [38]	2021	privacy, structured transparency, know-your-customer (KYC), zero-knowledge proof (ZKP), **SSI**
Silva and Moro [46]	2021	Blockchain, ..., Consumer transparency, Privacy, ...
Sousa [47]	2021	-
Bernabe et al. [6]	2022	-
Bosk et al. [9]	2022	**AC**, Unlinkability, ..., Hidden Issuer, Privacy
Connolly et al. [16]	2022	**AC**, Mercurial signatures, SPS-EQ
Esposito et al. [20]	2022	..., Social linked data, Solid, Privacy, Security, Personal data
Ghosh et al. [26]	2022	-
Jaques et al. [30]	2022	-
Kalos and Polyzos [32]	2022	**SSI**, privacy, ..., linked data, JSON LD, **AC**
Muth et al. [36]	2022	blockchains, **AC**, zero knowledge
Schlatt et al. [44]	2022	Distributed ledger, healthcare, token, privacy, **SSI**
Sedlmeir et al. [45]	2022	Certificate, digital wallet, distributed ledger, ..., VC
Yildiz et al. [51]	2022	**SSI**, Interoperability, Identity
Fu et al. [22]	2023	..., Privacy preservation, Single Sign-On, ...
Takaragi et al. [48]	2023	..., Privacy, PKI, **AC**, Zero-knowledge range proof
Zhu et al. [52]	2023	**DID**, range proofs, blockchain, social network, ...

system" [49, p.1] meaning that anyone and anything can establish an identity relationship with any other entity without requiring an authority.

Thus, all sources identified user control without a trusted third party as a defining feature since this is also a predominant concept in the articles from Allen [3] and Mühle et al. [35], and the goal Sovrin is trying to achieve. In SSI-Research, the interpretations from Allen [3] and Mühle et al. [35] are also predominant, focusing on user control, privacy and no trusted third party. However, a wider variety of definitions can be found, such as an implementation of the W3C standard called *decentralised identity* [52], with a focus on other aspects such as *portability* [44], or very general as "a set of technologies, tools, and governance models" [1].

Security researchers' view on AC

Papadopoulos et al. [37] and Zhu et al. [52] reference the original idea from Chaum [15]. Mostly a concrete implementation or project such as Idemix [19], Hyperledger AnonCreds [51] or ABC4Trust [25] is referenced. Only three sources out of ten provided additional information about the purpose of an Anonymous Credential System (ACS). The interpretations identify an ACS as a credential system with selective disclosure [27,44,52]. Often, AC are used as a black box building block, enhancing privacy within the described system. In contrast, the explanations given within AC-Research are often quite exhaustive, providing lists of other sources [8,16], the underlying used signatures or schemes [36,42,48] and the main idea: enhancing the users' privacy [9,24]. Sanders [41] states that there is no unique definition for Anonymous Credential but rather a set of similar interpretations of the same concept. There might be more awareness of this fact within the cryptographers' community, leading to more exhaustive definitions.

4 Conclusion

Self-Sovereign Identity and *Anonymous Credentials* are concepts with the same goal, compatible architectures and intertwined projects.

So far, research on *Anonymous Credentials* conducted by cryptographers is often hard to understand and research on *Self-Sovereign Identity* is often seen together with blockchains and cryptocurrencies considered critically by many cryptographers. This work is a first step towards a structured analysis of the intersection between the two research topics. However, the ambiguity of both terms complicates the search for relevant literature. Examples like Heiss et al. [29], which does not appear in the above analysis since neither *SSI* nor *Self-Sovereign Identity* is mentioned explicitly, show that the literature review as conducted is a starting point but does not suffice to derive the whole picture. However, their W3C verifiable credential-compliant solution is an SSI solution compared against a CL-signature-based approach conducted by Muth et al. [36]. Consequently, searching for related concepts such as *W3C verifiable credentials* or through all papers referencing Allen [3] as a source would create a more complete picture. The same is true for *Anonymous Credentials*, which can also be hidden behind terms such as *CL-signature based scheme*, *p-ABC system* or *Idemix*.

Despite these limitations, we were able to show that understanding in the research communities differs and would benefit from a more active exchange. Our work bridges the gap between both worlds, providing a consistent vocabulary and aligning understanding.

Acknowledgements. I would like to thank Tarek Galal, Alexander Mühle and Inga Stumpp for their comments and feedback and Professor Andreas Polze for his overall support.

References

1. Abramson, W., Hickman, N., Spencer, N.: Evaluating trust assurance in Indy-based identity networks using public ledger data. Front. Blockchain (2021)
2. Adane, P., Jadhav, A., Kodgire, S., Agrawal, A., Kumar, S.: Integration of self sovereign identity in security systems. Int. J. Next-Gener. Comput. (2021)
3. Allen, C.: The Path to Self-Sovereign Identity. Life with Alacrity (2016)
4. Au, M.H., Susilo, W., Mu, Y.: Constant-size dynamic k-TAA. In: De Prisco, R., Yung, M. (eds.) SCN 2006. LNCS, vol. 4116, pp. 111–125. Springer, Heidelberg (2006). https://doi.org/10.1007/11832072_8
5. Bernabe, J.B., David, M., Moreno, R.T., Cordero, J.P., Bahloul, S., Skarmeta, A.: Aries: evaluation of a reliable and privacy-preserving European identity management framework. Futur. Gener. Comput. Syst. **102**, 409–425 (2020)
6. Bernabe, J.B., et al.: An overview on ARIES: reliable European identity ecosystem. In: Challenges in Cybersecurity and Privacy-The European Research Landscape, pp. 231–254 (2022)
7. Bichsel, P., et al.: An architecture for privacy-ABCs. In: Rannenberg, K., Camenisch, J., Sabouri, A. (eds.) Attribute-Based Credentials for Trust, pp. 11–78. Springer, Cham (2015). https://doi.org/10.1007/978-3-319-14439-9_2
8. Bobolz, J., Eidens, F., Krenn, S., Ramacher, S., Samelin, K.: Issuer-hiding attribute-based credentials. In: Conti, M., Stevens, M., Krenn, S. (eds.) CANS 2021. LNCS, vol. 13099, pp. 158–178. Springer, Cham (2021). https://doi.org/10.1007/978-3-030-92548-2_9
9. Bosk, D., Frey, D., Gestin, M., Piolle, G.: Hidden issuer anonymous credential. In: Proceedings on Privacy Enhancing Technologies 2022, pp. 571–607 (2022)
10. Brands, S.: Rethinking Public Key Infrastructures and Digital Certificates: Building in Privacy. MIT Press (2000)
11. Camenisch, J., Lysyanskaya, A.: An efficient system for non-transferable anonymous credentials with optional anonymity revocation. In: Pfitzmann, B. (ed.) EUROCRYPT 2001. LNCS, vol. 2045, pp. 93–118. Springer, Heidelberg (2001). https://doi.org/10.1007/3-540-44987-6_7
12. Camenisch, J., Lysyanskaya, A.: Dynamic accumulators and application to efficient revocation of anonymous credentials. In: Yung, M. (ed.) CRYPTO 2002. LNCS, vol. 2442, pp. 61–76. Springer, Heidelberg (2002). https://doi.org/10.1007/3-540-45708-9_5
13. Camenisch, J., Lysyanskaya, A.: A signature scheme with efficient protocols. In: Cimato, S., Persiano, G., Galdi, C. (eds.) SCN 2002. LNCS, vol. 2576, pp. 268–289. Springer, Heidelberg (2003). https://doi.org/10.1007/3-540-36413-7_20
14. Cameron, K.: The laws of identity. Microsoft Corp **12**, 8–11 (2005)

15. Chaum, D.: Security without identification: transaction systems to make big brother obsolete. Commun. ACM **28**(10), 1030–1044 (1985)
16. Connolly, A., Lafourcade, P., Perez Kempner, O.: Improved constructions of anonymous credentials from structure-preserving signatures on equivalence classes. In: Hanaoka, G., Shikata, J., Watanabe, Y. (eds.) Public-Key Cryptography–PKC 2022, vol. 13177, pp. 409–438. Springer, Cham (2022). https://doi.org/10.1007/978-3-030-97121-2_15
17. Čučko, Š, Turkanović, M.: Decentralized and self-sovereign identity: systematic mapping study. IEEE Access **9**, 139009–139027 (2021)
18. Der, U., Jähnichen, S., Sürmeli, J.: Self-sovereign identity – opportunities and challenges for the digital revolution. arXiv preprint arXiv:1712.01767 (2017)
19. Dündar, Y., Sertkaya, I.: Self sovereign identity based mutual guardianship. J. Mod. Technol. Eng. **5**(3), 189–211 (2020)
20. Esposito, C., Hartig, O., Horne, R., Sun, C.: Assessing the solid protocol in relation to security & privacy obligations. arXiv preprint arXiv:2210.08270 (2022)
21. Ferreira, J.C., Ferreira da Silva, C., Martins, J.P.: Roaming service for electric vehicle charging using blockchain-based digital identity. Energies **14**(6), 1686 (2021)
22. Fu, Y., et al.: Non-transferable blockchain-based identity authentication. Peer-to-Peer Network. Appl. **16**, 1354–1364 (2023)
23. Fuchsbauer, G., Hanser, C., Slamanig, D.: Structure-preserving signatures on equivalence classes and constant-size anonymous credentials. J. Cryptol. **32**, 498–546 (2019)
24. García-Rodríguez, J., Moreno, R.T., Bernabe, J.B., Skarmeta, A.: Implementation and evaluation of a privacy-preserving distributed ABC scheme based on multi-signatures. J. Inf. Secur. Appl. **62**, 102971 (2021)
25. García-Rodríguez, J., Moreno, R.T., Bernabé, J.B., Skarmeta, A.: Towards a standardized model for privacy-preserving verifiable credentials. In: Proceedings of the 16th International Conference on Availability, Reliability and Security, pp. 1–6 (2021)
26. Ghosh, B.C., Patranabis, S., Vinayagamurthy, D., Ramakrishna, V., Narayanam, K., Chakraborty, S.: Private certifier intersection. Cryptology ePrint Archive (2022)
27. Gross, J., Sedlmeir, J., Babel, M., Bechtel, A., Schellinger, B.: Designing a central bank digital currency with support for cash-like privacy. In: PSN: Exchange Rates & Currency (Comparative) (Topic) (2021)
28. Halpin, H.: Vision: a critique of immunity passports and W3C decentralized identifiers. In: van der Merwe, T., Mitchell, C., Mehrnezhad, M. (eds.) SSR 2020. LNCS, vol. 12529, pp. 148–168. Springer, Cham (2020). https://doi.org/10.1007/978-3-030-64357-7_7
29. Heiss, J., Muth, R., Pallas, F., Tai, S.: Non-disclosing credential on-chaining for blockchain-based decentralized applications. In: Troya, J., Medjahed, B., Piattini, M., Yao, L., Fernández, P., Ruiz-Cortés, A. (eds.) Service-Oriented Computing: 20th International Conference, ICSOC, pp. 351–368. Springer, Cham (2022). https://doi.org/10.1007/978-3-031-20984-0_25
30. Jaques, S., Lodder, M., Montgomery, H.: ALLOSAUR: accumulator with low-latency oblivious sublinear anonymous credential updates with revocations. Cryptology ePrint Archive (2022)
31. Kakvi, S.A., Martin, K.M., Putman, C., Quaglia, E.A.: SoK: anonymous credentials. In: Security Standardisation Research: 8th International Conference, SSR, vol. 13895, pp. 129–151. Springer, Cham (2023). https://doi.org/10.1007/978-3-031-30731-7_6

32. Kalos, V., Polyzos, G.C.: Requirements and secure serialization for selective disclosure verifiable credentials. In: Meng, W., Fischer-Hübner, S., Jensen, C.D. (eds.) ICT Systems Security and Privacy Protection: 37th IFIP TC 11 International Conference, vol .648, pp. 231–247. Springer, Cham (2022). https://doi.org/10.1007/978-3-031-06975-8_14
33. Karatas, R., Sertkaya, I.: Self sovereign identity based E-petition scheme. Int. J. Inf. Secur. Sci. **9**(4), 213–229 (2020)
34. Moreno, R.T., et al.: The Olympus architecture: oblivious identity management for private user-friendly services. Sensors **20**(3), 945 (2020)
35. Mühle, A., Grüner, A., Gayvoronskaya, T., Meinel, C.: A survey on essential components of a self-sovereign identity. Comput. Sci. Rev. **30**, 80–86 (2018)
36. Muth, R., Galal, T., Heiss, J., Tschorsch, F.: Towards smart contract-based verification of anonymous credentials. In: Matsuo, S., et al. (eds.) Financial Cryptography and Data Security. FC 2022 International Workshops, vol. 13412, pp. 481–498. Springer, Cham (2023). https://doi.org/10.1007/978-3-031-32415-4_30
37. Papadopoulos, P., Abramson, W., Hall, A.J., Pitropakis, N., Buchanan, W.J.: Privacy and trust redefined in federated machine learning. Mach. Learn. Knowl. Extr. **3**(2), 333–356 (2021)
38. Pauwels, P.: zkKYC: a solution concept for KYC without knowing your customer, leveraging self-sovereign identity and zero-knowledge proofs. Cryptology ePrint Archive (2021)
39. Reed, D., Law, J., Hardman, D.: The technical foundations of Sovrin. The Technical Foundations of Sovrin (2016)
40. Richter, M., Bertram, M., Seidensticker, J., Margraf, M.: Cryptographic requirements of verifiable credentials for digital identification documents. In: COMPSAC 2023, Torino, Italy, 26–30 June 2023, pp. 1663–1668. IEEE (2023)
41. Sanders, O.: Efficient redactable signature and application to anonymous credentials. In: Kiayias, A., Kohlweiss, M., Wallden, P., Zikas, V. (eds.) PKC 2020. LNCS, vol. 12111, pp. 628–656. Springer, Cham (2020). https://doi.org/10.1007/978-3-030-45388-6_22
42. Schanzenbach, M., Kilian, T., Schütte, J., Banse, C.: ZKlaims: privacy-preserving attribute-based credentials using non-interactive zero-knowledge techniques. In: International Conference on E-Business and Telecommunication Networks (2019)
43. Schardong, F., Custódio, R.: Self-Sovereign identity: a systematic review, mapping and taxonomy. Sensors **22**(15), 5641 (2022)
44. Schlatt, V., Sedlmeir, J., Traue, J., Völter, F.: Harmonizing sensitive data exchange and double-spending prevention through blockchain and digital wallets: the case of E-prescription management. In: Distributed Ledger Technologies, Research and Practice (2022)
45. Sedlmeir, J., Barbereau, T., Huber, J., Weigl, L., Roth, T.: Transition pathways towards design principles of self-sovereign identity. In: International Conference on Interaction Sciences (2022)
46. da Silva, C.F., Moro, S.: Blockchain technology as an enabler of consumer trust: a text mining literature analysis. Telematics Inform. **60** (2021)
47. Sousa, P.R.R.: Privacy preserving middleware platform for IoT. Ph.D. thesis, Universidade do Porto (Portugal) (2021)
48. Takaragi, K., Kubota, T., Wohlgemuth, S., Umezawa, K., Koyanagi, H.: Secure revocation features in eKYC-privacy protection in central bank digital currency. IEICE Trans. Fundam. Electron. Commun. Comput. Sci. **106**(3), 325–332 (2023)
49. Tobin, A., Reed, D.: The inevitable rise of self-sovereign identity. Sovrin Found. **29**(2016), 18 (2016)

50. W3 Consortium: Verifiable credentials data model 1.0: expressing verifiable information on the web (2019). https://www.w3.org/TR/vc-data-model/?#core-data-model
51. Yildiz, H., Küpper, A., Thatmann, D., Göndör, S., Herbke, P.: A tutorial on the interoperability of self-sovereign identities. arXiv preprint arXiv:2208.04692 (2022)
52. Zhu, X., He, D., Bao, Z., Luo, M., Peng, C.: An efficient decentralized identity management system based on range proof for social networks. IEEE Open J. Comput. Soc. (2023)

Network Intrusion Detection with Incremental Active Learning

Münteha Nur Bedir Tüzün and Pelin Angin(✉)

Middle East Technical University, Ankara, Turkey
pangin@ceng.metu.edu.tr

Abstract. Increasing Internet usage in recent years has correspondingly increased the prevalence of cyber threats, emphasizing the necessity for robust intrusion detection systems (IDS). The efficacy of these systems is crucially dependent on their ability to adapt promptly to the continuously evolving types of cyber-attacks. Nonetheless, achieving the desired performance levels is often hindered by the scarcity of labeled data for newly emerging threats and the complexities associated with implementing incremental learning within machine learning frameworks. In this research, we introduce an IDS that employs active learning techniques for class incremental learning, aimed at adapting to the dynamic cyber security landscape while requiring fewer labeled data instances. The results from our experiments demonstrate that the proposed method significantly reduces the need for labeled training data while effectively incorporating new attack classes incrementally.

1 Introduction

The growth, diversity, and complexity of computer networks have surged dramatically, resulting in an unprecedented increase in cyber-attacks targeting web-based infrastructures. In the contemporary web environment, new types of attacks are rapidly developed, necessitating intrusion detection systems (IDS) to be highly adaptive to these rapidly changing conditions. The application of machine learning and deep learning techniques has become increasingly prevalent in the domain of IDS over the past decade, delivering satisfactory results. However, these methods inherently lack the capability for sequential learning, which is essential for adapting to new threats. The issue of lifelong learning has been a focus of research for the past two decades, yet a comprehensive solution remains elusive. A major challenge in incremental learning is the phenomenon of catastrophic forgetting, wherein newly acquired knowledge supersedes and eradicates previously learned information, necessitating the model to be retrained from scratch with each new data introduction. Despite the development of several incremental learning strategies, none have yet achieved the performance level of retraining the model with the entire dataset. Hence, maintaining a small training dataset enhances the efficiency of the continuous learning process. The lack of labeled data in the intrusion detection domain poses another significant challenge for achieving high-performance IDS. Active learning emerges as a promising solution to these challenges, where the active learner selects the most informative instances from an

L. Barolli (Ed.): AINA 2024, LNDECT 204, pp. 344–353, 2024.
https://doi.org/10.1007/978-3-031-57942-4_33

unlabeled dataset, subsequently querying these instances to an oracle for labeling. This approach achieves similar, if not superior, performance with significantly less training data.

In this study, we propose a signature-based network intrusion detection system that incorporates active learning to efficiently learn new types of attacks incrementally. This system requires fewer labeled data instances and offers a shorter training period, as the training dataset is minimized as much as possible.

2 Related Work

Security researchers have applied various machine learning algorithms combined with active learning techniques to develop IDS, and have compared the performance of these systems across different datasets. Despite the limited research combining active learning with IDS, the results from these studies are promising for achieving high-performance IDS through active learning.

Yang et al. [11] conducted the initial study applying active learning to the Internet of Things (IoT) IDS. Their system first identifies outliers in an unsupervised manner, then iteratively selects and labels data from the unlabeled dataset until the desired performance in terms of precision and recall is achieved. They demonstrated that an active learning-based method converges faster to the desired performance compared to random selection methods using the KDD'99 dataset.

Li et al. [4] introduced an IDS based on active transfer learning, named ACTrAdaBoost. Their experiments on the DARPA1998, KDD'99, and ISCX2012 datasets showed improved performance with the incorporation of active learning.

Boukela et al. [2] proposed a hybrid IDS that is adaptable to unseen attacks, consisting of a supervised module using Deep Neural Network (DNN) and an unsupervised module employing K-Nearest Neighbors (KNN). They analyzed the effect of the queried sample size in each window on system performance, comparing it with a baseline trained with the full training dataset. They found that as the queried sample size increases, the DNN performance improves.

In studies by McElwee et al. [7,8], the performance of attack classifiers was progressively enhanced with active learning, although new classes of attack were not incorporated. These studies primarily analyzed the effect of selected sample size for labeling and the minimum confidence threshold, without delving into other parameters. Boukela et al. [2] were the first to use active learning to achieve an adaptive IDS, though their system was limited to binary classification and could not categorize attacks as new types due to its non-multi-class classification nature.

Martina et al. [6] introduced an IDS named Soft-Forgetting Self-Organizing Incremental Neural Network (SF-SOINN) with the ability for incremental learning. They demonstrated short classification times and high performance. Amalapuram et al. [1] applied popular incremental learning methods, Elastic Weight Consolidation (EWC) and Gradient Episodic Memory (GEM), to IDS. They evaluated the performance of these methods across different tasks on the CICIDS-2017, CICIDS-2018, and KDD'99 datasets, showing that the proposed architecture provides a scalable solution. Lin et al. [5] proposed a two-stage IDS utilizing DNN and Incremental Learning, trained with

CAN data offline and updated with new data using incremental learning in the online stage, achieving high performance in experiments.

Our work is distinctive as it applies active learning to the multi-class classification problem within the intrusion detection domain for class incremental learning, a novel approach not previously explored to the best of our knowledge.

3 Methodology

3.1 System Design and Operational Framework

The design of our proposed intrusion detection system [10] is modular, comprising four critical components: the classification module, the incremental learning module, the active learning module, and the simulated oracle module. This section elucidates the interplay between these components and the overall operational mechanism of the system, as illustrated in Fig. 1.

Initially, the classification module processes input feature sets, assigning the most probable label based on the neural network model's predictions. Should the confidence level of a label fall below a predefined threshold, the corresponding input feature set is relegated to an unlabeled data repository. It is important to note that the neural network model undergoes preliminary training with a subset of the training dataset prior to deployment.

Activation of the incremental learning module occurs when the volume of data within the unlabeled repository surpasses a specified threshold. This module then interfaces with the active learning module, transferring the unlabeled data for selection and labeling. Introduction of a new attack category necessitates the expansion of the neural network model's output layer to accommodate additional classes, achieved by integrating new neurons corresponding to the number of new categories identified. Crucially, to prevent the obliteration of previously acquired knowledge, the existing weights of neurons in the hidden layers are preserved. Subsequent to the augmentation of the training dataset with newly labeled instances, the neural network model undergoes retraining. Upon completion, the unlabeled data repository is cleared, preparing the system for subsequent cycles of learning and adaptation. This iterative process of model refinement and expansion continues throughout the operational lifespan of the IDS, with intervals between cycles expected to lengthen as the model stabilizes and matures.

3.2 Classification Module

For the classification task within this work, we employ artificial neural networks (ANN) due to their proficiency in handling complex patterns. The configuration of the input layer's neurons directly corresponds to the number of features present in the dataset under consideration. Through our empirical analysis, we determined that ANNs configured with either a singular hidden layer comprising 64 neurons or a more complex structure with four hidden layers (256-128-64-32 neurons) deliver comparable levels of performance. Given this equivalence, our model adopts the simpler architecture of a single hidden layer with 64 neurons for efficiency. The Rectified Linear Unit (ReLU)

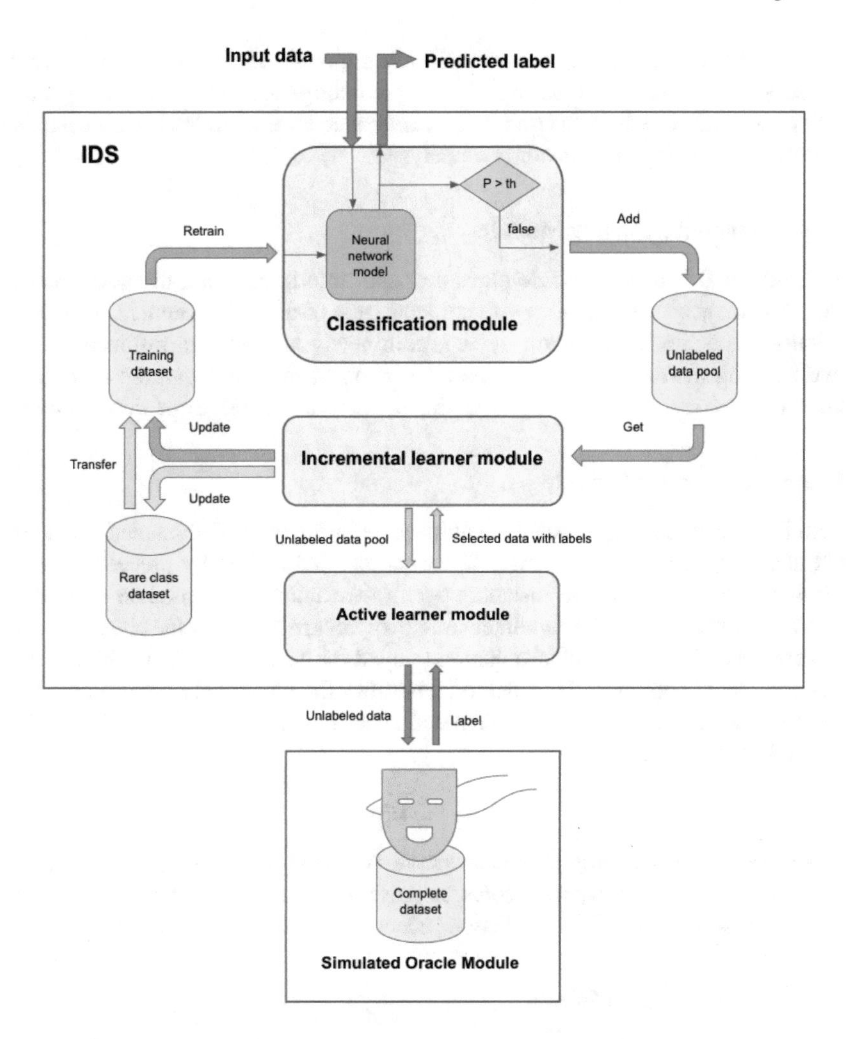

Fig. 1. Schematic of the Proposed Intrusion Detection System

function is utilized for activation within the hidden layer to introduce non-linearity, enhancing the model's learning capability. Adjustments to the output layer's neuron count are made to reflect the variety of classes within the training dataset, employing the softmax function for activation to facilitate classification across multiple categories. To mitigate the risk of overfitting, we incorporate dropout and batch normalization techniques, setting the dropout ratio at 0.20. The model employs a cross-entropy loss function optimized by the Adam algorithm, a choice driven by its effectiveness in converging to the optimal solution efficiently. Acknowledging the challenge posed by the imbalanced distribution of classes typical of intrusion detection datasets, we implement class weighting strategies. These weights are dynamically calculated based on the frequency of each class within the training data, ensuring a balanced consideration during the model training phase. This classification module serves as the cornerstone of our system, tasked with the dual functions of predicting the labels of incoming data and identifying instances where the prediction confidence level falls short of the established threshold, earmarking them for further analysis.

3.3 Incremental Learning Module

The incremental learning module plays a crucial role in ensuring the neural network model remains responsive to the evolving landscape of data it encounters. This module facilitates the model's adaptation by acquiring newly labeled data instances from the active learning module. It oversees the integration of this fresh data into the existing dataset and oversees the process of updating the neural network through retraining.

3.4 Active Learning Module

Within the active learning module, each instance in the pool of unlabeled data is evaluated and assigned a distinct score. These scores, determined by predefined criteria, guide the selection of certain instances for further action. This module establishes a communication link with the simulated oracle to procure labels for the chosen instances lacking labels. The scoring mechanism employed for each data instance hinges on the "least-confident" approach. This method quantifies the uncertainty associated with the predicted label for an instance x, calculated according to Eq. 1 below, where $\hat{y}$ is the most likely label for x.

$$\mathbf{S}_x = 1 - \mathbf{P}(\hat{\mathbf{y}}|\mathbf{x}) \tag{1}$$

The strategy for selecting instances operates on two principles: opting for either the top n instances with the highest scores or those with the lowest, thereby focusing on either the most or least certain predictions according to the system's assessment.

3.5 Simulated Oracle Module

Each piece of data within the dataset is uniquely tagged with an identifier number, which is also shared with the simulated oracle module alongside the data labels. Upon request for the label of a specific data instance, the simulated oracle employs this identifier number to locate and provide the corresponding label. The implementation of a simulated oracle module is primarily for experimental validation, compensating for the unavailability of a security expert's input within this study's framework. In practical applications, this module's role would be fulfilled by actual security professionals who would undertake the task of data labeling.

4 Experiments and Analysis

4.1 Dataset

For this study, the CICIDS-2017 intrusion detection dataset [9] served as the basis for our evaluation. This dataset, encompassing data from July 3 to July 7, 2017, includes

80 flow-based traffic features, with two being text-based. It is comprehensively labeled, featuring 15 distinct labels spanning seven prevalent attack types: Botnet, Brute Force, DoS, Heartbleed, Infiltration, DDoS, and Web Attacks, alongside benign traffic simulated based on the activities of over 20 users across FTP, HTTPS, HTTP, SSH, and email protocols, mirroring contemporary traffic patterns.

Data pre-processing: In this study, we focused exclusively on numeric features, resulting in a dataset with 78 attributes after preprocessing. Data corresponding to three labels were excluded due to insufficient sample sizes. Referencing findings from prior research [3], which demonstrated equivalent performance between the full dataset and a 10% subset when class distributions are maintained, we applied this strategy to reduce our dataset to 10% of its original size. Subsequently, we allocated 20% of this condensed dataset for testing purposes. A portion of what remained was then set aside for the initial phase of model training, with the specific proportion adjusted based on experimental needs. This preliminary training dataset was limited to three categories: BENIGN, PortScan, and DDos, while data from other categories were omitted. The residual dataset was processed by the IDS in batches of 100 instances each. This batch processing approach was adopted to enhance the efficiency of the experiments, acknowledging that individual data processing by the neural network would be considerably more time-consuming. Furthermore, this method mirrors the real-world operational environment of an IDS, where multiple data streams are analyzed concurrently.

4.2 Experimental Setup

The IDS prototype was developed in Python, utilizing the Tensorflow and Keras libraries to construct the ANN. A baseline model, trained on the entire dataset, was established for comparative purposes, with training parameters set to a batch size of 256 and 150 epochs. Key experimental parameters and their values include:

- **Initial Data Split Ratio:** The proportion of the dataset designated for initial model training, set at 0.05.
- **Batch Size:** The quantity of data instances processed per training batch, established at 256.
- **Epochs:** The total number of training cycles through the dataset, fixed at 10.
- **Selection Count:** The number of instances selected from the unlabeled pool for labeling, determined to be 100.
- **Pool Size Threshold:** The critical size of the unlabeled data pool that triggers the active learning process, set at 500.

4.3 Results and Discussion

Figure 2 depicts the classifier's performance through various training phases, with "train-0" indicating the initial pre-training cycle and subsequent cycles labeled as train-x. Throughout these cycles, newly identified data instances by the active learning module are incorporated into the training dataset, prompting updates to the classifier.

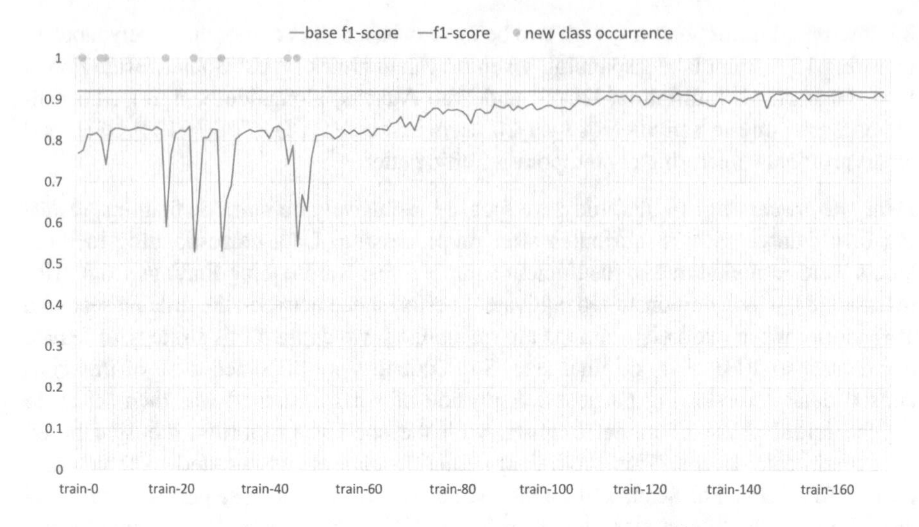

Fig. 2. Performance Analysis Across Training Cycles

Table 1. General performance table

name	accuracy	accuracy balanced	f1-score weighted	f1-score macro	precision weighted	precision macro	recall weighted	recall macro
train-0	0.833	0.149	0.770	0.125	0.717	0.109	0.833	0.149
train-1	0.835	0.160	0.776	0.134	0.745	0.144	0.835	0.160
train-5	0.836	0.263	0.807	0.227	0.788	0.213	0.836	0.263
train-6	0.712	0.261	0.742	0.197	0.797	0.174	0.712	0.261
train-19	0.517	0.321	0.590	0.157	0.829	0.174	0.517	0.321
train-25	0.439	0.370	0.529	0.179	0.797	0.183	0.439	0.370
train-31	0.445	0.279	0.531	0.163	0.794	0.160	0.445	0.279
train-45	0.716	0.386	0.744	0.201	0.802	0.171	0.716	0.386
train-47	0.458	0.436	0.543	0.190	0.836	0.193	0.458	0.436
train-172	**0.894**	**0.723**	**0.908**	**0.541**	**0.927**	**0.530**	**0.894**	**0.723**
baseline	**0.913**	**0.620**	**0.920**	**0.527**	**0.934**	**0.560**	**0.913**	**0.620**

The graph highlights moments where new attack categories are introduced (marked by yellow dots). Introduction of new classes initially leads to a slight decline in performance due to the increased complexity and potential confusion among classes. However, as the model undergoes further training cycles, incorporating data from these new classes, a noticeable improvement in performance is observed. The baseline performance, represented by the green line, is achieved by training with the entire dataset. The classifier's performance gradually aligns with this baseline, reaching parity around the 110th training cycle.

Table 1 outlines variations in performance metrics across different classifier stages. Metrics are evaluated in two forms: average and weighted average. The average met-

Table 2. F1-score table for each class

name	BENIGN	Bot	DDoS	DoS Golden Eye	DoS Hulk	DoS Slowhttp test	DoS slowloris	FTP-Patator	PortScan	SSH-Patator	Web Attack Brute Force	Web Attack XSS
train-0	0.926	0.000	0.569	0.000	0.000	0.000	0.000	0.000	0.000	0.000	0.000	0.000
train-1	0.927	0.000	0.607	0.000	0.000	0.000	0.000	0.000	0.073	0.000	0.000	0.000
train-5	0.924	0.000	0.555	0.000	0.000	0.563	0.000	0.000	0.686	0.000	0.000	0.000
train-6	0.835	0.000	0.603	0.000	0.089	0.198	0.000	0.000	0.639	0.000	0.000	0.000
train-19	0.652	0.039	0.535	0.000	0.307	0.057	0.000	0.000	0.301	0.000	0.000	0.000
train-25	0.564	0.004	0.553	0.000	0.144	0.123	0.052	0.019	0.692	0.000	0.000	0.000
train-31	0.579	0.001	0.521	0.067	0.064	0.000	0.035	0.035	0.651	0.000	0.000	0.000
train-45	0.853	0.065	0.374	0.046	0.000	0.132	0.000	0.000	0.729	0.213	0.000	0.000
train-47	0.583	0.013	0.531	0.192	0.170	0.039	0.031	0.047	0.628	0.005	0.041	0.000
train-172	**0.936**	**0.078**	**0.904**	**0.534**	**0.708**	**0.534**	**0.415**	**0.821**	**0.887**	**0.622**	**0.051**	**0.000**
baseline	**0.947**	**0.006**	**0.914**	**0.907**	**0.879**	**0.437**	**0.183**	**0.656**	**0.695**	**0.620**	**0.078**	**0.000**

Table 3. Train data table

name	BENIGN	Bot	DDoS	DoS Golden Eye	DoS Hulk	DoS Slowhttp test	DoS slowloris	FTP-Patator	PortScan	SSH-Patator	Web Attack Brute Force	Web Attack XSS	total
train_data-0	9115	0	494	0	0	0	0	0	616	0	0	0	10225
train_data-1	9160	0	499	0	4	0	0	0	662	0	0	0	10325
train_data-5	9419	0	509	0	43	2	0	0	749	0	0	0	10722
train_data-6	9493	0	509	0	52	2	0	2	765	0	0	0	10823
train_data-19	10429	2	533	0	158	4	0	11	985	0	0	0	12122
train_data-25	10983	2	536	0	166	4	2	12	1018	0	0	0	12723
train_data-31	11490	3	537	2	193	5	5	15	1074	0	0	0	13324
train_data-45	12622	4	553	3	242	5	10	27	1257	2	0	0	14725
train_data-47	12801	4	560	3	251	5	11	28	1257	2	3	0	14925
train_data-172	**19728**	**14**	**1287**	**27**	**2509**	**20**	**74**	**85**	**3542**	**130**	**9**	**0**	**27425**
base train data	**181705**	**157**	**10242**	**823**	**18409**	**440**	**464**	**635**	**12704**	**472**	**121**	**52**	**226224**

ric calculates the mean performance across all classes, while the weighted average takes into account the proportional representation of each class in the test data, offering insight into the impact of learning new classes. From the initial pre-training cycle ("train-0") through various stages of introducing new attack types, a progressive enhancement in all performance metrics is observed, evidencing a convergence towards baseline performance. Notably, the classifier exhibits superior performance in certain metrics under the average calculation method, suggesting a more effective learning of some classes compared to the baseline. Table 2 presents the evolution of f1-scores for individual classes from the pre-training phase through successive cycles, highlighting the classifier's capacity to incrementally learn and integrate new classes. In several cases, the classifier outperforms the baseline, showing its adeptness at adapting to and recognizing a broad spectrum of attacks.

Table 3 tracks the composition of the training dataset from its initial formation, concentrating initially on three classes, to its final state, where it encompasses nearly all targeted attack categories. A comparison between the baseline and the classifier's final training dataset reveals a more balanced representation of attack classes in the latter, alongside a significant reduction in the overall volume of labeled data required for training, showing the effectiveness of the incremental learning approach.

5 Conclusion

In this work, we introduced an IDS that utilizes active learning to incrementally identify new types of cyber attacks with minimal reliance on labeled data. This work marks a pioneering application of active learning for addressing the multi-class classification challenges within intrusion detection, specifically tailored for class incremental learning. Our experiments utilized the CICIDS-2017 dataset, effectively simulating contemporary network traffic conditions. Through detailed experimentation, we tested various parameters impacting the IDS's performance, revealing that our active learning approach requires less than 15% of the entire dataset for training to achieve comparable results to established benchmarks. This reduction in data requirement signifies a promising advancement towards crafting adaptive IDS solutions capable of evolving in tandem with the dynamic landscape of cyber threats.

Acnowledgements. This research has been supported by the TÜBİTAK 3501 Career Development Program under grant number 120E537 and the TÜBA GEBİP Program.

References

1. Amalapuram, S., Tadwai, A., Vinta, R., Channappayya, S., Tamma, B.: Continual learning for anomaly based network intrusion detection. In: 2022 14th International Conference on COMmunication Systems and NETworkS, COMSNETS 2022, pp. 497–505 (2022). https://doi.org/10.1109/COMSNETS53615.2022.9668482
2. Boukela, L., Zhang, G., Yacoub, M., Bouzefrane, S.: A near-autonomous and incremental intrusion detection system through active learning of known and unknown attacks. In: Conference Digest - 2021 International Conference on Security, Pattern Analysis, and Cybernetics, SPAC 2021, pp. 374–379 (2021). https://doi.org/10.1109/SPAC53836.2021.9539947
3. Gamage, S., Samarabandu, J.: Deep learning methods in network intrusion detection: a survey and an objective comparison. J. Network Comput. Appl. **169**, 102767 (2020). https://doi.org/10.1016/j.jnca.2020.102767
4. Li, J., Wu, W., Xue, D.: An intrusion detection method based on active transfer learning. Intell. Data Anal. **24**, 263–283 (2020). https://doi.org/10.3233/IDA-194487
5. Lin, J., Wei, Y., Li, W., Long, J.: Intrusion detection system based on deep neural network and incremental learning for in-vehicle CAN networks. In: Wang, G., Choo, KK.R., Ko, R.K.L., Xu, Y., Crispo, B. (eds.) Ubiquitous Security. UbiSec 2021. Communications in Computer and Information Science, vol. 1557 CCIS. Springer, Singapore (2022). https://doi.org/10.1007/978-981-19-0468-4_19
6. Martina, M., Foresti, G.: A continuous learning approach for real-time network intrusion detection. Int. J. Neural Syst. **31**, 2150060 (2021). https://doi.org/10.1142/S012906572150060X
7. McElwee, S.: Active learning intrusion detection using k-means clustering selection. In: Conference Proceedings - IEEE SOUTHEASTCON, pp. 1–7 (2017). https://doi.org/10.1109/SECON.2017.7925383

8. McElwee, S., Cannady, J.: Cyber situation awareness with active learning for intrusion detection. In: Conference Proceedings - IEEE SOUTHEASTCON, vol. 2019-April, pp. 1–7 (2019). https://doi.org/10.1109/SoutheastCon42311.2019.9020599
9. Sharafaldin, I., Lashkari, A., Ghorbani, A.: Toward generating a new intrusion detection dataset and intrusion traffic characterization. In: ICISSP 2018 - Proceedings of the 4th International Conference on Information Systems Security and Privacy, vol. 2018-January, pp. 108–116 (2018). https://doi.org/10.5220/0006639801080116
10. Tüzün, M.N.B.: Network intrusion detection system with incremental active learning. Master's thesis, Middle East Technical University (2022)
11. Yang, K., Ren, J., Zhu, Y., Zhang, W.: Active learning for wireless IoT intrusion detection. IEEE Wirel. Commun. **25**, 19–25 (2018). https://doi.org/10.1109/MWC.2017.1800079

Quantum Advancements in Securing Networking Infrastructures

Hadi Salloum[1,2(✉)], Murhaf Alawir[1,2], Mohammad Anas Alatasi[1,2], Saleem Asekrea[1,2], Manuel Mazzara[2], and Mohammad Reza Bahrami[2,3]

[1] QDeep, Innopolis, Russia
[2] Innopolis University, Innopolis, Russia
h.salloum@qdeep.net
[3] Samarkand International University of Technology, Samarkand, Uzbekistan

Abstract. This paper presents a exploration into the integration of Quantum Key Distribution (QKD) and Post-Quantum Cryptography (PQC) within networking infrastructures, marking a groundbreaking advancement in network security. It meticulously examines the vulnerabilities inherent in classical and post-quantum cryptographic methods, underlining the pressing necessity for a shift towards quantum-based security paradigms. The crux of novelty lies in the comprehensive analysis of implementing QKD and PQC strategies within networking systems, providing innovative insights into their fusion for heightened security measures. This integration not only addresses existing cryptographic vulnerabilities but also spearheads a transformative approach to fortify networking infrastructures against evolving threats. Furthermore, the paper anticipates and scrutinizes the forthcoming threats and opportunities in the quantum era, offering a forward-thinking perspective to navigate the dynamic landscape of network security.

Keywords: Post-Quantum Cryptography · Quantum Key Distribution · Networking Infrastructures · Security · Quantum Computing

1 Introduction

The evolution of computing technologies has consistently driven innovation across various domains, particularly in securing advanced networking infrastructures. Quantum computing, an emerging paradigm, has garnered significant attention owing to its potential to revolutionize computational capabilities [9] and address complex challenges in cryptography and network security. The roots of utilizing quantum computing advancements in securing advanced networking infrastructures trace back to the conceptualization and pioneering works of scientists such as Stephen Wiesner, who, in the 1960s, laid the theoretical groundwork for quantum cryptography. Wiesner's propositions of quantum money and quantum conjugate coding stimulated contemplation about the unique properties of quantum mechanics that could be harnessed for cryptographic purposes [6,18,22].

However, the concrete proposal to employ quantum computing advancements specifically for securing advanced networking infrastructures gained prominence in the

L. Barolli (Ed.): AINA 2024, LNDECT 204, pp. 354–363, 2024.
https://doi.org/10.1007/978-3-031-57942-4_34

late 20th century. Notably, the seminal work by Peter Shor in 1994 demonstrated the potential of quantum computers to efficiently solve integer factorization and discrete logarithm problems. Shor's algorithm posed a significant threat to widely-used cryptographic schemes, including RSA and ECC, by leveraging quantum principles to solve these problems exponentially faster than classical computers [5, 11, 12].

The pivotal suggestion to employ quantum computing advancements for bolstering the security of advanced networking infrastructures came as a response to the impending vulnerabilities posed by Shor's algorithm [21]. It became evident that the advent of scalable quantum computers could render traditional encryption protocols obsolete, necessitating the exploration and development of quantum-resistant cryptographic algorithms.

Subsequently, researchers, cryptographers, and computer scientists worldwide embarked on rigorous investigations into quantum-resistant cryptographic techniques and quantum key distribution protocols. Their endeavors aimed to fortify network security against the potential threat posed by quantum computing's computational capabilities. This intersection of quantum computing and network security paved the way for a new era of cryptographic mechanisms and communication protocols designed to withstand quantum-based attacks.

The paper aims to explore the convergence of Quantum Key Distribution and Post-Quantum Cryptography strategies within networking systems, presenting innovative insights to enhance security measures by addressing cryptographic vulnerabilities. This integration fortifies current infrastructures and anticipates future threats in the quantum era, offering a forward-thinking approach to navigate evolving security landscapes.

2 Core Principles of Quantum Mechanics: The Physics and Mathematical Foundations

The foundational principles of quantum mechanics provide the basis for understanding the intriguing behaviors exhibited by quantum systems. These include superposition, entanglement, and uncertainty [7, 8, 14], as detailed in the subsequent sections.

2.1 Superposition and Quantum States

In quantum mechanics, a system's state $|\psi\rangle$ is described by a complex-valued vector in a Hilbert space $\mathscr{H}$. The superposition principle allows for the representation of this state as a linear combination of basis states:

$$|\psi\rangle = \sum_i c_i |\psi_i\rangle, \tag{1}$$

where c_i are complex probability amplitudes and $|\psi_i\rangle$ are the basis states forming an orthonormal set. The coefficients c_i satisfy normalization conditions $\sum_i |c_i|^2 = 1$ to ensure probabilistic interpretation.

The evolution of a quantum state is governed by the Schrödinger equation:

$$i\hbar \frac{d}{dt}|\psi\rangle = \hat{H}|\psi\rangle, \tag{2}$$

where $\hat{H}$ is the Hamiltonian operator representing the system's energy, and $\hbar$ is the reduced Planck constant.

2.2 Entanglement and Quantum Correlation

For a composite system with subsystems A and B described by the joint state $|\psi\rangle_{AB}$, entanglement arises when the system cannot be written as a product state $|\psi\rangle_A \otimes |\psi\rangle_B$. Mathematically, this non-factorizability demonstrates entanglement's non-separability.

Quantum entanglement is often quantified using measures like the von Neumann entropy

$$S(\rho_A) = -\mathrm{Tr}(\rho_A \log \rho_A), \tag{3}$$

where ρ_A is the reduced density matrix obtained by tracing out subsystem B. For bipartite systems, entanglement measures reveal the extent of correlations that transcend classical correlations.

2.3 Quantum Uncertainty

Heisenberg's uncertainty principle is expressed mathematically as:

$$\Delta A \cdot \Delta B \geq \frac{1}{2}|\langle [A,B] \rangle|, \tag{4}$$

where ΔA and ΔB are the standard deviations of measurements of non-commuting observables A and B, and $\langle [A,B] \rangle$ is the expectation value of their commutator $[A,B] = AB - BA$.

For example, the position x and momentum p operators do not commute: $[x,p] = i\hbar$. This non-commutativity leads to the uncertainty relation

$$\Delta x \cdot \Delta p \geq \frac{\hbar}{2}, \tag{5}$$

highlighting the inherent trade-off between the precision of position and momentum measurements.

3 Security Innovations Through Quantum Cryptography

The advent of quantum computing has posed substantial challenges to classical cryptographic methods, particularly due to the capabilities of Shor's Algorithm, which can efficiently solve problems integral to classical encryption. Quantum computers' exponential computational advantage, notably in factoring large numbers and solving discrete logarithm problems, undermines the security foundations of classical encryption, endangering data integrity in networked communications. Consequently, the development of Post-Quantum Cryptography, encompassing strategies like lattice-based, hash-based, multivariate, and code-based cryptography, aims to fortify encryption against potential quantum threats. A pivotal aspect in securing communications amidst quantum advancements is Quantum Key Distribution, leveraging quantum principles to establish secure cryptographic key exchange, forming a critical component in safeguarding subsequent communications.

In this section, we will delve into PQC and QKD, two critical domains reshaping security paradigms amidst the emergence of quantum computing.

3.1 Vulnerabilities of Classical Cryptography and Post-Quantum Cryptography

The vulnerabilities inherent in classical cryptography have assumed a critical dimension with the advent of quantum computing. Classical cryptographic methods, long deemed secure against traditional computing capabilities, now face significant threats due to the unprecedented computational power exhibited by quantum algorithms [2].

Central to this shift is Shor's Algorithm, a quantum algorithm that fundamentally challenges the security pillars of classical encryption methods. These classical algorithms, like RSA and ECC, rely on the complexity of factoring large numbers or solving discrete logarithm problems for ensuring data security. However, Shor's Algorithm has the unique ability to efficiently perform these computations on quantum computers, posing a substantial threat to the presumed security of these classical cryptographic primitives.

The quantum advantage lies in the potential exponential speedup of quantum computers compared to classical counterparts when solving specific problems [13]. This notable advantage undermines the assumed security of classical encryption methods, opening doors to efficient brute-force attacks that compromise the confidentiality and integrity of data transmitted over networks.

Consequently, there's an increasing necessity to reassess cryptographic protocols and algorithms in the wake of quantum computing's advancements. This urgent need arises from the looming risk of classical encryption methods becoming vulnerable to quantum-enabled attacks.

In response to this paradigm shift, there's a growing focus on developing and implementing quantum-resistant cryptographic algorithms, collectively referred to as Post-Quantum Cryptography. PQC encompasses diverse approaches such as lattice-based, hash-based, and multivariate cryptography, aiming to establish encryption schemes resilient against quantum threats.

3.1.1 Diverse Approaches in Post-Quantum Cryptography

1. Lattice-Based Cryptography

Lattice-based cryptography relies on the computational complexity of lattice problems [10], such as the Shortest Vector Problem (SVP) and Learning with Errors (LWE). These problems serve as the foundation for constructing cryptographic primitives like encryption, digital signatures, and key exchange protocols. Notable lattice-based schemes include NTRUEncrypt, NTRUSign, and Ring-LWE-based systems, offering promising alternatives resilient to quantum attacks.

2. Hash-Based Signatures

Hash-based signature schemes leverage cryptographic hash functions to create digital signatures resistant to quantum attacks. The Merkle signature scheme, including the Lamport and Winternitz one-time signature schemes, provides unforgettable signatures based on the security of hash functions. However, their reliance on large signature sizes and key management complexities present practical challenges for deployment in networked environments [15].

3. Multivariate Cryptography

Multivariate cryptography constructs cryptographic primitives based on solving systems of multivariate polynomial equations. These schemes, including the Rainbow and Unbalanced Oil and Vinegar (UOV) schemes, exhibit resistance against quantum algorithms due to their reliance on mathematical problems not easily solvable by quantum computers. However, their efficiency and scalability remain subjects of ongoing research and optimization efforts [23].

4. Code-Based Cryptography

Code-based cryptographic systems rely on error-correcting codes to create cryptographic primitives resilient to quantum attacks. Schemes like McEliece and Niederreiter encryption offer security based on the difficulty of decoding certain linear codes. These systems have shown promising resistance against quantum algorithms but may require larger key sizes [4].

5. Other Approaches

Additional PQC approaches include lattice-based fully homomorphic encryption, isogeny-based cryptography (e.g., SIDH), and quantum-resistant digital signature schemes like hash-based signatures or those based on error-correcting codes [4].

3.2 Quantum Key Distribution for Secure Communication

In modern networking, the exponential surge in data transmission across intricate infrastructures accentuates the critical necessity for robust security measures. Traditional cryptographic methods, while effective, face vulnerabilities from emerging threats posed by quantum computing capabilities. Quantum Key Distribution (QKD) emerges as a groundbreaking solution rooted in the principles of quantum mechanics to fortify the security of advanced networking infrastructures [1, 16, 26].

3.2.1 Mechanics of Quantum Key Distribution for Secure Communication

QKD harnesses principles from quantum mechanics, describing particle behavior at the quantum level, where properties such as photon polarizations exist in quantized states. Quantum entities like photons facilitate secure cryptographic key exchange between entities, typically referred to as Alice and Bob. Qubits, the information carriers, differentiate themselves from classical bits due to their ability to exist in multiple states simultaneously, known as superposition. Photon polarization, often used in QKD, serves as qubit states, where, for instance, horizontal polarization might denote '0', and vertical polarization '1'.

Qubits, upon encoding, traverse a communication channel from Alice to Bob, susceptible to interception or measurement by an eavesdropper, posing potential security threats. The primary objective of QKD is to establish a shared secret key between Alice and Bob while actively detecting potential eavesdropping attempts.

Upon reception, Bob randomly measures the qubits' quantum states using chosen bases, while Alice records the encoding bases without disclosing specific qubit values.

Interference by an eavesdropper, such as Eve, disrupts the qubits' states, causing errors or inconsistencies detected through discussions between Alice and Bob concerning the encoding and measurement bases used.

Public discussions aid in identifying inconsistencies or errors arising from eavesdropping attempts, as any tampering during transmission alters the qubits' states, hinting at potential security breaches. By discarding inconsistent qubits and retaining matching ones, Alice and Bob derive a subset constituting their shared secret key. This key is instrumental in conventional cryptographic algorithms for secure encryption and decryption of subsequent communications, ensuring confidentiality and integrity.

QKD amalgamates principles from quantum mechanics into cryptography, providing a framework for secure communication. It not only generates shared secret keys but actively monitors and detects potential eavesdropping attempts, ensuring robust security for sensitive data exchanges between entities.

3.2.2 Security in Quantum Key Distribution

QKD rests on foundational principles in quantum mechanics, leveraging quantum indeterminacy and the no-cloning theorem to fortify communication security.

The Heisenberg Uncertainty Principle, a cornerstone of quantum mechanics, stipulates the inherent uncertainty when measuring certain paired properties of a quantum system. In QKD, attempts by an eavesdropper (Eve) to measure qubits in transit unavoidably disturb their quantum states due to measurement-induced uncertainties. This disturbance triggers detectable interference, signaling potential tampering to the communicating parties.

Concurrently, the no-cloning theorem asserts the impossibility of perfectly replicating an unknown quantum state. In the context of QKD, this implies that unauthorized cloning of transmitted qubits without altering their states is unfeasible. Any endeavor by Eve to replicate quantum information inevitably disrupts the quantum state, rendering it detectable by the intended recipients, typically represented as Alice and Bob.

Security assurances in QKD arise from these quantum principles, constraining eavesdropping attempts due to inherent quantum mechanics constraints. Efforts to eavesdrop or clone qubits disrupt the transmitted qubits' quantum states, leading to observable disruptions.

To bolster security, QKD protocols, such as BB84 [20], incorporate detection mechanisms. Through these protocols, legitimate parties validate quantum states to ensure coherence between transmitted and received qubits. Any discrepancies in the quantum state between transmission and reception phases serve as indicators of potential unauthorized interventions, facilitating the rejection of compromised data.

The security underpinnings of QKD hinge on the inherent uncertainty and non-clonability of quantum states, governed by the Heisenberg Uncertainty Principle and the no-cloning theorem. These principles deter covert eavesdropping attempts by ensuring that any endeavor to intercept or clone quantum information inevitably disrupts the quantum states, enabling legitimate parties to detect and discard compromised data.

3.3 Implementation of PQC and QKD in Networking Infrastructure

Securing networking infrastructure against potential quantum threats necessitates a systematic and comprehensive approach towards the implementation of PQC and QKD. This endeavor comprises several strategic phases:

1. **Assessment of Networking Infrastructure:** Initiate the process with a meticulous assessment of the existing networking architecture. Conduct an in-depth analysis to identify inherent vulnerabilities susceptible to quantum attacks. It is imperative to categorize these vulnerabilities and tailor specific security requirements, aligning them with data sensitivity and regulatory compliance standards.
2. **Integration of PQC:** Proceed by judiciously selecting PQC algorithms, such as lattice-based, hash-based, multivariate, or code-based algorithms, considering their compatibility and resilience against quantum threats. Formulate a meticulous integration plan, delineating a seamless integration of the chosen PQC algorithms into the existing cryptographic protocols. This integration must maintain operational efficiency while enhancing security.
3. **Adoption of QKD:** Identify critical communication links within the network infrastructure where the integration of QKD can significantly fortify security. Strategically deploy QKD devices at these pivotal points to facilitate secure quantum-based key distribution among communicating entities. This step ensures the establishment of robust cryptographic keys, resistant to potential quantum attacks.
4. **Enhancement of Network Security Measures:** Augment the implemented PQC and QKD solutions with a comprehensive suite of complementary security measures. This entails the integration of sophisticated intrusion detection systems, robust firewalls, and stringent access controls. Additionally, implement continuous monitoring systems to vigilantly track network activities and promptly identify any anomalies or potential quantum threats.

The integration of PQC and QKD into networking infrastructure demands meticulous planning, precise execution, and ongoing evaluation to mitigate the evolving landscape of quantum threats. This holistic approach fortifies the resilience of network security against the formidable challenges posed by quantum computing advancements [16,17,28].

4 Threats and Opportunities in Quantum Era Network Security

The evolution of technology has ushered in the quantum era, significantly altering the landscape of network security. This transformation brings forth both unprecedented threats and opportunities, necessitating a reevaluation of conventional security paradigms [3,19,24,25,27].

4.1 Threat Landscape

The advent of quantum computing introduces critical vulnerabilities to existing encryption methods, particularly the widely-utilized RSA and ECC. Quantum algorithms like

Shor's algorithm significantly elevate the risk of cryptographic breaches, undermining the confidentiality of sensitive data transmitted across networks.

Moreover, the transitional phase from classical to post-quantum cryptographic systems poses inherent risks. During this migration, systems remain susceptible to quantum attacks, potentially compromising network security until newer, quantum-resistant algorithms are fully integrated.

4.2 Leveraging Opportunities

Amidst these threats, innovative advancements in post-quantum cryptography offer promising avenues for bolstering network security. New cryptographic methodologies such as lattice-based cryptography, hash-based cryptography, and code-based cryptography emerge as robust solutions within the quantum computing paradigm. These methodologies fortify data protection, countering the looming quantum threats effectively.

Furthermore, the integration of Quantum Key Distribution into network infrastructures presents an opportunity to establish highly secure communication channels. Leveraging quantum principles, QKD promises substantial enhancements in data confidentiality and integrity, reinforcing the overall security posture.

4.3 Addressing Challenges and Forging Ahead

The implementation of PQC algorithms and QKD encounters significant technical complexities. Challenges related to scalability, compatibility issues across diverse network environments, and the standardization of protocols are critical hurdles that need addressing.

Despite the potential solutions that post-quantum cryptographic schemes offer, their widespread adoption faces multiple challenges. These include the computational overhead of many algorithms compared to their classical counterparts, larger key sizes, increased storage requirements, and the necessity to ensure interoperability and compatibility with existing systems. The transition period from classical to post-quantum cryptographic algorithms requires meticulous planning and integration, posing practical challenges and complexity.

Amidst these challenges, collaborative interdisciplinary research efforts and increased investments in R&D stand as pivotal strategies. Standardized frameworks for quantum-safe communication must be established to pave the way for quantum-resilient network architectures.

5 Conclusion

The convergence of quantum computing with advanced networking introduces a new era of unprecedented capabilities, challenging traditional networking paradigms and unlocking opportunities for innovation. Embracing this transformative potential requires concerted efforts in research, development, standardization, and practical implementations to usher in the era of quantum-powered networking infrastructures.

Navigating the imminent threats posed by quantum computing necessitates a proactive stance in fortifying network security. Embracing post-quantum cryptography, integrating QKD, and addressing technical challenges through collaborative efforts are pivotal in steering network architectures towards resilience in the face of quantum threats. Crafting secure networks in the quantum era demands a cohesive and forward-looking strategy encompassing innovation, collaboration, and robust standards.

References

1. Nurhadi, A.I., Syambas, N.R.: Quantum key distribution (QKD) protocols: a survey. In: 2018 4th International Conference on Wireless and Telematics (ICWT), pp. 1–5. IEEE (2018)
2. Huang, A., Barz, S., Andersson, E., Makarov, V.: Implementation vulnerabilities in general quantum cryptography. New J. Phys. **20**(10), 103016 (2018)
3. Vaishnavi, A., Pillai, S.: Cybersecurity in the quantum era-a study of perceived risks in conventional cryptography and discussion on post quantum methods. J. Phys. Conf. Ser. **1964**(4), 042002 (2021)
4. Balamurugan, C., Singh, K., Ganesan, G., Rajarajan, M.: Post-quantum and code-based cryptography-some prospective research directions. Cryptography **5**(4), 38 (2021)
5. Ugwuishiwu, C.H., Orji, U.E., Ugwu, C.I., Asogwa, C.N.: An overview of quantum cryptography and Shor's algorithm. Int. J. Adv. Trends. Comput. Sci. Eng. **9**(5), 1–9 (2020)
6. Van Assche, G.: Quantum Cryptography and Secret-Key Distillation. Cambridge University Press, Cambridge (2006)
7. Naber, G.: Foundations of Quantum Mechanics (2016)
8. Mackey, G.W.: Mathematical Foundations of Quantum Mechanics. Courier Corporation (2013)
9. Salloum, H., et al.: Integration of machine learning with quantum annealing. In: International Conference on Advanced Information Networking and Applications. Springer (2024)
10. Nejatollahi, H., Dutt, N., Ray, S., Regazzoni, F., Banerjee, I., Cammarota, R.: Post-quantum lattice-based cryptography implementations: a survey. ACM Comput. Surv. (CSUR) **51**(6), 1–41 (2019)
11. Charjan, S., Kulkarni, D.H.: Quantum key distribution by exploitation public key cryptography (ECC) in resource constrained devices. Int. J. **5**, 1–8 (2015)
12. Monz, T., et al.: Realization of a scalable Shor algorithm. Science **351**(6277), 1068–1070 (2016)
13. Rønnow, T.F., et al.: Defining and detecting quantum speedup. Science **345**(6195), 420–424 (2014)
14. McMahon, D.: Quantum Computing Explained. John Wiley & Sons, New York (2007)
15. Mozaffari-Kermani, M., Azarderakhsh, R.: Reliable hash trees for post-quantum stateless cryptographic hash-based signatures. In: 2015 IEEE International Symposium on Defect and Fault Tolerance in VLSI and Nanotechnology Systems (DFTS), pp. 103–108. IEEE (2015)
16. Mehic, M., et al.: Quantum key distribution: a networking perspective. ACM Comput. Surv. (CSUR) **53**(5), 1–41 (2020)
17. Stanley, M., Gui, Y., Unnikrishnan, D., Hall, S.R.G., Fatadin, I.: Recent progress in quantum key distribution network deployments and standards. J. Phys. Conf. Ser. **2416**(1), 012001 (2022)
18. Torres, N.N., Garcia, J.C.S., Guancha, E.A.V.: Systems security affectation with the implementation of quantum computing. Int. J. Adv. Comput. Sci. App. **12**(4), 1–11 (2021)
19. Althobaiti, O.S., Dohler, M.: Cybersecurity challenges associated with the Internet of Things in a post-quantum world. IEEE Access. **8**, 157356–157381 (2020)

20. Shor, P.W., Preskill, J.: Simple proof of security of the BB84 quantum key distribution protocol. Phys. Rev. Lett. **85**(2), 441 (2000)
21. Shor, P.W.: Polynomial-time algorithms for prime factorization and discrete logarithms on a quantum computer. SIAM Rev. **41**(2), 303–332 (1999)
22. Shor, P.W., Farhi, E., Gosset, D., Hassidim, A., Lutomirski, A.: Quantum Money, vol. 19. MIT, Cambridge (2012)
23. Kuang, R., Perepechaenko, M., Barbeau, M.: A new post-quantum multivariate polynomial public key encapsulation algorithm. Quant. Inf. Process. **21**(10), 360 (2022)
24. Eddin, S., et al.: Quantum microservices: transforming software architecture with quantum computing. In: International Conference on Advanced Information Networking and Applications. Springer (2024)
25. Bajrić, S.: Enabling secure and trustworthy quantum networks: current state-of-the-art, key challenges, and potential solutions. IEEE Access **11**, 128801–128809 (2023)
26. Scarani, V., et al.: The security of practical quantum key distribution. Rev. Mod. Phys. **81**(3), 1301 (2009)
27. Usenko, V.C., Filip, R.: Trusted noise in continuous-variable quantum key distribution: a threat and a defense. Entropy **18**(1), 20 (2016)
28. Yang, Y.-H., et al.: All optical metropolitan quantum key distribution network with post-quantum cryptography authentication. Opt. Express. **29**(16), 25859–25867 (2021)

In-Network Monitoring Strategies for HPC Cloud

Masoud Hemmatpour[1(✉)], Tore Heide Larsen[2], Nikshubha Kumar[3], and Ernst Gunnar Gran[4]

[1] Arctic University of Norway (UiT), Tromsø, Norway
mhe222@uit.no

[2] Simula Research Laboratory (SRL), Oslo, Norway
torel@simula.no

[3] Norwegian Business School (BI), Oslo, Norway
nikshubha.kumar@bi.no

[4] Norwegian University of Science and Technology (NTNU), Trondheim, Norway
ernst.g.gran@ntnu.no

Abstract. The optimized network architectures and interconnect technologies employed in high-performance cloud computing environments introduce challenges when it comes to developing monitoring solutions that effectively capture relevant network metrics. Moreover, network monitoring often involves capturing and analyzing a large volume of network traffic data. This process can introduce additional overhead and consume system resources, potentially impacting the overall performance of HPC applications. Balancing the need for monitoring with minimal disruption to application performance is a key challenge. In this paper, we study different strategies to enable a low-overhead monitoring system utilizing emerging programmable network devices.

1 Introduction

Cloud service providers strategically design High Performance Computing (HPC) systems with a focus on fault tolerance to guarantee uninterrupted application execution in the face of potential hardware or software failures. Undoubtedly, monitoring constitutes a fundamental element in establishing fault tolerance within HPC systems, affording timely insights into the health, performance, and behaviors of diverse components within a HPC cloud infrastructure. The network stands out as a vital constituent within HPC applications. This prominence arises from the prevalent utilization of parallel processing and distributed computing in HPC applications, entailing task division across multiple processors or nodes.

By continuously monitoring network traffic, bandwidth usage, and device performance, administrators can identify areas of congestion, predict future growth requirements, and make informed decisions regarding network upgrades, expansions, or optimization strategies. Furthermore, communication-intensive HPC applications can

The research project was initiated at SRL and subsequently continued at UiT.

L. Barolli (Ed.): AINA 2024, LNDECT 204, pp. 364–373, 2024.
https://doi.org/10.1007/978-3-031-57942-4_35

exploit network monitoring information to improve their performance by efficiently utilizing the network resources [11].

In the early stages of network monitoring, only basic tools like packet sniffers were accessible. These tools enabled network administrators to capture and analyze network traffic, aiding in the identification of connectivity problems and abnormal behavior. Over time, specialized monitoring protocols, such as the Simple Network Management Protocol (SNMP) introduced in the 1990s [5], emerged, specifically tailored to cater to the needs of network monitoring. SNMP enabled the monitoring of network devices and provided a standardized way to collect and manage information about their performance and health. With the advent of Software Defined Networking (SDN) network monitoring has significantly increased due to the easy network observation with a centralized control plane. One of the key benefits of SDN is the centralized control and management of network devices through a software-based controller [9]. This centralized control enables network administrators to have a holistic view of the entire network.

In spite of significant advancements in network monitoring, the task of efficiently monitoring HPC networks remains a challenging endeavor, primarily due to the unique characteristic of HPC network, which necessitates specialized monitoring tools tailored for their requirements. One of the primary factors contributing to this challenge arises from a specific characteristic exhibited by HPC protocols that bypass the processing unit (CPU) and operating system (OS) of a remote system when accessing data, rendering conventional monitoring approaches ineffective in capturing and analyzing network traffic within the HPC context at an end node. Moreover, because of the high cost associated with HPC devices allocating a dedicated processing unit solely for network monitoring is an inefficient and undesirable solution. These circumstances have compelled us to investigate low-overhead approaches for monitoring HPC networks through the utilization of emerging programmable network devices, effectively leveraging their computing power to analyze network traffic.

In Sect. 2, we present monitoring protocols and existing HPC monitoring tools. In Sect. 3, we propose and detail our in-network solutions for monitoring HPC networks. Section 4 evaluates the in-network monitoring solution. Finally, in Sect. 5 we conclude and indicate directions for future work.

2 Network Monitoring

The first step in network monitoring is data measurement. Data measurement includes obtaining and preserving data which is then subjected to various analysis for investigation. Measured data can also be visualized, e.g. to provide network operators with easy to interpret network status overviews. Certain control and monitoring protocols, such as the Internet Control Message Protocol (ICMP), are characterized by their simplicity and primarily serve the purpose of reporting communication errors between nodes. Nonetheless, there exist more sophisticated monitoring protocols specifically designed to cater to advanced monitoring requirements. Below, we discuss some of the commonly utilized monitoring protocols:

Simple Network Management Protocol (SNMP) is one of the most used protocols to monitor network status [5]. SNMP for instance can be used to request per-interface port-counters and overall node statistics from a switch. It is implemented in most network devices. Monitoring using SNMP is typically achieved by regularly polling the switch, though switch efficiency may degrade with frequent polling due to CPU overhead. Although vendors are free to implement their own SNMP counters, most switches are limited to counters that aggregate traffic for the whole switch and each of its interfaces, disabling insight into flow level statistics necessary for fine-grained traffic engineering. Therefore, SNMP is not suitable for flow based monitoring.

sFlow is an industry standard network traffic monitoring solution with sampling technology provided by a wide range of network equipment and software application vendors [18]. It is a scalable technique for measuring network traffic, collecting, storing, and analyzing traffic data. One of the advantages of sFlow is scalability of monitoring high speed links without adding significant network load.

Vendor supplied Software Development Kits (SDKs) provide APIs to implement desired functionality in a switch without compromising packet rate performance [14]. In particular, they allow to monitor several objects or performance counters of a network infrastructure. However, these SDKs are vendor specific and limited to the APIs provide by a vendor, thus they do not represent a standard monitoring approach.

2.1 HPC Monitoring Tools

Communications over traditional LAN technologies and the TCP/IP protocol stack are unfit for HPC networks because of inferior performance characteristics and significant CPU and memory usage. Consequently, a new category of network fabrics emerged, using a technology known as Remote Direct Memory Access (RDMA) [6]. RDMA is a high-performance networking technology that allows direct memory access between computers in a network, bypassing traditional network stack layers and reducing latency. In HPC environments, where data-intensive applications require fast and efficient data transfers, RDMA plays a crucial role. This makes the monitoring of RDMA protocols more fundamental to ensure optimal performance, to troubleshoot issues, and to maintain the overall health of the network.

Monitoring RDMA networks requires specialized tools and techniques designed to capture and analyze RDMA-specific traffic and performance metrics. These tools provide insights into key parameters such as bandwidth utilization, latency, and congestion [19]. In the following, some of the known tools are discussed:

ibdiagnet [13] is an advanced diagnostic tool designed to analyze and troubleshoot InfiniBand networks. The primary purpose of ibdiagnet is to perform in-depth tests and verification on various components of an InfiniBand network. It can diagnose potential problems related to link connectivity, network congestion, hardware failures, and other issues that may affect the overall performance and reliability of the fabric. ibdiagnet operates by utilizing the Subnet Manager (SM) component of the InfiniBand fabric, which is responsible for managing and configuring the network. It communicates with the SM to obtain information about the network's topology, nodes, and configuration. Based on this data, ibdiagnet performs a series of tests and analyses to evaluate the network's integrity and identify any abnormalities or potential bottlenecks. A significant drawback of utilizing ibdiagnet for network monitoring is the necessity to parse its output in order to extract the desired information. This additional step of parsing introduces complexity and may require specialized tools to effectively extract the relevant data for analysis and interpretation.

rdma_statistic [10] is part of the iproute2 package that provides a collection of utilities for controlling networking and traffic control in Linux. RDMA statistic shows information about counter configuration depending on the option and the command selected. However it is restricted to show information about the RDMA counters of the host machine. In effect, it cannot be extended to obtain information about RDMA communication happening in the network.

Performance Co-Pilot (PcP), developed by Silicon Graphics, ported to Linux from 1999 and eventually open-sourced in the mid-2000s, serves as a framework for supporting performance monitoring and management [16]. It is actively employed by companies such as Netflix and Red Hat to gather high-resolution performance metrics. It is available for all major Linux distributions. Within PCP, the Performance Metrics Collector Daemon (PMCD) acts as an agent running on hosts slated for monitoring. PMCD incorporates various Performance Metrics Domain Agents (PMDA), dynamically loaded libraries with a specified API. Each PMDA collects metrics from a performance metric domain, such as a kernel or a database server. PcP connects to PMCD, requests specific metrics, and PMCD directs the request to the relevant PMDA, subsequently returning the response.

3 HPC In-Network Monitoring Discussion

Effective communication plays a vital role in HPC applications, as a large problem is divided into smaller parts and distributed across multiple machines within an HPC cluster. Consequently, network utilization and congestion level emerge as critical metrics in HPC networks [1], attracting research attention to monitor and analyze these parameters [2,7]. In this section, we discuss the available approaches to monitor theses metrics through programmable network devices since there are few studies in this field [4].

Our research initially delved into the SDK monitoring approach, which entails utilizing SDKs provided by vendors for specific Smart Switches and Smart NICs, such as those offered by NVIDIA [14]. These SDKs grant access to hardware counters, enabling the monitoring and analysis of network performance. However, during our investigation, we encountered several limitations and challenges associated with this approach.

First and foremost, relying on an SDK-based monitoring approach is inherently vendor-specific, rendering it inapplicable to network devices from different vendors. This lack of cross-compatibility restricts the generalizability of the monitoring approach. Additionally, certain SDKs, including the NVIDIA Ethernet SDK, were not publicly accessible and necessitated specific circumstances for vendor-granted access. Hence, employing an SDK-based strategy is not the most optimal choice for developing an open and freely accessible cross-platform monitoring system, considering these constraints.

Furthermore, we found that some SDKs were still under development and did not offer complete functionality, despite their documentation promising otherwise. This lack of comprehensive functionality impeded our reliance on these SDKs for thorough monitoring purposes. Acknowledging these limitations, our research shifted its focus towards alternative monitoring approaches that provide more comprehensive and vendor-agnostic solutions for network monitoring. Therefore, we prioritized standard protocols that furnish a structured framework for collecting, analyzing, and visualizing network data. Furthermore, we focus on the RDMA RoCE protocol rather than InfiniBand since RoCE leverages standard Ethernet infrastructures, which is widely deployed in cloud infrastructure and familiar to network administrators.

3.1 Traffic Monitoring

Firstly, we examined SNMP in our research. However, our investigation revealed its limitations in monitoring RDMA traffic. Although certain Smart NICs, like BlueField, offer SNMP for monitoring specific counters [12], our evaluation indicated incomplete functionality of SNMP at the time of writing this paper. As a result, we explored alternative standard protocols to address these constraints.

sFlow as a standard monitoring protocol provides real-time visibility and insights into network traffic and performance [18]. It uses packet sampling technology which most of the switches support. It's a sampling mechanism which recently supports RDMA RoCE traffic monitoring which fulfills all the necessary conditions for a HPC network traffic monitoring solution. sFlow has two components: a sFlow agent, which is a software process, needs to be enabled within the switch, and a collector runs on a host that collects sFlow datagrams. The packet sampling is typically performed by the switching/routing ASICs, providing wire-speed performance. Packet sampling uses randomness in the sampling process to prevent synchronization with any periodic patterns in the traffic. On average, 1 in every N packets is captured and analyzed. While this type of packet sampling does not provide a 100% accurate result, it does provide a result with quantifiable accuracy. The sFlow packets are encapsulated and sent in UDP over IP (port 6343 by default). The sFlow collector receives sFlow data, and generates either a simple-to-parse tagged-ASCII output, or binary output in tcpdump format. To display a real-time trend chart of network traffic, sFlow provides an open source tools called sFlow-RT/sFlowTrend.

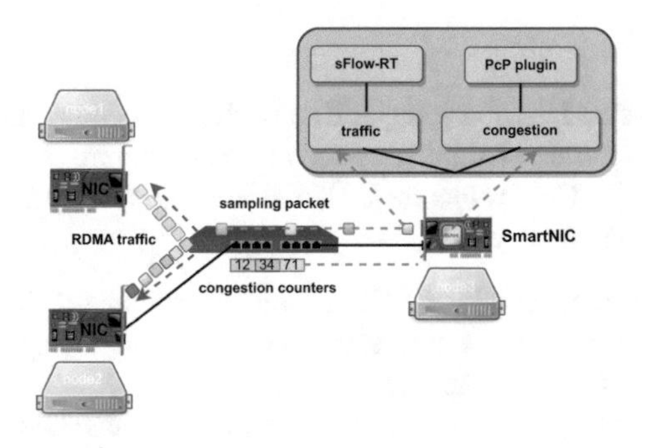

Fig. 1. Testbed consists of three main components to collect and visualize the observed RDMA RoCE traffic.

3.2 Congestion Monitoring

Detecting congestion in a network is critical for maintaining a stable and reliable network environment. There are different techniques to manage congestion when the congestion is detected such as the use of Quality of Service (QoS), which prioritize certain types of traffic, or injection throttling. Congestion counters allow us to measure and monitor the level of congestion in a network. Congestion counters in RDMA protocols may vary depending on the implementation and hardware. In order to monitor congestion counters in the switch, we exploit JSON APIs to perform commands remotely from a node in the HPC cluster.

Figure 1 illustrates our in-network monitoring strategy. We implement our monitoring and visualization in a SmartNIC. Network traffic is observed through the sFlow protocol, and congestion counters are observed through calling JSON API. We visualize the sFlow datagram through sFlow-RT. In order to visualize the congestion counters we implemented a plugin for PcP to support RoCE protocol. So, firstly we implemented new metrics and created PMCD to extract the congestion counters from the switch and visualize them in PcP. Our implementation is publicly available at https://github.com/niks16/iNet.

4 Evaluation

Relocating a monitoring system to network devices poses a significant challenge in terms of allocating resources effectively to optimize the use of resources. To assess the monitoring strategy and its associated overhead, we devised various scenarios for evaluation.

We established a test environment compromising three machines, each equipped with a BlueField-2 operating on Ubuntu 18.04.6. These machines are interconnected by an NVIDIA SN2100 switch through 100Gbps RoCE links, as shown in Fig. 1.

4.1 Monitoring

Open Fabrics Enterprise Distribution (OFED) performance test, also called perftest, is a collection of tests intended for performance micro-benchmark of RDMA operations over infiniBand and RoCE protocols [17]. We evaluated the followings tests in our environment between two nodes:

- **ib_write_bw** generates *RDMA WRITE* traffic and calculates the bandwidth. *RDMA WRITE* operation writes to the memory of a remote machine bypassing the CPU and operating system of this remote node.
- **ib_read_bw** generates *RDMA READ* traffic and calculates the bandwidth between a pair of machines. *RDMA READ* reads from the memory of a remote machine, again bypassing the CPU and operating system of the remote node.

We compared the sFlow bandwidth with the OFED performance tests. As can be seen in Table 1, sFlow reports very similar result to perftest.

Table 1. sFlow and perftest comparison.

RDMA operation	sFlow (Gbps)	perftest (Gbps)
ib_write_bw	78.0	79.79
ib_read_bw	80.0	80.85

Furthermore, we conducted testing of our in-network monitoring strategy utilizing the HPC message passing protocol, which leverages RDMA in its underlying layer. Message Passing Interface (MPI) is a popular parallel programming model for scientific applications. The Open MPI is an open-source implementation of the MPI that is developed and maintained by a consortium of academic, research, and industry partners [15]. The Open MPI supports the RDMA UCX library. UCX is an open-source optimized communication library allowing to choose the communication protocols by the runtime characteristics of MPI [21]. Open MPI itself supports different transport protocols to handle communication with different messages sizes [20]. For example, large messages are sent over *Rendezvous protocol*, where the sender first negotiates the buffer availability at the receiver side before the message is actually transferred. In our study, we conducted experiments utilizing the OSU micro-benchmarks [3], varying the *UCX_RNDV_THRESH* parameter and closely monitoring network traffic. Our observations indicate that UCX adapts its RDMA operations based on the threshold setting. For instance, when setting the threshold to a higher value (e.g., 20M), all messages are transmitted using *RDMA SEND*. *RDMA SEND* represents a two-sided communication approach, necessitating acknowledgment from the receiver before data transmission. This analysis are important as UCX exhibits the ability to dynamically adapt its transport protocol and network interface in response to varying circumstances and not only message size. The utilization of lightweight in-network monitoring facilitates a comprehensive understanding of the underlying MPI communication without imposing any burden on the processes running on the host. Further analysis of *UCX_RNDV_THRESH* can be found on our GitHub repository.

4.2 Monitoring Overhead

We evaluated the overhead of our monitoring strategy from augmented network traffic, CPU and memory consumption:

- The network traffic of monitoring depends on the sFlow parameters settings such as sampling rate and sFlow counter-poll-interval. We generated *RDMA READ* traffic through perftest while systematically varying the mentioned parameters and recorded the obtained results. As Fig. 2 (a) and (b) show by increasing the counter-poll-interval and sampling rate the monitoring traffic caused by sFlow decreases.
- The processing overhead of the collector on SmartNIC is impressively low, with less than 5% of CPU usage. As for the switch, packet sampling for sFlow is primarily handled by the high-speed switching/routing ASICs, ensuring efficient and swift processing.
- The memory consumed by collector on the SmartNIC was found to be less than 3% on average, which is considerably low.
- When using PcP, the system overhead depends on various factors. These include the number of managed hosts, frequency of logging interval, duration of historical data retention, number of performance metrics measured and application performance [8].

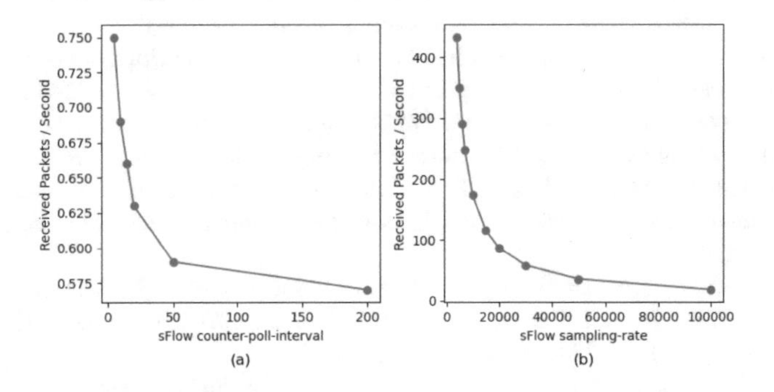

Fig. 2. Network load when using sFlow.

Following the analysis of the monitoring accuracy and its overhead, we have established that the implementation of an accurate lightweight in-network strategy for monitoring RoCE RDMA is feasible and achievable through the utilization of standard protocols. We efficiently utilized only one core of the ARM CPU, leaving a substantial amount of additional resources. Furthermore, the SmartNIC's connection through PCIe to the host enables us to store the observed data at high speeds. This combination of factors enhances our capabilities for in-depth data analysis and processing of observed network traffic.

5 Conclusion and Future Work

In this research, different strategies for monitoring RDMA RoCE protocol have been discussed and evaluated to implement an accurate and lightweight in-network monitoring in an HPC cloud setting. In this research, network traffic and congestion from a switch have been monitored. However, as future work we plan to evaluate monitoring of multiple network devices. Moreover, we plan to provide actions based on the network status through some machine learning methods. Additionally, we intend to extend our monitoring scope to include both the network and end nodes using SmartNICs.

Acknowledgment. The work was supported by European Commission under MISO project, (grant 101086541), EEA grants under the HAPADS project (grant NOR/ POLNOR/ HAPADS/ 0049/2019-00) and Research Council of Norway under eX3 project (grant 270053) and Sigma2 (grant NN9342K).

References

1. Bhatele, A., et al.: Identifying the culprits behind network congestion. In: 2015 IEEE International Parallel and Distributed Processing Symposium, pp. 113–122. IEEE (2015)
2. Cascajo, A., et al.: Monitoring infiniband networks to react efficiently to congestion. IEEE Micro **43**(2), 120–130 (2023)
3. Dhabaleswar, K., Panda, D.K.: Mvapich: MPI over infiniband, omni-path, ethernet/iWarp, roce, and slingshot, Network-Based Computing Laboratory (2023)
4. Ding, D., Savi, M., Pederzolli, F., Siracusa, D.: Design and development of network monitoring strategies in p4-enabled programmable switches. In: NOMS 2022-2022 IEEE/IFIP Network Operations and Management Symposium, pp. 1–6. IEEE (2022)
5. Network Working Group. https://www.rfc-editor.org/rfc/rfc1157. Accessed 12 Oct 2023
6. Hemmatpour, M., Montrucchio, B., Rebaudengo, M.: Communicating efficiently on cluster-based remote direct memory access (RDMA) over infiniband protocol. Appl. Sci. **8**(11), 2034 (2018)
7. Hintze, K., Graham, S., Dunlap, S., Sweeney, P.: Infiniband network monitoring: challenges and possibilities. In: ICCIP 2021. IAICT, vol. 636, pp. 187–208. Springer, Cham (2022). https://doi.org/10.1007/978-3-030-93511-5_9
8. Jin, E.: Common questions about performance co-pilot. https://www.redhat.com/en/blog/common-questions-about-performance-co-pilot. Accessed July 2023
9. Kreutz, D., Ramos, F.M.V., Verissimo, P.E., Rothenberg, C.E., Azodolmolky, S., Uhlig, S.: Software-defined networking: a comprehensive survey. Proc. IEEE **103**(1), 14–76 (2014)
10. Ostrovsky, N., Zhang, M., Alfasi, E.: https://www.man7.org/linux/man-pages/man8/rdma-statistic.8.html (2019)
11. Mishra, P., Agrawal, T., Malakar, P.: Communication-aware job scheduling using slurm. In: Workshop Proceedings of the 49th International Conference on Parallel Processing, pp. 1–10 (2020)
12. NVIDIA (2023). https://docs.nvidia.com/networking/display/bluefielddpuosv380/performance+monitoring+counters#heading-SNMPSubagent
13. NVIDIA. https://network.nvidia.com/products/adapter-software/infiniband-management-and-monitoring-tools/. Accessed July 2023
14. NVIDIA. Nvidia ethernet switch SDK. https://developer.nvidia.com/networking/ethernet-switch-sdk. Accessed July 2023

15. OpenMPI. https://docs.open-mpi.org/en/v5.0.x/index.html. Accessed July 2023
16. Performance Co-Pilot (PcP). https://pcp.io/. Accessed July 2023
17. Linux RDMA and InfiniBand. Open fabrics enterprise distribution (OFED) performance tests readme. https://github.com/linux-rdma/perftest. Accessed July 2023
18. sFlow. Traffic monitoring using sFlow (2003). https://sflow.org/sFlowOverview.pdf
19. Stefanov, K.S., Pawar, S., Ranjan, A., Wandhekar, S., Voevodin, V.V.: A review of supercomputer performance monitoring systems. Supercomput. Front. Innov. **8**(3), 62–81 (2021)
20. Sur, S., Jin, H.-W., Chai, L., Panda, D.K.: RDMA read based rendezvous protocol for MPI over infiniband: design alternatives and benefits. In: Proceedings of the Eleventh ACM SIGPLAN Symposium on Principles and Practice of Parallel Programming, pp. 32–39 (2006)
21. Unified Communication X. https://openucx.org/. Accessed July 2023

Optimizing Network Latency: Unveiling the Impact of Reflection Server Tuning

Jan Marius Evang[1,2](✉) and Thomas Dreibholz[1]

[1] Simula Metropolitan Centre for Digital Engineering, Pilestredet 52, 0167 Oslo, Norway
{marius,dreibh}@simula.no

[2] Oslo Metropolitan University, Postboks 4 St. Olavs plass, 0130 Oslo, Norway
https://www.simula.no/people/marius , https://www.simula.no/people/dreibh

Abstract. This study investigates the dynamics of network latency optimizations, with a focus on the role of reflection server tuning. In an era marked by the demand for precise and low-latency network measurements, our exploration unveils the interplay of diverse parameters in achieving optimal performance. Notably, the implementation of a tuned profile on Linux emerges as a standout strategy, showcasing significant rewards in network efficiency. We highlight the importance of early acceptance of latency-critical traffic in the firewall chain and emphasize the cumulative impact of various optimizations. These findings have practical implications for network administrators and system architects, providing valuable insights for the deployment of efficient and low-latency network infrastructures, essential in the landscape of emerging technologies such as 5G networks and edge computing solutions.

1 Introduction

Effective management of network latency is increasingly important, especially in our era of technologies such as 5G networks, cloud-based Radio Access Networks (RAN) and low-latency edge computing. HiPerConTracer [4,5] is a valuable tool for high-precision short- and long-term network latency measurements through the utilization of hardware timestamping.

Traditional latency measurements involve the transmission of Internet Control Message Protocol (ICMP) [3,10] Echo Request packets across network paths. Upon reception of an ICMP Echo Request, the receiver's network stack, i.e. usually the operating system kernel, replies with an ICMP Echo Reply (by default, unless deactivated or firewalled). The time between sending an Echo Request and receiving the corresponding Echo Reply is the Round-Trip Time (RTT). ICMP therefore provides relatively simple and reliable means of measuring latency, without any need for an additional service at the remote site.

While ICMP is convenient for measuring, it may be subject to varying treatments by networking and security devices. In contrast, the User Datagram Protocol (UDP), though less accurate, is the preferred protocol for quality measurements in modern network environments, because the high-demanding services

L. Barolli (Ed.): AINA 2024, LNDECT 204, pp. 374–384, 2024.
https://doi.org/10.1007/978-3-031-57942-4_36

often use UDP payloads [1]. Similar to ICMP, the UDP Echo protocol [11] can be used for UDP-based RTT measurements. However, unlike for ICMP, a UDP Echo service has to be deployed at the remote endpoint, replying to incoming UDP packets. That is, it requires explicit support by the remote endpoint, either by a UDP Echo service in user-space, or by a special router setup for a reply in hardware. This is denoted as "reflector", since it simply echoes the UDP packet back to the sender.

This paper uses HiPerConTracer [4,5] to measure round-trip time using reflectors with varying configurations, with the aim to optimize reflectors and to provide recommendations for configuring servers and network devices. By addressing this trade-off between measurement ease and accuracy associated with ICMP and UDP, our research contributes to enhancing latency- and jitter-sensitive production services in contemporary networks.

While recent efforts have been dedicated to achieving low latency by modifying the Linux kernel [2,7,9], our approach in this paper is distinct, as we only leverage standard Linux kernel tuning. Focusing on the lightweight UDP reflector service, involving minimal CPU processing, allows for a detailed examination of network performance factors induced by underlying hardware, drivers, and the kernel.

A comprehensive investigation into the underlying mechanisms contributing to the observed result variations is deferred to future research.

2 Methodology: The HiPerConTracer Framework

The measurements in this paper utilized the High-Performance Connectivity Tracer (HiPerConTracer) framework[1] [4,5]. The HiPerConTracer architecture comprises measurement vantage points (clients) sending ICMP and/or

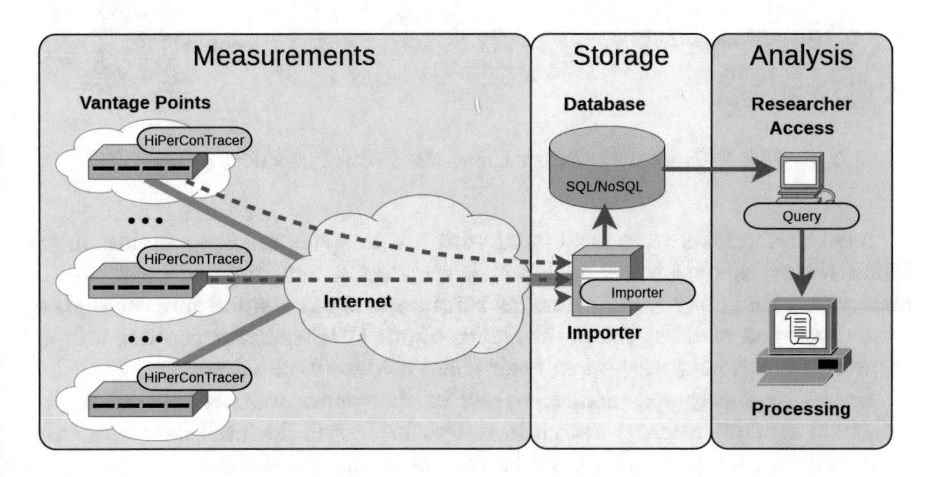

Fig. 1. An Overview of the Architecture of HiPerConTracer.

[1] HiPerConTracer: https://www.nntb.no/~dreibh/hipercontracer/.

UDP packets to echo-servers (reflectors), alongside the associated storage and analysis devices, as shown in Fig. 2 (Fig. 1).

HiPerConTracer's precise timestamping [5] enables both short- and long-term latency measurements. Measurement vantage points act as clients towards echo-servers (reflectors) to perform latency measurements. The modular architecture of the framework enables flexibility in configuring measurement scenarios, making it well-suited for our investigation into round-trip time optimization.

3 Measurements Infrastructure

For our experiments, we employed a single HiPerConTracer client composed of a server equipped with a 24-thread 3.2 GHz Intel i9-12900K CPU, 32 GiB RAM, and an Intel I225-V Ethernet card. The reflector server featured a 4-core 3.4 GHz Intel Xeon E-2224 CPU and was equipped with five different Ethernet cards. Both servers operated on Ubuntu 22.04. For some tests the setup was configured as in Fig. 2 with a Linux kernel firewall on the reflector server. For other tests, a Juniper MX80 was configured as a routing firewall as illustrated in Fig. 3, addressing typical real-world scenarios.

Each test consisted of 20 iterations, with packets sent in bursts of 1, 10, and 50 as rapidly as possible. The measurement client remained static, while the configuration of the reflector server varied as described below.

As a measure of latency, HiPerConTracer records the RTT, which is used in this paper. As a measure of jitter we use the Inter-Quartile Range (IQR), defined as the difference between the 75th percentile (Q3) and the 25th percentile (Q1) of the RTT values for all packets in a test.

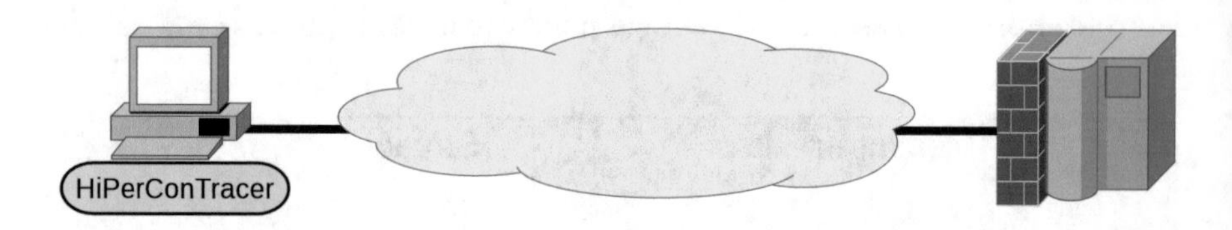

Fig. 2. Scenario 1: Direct Connection with Firewalled Server

Most of the tests were performed with a user-space process running on the Linux server as reflector. The Linux kernel has a multitude of configuration parameters that may be adjusted at runtime using the *sysctl* interface. Many of the parameters affect the performance of our UDP reflector process, and are adjusted by various tools to test their effect as described below.

Unless explicitly specified, the tests in this paper were conducted with a 100/1000 Mbit/s Ethernet switch in the path. Test G deviated from this standard, utilizing a direct cable between the client and the server.

The tests were executed with both, iptables and nftables modules loaded, employing an "accept all" configuration, with a tuned profile of "balanced". The Intel I350 card with ID 0.1 was used, unless stated otherwise in Sect. 4.

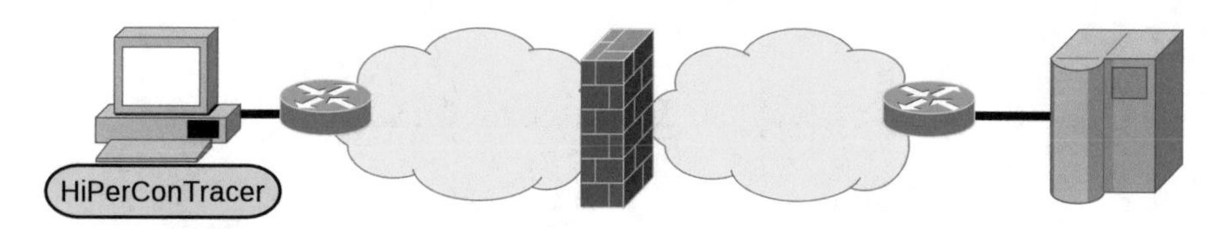

Fig. 3. Scenario 2: Routed Connection with Firewall in the Network

4 Results

We conducted a series of tests on the established infrastructure, exploring diverse aspects of potential tuning options for the reflector server system, as detailed in the following subsections. The used HiPerConTracer source branch is available via GitHub[2], the dataset of the following experiments is available as well [6].

4.1 Test A: Tuned Profile

Tuned[3] is a tuning daemon designed to enhance the performance of the operating system under specific workloads by applying tuning profiles. It can dynamically respond to changes in CPU and network utilization, adjust `sysctl` settings to optimize performance for active devices and conserve power for inactive. We explored three distinct Tuned profiles:

1. **network-latency:** This server profile prioritizes reducing network latency, emphasizing performance over power savings. It configures `sysctl` parameters such as setting intel_pstate, min_perf_pct=100, disabling transparent huge pages, and turning off automatic Non-Uniform Memory Access (NUMA) balancing. Additionally, it utilizes cpupower to set the performance cpufreq governor, requests a cpu_dma_latency value of 1, and adjusts busy_read, bus_poll times to 50 μs, and tcp_fastopen to 3.[4]
2. **balanced:** This profile strikes a balance between performance and power consumption, employing automatic scaling and tuning when feasible. However, it may result in slightly increased latency.
3. **powersave:** Geared towards maximizing power savings, this profile can minimize actual power consumption by throttling performance.

As depicted in Fig. 4, employing the kernel Tuned profile for network-latency results in a notable enhancement, reducing latency by a factor of 7.0 to 15.9, and jitter by a factor of 4.0 to 44.8.

---

[2] HiPerConTracer version 2.0.0~beta4 sources:
Git repository: https://github.com/dreibh/hipercontracer, branch "dreibh/udpping".

[3] Tuned: https://access.redhat.com/documentation/en-us/red_hat_enterprise_linux/7/html/performance_tuning_guide/chap-red_hat_enterprise_linux-performance_tuning_guide-tuned.

[4] A description of the options can be found in https://access.redhat.com/sites/default/files/attachments/201501-perf-brief-low-latency-tuning-rhel7-v2.1.pdf.

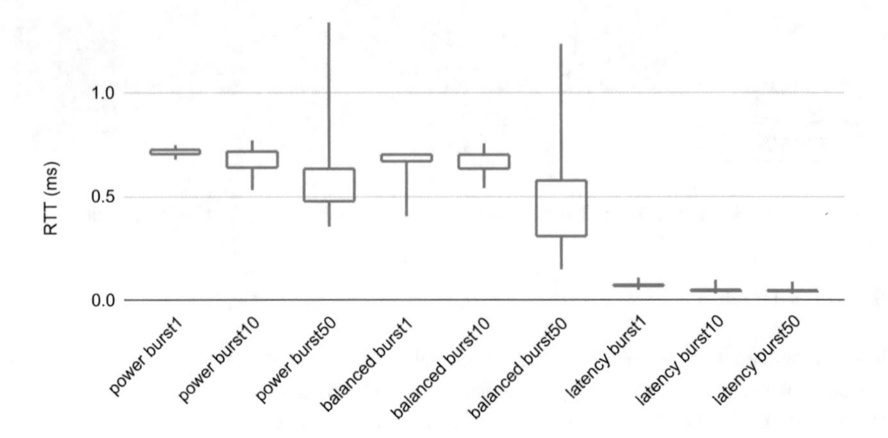

Fig. 4. Test A: Effects of TUNED profiles: Power-optimized, balanced and latency-optimized.

4.2 Test B: Firewall Rules

Firewalls can introduce variability to network latency. In this test, we measured the latencies imposed by the Linux kernel firewall running on the reflector server, utilizing the NFTABLES module in IPTABLES compatibility mode.

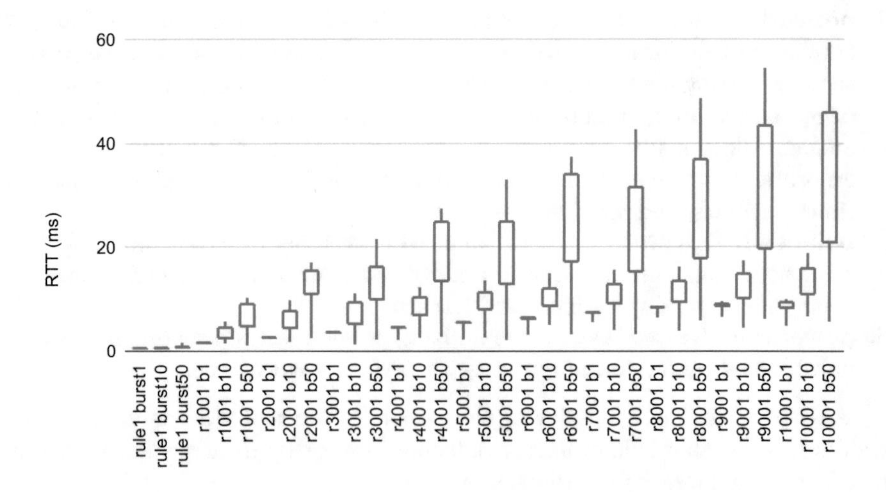

Fig. 5. Test B2: The Effect of a Large Linux Firewall Table

Figure 5 illustrates the impact of incorporating firewall rules into the Linux kernel. For this experiment, the reflection server was equipped with 10001 firewall rules, each accepting a single UDP port. Subsequent tests were conducted, where the traffic matched the 1st, 1001st, ..., 10001st firewall rule to gauge differences in processing time within the Linux firewall.

With just one firewall rule, we observed an almost 0.4 ms latency increase for a burst of 50 packets. Furthermore, we noticed a nearly linear latency escalation, with an additional latency of approximately 2 µs per firewall rule. It is worth noting that the measurement with 10001 rules, while perhaps unrealistically large, resulted in an almost 60 ms latency increase.

4.3 Test C: Network Interface Cards

The reflection server has three different network cards, with altogether five Ethernet interfaces. One Intel I350 with two interfaces identified with ID 0 and ID 0.1, one Broadcom (BCM) with interfaces ID 0 and ID 0.1, and one Intel 82574L with one interface. Four of the interfaces were tested to observe any differences.

Some notable differences were observed. Particularly intriguing was the variance between I350-0 and I350-0.1, where the ID0 interface exhibited significantly lower latency for burst50, while the ID0.1 interface demonstrated superior RTT albeit with a single high outlier displaying extended latency. There was no similar pattern for BCM, and the cause is unknown.

4.4 Test D: Coalesce

The network cards support network coalescing. This is a mechanism where the Linux kernel does not receive an interrupt (IRQ) for every packet. Instead, the network card queues up a number of packets before sending an interrupt to the kernel to handle a batch of packets. While this approach improves throughput, it introduces latency and jitter. The default coalescing setting for our server is 3 µs.

Surprisingly, disabling coalescing has a modest impact on RTT and jitter. For burst1, a coalescing setting of 0 µs shows a slight improvement. However, for bursts of 10 or 50 packets, the default 3 µs setting has the best results.

4.5 Test E: Kernel Options

Ubuntu Linux offers various kernel options, including the standard kernel, the low-latency kernel, and the real-time kernel. Test E measures the impact of using the low-latency kernel, without modifying the used `udp-echo-server` program (which is the UDP Echo server provided by the HiPerConTracer framework). By enabling CONFIG_PREEMPT, the low-latency Ubuntu kernel disables preemption only at critical locations where the kernel must protect data from concurrent access. This means that the UDP reflector will send the response immediately, unless a higher-priority task is being executed. The real-time kernel imposes even stricter constraints but is generally not recommended for standard servers unless precise timing is a requirement.

The results of our tests show a latency improvement ranging from 0.02 ms to 0.2 ms with the low-latency kernel. However, this improvement comes at the cost of slightly increased jitter.

4.6 Test F: Process Affinity for the UDP Echo Server

Since we suspected that CPU scheduling might contribute to latency and jitter in `udp-echo-server`, we investigated the impact of CPU affinity. To explore this, we locked the `udp-echo-server` process to a specific CPU core using CPU affinity.

Applying CPU affinity resulted in a slight reduction of jitter by 0.02 ms for the burst1 tests. For the other cases, CPU affinity did not significantly impact the results.

4.7 Test G: Direct Connection

All previous tests were conducted with a single 1 Gbit/s switch between the HiPerConTracer client and the UDP Echo reflection server. In order to evaluate whether the switch significantly influenced our measurements, we conducted Test G with a direct cable connection.

The results show a reduction in RTT varying from 0.01 ms to 0.1 ms. Additionally, there is a decrease in jitter for burst1, while jitter increases for burst10 and burst50.

4.8 Test H: Juniper MX80 Router as Reflector

While the standard measurement setup with HiPerConTracer's UDP Ping involves using a user-space UDP Echo server on a Linux server, certain router-firewalls like Juniper's MX80 offer a built-in system for latency measurement called Real-Time Performance Monitoring (RPM). In this test, we compare the performance of HiPerConTracer with the RPM probe-server against the Linux UDP Echo server.

The results depicted in Fig. 6 reveal that the RPM probe-server significantly lags behind the Linux UDP Echo server, causing RTT and jitter to increase by 16–120 times. It is noteworthy that in the RPM architecture, hardware timestamping is supported, but only on the client-side.

4.9 Test I: Inline MX80 Firewall

This test explores the impact of the configuration depicted in Fig. 3, where a dedicated routing firewall (Juniper MX80) is positioned between the HiPerConTracer client and the UDP Echo server. The firewall is specifically configured with 1 to 10001 firewall rules, and the corresponding results are visualized in Fig. 7.

The Juniper MX80 firewall is set up with a stateless firewall configuration. Comparing these results to those presented in Fig. 5 yields intriguing insights. The data indicates that the MX80 firewall is notably optimized for large firewall filters. Remarkably, aside from a few outliers (primarily the first packet in a burst), there is no discernible difference in performance, irrespective of whether a packet matches the first or the 10001st filter rule. This pattern is consistent across most tests and suggests that the MX80 firewall employs some form of connection tracking optimization, even in stateless mode.

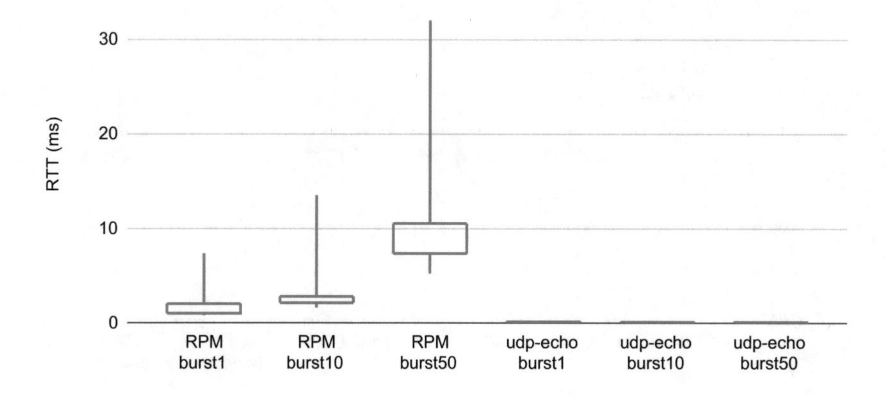

Fig. 6. Test H: Comparison Between RPM Probe-Server and UDP Echo Server.

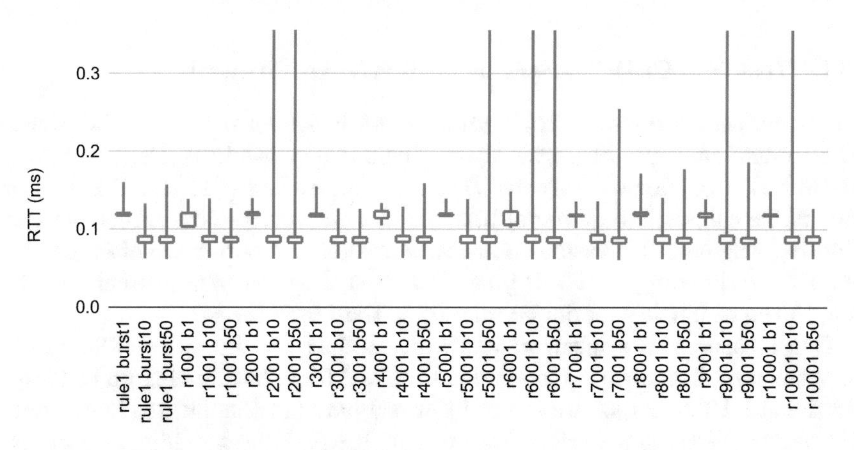

Fig. 7. Test I: The impact of an in-line MX80 firewall.

4.10 Test J: UDP vs. ICMP

Testing with ICMP is often simpler to implement than testing with UDP, as any device can be used to reflect ICMP packets. This method may offer higher accuracy, as packets are reflected in kernel-space. However, considering that latency-sensitive production traffic typically involves UDP, and ICMP packets may be treated differently by network devices, conducting tests with UDP provides more representative measurements.

In Fig. 8, the contrast between UDP and ICMP is examined using the TUNED (latency-optimized) Linux server as a reflector. The results indicate that, for burst1, ICMP exhibits significantly lower RTT. In all cases, ICMP demonstrates approximately half the jitter compared to UDP.

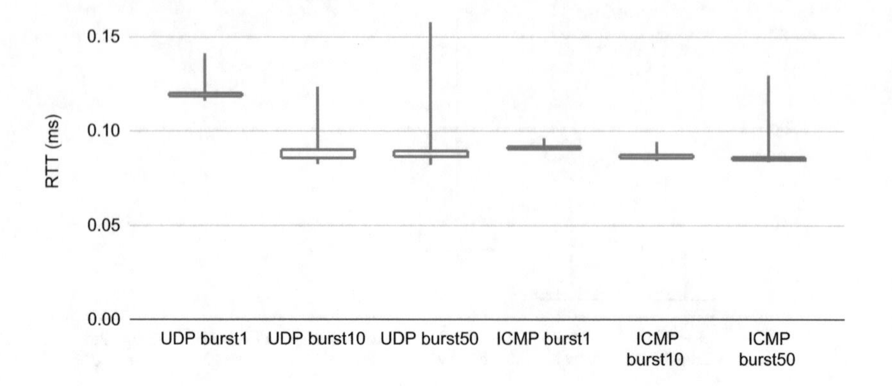

Fig. 8. Test J: Comparison Between UDP and ICMP.

4.11 Test K: ICMP to Server vs. ICMP to Firewall

The Linux server responds to ICMP packets in kernel-space, but the Juniper MX80 offers multiple response layers, including a Modular Port Concentrator (MPC), two Flexible Physical Interface Controllers (PIC), a Trio [12] programmable chipset, a bare-metal Linux kernel and a virtualized FreeBSD kernel (Routing Engine), plus various user-space processes. Routers are often not optimized for responding to ICMP ping. This test aims to evaluate whether ICMP Ping to the router is superior or inferior to the Linux server.

Our measurements unsurprisingly reveal that the Juniper MX80 exhibits suboptimal performance in responding to ICMP Ping, introducing up to 18 ms of additional RTT and 1000 times the jitter compared to the TUNED Linux kernel. The reason for the low performance is that ICMP Echo handling is performed in software and is a low priority task.

4.12 Test L: Linux Kernel-Space Reflector

Since HIPERCONTRACER supports specifying the source UDP port, it is possible to use the Linux kernel's Network Address Translation (NAT) functionality to enable UDP reflection in kernel-space. Our tests show that this further reduces jitter for the bursty measurements by 40% to 50% down to the minimal attained 2.4 μs to 2.9 μs IQR.

4.13 Test Z: Multiple Optimizations

The preceding tests measured the individual variables affecting RTT and jitter. This test, however, examines the collective impact of applying the optimal values for all parameters: It was conducted using a Linux kernel with a latency-optimized TUNED profile, the first firewall rule matching packets, the optimal interface card (BCM-0), zero coalescing, a low-latency kernel, UDP tests, no inline firewall, and utilizing the UDP Echo server.

The improvements achieved through these optimizations are significant, median RTT was reduced from 0.66 ms to 0.04 ms and jitter was reduced by 99%.

5 Conclusion

In this study, we delved into the intricate landscape of network latency optimizations, unveiling the critical importance of reflection server tuning in achieving precise and low-latency network measurements. Our findings highlight the synergistic effects of various parameters, emphasizing the need for a holistic approach to network optimization.

The implementation of a TUNED profile on Linux emerged as a standout strategy, showcasing substantial rewards in network performance, in particular for a kernel-space reflector. Early acceptance of latency-critical traffic in the firewall chain proved pivotal. While the impact of individual factors might seem modest, the cumulative effect of optimizations became pronounced, especially in the context of stringent latency requirements, such as those dictated by emerging technologies like 5G networks.

Our study underscores the practical implications for system architects and network administrators, emphasizing the need to consider multiple factors collectively for optimal performance. As technology continues to advance, these insights contribute to the ongoing pursuit of efficient and low-latency network infrastructures, which are essential for the success of latency-sensitive applications and edge computing solutions.

Future research should probably delve more into changing the software, exploring socket options and further Linux kernel scheduling features, as well as TWAMP Light (Two-Way Active Measurement Protocol as defined in [8]).

Another topic for future research is the observation that the differences in firewall behaviour between the Linux firewall and the Juniper MX80 together with the achieved level of accuracy (jitter) has the potential to allow fingerprinting of firewalls based on similar network measurements.

References

1. Arouna, A., Bjørnstad, S., Ryan, S.J., Dreibholz, T., Rind, S., Elmokashfi, A.M.: Network path integrity verification using deterministic delay measurements. In: Proceedings of the 6th IEEE/IFIP Network Traffic Measurement and Analysis Conference (TMA), Enschede, Overijssel/Netherlands (2022)
2. Beifuß, A., Runge, T.M., Raumer, D., Emmerich, P., Wolfinger, B.E., Carle, G.: Building a low latency Linux software router. In: Proceedings of the 28th International Teletraffic Congress (ITC 28), vol. 01, pp. 35–43 (2016). https://doi.org/10.1109/ITC-28.2016.114
3. Conta, A., Deering, S.E., Gupta, M.: Internet Control Message Protocol (ICMPv6) for the Internet Protocol Version 6 (IPv6) Specification. Standards Track RFC 4443, IETF (2006). https://doi.org/10.17487/RFC4443

4. Dreibholz, T.: HiPerConTracer - a versatile tool for IP connectivity tracing in multi-path setups. In: Proceedings of the 28th IEEE International Conference on Software, Telecommunications and Computer Networks (SoftCOM), pp. 1–6. Hvar, Dalmacija/Croatia (2020). https://doi.org/10.23919/SoftCOM50211.2020.9238278
5. Dreibholz, T.: High-precision round-trip time measurements in the internet with HiPerConTracer. In: Proceedings of the 31st International Conference on Software, Telecommunications and Computer Networks (SoftCOM). Split, Dalmacija/Croatia (2023). https://doi.org/10.23919/SoftCOM58365.2023.10271612
6. Evang, J.M., Dreibholz, T.: Reflection server tuning dataset. IEEE DataPort (2024). https://doi.org/10.21227/8wbz-7e29
7. Fujimoto, K., Natori, K., Kaneko, M., Shiraga, A.: Energy-efficient KBP: kernel enhancements for low-latency and energy-efficient networking. IEICE Trans. Commun. **Vol.E105-B** (2022). https://doi.org/10.1587/transcom.2021EBP3194
8. Hedayat, K., Krzanowski, R.M., Morton, A., Yum, K., Babiarz, J.Z.: A Two-Way Active Measurement Protocol (TWAMP). Standards Track RFC 5357, IETF (2008). https://doi.org/10.17487/RFC5357
9. Heursch, A.C., Rzehak, H.: Rapid reaction Linux: Linux with low latency and high timing accuracy. In: Proceedings of the 5th Annual Linux Showcase & Conference (ALS) (2001)
10. Postel, J.B.: Internet Control Message Protocol. RFC 792, IETF (1981). https://doi.org/10.17487/RFC0792
11. Postel, J.B.: Echo Protocol. RFC 862, IETF (1983). https://doi.org/10.17487/RFC0862
12. Yang, M., et al.: Using trio: juniper networks' programmable chipset - for emerging in-network applications. In: Proceedings of the ACM SIGCOMM Conference. Association for Computing Machinery, New York (2022). https://doi.org/10.1145/3544216.3544262

Extension of Resource Authorization Method with SSI in Edge Computing

Ryu Watanabe[1(✉)], Ayumu Kubota[1], Jun Kurihara[2], and Kouichi Sakurai[3]

[1] KDDI Research, Inc., 2–1–15 Ohara, Fujimino, Saitama 356–8502, Japan
ry-watanabe@kddi.com

[2] Graduate School of Information Science, University of Hyogo, 7–1–28 Minatojima-Minamimachi, Chuo, Kobe, Hyogo 650–0047, Japan

[3] Faculty of Information Science and Electrical Engineering, Kyushu Univercity, 744 Nishiku Motooka, Fukuoka-shi, Fukuoka 819–0395, Japan

Abstract. The authors have previously confirmed that the flow of OAuth, a traditional authorization method on the Internet, can be managed appropriately by applying it to allocating computational resources in edge computing. In addition, in recent years, the concept of self-sovereign identity (SSI), which respects the rights of individuals, and technologies for this purpose have been studied in examining digital identities responsible for authentication and authorization. In this paper, we report on a study of the suitability and security of introducing the concept of SSI as an extension of resource authorization for edge computing.

1 Introduction

One type of mobile networking is edge computing, in which processing that is conventionally performed by a central server is performed by edge servers on the network periphery, near devices that have communication capabilities and can connect to the network, to distribute the processing load and ensure immediate response. Since the capacity of individual edge servers is weaker and scarcer than that of the central server, it is necessary to properly manage various resources, such as computing power, to operate them in a qualified manner. For this reason, it is essential to dynamically assign usage privileges to the resources of edge servers so that only authorized users or service providers can use the computing resources and also to avoid depletion of computing resources. Authorization methods can be applied to manage this usage authority. The authors have focused on the relationship among participating entities in edge computing, categorized the forms in which authorization methods can be applied, clarified the requirements for such methods, selected nearby edge servers, and clarified the sequence in which OAuth, a typical authorization method used on the Internet, is applied to realize this method. It also clarifies the application of OAuth sequence, a standard authorization method used on the Internet, by selecting nearby edge servers [16, 17].

On the other hand, in research on ID management technology for authentication and authorization, studies on self-sovereign style management methods that delegate authority to users have been increasing rapidly in recent years, replacing conventional

L. Barolli (Ed.): AINA 2024, LNDECT 204, pp. 385–394, 2024.
https://doi.org/10.1007/978-3-031-57942-4_37

decentralized management methods [6,13,15]. This concept, called self-sovereign identity (SSI), solves issues such as privacy leaks and a problem with centralized management by making the users themselves the management entity. In addition, problems such as a single point of failure can be solved by utilizing distributed ledger technology. The OpenID Foundation, which is working on OpenID standardization, is following this trend and is studying OpenID Connect for SSI, a specification following the SSI concept [20]. Therefore, the authors are studying the introduction of the SSI concept as an expansion of resource management in edge computing. The authors have confirmed the suitability and safety of applying OpenID Connect for SSI. This paper describes the results of these studies.

2 Related Work

Research on edge computing security [11,12] identify malicious actions such as Distributed Denial of Service (DDoS) attacks, side-channel attacks, and malware infections. In addition, essential security challenges in the edge computing environment are issued. Regarding authentication and authorization, many studies are done. There is also a lot of research on access control and authorization for IoT terminals in an IoT environment, with use of extended Role Based Access Control (RBAC) or Attribute Based Access Conrtol (ABAC), and capability based access control method [1,8,10]. There are aslo proposals to access control edge devices using the SSI technique described below, but they are unrelated to edge server resource authorization [2,3]. However, prior studies of edge server resource authorization in edge computing are rare.

There are many researche articles about the management of devices, nodes ,and servers in edge computing environment (ex., [4,9,14]). However, in edge computing

Table 1. Teaminology of OAuth

Team	Description
Protected resource	Information on a resource server.
Resource server	A server that holds protected resources and provides protected resources upon request.
Client	A service or application that accesses resource servers
Resource owner	The entity that owns protected resources. Also known as end users. Resource owner authorizes access to its protected resources.
Authorization server	A server that authenticates the resource owner, issues access token, and manages authorization flow. It has an authorization endpoint that performs the actual authorization process and a token endpoint that issues and manages access tokens.
Access token	A token for an authenticated request, presented by the client to the resource server.
Scope	Scope of access privileges to be delegated (authorization scope)

environments, users or devices want to use the servers with available resources. Therefore, the resource authorization on edge servers must be integrated with resource servers and device management methods.

2.1 OpenID

OpenID is a standard specification for authentication and authorization developed by the OpenID Foundation, a non-profit organization. The SP (Service Provider), which provides services to users, verifies the signature on the security token and confirms the validity of the token to authenticate users who present the token. In the past, the first author has proposed a system that uses cell phones and certificates for authentication in OPs to ensure the reliability of identity verification [18, 19]. The current OpenID specification uses the OpenID Connect specification [7], an extension of OAuth, an authorization specification developed by the same organization after the OpenID 2.0 specification. It is possible to authenticate a user based on the consent between a user and a service provider without prior trust confirmation, quickly realizing a single sign-on environment. Although multiple OPs can be utilized, users use a single OP, and there is a concern that this may lead to a single point of failure.

2.1.1 OAuth Consider the case where a site (site A) provides user information to another site (site B) on a limited basis. In this case, if the user's authentication information on site A is used by site B, site B can view all the information that site A possesses. Site B can read all the information about the user. So, there is a leakage of information that is not necessary or that the user does not want to disclose. For this reason, authorization is a mechanism to allow site A to provide information that site B needs on site A. OAuth [5] is one of the open standards for authorization. The current state-of-the-art is OAuth2.0 (hereafter, OAuth refers to OAuth2.0). Table 1 summarizes the terminology used in OAuth. The target information of disclosure is called the scope in OAuth and is provided to Site B securely in encrypted form. Therefore, an entity other than Site B cannot obtain the information described in the scope from Site A.

2.1.2 OpenID Connect for SSI[20] The OpenID Foundation is working on the OpenID Connect specification for SSI. Its specifications support Verifiable Credential (VC) and Verifiable Presentation (VP), data model standards for digital credentials, and DID, an identifier managed using a distributed ledger. These three specifications, Self-Issued OpenID Provider v2 (SIOPv2), OpenID for Verifiable Presentation, and OpenID for Verifiable Credential are included. The draft specifications apply VC and DID standardized by W3C.

Self-Issued OpenID Provider (SIOP)
SIOPv2 is an extension of OpenID Connect; this specification expands the end user's scope of control. Users issue their ID tokens encrypted with user-controlled encryption keys and use them for their authentication. This specification allows the user to control the identifier issued to the client RP and the disclosure of digital identities.

OpenID for Verifiable Presentation
This specification is used for requesting token disbursement as a VP form in the OAuth flow. When combined with SIOP, it is possible to authenticate with SIOP and obtain a token in VP format. Besides W3C's VC, In addition to W3C's VC, it also supports ISO/IEC 18013-5:2021 mDL, a standard for mobile driver's licenses, and AnonCreds, credentials standardized by the Hyperledger foundation.

OpenID for Verifiable Credential Issuance
This specification allows an end user to request an ID provider to issue VC. Thus, users can use credentials issued and managed by themselves using SIOPv2 or credentials issued by a trusted third party (OP) in the form of a VC, as in the past.

3 Authors' Previous Consideration on Resource Authorization

This section presents the edge computing model envisioned by the authors and then presents the authors' previous studies [16,17] on resource authorization in edge computing.

Table 2. Terms definition in our study

Term	Definition
Edge Provider (EP)	The entity that owns edge servers.
Edge Server (ES)	A Server with computational resources and performs computation function.
Service Provider (SP)	An entity that provides a service with ES.
Computational resource	Resources held by an edge server (CPU, memory, and so on.).
Instance	A computation service performed with computational resources on ES(s) by an SP.
User	An entity who is provided services by SPs

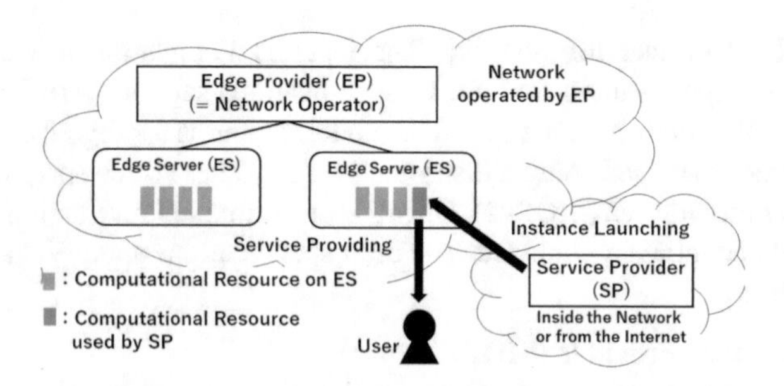

Fig. 1. Edge computing model

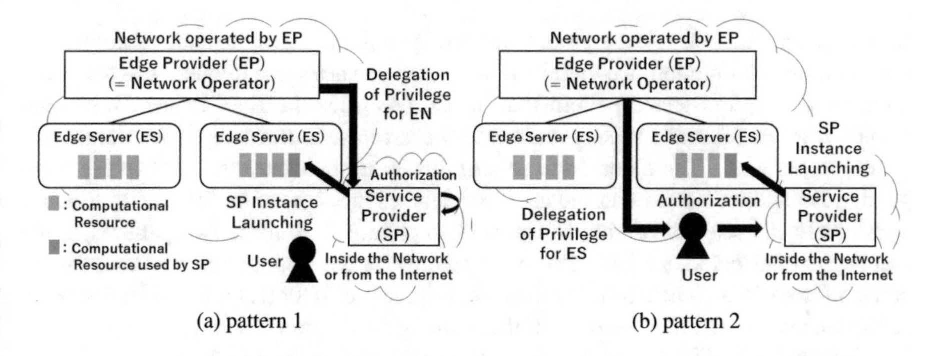

Fig. 2. Patterns in Edge computing authorization

3.1 Edge Computing Models

The terms on our edge computing models are summarized in table 2, and our assumed models are shown in Fig. 1 The details of the model are summarized in the article [16]. In our model, a network operator (=EP) provides ESs, and computational resources are used to provide service to the subscribers (=users). For the authorization in the model, the appropriate authorization flows are expected for the use of resources on edge services (ESs), and they are different based on the respective interrelationships among EP, SP, and user.

Under the assumption that edge computing is the provision of ESs and their computational resources owned by an EP to another participant, We defined the applicable patterns by considering the authorized participants to whom the computational resources

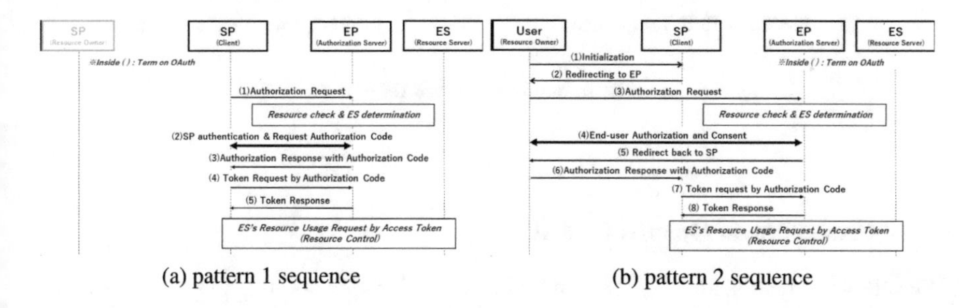

Fig. 3. Sequence in each pattern

Table 3. Summary of Authorization Patterns

Pattern	Rights	Note
1	SP	Difficult to determine the location of a user
2	user	Easy to determine the location of a user
1 derivative	SP	Pattern 1 + Need coordination between EPs
2 derivative	user	Pattern 2 + Need coordination between EPs

are rented and the role of each participant in edge computing in the authorization flow. Based on this assumption, we classified the patterns. They were divided into two main categories. The first is the pattern that an EP delegates the right to use computing resources to an SP, and the SP provides services to users as a tenant. We define this pattern as pattern 1. The other is the pattern that an EP delegates the right to use computing resources to a user, and the user re-delegates the right to an SP, which provides a service to the user. We define this pattern as pattern 2. In these two patterns, when roaming is considered, we have also confirmed the usage form of authorization for the derived form in which EPo (= Edge Provider original) and EPr (Edge Provider roaming) cooperate and use the ESs' resouce of the roaming destination. The main patterns 1 and 2 are shown in the Fig. 2, and the sequences are shown in the Fig. 3. The overview of patterns is summarized in Table 3.

4 Application of SSI Concept

This section describes the case where the SSI concept is introduced for resource authorization in edge computing, as described earlier.

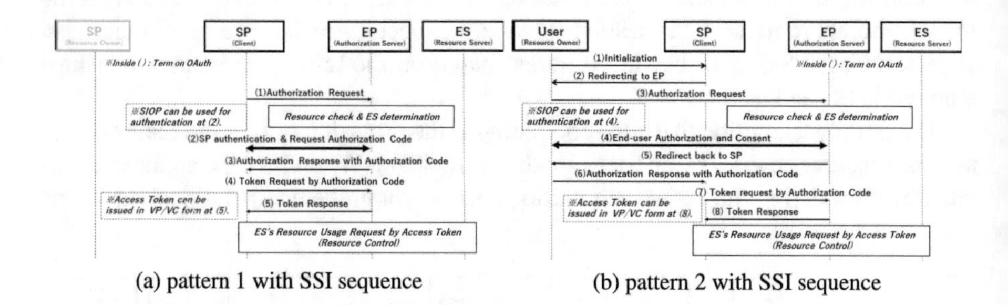

(a) pattern 1 with SSI sequence

(b) pattern 2 with SSI sequence

Fig. 4. Sequence in SSI pattern

4.1 Introduction of OpenID for SSI

The following results are confirmed with this consideration about applying the SSI concept for resource authorization at edge servers. In patterns 1 and 2, the OpenID connect for SSI can be allied without any problem. SIOPv2 can be used for authentication on EP, and Access tokens are issued in VP/VC format (Fig. 4).

When OpenID for SSI is adapted to a derivative of the authorization pattern in a roaming situation, the VP can be used as authentication information at the EP on the roaming destination (Figs. 5 and 7). Therefore, the VP is requested by an SP or a user before the authorization process with EPr (P1 and P2 in Figs. 6 and 8).

4.2 Consideration

Since the OpenID Conncet for SSI specification is an extension of OAuth2.0, it can be easily applied to the authors' study, and the benefits of the SSI concept can be provited as is. By issuing out access tokens in the form of VP/VC, verification and revocation of authorization information can be performed on a distributed ledger, thus increasing management flexibility. However, in Pattern 1, the SP has the right to use the ES, and the SP is authenticated by the EP, so the SP is not an individual user but an organization that provides the service. Therefore, it does not lead to the protection of the user's own rights, which is the concept of SSI. In the case of roaming, which is a derivative of the patterns, VP/VC can be used for authentication at the roaming destination, and thus privacy protection can be strengthened when using a derivative of pattern 2, in which the user retains the right to use ES.

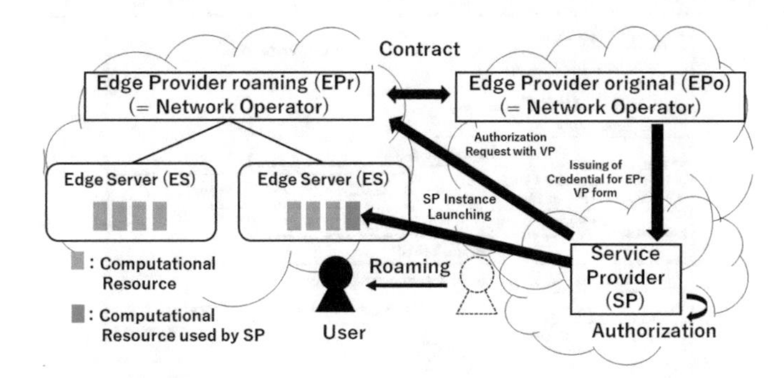

Fig. 5. Concept of Pattern1 derivative with SSI

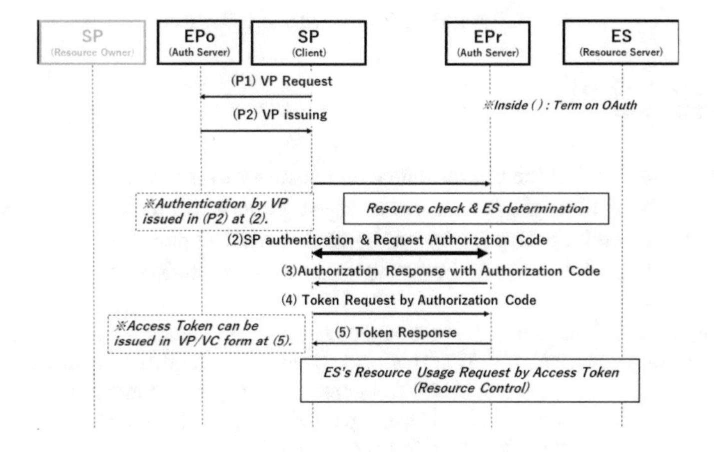

Fig. 6. Pattern1 derivative with SSI: Sequence

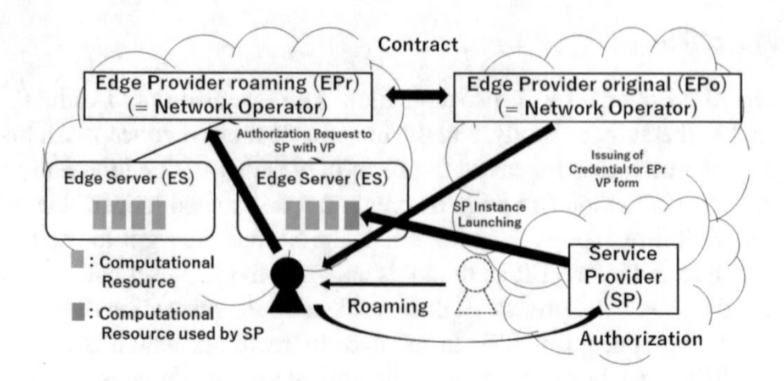

Fig. 7. Concept of Pattern2 derivative with SSI

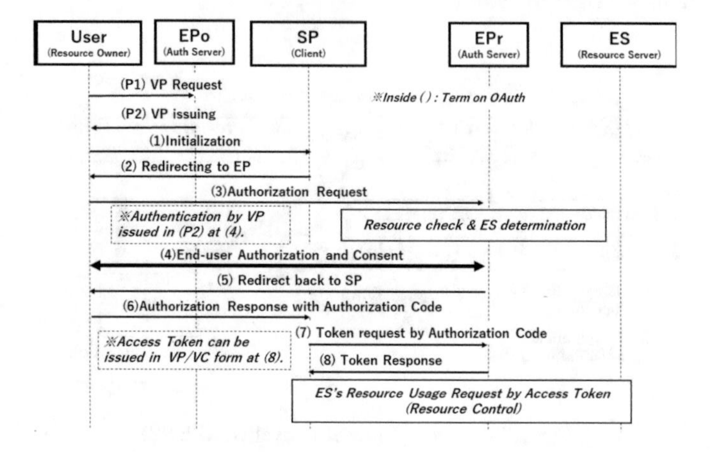

Fig. 8. Pattern2 derivative with SSI: Sequence

5 Conclusion

In this paper, we studied the conformance and security of applying the concept of SSI to resource authorization of edge servers in edge computing. Specifically, we confirmed that the benefits can be promoted by applying OpenID Connect for SSI. In the future, we plan to verify performance and other factors using simulation and other methods.

Acknowledgements. The fourth author is supported by support center for advaced telecommunications technology research (SCAT). This work was supported in part by the National Institute of Information and Communications Technology, Japan [Grant Number NICT22401], the Japan Society for the Promotion of Science, Japan [KAKENHI Grant Numbers JP22K11994, JP21H03442], and the KDDI Foundation Research Grant.

References

1. Bernal Bernabe, J., Hernandez Ramos, J.L., Skarmeta Gomez, A.F.: Taciot: multidimensional trust-aware access control system for the internet of things. Soft. Comput. **20**(5), 1763–1779 (2016). https://doi.org/10.1007/s00500-015-1705-6
2. Fotiou, N., Pittaras, I., Chadoulos, S., Siris, V.A., Polyzos, G.C., Ipiotis, N., Keranidis, S.: Authentication, authorization, and selective disclosure for iot data sharing using verifiable credentials and zero-knowledge proofs. In: Saracino, A., Mori, P. (eds.) ETAA 2022, vol. 13782, pp. 88–101. Springer, Cham (2023). https://doi.org/10.1007/978-3-031-25467-3_6
3. Fotiou, N., Siris, V.A., Polyzos, G.C., Kortesniemi, Y., Lagutin, D.: Capabilities-based access control for iot devices using verifiable credentials. In: 2022 IEEE Security and Privacy Workshops (SPW), pp. 222–228 (2022). https://doi.org/10.1109/SPW54247.2022.9833873
4. Hong, C.H., Varghese, B.: Resource management in fog/edge computing: a survey on architectures, infrastructure, and algorithms using edge computing. ACM Comput. Surv. **52**(5), 1–37 (2019)
5. IETF OAuth Working Group. https://tools.ietf.org/wg/oauth/
6. Lim, S.Y., et al.: Blockchain technology the identity management and authentication service disruptor: a survey. Int. J. Adv. Sci. Eng. Inf. Technol. **8**, 1735 (2018). https://doi.org/10.18517/ijaseit.8.4-2.6838
7. OpenID Foundation: Openid connect 1.0 (2014). https://openid.net/wg/connect/specifications/
8. Pal, S., Rabehaja, T., Hill, A., Hitchens, M., Varadharajan, V.: On the integration of blockchain to the internet of things for enabling access right delegation. IEEE Internet Things J. **7**(4), 2630–2639 (2020). https://doi.org/10.1109/JIOT.2019.2952141
9. Pradhan, M., Poltronieri, F., Tortonesi, M.: Dynamic resource discovery and management for edge computing based on SPF for HADR operations. In: Proceedings of IEEE ICMCIS 2019 (2019)
10. Putra, G.D., Dedeoglu, V., Kanhere, S.S., Jurdak, R.: Trust management in decentralized iot access control system. In: 2020 IEEE International Conference on Blockchain and Cryptocurrency (ICBC), pp. 1–9 (2020). https://doi.org/10.1109/ICBC48266.2020.9169481
11. Román, R., López, J., Mambo, M.: Mobile edge computing, fog et al.: a survey and analysis of security threats and challenges. Future Gener. Comput. Syst. **78**, 680–698 (2016). https://api.semanticscholar.org/CorpusID:3767252
12. Soni, N., Malekian, R., Thakur, A.: Edge computing in transportation: security issues and challenges. arXiv:2012.11206 (2020)
13. Tobin, A., Reed, D., Windley, P.J.: The inevitable rise of self-sovereign identity. The Sovrin Foundation (2017)
14. Toczé, K., Nadjm-Tehrani, S.: A taxonomy for management and optimization of multiple resources in edge computing. Wirel. Commun. Mob. Comput. (2018)
15. Toth, K.C., Anderson-Priddy, A.: Self-sovereign digital identity: a paradigm shift for identity. IEEE Secur. Priv. **17**(3), 17–27 (2019). https://doi.org/10.1109/MSEC.2018.2888782
16. Watanabe, R., Kubota, A., Kurihara, J.: Resource authorization methods for edge computing. In: Barolli, L., Hussain, F., Enokido, T. (eds.) AINA 2022. LNNS, vol. 449, pp. 167–179. Springer, Cham (2022). https://doi.org/10.1007/978-3-030-99584-3_15
17. Watanabe, R., Kurihara, A., Kurihara, J.: Resource authorization patterns on edge computing. IN2020-68 **120**(414), 85–90 (2021)
18. Watanabe, R., Miyake, Y.: User authentication method with mobile phone as secure token. In: SECURWARE 2013, The Seventh International Conference on Emerging Security Information, Systems and Technologies (2013). https://api.semanticscholar.org/CorpusID:197645242

19. Watanabe, R., Nakano, Y., Tanaka, T.: Single sign-on techniques with pki-based authentication for mobile phones. In: Security and Management (2010). https://api.semanticscholar.org/CorpusID:38455654
20. Yasuda, K., Loddersted, T.: Openid Connect for SSI (2021). https://openid.net/wordpress-content/uploads/2021/09/OIDF_OIDC4SSI-Update_Kristina-Yasuda-Torsten-Lodderstedt.pdf

An AI-Driven System for Identifying Dangerous Driving Vehicles

Hibiki Tanaka[1], Kazuki Shimomura[2], Naoki Tanaka[1], Makoto Ikeda[2(✉)], and Leonard Barolli[2]

[1] Graduate School of Engineering, Fukuoka Institute of Technology, 3-30-1 Wajiro-Higashi, Higashi-Ku, Fukuoka 811-0295, Japan
mgm23107@bene.fit.ac.jp

[2] Department of Information and Communication Engineering, Fukuoka Institute of Technology, 3-30-1 Wajiro-Higashi, Higashi-Ku, Fukuoka 811-0295, Japan
mgm22101@bene.fit.ac.jp, makoto.ikd@acm.org, barolli@fit.ac.jp

Abstract. As the world faces aging population, traffic accidents caused by elderly drivers have become a social issue. The decline in driving skills and judgment can lead to deviation of vehicles from their lanes while they are in motion resulting in accidents. In this paper, we propose an intelligent safety drive assisting system that identifies dangerous vehicles. The proposed system has an AI application that identifies vehicles engaging in risky driving and notifies both the control center and the driver. We present the detection performance of four different datasets. From the evaluation results, we observed that datasets based on remote-controlled cars and actual vehicle images reduced misrecognition rate and improved detection accuracy.

Keywords: Dangerous driving vehicles · Dataset · YOLO

1 Introduction

The world is facing the aging population and the traffic accidents caused by elderly drivers have become a social issue. Data published from 2009 to 2019 [4] show that the number of driver license holders over 75 years old has approximately increased by 1.8 times, while for people aged over 80 years is approximately 1.9 times. Compared to 2009 in 2022 the number of driver license holders aged over 75 years has approximately increased by 1.9 times, while for people aged over 80 years by approximately 2.2 times, indicating a continuing upward trend.

Because of declining driver skills, various advanced technologies have been developed to ensure automotive traffic safety. However, there are only a few works that aim to add safety features to existing vehicles. In addition, there are cases of dangerous driving by inexperienced young drivers, resulting in unfortunate accidents.

Artificial intelligence (AI) systems have received a lot of interest in several industries [5,9,11,16,19]. Edge-focused AI systems employ cloud-based training for regular tasks, whereas real-time detection operates at the edge [10,13,14,18]. The presence of

L. Barolli (Ed.): AINA 2024, LNDECT 204, pp. 395–400, 2024.
https://doi.org/10.1007/978-3-031-57942-4_38

open datasets, models and platforms has accelerated the process of application development by minimizing the time needed for creating applications [1,2].

The decline in driving skills and judgment can lead to deviation of vehicles from their lanes while they are in motion resulting in accidents. In this paper, we propose an intelligent safety drive assisting system that identifies dangerous vehicles. We have presented the detection performance of four datasets. From evaluation results, we observed that datasets based on remote-controlled cars and actual vehicle images reduced misrecognition rate and improved detection accuracy.

The structure of this paper is organized as follows: Sect. 2 presents an overview of AI models. Section 3 outlines the system designed to support drivers in detecting vehicles exhibiting potentially dangerous driving vehicle behaviors. The evaluation results are discussed in Sect. 4. Finally, the paper concludes with Sect. 5.

2 Deep Neural Networks and YOLO Models

Deep Neural Networks (DNNs) have many applications and they have drawing significant interest because of the multi-layered architectures, which can be used in feature detection and data representation learning are crucial for gleaning essential insights from observed data [6,8,15].

There are different version of YOLO algorithms. The pioneering YOLO algorithm [12] by Joseph Redmon's team halted the development over concerns of military use and privacy. Then, the work was taken up by Alexey Bochkovskiy with YOLOv4's introduction [3], followed by Glenn Jocher's YOLOv5 [7], which incorporated instance segmentation. The series evolved with YOLOv6 and YOLOv7, culminating in YOLOv8's launch [17]. Each version has different features and can be used for specific computer vision tasks.

YOLO's methodology (processing images in one fell swoop) segments the images into a grid to predict across the entire image, thereby decreasing the background error.

3 Intelligent Dangerous Vehicle Detection and Aid System

3.1 System Architecture and Design

Figure 1 depicts our smart driver assistance system architecture. This system includes an AI application that identifies vehicles engaging in risky driving and notifies both the control center and the driver.

Figure 2 shows the graphical user interface of the application, which will be able to store information on dangerous vehicles, which will be detected in real time and report the information to the central system.

To optimize the system, we have integrated an edge computing device called the Jetson Xavier NX, which is connected to the Internet to store training images. We are developing a feature to enable in-vehicle image training when the device is not connected to the Internet, with the goal of improving training accuracy. For object detection, the system employs the YOLOv5 model.

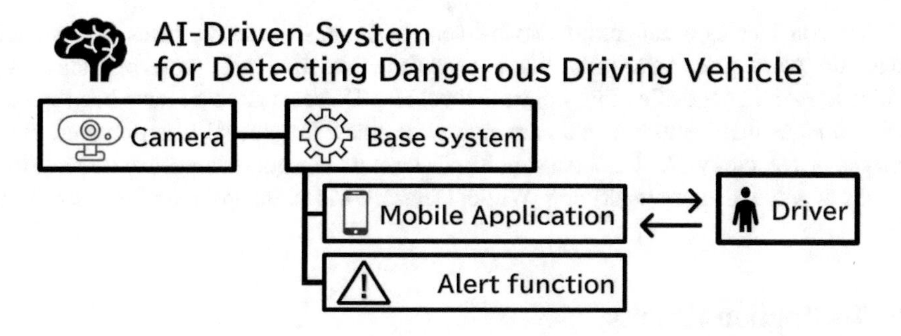

Fig. 1. System model.

Fig. 2. Application images.

3.2 Dataset and Training Models

In our research, we collect images for dangerous driving vehicle behaviors to build a distinct dataset, which includes a total of 696 images. Table 1 summarizes the contents of the four datasets. We chose YOLOv5x model for the network model, which is used up to 200 epochs. The images are processed at a resolution of 300 and the training is conducted for 16 batches.

Table 1. Dataset comparison.

Label	Number of Images	Vehicle Type	Traffic Lane
Dataset #1	139	Radio-controlled Car	White Tape
Dataset #2	303	Actual Car (Parked)	Actual
Dataset #3	254	Actual Car (Consider DL)	Actual
Dataset #4	696	Radio-controlled Car and Actual Car	Actual

We consider as a dangerous driving vehicle the case when the tires of a vehicle touch the white line. As a traffic line is used the white tape in the case of Dataset #1, which uses the image of a radio-controlled vehicle. Datasets #2 to #4 have been trained using images that capture actual cars and white lines. Dataset #2 was prepared using images of vehicles parked in a parking lot. Dataset #3 considers images extracted from videos of actual dangerous driving. While, Dataset #4 is an integration of datasets #1 to #3.

4 Evaluation Results

The learning models derived from each dataset were validated using multiple test videos. The image in Fig. 3 represents a frame from one of the test videos.

Considering Dataset #1, there were no detections of tires on white lines, but there were numerous instances of false detections of the car body and cases where no detection occurred (see Fig. 3(a)). In Dataset #2, there were detections of tires on white lines. However, the instances of accurately identifying tires on white lines were very few and false detections were observed. As can be seen in the video of Fig. 3(a), nothing was detected. In Dataset #3, the test video of Fig. 3(b) detected tires on white lines with an accuracy of 0.91, but there were some false detections. In Dataset #4, all areas were accurately identified without any false detections, achieving better accuracy than Dataset #3. In the frame shown in Fig. 3(c) 3% improvement was observed compared to Dataset #3. This improvement is because we considered various vehicle types, which contributed to the enhanced accuracy.

Based on the above results, Table 2 categorizes each dataset in terms of accuracy level, false positive level and overall level. In the table, a greater number of stars indicates a higher level. We observed that datasets based on remote-controlled cars and actual vehicle images reduced misrecognition rate and improved detection accuracy.

Table 2. Evaluation of each dataset.

Label	Accuracy Level	False Positive Level	Overall Level
Dataset #1	⋆	⋆	⋆
Dataset #2	⋆⋆	⋆	⋆⋆
Dataset #3	⋆⋆⋆	⋆⋆⋆	⋆⋆⋆
Dataset #4	⋆⋆⋆⋆⋆	⋆⋆	⋆⋆⋆⋆

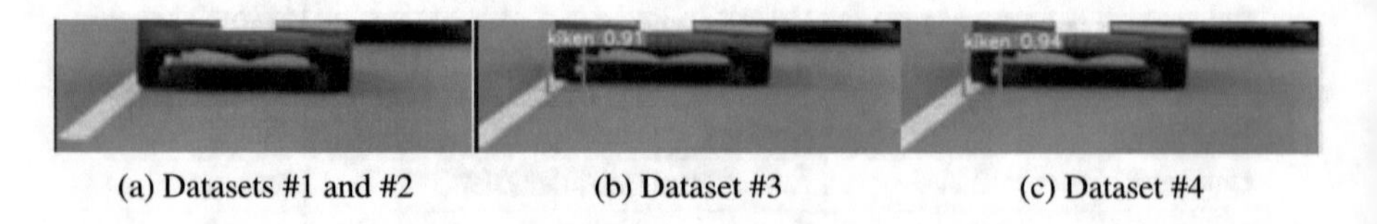

(a) Datasets #1 and #2 (b) Dataset #3 (c) Dataset #4

Fig. 3. Comparison of detection performance for each dataset.

5 Conclusions

In this paper, we proposed an intelligent safety drive assisting system that identifies dangerous vehicles. We presented an AI application and its interface, which can identify vehicles engaging in risky driving and notifies both the control center and the driver. We discussed the detection performance of four different datasets. From the evaluation results, we observed that datasets based on remote-controlled cars and actual vehicle images reduced misrecognition rate and improved detection accuracy.

In the future research, we aim to enhance the proposed system by incorporating other dangerous driving behaviors and parameter tuning.

References

1. Kaggle: Data science community. https://www.kaggle.com/
2. Roboflow: The world's largest collection of open source computer vision datasets and apis. https://universe.roboflow.com/
3. Bochkovskiy, A., Wang, C.Y., Liao, H.Y.M.: YOLOv4: Optimal speed and accuracy of object detection. Computer Vision and Pattern Recognition (cs.CV), April 2020. https://arxiv.org/abs/2004.10934
4. Cabinet Office, G.o.J.: 2023 white paper on traffic safety (2023). https://www8.cao.go.jp/koutu/taisaku/r05kou_haku/index_zenbun_pdf.html
5. Chen, G., Wang, F., Li, W., Hong, L., Conradt, J., Chen, J., Zhang, Z., Lu, Y., Knoll, A.: NeuroIV: neuromorphic vision meets intelligent vehicle towards safe driving with a new database and baseline evaluations. IEEE Trans. Intell. Transp. Syst. **23**(2), 1171–1183 (2022). https://doi.org/10.1109/TITS.2020.3022921
6. Hinton, G.E., Osindero, S., Teh, Y.W.: A fast learning algorithm for deep belief nets. Neural Comput. **18**(7), 1527–1554 (2006)
7. Jocher, G.: The project page of Ultralytics YOLOv5 (2020). https://github.com/ultralytics/yolov5/wiki
8. Le, Q.V.: Building high-level features using large scale unsupervised learning. In: Proceedings of IEEE International Conference on Acoustics, Speech and Signal Processing 2013 (ICASSP-2013), pp. 8595–8598, May 2013.https://doi.org/10.1109/ICASSP.2013.6639343
9. Li, B., Chen, J., Huang, Z., Wang, H., Lv, J., Xi, J., Zhang, J., Wu, Z.: A new unsupervised deep learning algorithm for fine-grained detection of driver distraction. IEEE Trans. Intell. Transp. Syst., 1–13 (2022) https://doi.org/10.1109/TITS.2022.3166275
10. Mnih, V., Kavukcuoglu, K., Silver, D., Rusu, A.A., Veness, J., Bellemare, M.G., Graves, A., Riedmiller, M., Fidjeland, A.K., Ostrovski, G., Petersen, S., Beattie, C., Sadik, A., Antonoglou, I., King, H., Kumaran, D., Wierstra, D., Legg, S., Hassabis, D.: Human-level control through deep reinforcement learning. Nature **518**, 529–533 (2015). https://doi.org/10.1038/nature14236
11. Poon, Y.S., Lin, C.C., Liu, Y.H., Fan, C.P.: YOLO-based deep learning design for in-cabin monitoring system with fisheye-lens camera. In: Proceedings of the IEEE International Conference on Consumer Electronics (ICCE-2022), pp. 1–4, January 2022. https://doi.org/10.1109/ICCE53296.2022.9730235
12. Redmon, J., Divvala, S., Girshick, R., Farhadi, A.: You only look once: unified, real-time object detection. In: Proceedings of the IEEE Conference on Computer Vision and Pattern Recognition (CVPR-2016). pp. 779–788, June 2016. https://doi.org/10.1109/CVPR.2016.91
13. Silver, D., et al.: Mastering the game of Go with deep neural networks and tree search. Nature **529**, 484–489 (2016). https://doi.org/10.1038/nature16961

14. Silver, D., et al.: Mastering the game of Go without human knowledge. Nature **550**, 354–359 (2017). https://doi.org/10.1038/nature24270
15. Simonyan, K., Zisserman, A.: Very deep convolutional networks for large-scale image recognition. In: Proceedings of the 3rd International Conference on Learning Representations (ICLR-2015), May 2015. https://doi.org/10.48550/arXiv.1409.1556
16. Ugli, I.K.K., Hussain, A., Kim, B.S., Aich, S., Kim, H.C.: A transfer learning approach for identification of distracted driving. In: Proceedings of the 24th International Conference on Advanced Communication Technology (ICACT-2022), pp. 420–423, February 2022. https://doi.org/10.23919/ICACT53585.2022.9728846
17. Ultralytics: The project page of Ultralytics YOLOv8 (2023). https://github.com/ultralytics/ultralytics
18. Vicente, F., Huang, Z., Xiong, X., la Torre, F.D., Zhang, W., Levi, D.: Driver gaze tracking and eyes off the road detection system. IEEE Trans. Intell. Transp. Syst. **16**(4), 2014–2027 (2015). https://doi.org/10.1109/TITS.2015.2396031
19. Xing, Y., Lv, C., Wang, H., Cao, D., Velenis, E., Wang, F.Y.: Driver activity recognition for intelligent vehicles: a deep learning approach. IEEE Trans. Veh. Technol. **68**(6), 5379–5390 (2019). https://doi.org/10.1109/TVT.2019.2908425

Multiclassification Analysis of Volumetric, Protocol, and Application Layer DDoS Attacks

Eric Brown[1,2], John Fisher[1,2], Aaron Hudon[1,2], Erick Colston[1,2], and Wei Lu[1,2](✉)

[1] Department of Computer Science, Keene State College, Keene, NH, USA
wlu@usnh.edu

[2] The University System of New Hampshire, Concord, USA

Abstract. In today's digital landscape, Distributed Denial of Service (DDoS) attacks represent a constantly evolving and significant threat, interrupting online services via many attack vectors. These attacks universally aim to render websites and services inoperable. Developing effective detection strategies is imperative as DDoS attacks become more frequent, varied, and destructive. This paper introduces an in-depth multiclassification analysis of DDoS attacks, categorizing them into Volumetric, Protocol, and Application Layer attacks. Utilizing the extensive CICDDoS2019 dataset, which includes eleven specific DDoS attack variations, we systematically classify these variations. Our analysis employs six diverse Machine Learning (ML) models: Logistic Regression (LR), Decision Tree (DT), Random Forest (RF), Support Vector Machine (SVM), Neural Networks (NN), and Extreme Gradient Boosting (XGB). We aim to identify the most effective model for predicting DDoS traffic across each attack category and to ascertain the model with superior overall performance. The findings of this study provide valuable insights into the effectiveness of various ML techniques in countering DDoS attacks, thereby contributing to the fortification of digital infrastructures against this pervasive threat.

1 Introduction

Recently, there has been much growth in networked devices in use globally and an apparent rise in demand. People, businesses, and governments are increasingly relying on this technology. While many benefits come from this, it opens the door to many potential vulnerabilities and attack types [1]. One of the most prominent attack types seen regularly is distributed denial-of-service (DDoS) attacks [2, 3]. A distributed denial of service (DDoS) attack constitutes a malicious effort to render an online service inaccessible to users, typically achieved by temporarily disrupting or suspending the functions of its hosting server [4]. While these attacks have a simple enough concept, their effects can be detrimental to a person or company. With a few compromised networked devices, malicious traffic can be sent to disrupt and crash web-based or network resources. Moreover, DDoS attacks can be difficult to detect manually as there are many possible strategies for carrying out these attacks. They can be implemented at different layers within the TCP/IP model, such as application, transport, and network layers) and exploit several protocols, such as UDP, TCP, SNMP, and LDAP.

L. Barolli (Ed.): AINA 2024, LNDECT 204, pp. 401–413, 2024.
https://doi.org/10.1007/978-3-031-57942-4_39

As a result, these attacks can swiftly render the services of a target effectively useless until a remediation process can be completed. This can be extremely difficult depending on the severity, type of DDoS attack, and the victim. Mirai is a malicious software that infects smart devices using ARC processors, transforming them into remotely controlled bots or "zombies." These infected devices form a network known as a botnet, commonly utilized for initiating DDoS attacks [5, 6]. In September 2016, Mirai executed its inaugural major assault targeting a French technology firm, OVH. The attack reached an unparalleled peak of 1 terabit per second (Tbps) and is approximated to have engaged approximately 145,000 devices in the offensive maneuver [7]. AWS reported mitigating a massive DDoS attack in February 2020. At its peak, this attack saw incoming traffic at 2.3 terabits per second (Tbps) [8]. A notable example of a severe DDoS attack against Google Cloud occurred in August 2023 [9, 10]. This attack is noted to be one of the largest attacks in history, with its peak traffic volume of 398 million requests per second (RPS), indicating that the capabilities of attackers are growing rapidly, which must be addressed by creating new DDoS detection methods with fast and accurate performance. By contrast, in 2022, the largest recorded DDoS attack peaked at 46 million reps [11].

This paper performs two multi-class classification experiments based on this growing concern. Using the CICDDoS2019 dataset, we have grouped the available attack data into Volumetric, Protocol, and Application Layer DDoS attacks [22]. Within these three categories, there are eleven different types of DDoS attacks, which are DNS Flood, UDP Flood, UDP Lag, SYN Flood, TFTP, LDAP, MSSQL, NetBIOS, NTP, SNMP, and SSDP Attacks. Before analyzing our ML models, we performed feature selection to identify some of the most influential and relevant features contributing to detecting an attack. After selecting these features, we then analyze and compare the performance of six ML algorithms on our datasets to find three answers to two fundamental questions: (1) which algorithm can perform best for each of the three DDoS attack categories; (2) which algorithm can perform best when classifying based on the eleven individual attack types and then explore how these happen. The ML algorithms we use are Logistic Regression (LR) [25], Decision Tree (DT) [26], Random Forest (RF) [27], Support Vector Machine (SVM) [28], K-Nearest Neighbors (KNN) [29] and Extreme Gradient Boosting (XGB) [30].

2 Related Work

Recent studies have explored various machine learning techniques for detecting DDoS attacks, emphasizing accuracy, efficiency, and robustness [12, 13]. Approaches include Higher Order Singular Value Decomposition, Random Forest, and Gradient Boosting with datasets like CICDDoS2019 and BoT-IoT [23]. Some focus on IoT networks, employing unique feature extraction methods and classifiers. Others explore neural networks and advanced algorithms like CatBoost for traffic classification [24]. In this section, we highlight the evolving nature of DDoS attack detection, with a trend towards high-accuracy, real-time capable models.

The study in [14] presents a new DDoS attack detection method using higher-order singular value decomposition and machine learning techniques like decision trees, achieving better accuracy and lower false positives. It highlights the method's robustness against errors but notes its higher computational demand, suggesting a focus on real-time application viability and future exploration of deep learning algorithms. In [16], Batchu and Seetha's research focuses on improving DDoS attack detection in the face of increasing network traffic and sophisticated attack methods. Addressing data quality issues affecting model accuracy, they utilize the CICDDoS2019 dataset with KNORA-E and KNORA-U algorithms, achieving exceptional accuracy between 99.9878% and 99.9909%. Their study thoroughly compares their model with other methods, demonstrating its superior accuracy, recall, precision, and efficiency, positioning it as an effective solution for DDoS attack detection in diverse. Chu et al. [15] utilize a Random Forest model to detect DDoS attacks in vulnerable network servers, focusing on the transport layer. This model, featuring 24 key attributes, outperforms previous ones with a 97% accuracy rate across various attack types. It shows high precision, recall, and F1 scores, particularly effective in HTTP attack detection. The study emphasizes the model's simplicity and accuracy, illustrating the effectiveness of their approach in DDoS attack detection.

In [17], Ma et al. developed a comprehensive DDoS detection model termed "feature and model selection" (FAMS). This methodology, divided into data preprocessing, feature selection, machine model selection, and parameter optimization, resulted in a Random Forest model with a notable 99.99% accuracy and swift processing time (0.216050 s). FAMS's structured, multi-stage approach ensures high accuracy and offers adaptability for future models and datasets, showcasing its practicality and efficiency in optimizing machine-learning models for DDoS detection. Parfenov et al. [18] focus on a machine learning method to classify network traffic in IoT networks, distinguishing benign activities from DDoS attacks. The study introduces a novel approach using altered hash values in device configuration files for identifying malicious activities, highlighting their method's effectiveness, especially with CatBoost and Gradient Boosting algorithms, achieving high scores in precision, recall, and F1 metrics. The research concludes successfully, with plans to refine the accuracy for multiclass classification scenarios.

In [19], Talaei Khoeiand and Kaabouch thoroughly evaluated nine machine learning models to identify the most effective for specific tasks. The study highlights the total length of forward packets, flow byte, flow packet, flow IAT mean, and flow IAT standard deviation as key features. Among the models, Alex Neural Networks (ANN) and Variational Autoencoder (VA) stand out, with ANN achieving a higher accuracy of 98.71% and faster processing time (1.29 s) compared to VA's 96.7% accuracy and 3.3 s processing time. In [20], Li et al. developed a strategy for detecting volumetric DDoS attacks, focusing on volume-based attacks like HTTP and TCP/UDP floods. It identifies key characteristics such as traffic size, source information, and response time to malicious traffic. The approach uses sliding time windows and information entropy for accuracy. The model, employing a Long Short-Term Memory (LSTM) system, undergoes a three-stage process of traffic analysis, entropy calculation, and attack detection. The results show high accuracy and low false positives, with future research directions including IoT network applications and varied DDoS attack types. In [19], Navruzov and Kabulov tackle detecting DDoS attacks in large networks, highlighting difficulties due to device diversity and high traffic. It proposes using data mining for enhanced security models, focusing on big data preprocessing and machine learning. The study uses the CICDDoS2019 dataset to analyze relationships between attack types and normal traffic, underscoring the importance of feature stability. Findings indicate that Random Forest and K-NN are effective for DDoS detection, especially when feature selection is stability-focused.

3 Volumetric, Protocol, and Application Attack Taxonomy

Utilizing the CIC-DDoS2019 dataset, we introduce a novel taxonomy to categorize eleven types of attack packets into three distinct classes: volumetric, application, and protocol attacks [22], which are explained in the following.

Volumetric Attacks: This category includes DNS, NTP, UDP, UDP-Lag, and Syn attacks. The unifying characteristic of these attacks is their strategy to overwhelm a network with a deluge of illegitimate packets. By grouping these attacks, we aim to refine the model's capability to effectively classify volumetric attacks, which might be less discernible when treated as separate entities.

Application Layer Attacks: This classification encompasses TFTP and MSSQL attacks. MSSQL attacks specifically exploit a system's SQL server to deplete resources, while TFTP attacks manipulate the trivial file transfer protocol. Both target the application layer of a computer's OSI model. Grouping these attacks enhances the model's precision in identifying threats targeting the application layer.

Protocol Attacks: This category includes SSDP, SNMP, NetBIOS, and LDAP attacks. Although targeting the application layer, they are distinguished by their method of monopolizing system resources using protocols, resulting in a packet structure different from application layer attacks. This categorization is designed to improve the model's generalization capabilities in identifying new types of protocol-based attacks.

The SDG Classifier was employed in our analysis to extract feature significance across the established attack categories. Figure 1 presents an in-depth breakdown of how the machine-learning model interprets various attack types. This includes a mapping of attack-type labels to coefficients using a predefined dictionary. The top 5 features with the highest absolute weights are identified for each attack type. These weights are crucial as they indicate the features' substantial impact on the prediction accuracy of the attack. The absolute weight of each feature quantifies its contribution, with higher values signifying a more pronounced influence on the model's decision-making process. Furthermore, the mean values of these significant features in the training dataset are provided. These averages offer valuable insights into the typical characteristics associated with each type of attack, thereby enhancing our understanding of the attack patterns. This detailed exposition not only improves the transparency of the model but also serves as a vital resource for cybersecurity professionals. It aids them in comprehending the relevance and impact of specific features, thereby bolstering effective threat-detection strategies.

Attack Type	Feature	Weight	Mean Value
BENIGN	Fwd Packet Length Min	-0.573648	0.00316783
BENIGN	Fwd Packet Length Mean	-0.585541	0.00315004
BENIGN	Avg Fwd Segment Size	-0.585541	0.00315004
BENIGN	URG Flag Count	1.28976	-0.000393417
BENIGN	Bwd Packet Length Min	1.52892	-0.000483885
VOLUMETRIC	Bwd Packet Length Min	-0.372433	-0.000483885
VOLUMETRIC	Bwd Packet Length Std	-0.67398	-0.00036441
VOLUMETRIC	Init_Win_bytes_forward	-1.13436	0.00156095
VOLUMETRIC	act_data_pkt_fwd	1.14947	-0.00379195
VOLUMETRIC	ACK Flag Count	1.24143	-0.00138829
PROTOCOL	Average Packet Size	0.526802	0.00331103
PROTOCOL	Bwd Packet Length Min	-0.60519	-0.000483885
PROTOCOL	CWE Flag Count	-0.947994	-0.00123488
PROTOCOL	Flow IAT Mean	-1.26262	0.00224622
PROTOCOL	Init_Win_bytes_forward	-4.01162	0.00156095
APPLICATION LAYER	Bwd Packet Length Min	-0.802017	-0.000483885
APPLICATION LAYER	Idle Max	-1.31773	0.002554
APPLICATION LAYER	Init_Win_bytes_forward	-1.36022	0.00156095
APPLICATION LAYER	CWE Flag Count	-1.36688	-0.00123488
APPLICATION LAYER	Fwd Packet Length Std	-1.58547	-0.000270804

Fig. 1. Breakdown of the machine learning's interpretation for different attack types

4 Methodology

Various datasets are currently utilized in developing Machine Learning models for DDoS attack detection. Among these, the CIC-DDoS2019 dataset stands out due to its comprehensive nature. It encompasses a substantial amount of data, approximately 30 GB, and includes a detailed array of 88 distinct features. This dataset is particularly notable for having eleven different types of DDoS attacks, providing a broad spectrum for analysis and model training. The covered attack types are NTP, DNS, LDAP, MSSQL, NetBIOS, SNMP, SSDP, UDP, UDP-Lag, Syn, and TFTP. The diversity and volume of data in the CIC-DDoS2019 dataset make it an invaluable resource for developing robust and effective DDoS detection algorithms in ML.

The optimization of the CIC-DDoS2019 dataset was our initial focus, aiming to ensure a robust and reliable dataset for testing our machine learning models. This process involved several critical steps: (1) **Data Cleaning:** We began by removing all null values, duplicated entries, and infinite values from the dataset. This step was crucial to enhance the dataset's reliability and accuracy; (2) **Feature Selection and Omission:** Certain features were omitted due to their lack of relevance or potential to contribute meaningfully to our analysis. These include 'Unnamed: 0', 'Flow ID,' and 'Inbound.' Additionally, we excluded features containing source/destination IP and port numbers and the 'Timestamp' feature. The rationale was to prevent data leakage, which could otherwise result in model overfitting; **(3) Label Encoding:** For most of the dataset, label encoding was conducted. This was particularly essential for the 'Label' variable for classifying DDoS types. We programmatically assigned an integer representation to benign traffic (labeled as 0) and various attack types (marked as 1–11); and (4) **Data Scaling:** The final preprocessing step involved scaling our data using the StandardScaler method. This standardization ensures that the scaled features have a mean of 0 and a standard deviation of 1. Such scaling is beneficial for better handling outliers and improving the model's performance. These preprocessing steps were carried out to prepare the dataset for effective and accurate machine learning model application, laying a solid foundation for our subsequent analysis.

Two datasets were required, one for binary classification and one for multi-classification. This is because the data needs to be balanced to ensure no bias in the machine learning model. Each category must have the same number of entries as the others to mitigate the bias. To elaborate, binary classification has two classifications: benign and attack packets. The attack packets are made up of eleven types of attacks, and benign is made up of 60,000 entries. This means the attack category cannot total over 60,000. One can first find the smallest entries for the eleven attack types. This came to be UDP-Lag with 4,818 entries after data preprocessing. Next, 4,818 packets of the other attack types must be compiled together. After this was completed, the attack category totaled 52,998 entries. To create the final binary classification dataset, 52,998 benign entries were taken to balance the data. The final dataset for binary classification counted 105,996 entries. A binary label column was also added to represent the type of attack. Benign was represented by 0, and malicious packets were noted by 1.

Similar steps are being followed to extract the dataset for our multi-class classifications. However, because the categories differed, the data must be balanced differently. The smallest traffic category was benign at 60,000 entries, so the other categories must contain 60,000 entries. The volumetric category is made up of five attacks. This means that 12,000 entries from each specific attack needed to be included in the final category for the volumetric classification. Next, the protocol category is required to be created. It comprises four attacks, so 15,000 entries must be made from each attack. Then, the application layer category was formed by taking 30,000 packets from the TFTP and MSSQL attacks. Lastly, all the categories were combined into one CSV file, and a new column named 'Multi_Class' was appended. This column represented the packet type by assigning an integer 0–3 to a row. Benign was represented by 0, volumetric was represented by 1, protocol was represented by 2, and application layer attacks were denoted by 3. With the sub-datasets extracted, we moved on to further preprocessing and feature extraction.

One aspect to note in the classification based on DDoS types is that each traffic attack type has widely varying amounts of entries. Benign traffic consists of 60,000 out of 240,000 entries, while some attack types, such as NetBIOS, only contain 12,000. To resolve such a significant imbalance, we randomly dropped values from classes until each class had 12,000 values, leading to 144,000 entries.

The CIC2019DDoS dataset is notably large, comprising over 80 features across more than 200,000 traffic records. Not all features may significantly contribute to the detection of DDoS attacks. Considering the increased computational demands of using the entire feature set, it was essential to devise a strategy for identifying the ten most influential features. To address this challenge of high dimensionality, we employed the Sequential Forward Selection (SFS) method. SFS is a wrapper-based feature selection algorithm, and for our purposes, it was configured to utilize a Random Forest Classifier as its estimator. This algorithm starts from a baseline of zero selected features. It iteratively adds one feature at a time, basing the selection on each feature's improvement to the model's accuracy. Crucially, we integrated a 5-fold cross-validation within the SFS process. This approach helps prevent overfitting and offers a more accurate estimation of the model's performance as the feature selection unfolds. For consistency and to maintain a unified approach across our study, we decided to use the same top ten features identified by SFS for all classifications in our research, which includes one binary classification and two multi-class classifications. This systematic approach in feature selection is pivotal, as it ensures that our model is both efficient and effective, focusing only on the most impactful features for DDoS attack detection.

Now that our dataset is optimized and we have established our ten most influential features for DDoS detection, we can proceed to modeling strategies. We utilize the ML algorithms LR, DT, RF, KNN, SVM, and XGB, as mentioned. For each model, we use a Bayesian Search with 30 iterations to optimize our models' hyperparameters. The Bayesian Search algorithm explores various hyperparameter combinations for each given ML model, repeatedly training models until completion. Once completed, the best results are chosen based on the highest accuracy rating. The results will contain the top ten feature names and the training accuracy rating associated with that model. We then create the base model for each algorithm, where no parameters are given and default values are used. This will establish a baseline accuracy we can compare to our optimized model. The optimized model can be constructed with the hyperparameters given by Bayesian Search, from which we then test this model and compare its accuracy, recall, F1, and precision to the base. We repeat these steps to classify DDoS attacks into three categories and 11 attack types. Figure 2 below illustrates the flowchart of our methodology.

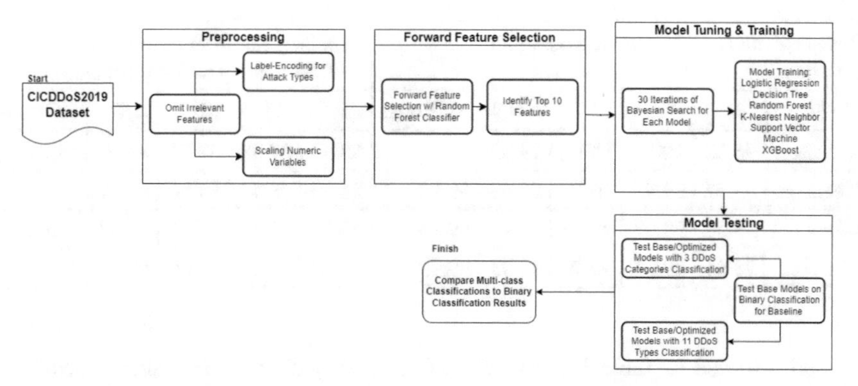

Fig. 2. Framework of the proposed methodology.

5 Result Analysis

The initial phase of our results analysis involves examining the critical features selected through our Sequential Forward Selection (SFS) implementation. These features are essential in differentiating between benign and DDoS-related traffic. We will delve into the first three features, which are ranked by their overall influence on the model:

Flow Bytes/s: This feature indicates the byte rate transmitted through a network flow. An abnormally high value in this metric could suggest involvement in a DDoS attack, particularly those of the Volumetric category. Such patterns are often observed in attacks where large volumes of data are sent to overwhelm the target.

Fwd Header Length: Representing the total length of a packet's header, this feature is particularly telling in specific DDoS strategies, such as DNS Amplification attacks. These attacks typically utilize disproportionate packet headers, where a small DNS query triggers a substantially larger response packet, overwhelming the victim's network. This feature is, therefore, a strong indicator of DDoS activity, especially in attacks like NTP and LDAP.

Min Packet Length: The shortest packet length within a flow is denoted by this feature. Its significance stems from certain DDoS attack types that use small, maliciously crafted packets to congest system resources. Examples include UDP and SYN Floods, categorized as Volumetric attacks in our dataset.

The patterns and behaviors identified by these features predominantly align with the detection of Volumetric DDoS attacks, likely due to the similar characteristics shared among these types of attacks. Following this feature analysis, we tested baseline binary classification with various ML algorithms. This testing yielded encouraging results across almost all algorithms, setting a positive foundation for further in-depth analysis and model refinement.

Table 1. Results given from base ML models using binary and 3-category multiclassification.

Results given from ML models using binary classification						Results given from base ML models with 3-category multiclassification				
Model	LR	DT	RF	KNN	XGBoost	LR	DT	RF	KNN	XGBoost
Accuracy	96.32	100	100	99.93	100	55.59	87.58	87.71	88.03	88.01
Precision	93.52	99.98	100	99.98	99.99	56.68	88.43	88.56	88.37	88.69
Recall	99.50	99.97	99.99	99.97	99.99	55.59	87.58	87.71	88.03	88.01
F1-Score	98.64	99.97	99.99	99.93	99.99	52.98	87.54	87.66	88.15	87.99

As illustrated in Table 1, the results of our binary classification show outstanding performance across all metrics displayed. While LR has the lowest performance out of our selection, a 96.32% accuracy is quite good. Despite these high scores, we believe there may be some issues about overfitting. This could be caused by errors within our dataset and preprocessing or potential data leakage during model training. All three of our tree-based algorithms score 100% accuracy, which is unusually high. Given that these were allowed to run with no specified parameters, these algorithms may have created trees with enough complexity to make perfect classifications, which could cause additional overfitting.

Upon testing our base models with no given hyperparameters, we see a good performance from the DT, RF, KNN, and XGBoost algorithms, as illustrated in the right part of Table 1. However, we noted that LR appears to perform poorly in this case. This makes sense, as LR tends to struggle with highly complex datasets. Additionally, LR is known to work under multi-class settings, which could also contribute to its poor performance. Granted, this model had no specified parameters, which further contributed to the poor accuracy, precision, recall, and F1 score. On the other hand, we see that of all algorithms, KNN scored the highest accuracy rating of 88.03%. We did not expect to see this, but ultimately, it is a sensible result. In short, KNN works by classifying incoming data entries based on the "k-nearest-neighbors." Depending on one entry's values, it will be classified based on which existing class holds many similar values. Since these DDoS attacks work similarly in their respective categories, it's sensible that KNN performs well here. Lastly, our tree-based algorithms (DT, RF, and XGB) have notable performances. DTs alone excel under these use cases of multi-class classification with highly complex data in large volumes. Given that RF and XGB are advancements of the DT algorithm, their improved performance over DT is expected.

Upon testing our optimized models with hyperparameters given through the Bayesian search, we do not see much improvement in any algorithms (for some algorithms, there is a decrease) except LR and KNN, as illustrated in Table 2. We were left fairly confused by these results, as our implementation of Bayesian Search allows for almost all algorithm parameters to be tested at large ranges. We see definite performance improvements in optimizing LR, albeit still achieving a low accuracy rating of 66.40%. One notable parameter contributing to the performance increase is 'multi_class,' in which we test with the values 'auto,' 'over,' and 'multinomial.' This parameter is dedicated to multi-class use cases in which 'ovr', or the One-vs-Rest approach, trains a binary classifier to predict which class a data entry will be selected. The choice 'multinomial' uses a different approach, where probabilities are calculated for each class, and the current data entry will be classified based on the highest probability. Beyond the improvements in LR, we see slight decreases in the performance of our optimized tree-based models. We are unsure exactly why this is, as it could stem from several reasons. The parameter ranges we specified could be too large or small, the models could be overfitted in training, and issues could exist in our dataset, to name a few. We did note a minimal improvement in the optimized KNN model's accuracy of 0.03%. We aren't sure of what caused this, but we assume that the tested hyperparameters contributed to this improvement. This could have led to minimal improvement since KNN has few parameters to test. Also, the default parameters in our base model could have ended up like those in our optimized model.

Table 2 illustrates the multiclassification results on DDoS attacks by individual type, in which there is a notable drop in the performance of all algorithms except LR. Compared to classifying based on the 3 DDoS categories, LR performs much better in this case of categorizing based on the DDoS types. The remaining algorithms perform worse in this classification compared to their former classification by attack categories. A potential reason behind this could be that when split by individual types, specific attacks may have more detailed, prominent patterns/behavior than other types. This could likely aid the LR model in identifying particular classes. However, we see that this LR model has low precision and F1-score, which probably means that this model also has difficulty in accurately classifying true/false positives for one or more attack types. The drop in performance metrics could also be caused by the increase in complexity, classifying into 12 classes instead of 4. We also see reduced performance metrics in our tree-based models and KNN, which we somewhat attribute to the increase in complexity. Ultimately, our RF model scored the highest testing accuracy of 81.61%, but we see that other performance metrics are subpar. With an F1-score of 68.84%, this tells us that this model is struggling to make accurate classifications.

Table 2. Results given from optimized ML models with 3- and 11-type multiclassification.

Results given from optimized ML models with 3-category multiclassification						Results given from base ML models with 11-type multiclassification				
Model:	LR	DT	RF	KNN	XGBoost	LR	DT	RF	KNN	XGBoost
Accuracy:	66.40	87.51	87.64	88.06	87.97	56.93	72.11	81.61	70.33	73.89
Precision	61.09	88.46	88.52	88.44	88.69	53.94	74.30	75.11	71.48	75.91
Recall	61.40	87.51	87.64	88.06	87.97	56.932	72.11	72.31	70.33	73.89
F1-Score	59.15	87.41	87.58	88.19	87.94	52.22	68.72	68.84	69.32	72.51

6 Conclusions

Our findings indicate that the most effective approach for multi-class classification of DDoS attacks involves categorizing based on the three broad attack categories rather than differentiating among the 11 specific attack types. This strategic shift has yielded superior performance metrics across all algorithms in our study. The models developed for the 3-category classification consistently outperform those designed for classifying the 11 individual DDoS attack types across various performance metrics. Notably, the K-Nearest Neighbors (KNN) model has emerged as the top performer. It excels particularly in recall and F1 score, albeit with marginally lower precision than our tree-based models. The effectiveness of KNN can be attributed to its proficiency in scenarios where classes exhibit closely related values and characteristics. This aligns well with the three-category classification approach, where each category embodies distinct, recognizable strategies and behaviors during an attack. While slightly overshadowed by KNN in this context, the XGBoost algorithm has also demonstrated commendable performance. It stands out as a robust choice for our multi-class classification needs, further substantiating the effectiveness of our chosen methodology.

These insights are instrumental in guiding our ongoing efforts to refine DDoS attack detection and classification. The distinct advantages of the 3-category classification strategy and the strengths of the KNN and XGBoost models offer promising avenues for future enhancements in cybersecurity. In the future, we will further refine and optimize the KNN and XGBoost models and explore other approaches for detecting, evaluating, and managing the alerts generated by our detection system [31–33].

Acknowledgments. This research is supported by New Hampshire - INBRE through an Institutional Development Award (IDeA), P20GM103506, from the National Institute of General Medical Sciences of the NIH.

References

1. Ghorbani, A.A., Wei, L., Tavallaee, M.: Network Attacks. In: Ghorbani, A.A., Wei, L., Tavallaee, M. (eds.) Network Intrusion Detection and Prevention, pp. 1–25. Springer US, Boston, MA (2010). https://doi.org/10.1007/978-0-387-88771-5_1
2. Garant, D., Lu, W.: Mining botnet behaviors on the large-scale web application community. In: Proceedings of 27th IEEE International Conference on Advanced Information Networking and Applications, Barcelona, Spain, March 25–28 (2013)
3. Lu, W., Miller, M., Xue, L.: Detecting command and control channel of botnets in cloud. In: Traore, I., Woungang, I., Awad, A. (eds.) ISDDC 2017. LNCS, vol. 10618, pp. 55–62. Springer, Cham (2017). https://doi.org/10.1007/978-3-319-69155-8_4
4. DDoS Attack Types & Mitigation Methods: Imperva. DDoS Attacks, 3 Oct. 2023. https://www.imperva.com/learn/ddos/ddos-attacks/
5. Mirai Botnet. What is the Mirai Botnet? https://www.cloudflare.com/learning/ddos/glossary/mirai-botnet/. Accessed 18 Dec 2023
6. Lu, W., Ghorbani, A.A.: Bots behaviors vs. Human behaviors on large-scale communication networks (Extended Abstract). In: Lippmann, R., Kirda, E., Trachtenberg, A. (eds.) Recent Advances in Intrusion Detection. RAID 2008. Lecture Notes in Computer Science, vol. 5230. Springer, Berlin, Heidelberg (2008). https://doi.org/10.1007/978-3-540-87403-4_33
7. CIS. Blog: The Mirai Botnet - Tips to Defend Your Organization. 30 July 2021. https://www.cisecurity.org/insights/blog/the-mirai-botnet-threats-and-mitigations
8. Famous DDoS Attacks. The largest DDoS attacks of all time. https://www.cloudflare.com/learning/ddos/famous-ddos-attacks/. Accessed 15 Dec 2023
9. Kiner, E, April, T.: Google Mitigated the Largest DDoS Attack to Date, Peaking above 398 Million RPS. Google cloud mitigated largest DDos attack, peaking above 398 million RPS, Google Cloud Blog, October 10, 2023. https://cloud.google.com/blog/products/identity-security/google-cloFud-mitigated-largest-ddos-attack-peaking-above-398-million-rps
10. Vaughan-Nichols, S.: Google Cloud, AWS, and Cloudflare report largest DDoS attacks ever, October 10, 2023. https://www.zdnet.com/article/google-cloud-aws-and-cloudflare-report-largest-ddos-attacks-ever/
11. Kiner, E, Konduru, S.: How Google Cloud Blocked Largest Layer 7 DDos Attack yet, 46 Million RPS. Google, August 18, 2022. https://cloud.google.com/blog/products/identity-security/how-google-cloud-blocked-largest-layer-7-ddos-attack-at-46-million-rps
12. Lu, W., Mercaldo, N., Tellier, C.: Characterizing command and control channel of mongoose bots over TOR. In: Woungang, I., Dhurandher, S.K. (eds.) WIDECOM 2020. LNDECT, vol. 51, pp. 23–30. Springer, Cham (2020). https://doi.org/10.1007/978-3-030-44372-6_2

13. Nunley, K., Lu, W.: Detecting network intrusions using a confidence-based reward system. In: 2018 32nd International Conference on Advanced Information Networking and Applications Workshops (WAINA), pp. 175–180 (2018). https://doi.org/10.1109/WAINA.2018.00083
14. Maranhão, A., et al.: Error-robust distributed denial of service attack detection based on an average common feature extraction technique. Sensors **20**(20), 5845 (2020). https://doi.org/10.3390/s20205845
15. Chu, T.S., Si, W., Simoff, S., Nguyen, Q.V.: A machine learning classification model using random forest for detecting DDoS attacks. In: 2022 International Symposium on Networks, Computers and Communications (ISNCC), Shenzhen, China, pp. 1–7 (2022).https://doi.org/10.1109/ISNCC55209.2022.9851797
16. Batchu, R.K., Seetha, H.: An integrated approach explaining the detection of distributed denial of service attacks. Comput. Netw. **216**, 109269 (2022). https://www.sciencedirect.com/science/article/pii/S1389128622003334
17. Ma, R.K., Chen, X.B., Zhai, R.: A DDos Attack Detection Method Based on Natural Selection of Features and Models. MDPI, February 20, 2023. https://www.mdpi.com/2079-9292/12/4/1059#:~:text=In%20this%20paper%2C%20we%20propose,divided%20into%20four%20main%20phases
18. Parfenov, D., Kuznetsova, L., Yanishevskaya, N, Bolodurina, I., Zhigalov, A., Legashev, L.: Research application of ensemble machine learning methods to the problem of multiclass classification of DDoS attacks identification. In: 2020 International Conference Engineering and Telecommunication (En&T), Dolgoprudny, Russia, pp. 1–7 (2020).https://doi.org/10.1109/EnT50437.2020.9431255
19. Talaei Khoei, T., Kaabouch, N.: A comparative analysis of supervised and unsupervised models for detecting attacks on the intrusion detection systems. Information **14**(2), 103 (2023). https://doi.org/10.3390/info14020103
20. Li, J., Liu, M., Xue, Z., Fan, X., He, X.: RTVD: a real-time volumetric detection scheme for DDoS in the Internet of Things. IEEE Access **8**, 36191–36201 (2020). https://doi.org/10.1109/ACCESS.2020.2974293
21. Navruzov, E., Kabulov, A.: Detection and analysis types of DDoS attack. In: 2022 IEEE International IOT, Electronics and Mechatronics Conference (IEMTRONICS), Toronto, ON, Canada, pp. 1–7 (2022).https://doi.org/10.1109/IEMTRONICS55184.2022.9795729
22. Sharafaldin, I., Lashkari, A.H., Hakak, S., Ghorbani, A.A.: Developing realistic distributed denial of service (DDoS) attack dataset and taxonomy. In: IEEE 53rd International Carnahan Conference on Security Technology, Chennai, India (2019)
23. Koroniotis, N., Moustafa, N., Sitnikova, E., Turnbull, B.: Towards the development of realistic botnet dataset in the internet of things for network forensic analytics: Bot-IoT dataset. Futur. Gener. Comput. Syst. **100**, 779–796 (2019)
24. Hancock, J.T., Khoshgoftaar, T.M.: CatBoost for big data: an interdisciplinary review. J Big Data **7**, 94 (2020). https://doi.org/10.1186/s40537-020-00369-8
25. Kleinbaum, D.G., Klein, M.: Logistic Regression. Springer New York, New York, NY (2010). https://doi.org/10.1007/978-1-4419-1742-3
26. Fürnkranz, J.: Decision Tree. In: Sammut, C., Webb, G.I. (eds.) Encyclopedia of Machine Learning, pp. 263–267. Springer US, Boston, MA (2010). https://doi.org/10.1007/978-0-387-30164-8_204
27. Breiman, L.: Random forests. Mach. Learn. **45**, 5–32 (2001)
28. Christmann, A., Steinwart, I.: Support Vector Machines, Springer (2008). https://doi.org/10.1007/978-0-387-77242-4
29. Mucherino, A., Papajorgji, P.J., Pardalos, P.M.: K-Nearest Neighbor Classification. In: Mucherino, A., Papajorgji, P.J., Pardalos, P.M. (eds.) Data Mining in Agriculture, pp. 83–106. Springer New York, New York, NY (2009). https://doi.org/10.1007/978-0-387-88615-2_4

30. Bartz-Beielstein, T., Chandrasekaran, S., Rehbach, F.: Case Study II: Tuning of Gradient Boosting (xgboost). In: Bartz, E., Bartz-Beielstein, T., Zaefferer, M., Mersmann, O. (eds.) Hyperparameter Tuning for Machine and Deep Learning with R: A Practical Guide, pp. 221–234. Springer Nature Singapore, Singapore (2023). https://doi.org/10.1007/978-981-19-5170-1_9
31. Ghorbani, A., Lu, W., Tavallaee, M.: Detection Approaches, Network Intrusion Detection and Prevention: Concepts and Techniques. Springer Publisher, pp. 27–53 (2009)
32. Ghorbani, A., Lu, W., Tavallaee, M.: Evaluation Criteria, In: Network Intrusion Detection and Prevention: Concepts and Techniques. Springer, pp. 161–183 (2009)
33. Ghorbani, A., Lu, W., Tavallaee, M.: Alert Management and Correlation, In: Network Intrusion Detection and Prevention: Concepts and Techniques. Springer, pp. 129–160 (2009)

A Classification Method of Image Feature Using Neural Metric Learning for Natural Environment Video

Yukito Seo[1(✉)], Rafly Arief Kanza[2], Nobuya Watanabe[3], Kin Fun Li[4], and Kosuke Takano[1]

[1] Department of Information and Computer Sciences, Kanagawa Institute of Technology, Atsugi, Japan
2021065@cco.kanagawa-it.ac.jp, takano@ic.kanagawa-it.ac.jp
[2] Politeknik Elektronika Negeri Surabaya, Surabaya, Indonesia
raflykanza@pasca.student.pens.ac.id
[3] Chubu Institute for Advanced Studies, Chubu University, Kasugai, Japan
nov@isc.chubu.ac.jp
[4] Department of Electrical and Computer Engineering, Faculty of Engineering, University of Victoria, Victoria, Canada
kinli@ece.uvic.ca

Abstract. This paper proposes an image feature classification method that applies a distance learning neural network to image feature vectors extracted from an autoencoder. There is active research on similar image retrieval methods using image feature vectors extracted from neural networks. If the image classification performance is not sufficient, it is possible to further improve it by applying a distance learning neural network to convert it into an image feature vector for obtaining appropriate ranking results. In the proposed method, by constructing a model that connects an autoencoder and a distance learning neural network, the reusability of image features extracted from the autoencoder is maintained. In addition, it allows the model to flexibly be combine the autoencoder and distance learning neural network for the model construction. In the experiment, we evaluate the image classification accuracy using an aerial photo dataset provided by the Geospatial Information Authority of Japan and confirm the feasibility of the proposed method.

1 Introduction

Climate change due to global warming, rising sea levels, desertification, and changes in ecosystems are urgent global-scale natural environmental problems to be solved. For this purpose, data analysis by constantly observing the natural environment and recording visual data such as movies and images plays an important role, since it allows us to understand natural phenomena over time and to extract similar natural phenomena. However, the amount of data required to record video data of observations of the natural environment in various regions of the world is extremely large, and the processing costs required for data analysis are expected to be very high. To tackle with this problem, we

L. Barolli (Ed.): AINA 2024, LNDECT 204, pp. 414–421, 2024.
https://doi.org/10.1007/978-3-031-57942-4_40

have proposed an image scene retrieval method using low-dimensional image feature vectors extracted by a neural network with input images [1]. However, when feature vectors are extracted using a neural network with many intermediate layers, such as a VGG convolutional neural network, we found that in many cases, there were unstable variations in the quality of the feature vectors obtained depending on the characteristics of intermediate layers.

In this study, we extend this method and propose an image feature classification method that applies an autoencoder as a network for extracting image features and a distance learning neural network. The reasons for using an autoencoder are that the network structure is not complex and easy to implement, and the number of dimensions of the image feature vectors obtained as the output of the encoder can be easily adjusted. However, since image classification accuracy is not always sufficient for the image feature vectors extracted from the autoencoder, a distance learning neural network is applied to extract more appropriate image feature vectors for the classification using positive and negative examples as training data, so that the accuracy of image classification can be improved. The proposed method, by constructing a model with a structure that connects an autoencoder and a distance learning neural network, makes it possible to maintain the reusability of image features extracted from the autoencoder and to realize flexible model construction by combining an autoencoder and a distance learning neural network.

In the experiments, we confirm the feasibility of the proposed method using an aerial photo dataset provided by the Geospatial Information Authority of Japan [6].

2 Related Research

Many researches have been conducted on applying autoencoders as image feature extractors where the extracted image feature vectors are incorporated into image retrieval.

In [2], for the purpose of image retrieval of Kaou, which is a symbol to authorize a writer of historical documents from medieval Japan, a machine learning-based shape analysis method for Kaou images is proposed. In [3], Hosoe et al. proposed a conditional autoencoder with a condition of character type label for handwriting analysis. By training only the image generation part of the decoder with the character type labels, the encoder extracts the writer's specific writing habits as features from the handwriting image, independent of the character type labels. In [4], Ishfaq proposed an autoencoder model that builds a neural network combining a variational autoencoder and triplet loss to learn embedded representations for extracting similarity concepts.

In addition, autoencoders are used as feature extractors for various data other than images, and methods such as [5], which constructs a neural network combining an autoencoder and triplet loss to detect attacks and intrusions from network traffic, have been proposed.

3 Proposed Method

The proposed method extracts compressed image features using an autoencoder and converts the image features into a classifiable feature vector by adapting a distance learning neural network. The converted feature vectors are suitable for classification and ranking.

The following describes the implementation procedure of the proposed method.

Step-1: Apply autoencoder to an image set $I = \{i_1, i_2,\ldots,i_n\}$ and train it to create a trained model *AE* from which image features can be extracted.

Step-2: Extract the image feature vector $V = \{v_1, v_2,...,v_n\}$ of the image set from the output of the intermediate layer of the encoder of *AE* constructed in Step-1 (Figure 1). The *m*-order tensor $T = \{t_1, t_2,...,t_m\}$, which is the output of the convolutional layer of *AE*, can be used as the image features (Figure 2). Where each t_x is a multidimensional vector. In this study, we call this *m*-order tensor the feature tensor.

Step-3: By applying the distance learning neural network *MLNN*, transform the image feature vector *V* or the image feature tensor *T* into a feature vector $V' = \{v'_1, v'_2,...,v'_n\}$ that is suitable for ranking and classification. Figure 3 show that the image feature vectors optimized into embedding vectors through the training of *MLNN* to adjust the distance relationship between similar feature vectors using positive and negative examples. In *MLNN*, the input layer is adapted according to the data format of the input image features. For example, a fully-connected layer is applied for the image feature vector input, and meanwhile, a convolution layer is applied for the image feature tensor input.

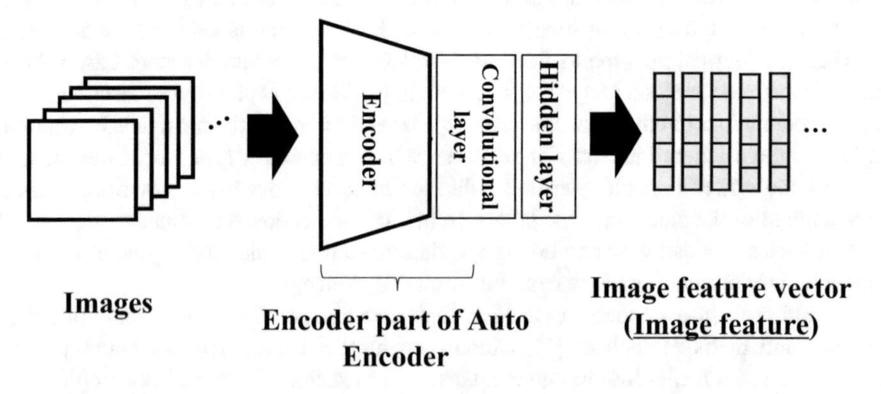

Fig. 1. Extraction of image feature vector from intermediate layer of autoencoder

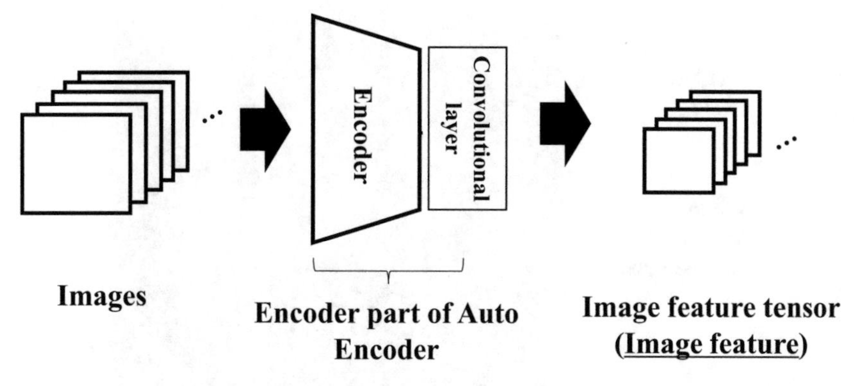

Fig. 2. Extraction of image feature tensor from convolutional layers of autoencoder

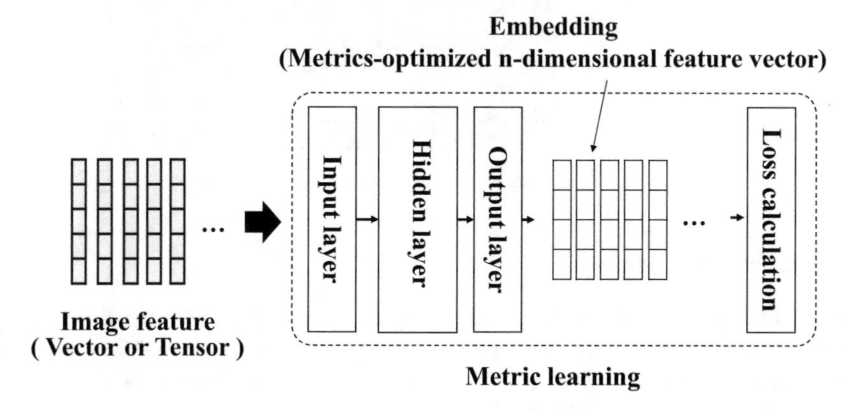

Fig. 3. Image feature optimization using distance learning neural networks

4 Experiment

We confirm the feasibility of the proposed method by evaluating the accuracy of image classification using the feature vectors extracted by the proposed method.

4.1 Experimental Environment

The aerial photo dataset provided by the Geospatial Information Authority of Japan [6] was used as the image dataset. The dataset consists of 576 aerial photographs, which are classified into four classes: "Fields," "Water Source Areas," "Forests," and "Buildings," with 144 images in each class. Figure 4 shows an example of photo data.

For comparison, we prepared for one type of auto encoder *AE* with a convolutional layer, and two types of distance learning neural networks: *NN*-1 without convolution layer, and *NN*-2 with a convolution layer. Autoencoder *AE* is trained using 576 aerial photographs. The distance learning neural network is trained using image features (image vectors or image tensors) or image data extracted from the autoencoder *AE* in a supervised manner.

Fig. 4. Example of aerial photo data (4 classes)

4.2 Experimental Method

We calculate the accuracy of classification by applying k-means for the following six models: $M1$-$M6$ and compare the results. In each neural network, we set the dimensions of the intermediate layer of the autoencoder and the output layer of the distance learning neural network to 1000, 500, and 100 dimensions.

$M1$: Consists of *AE* only: the output from the encoder of AE is used as a feature vector.
$M2$: Consists of *AE* only: the output of the convolutional layer in the encoder of *AE* is converted into a feature vector (32*32*64 dimensions).
$M3$: Consists of combination of *AE* of $M1$ and *NN*-1: feature vectors extracted from *AE* of $M1$ are optimized by *NN*-1.
$M4$: Consists of combination *AE* of $M2$ and *NN*-1: feature vectors are extracted from *AE* of $M1$ and optimized by *NN*-1.
$M5$: Consists of combination of *AE* of $M1$ and *NN*-2: feature tensors (32 dimensions, 32 dimensions, 64 dimensions) extracted from *AE* of $M1$ and optimized as a feature vector by *NN*-2.
$M6$: Consists of *NN*-2 only. The image is input as a feature tensor (256 dimensions, 256 dimensions, 3 dimensions), and optimized as a feature vector by NN-2.

We calculate recall and precision by the following equations.

$$\text{Recall} = N_c \big/ N_{rc} \tag{1}$$

$$\text{Precision} = N_c \big/ N_{pc} \tag{2}$$

where,

N_C: Number of correct answers

N_{TC}: Total number of correct answers

N_{PC}: Number of answers predicted as correct.

4.3 Experimental Results

The experimental results are shown in Figs. 5, 6 and 7 and Table 1. Figures 5 and 6 show the comparison results of the recall and precision rates for each model.

Models $M1$ and $M2$ consist of an autoencoder only, while models $M3$ and $M4$ consist of combination of an autoencoder and distance learning neural network. First, since models $M3$ and $M4$ have improved both recall and precision, it can be confirmed that the accuracy of image classification can be improved by applying the distance learning neural network.

Meanwhile, model $M5$, like $M3$ and $M4$, consists of combination of an autoencoder and a distance learning neural network, but it has a convolution layer in the distance learning neural network. On the other hand, models $M3$ and $M4$ are composed of a distance learning neural network with a fully-connected layer instead of a convolution layer. The improved recall and precision of model $M5$ indicate that using a convolutional layer in the distance learning neural network is more effective in improving classification accuracy in image classification.

The model $M6$ does not use an autoencoder, but only a distance learning neural network, and uses the image data itself as input instead of extracted image features. The recall and precision rates of $M6$ were improved when dimensions of the intermediate layer of autoencoder and the output layer of distance learning neural network were set to 100 and 1000 dimensions, respectively.

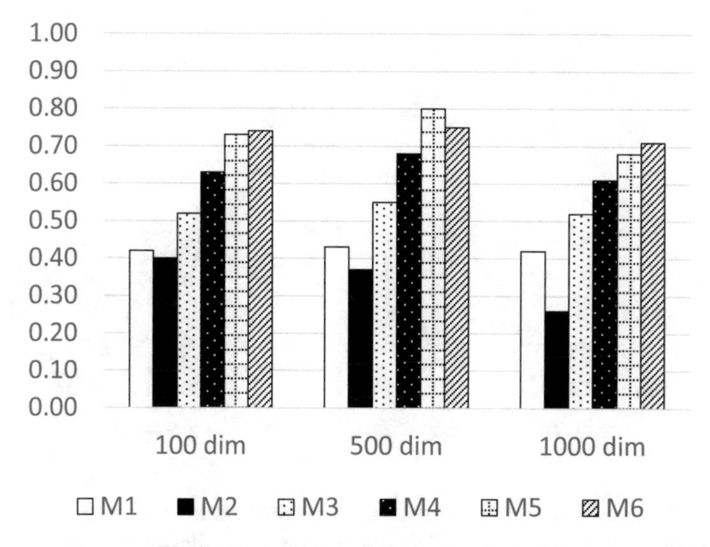

Fig. 5. Comparison of recall of each model

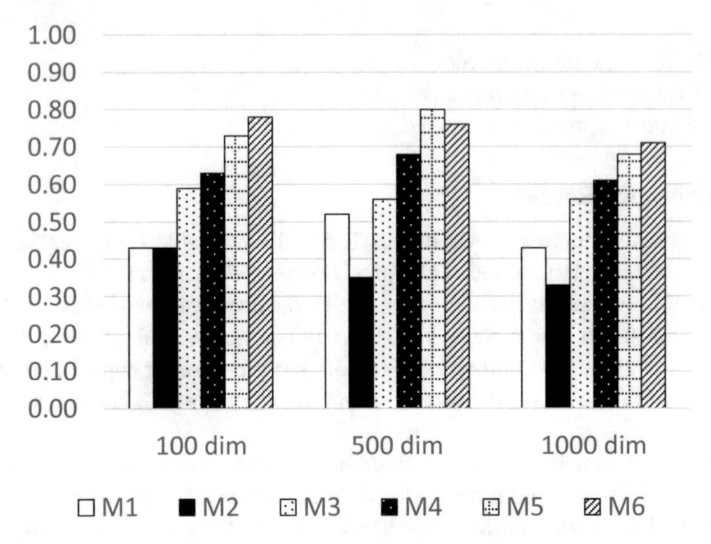

Fig. 6. Comparison of precision of each model

Table 1. Comparison of average recall and average precision ($M5$ and $M6$)

	$M5$	$M6$
Epochs	600	1000
Average recall	0.736	0.733
Average precision	0.736	0.750

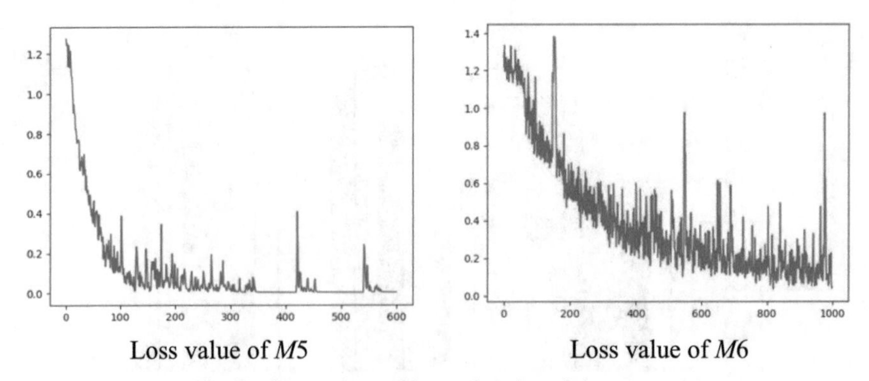

Loss value of $M5$
Loss value of $M6$

Fig. 7. Comparison of loss values in training process

Table 1 and Fig. 7 show the results of comparing the average recall and precision rates for models $M5$ and $M6$, as well as the change in loss during training process (100 dimensions). Table 1 shows that $M5$ has higher average recall in the 100, 500, and 1000 dimensions, and M6 has a slightly higher average precision, but the difference in average fit rate is small at 0.014. The results in Fig. 7 also confirm that the training process is more stable with smaller fluctuations in the loss values for $M5$. The loss values of $M5$ and *M6* both start at around 1.2 in the training process, indicating that the distance relationship is not sufficiently appropriate at the start of training process for both cases using the feature tensor extracted from the autoencoder ($M5$) and the image data ($M6$). The reason why the number of training epochs for $M5$ is small is considered to be that the image features appropriate for distance learning have been extracted by training with the autoencoder.

From these comparison results of the models $M5$ and $M6$, it is considered that the separation of the autoencoder and distance learning networks has certain effects in terms of classification accuracy and learning stability.

5 Conclusion

In this study, we proposed an image classification method that applies an autoencoder and a distance learning neural network for extracting and optimizing image features. Experiments using aerial photo data showed that the proposed method can improve the classification accuracy of image features by applying the distance learning neural network and confirmed the feasibility of the proposed method.

As our future work, we are considering to apply the proposed method to a video scene analysis system for a massive amount of video data recorded and accumulated from observation cameras, which will be installed at different locations in natural environment observation.

Acknowledgments. This research was supported by the Collaboration Research Program of IDEAS, Chubu University IDEAS202303, and by JSPS Grant-in-Aid for Scientific Research 23K11120.

References

1. Mimura, H., Tahara, M., Takano, K., Watanabe, N., Li, K.F.: Video Indexing for Live nature camera on digital earth, In: International Conference on Advanced Information Networking and Applications, pp. 660–667 (2023)
2. Onitsuka, Y., Ohyama, W., Yamada, T., Inoue, S., Uchida, S.: Convolutional feature extraction for kaou image retrieval. Proc. IPSJ Comput. Humanit. Symp. (Jinmoncon) **2018**, 257–262 (2018)
3. Hosoe, M., Yamada, T., Kato, K., Yamamoto, K.: A proposal of method for extraction of handwriting feature using conditional AutoEncoder. In: The 80th National Convention of IPSJ, vol. 2C-06, No. 2, pp. 37–38 (2018)
4. Ishfaq, H., Hoogi, A., Rubin, D.: TVAE: Triplet-Based Variational Autoencoder using Metric Learning (2018). arXiv:1802.04403
5. Andresini, G., Appice, A., Malerba, D.: Autoencoder-based deep metric learning for network intrusion detection. Inf. Sci. **569**, 706–727 (2021)
6. Geospatial Information Authority of Japan: Teacher image data for paddy field extraction using CNN, Geospatial Information Authority of Japan technical data, H1-No. 26 (2023)

An Imaging Camera Anomaly Detection System Based on Optical Flow

Chihiro Yukawa[1], Tetsuya Oda[2(✉)], Yuki Nagai[1], Kyohei Wakabayashi[1], and Leonard Barolli[3]

[1] Graduate School of Engineering, Okayama University of Science (OUS), 1-1 Ridaicho, Kita-ku, Okayama 700-0005, Japan
{t22jm19st,t22jm24jd,t22jm23rv}@ous.jp

[2] Department of Information and Computer Engineering, Okayama University of Science (OUS), 1-1 Ridaicho, Kita-ku, Okayama-shi 700-0005, Japan
{t19j061nm,t20j091yy}@ous.jp, oda@ous.ac.jp

[3] Department of Information and Communication Engineering, Fukuoka Insitute of Technology, 3-30-1 Wajiro-Higashi-ku, Fukuoka 811-0295, Japan
barolli@fit.ac.jp

Abstract. In the manufacturing industry, automation is crucial for increasing the efficiency of production processes. The robot-based automation is being used in all tasks required in the factory such as transport, processing, inspection and testing. Robot vision, in which a camera is mounted on a robot arm, is actively used in automated inspection. However, during automated inspection, not only the object to be inspected but also the robot vision itself may be damaged, stopped or collided. Therefore, the anomaly detection is required to deal with these problems. In recent research works, for automatic inspection is used deep learning to recognise the image of the object. However, sometime it is not possible to properly get the imaging target due to camera malfunction or dust on the lens. In this paper, we propose a camera anomaly detection system based on optical flow to detect anomalies such as camera failures and object adhesion in lens. The experimental results show that the proposed system can detect anomalies based on optical flow.

1 Introduction

In the manufacturing industry, the automation is very important to improve the efficiency of the production process [1–8]. The automation using robots is being promoted in many tasks of factories such as transportation, processing, inspection and testing. In some research works is used the robot vision for inspection and testing [9–12]. There are various types for image recognition using robot vision such as a robot arm attached to imaging camera and a vertical or horizontal direction fixed imaging camera over a belt conveyor [13–16].

L. Barolli (Ed.): AINA 2024, LNDECT 204, pp. 422–431, 2024.
https://doi.org/10.1007/978-3-031-57942-4_41

In automatic inspections and tests using robot vision, there are cases where it is not possible to properly get the imaging target due to camera malfunction or dust on the lens. Currently, the deep learning is used for detecting anomaliies in automatic inspections and tests [17–19]. On the other hand, executing deep learning requires a lot of computational resources. In addition, annotations for creating datasets are required to expend effort when performing anomaly detection using image recognition based on deep learning. Also, when there is noise they can not detect anomalies because of dust adhering in lens. Therefore, there are needed new methods for detecting anomalies with less computational resources and dataset.

We previously proposed automatic inspection methods such as a robot vision for micro-roughness recognition based on a convolutional neural network, a robot arm control for vibration suppression based on fuzzy control and the effect of lighting of metal surfaces by different colors for a robotic vision system.

In this paper, we propose an anomaly detection system for imaging camera using optical flow. In order to detect anomalies the robot arm is attached in imaging camera. The proposed system can be applied in belt conveyor for inspection system or surveillance camera. To evaluate the proposed system, we performed an experiment and considered a scenario to detect an anomaly when an object is artificially adhering to the lens of the imaging camera and there is a black noise in the imaging camera. From experimental results we found that the proposed system can detect anomalies in the imaging camera using optical flow.

The structure of the paper is as follows. In Sect. 2, we present the proposed system. In Sect. 3, we discuss the experimental results. Finally, conclusions and future work are given in Sect. 4.

2 Proposed System

The proposed system is shown in Fig. 1. The proposed system can detect the anomaly in a imaging camera by optical flow [20,21].

2.1 Anomaly Detection for Imaging Camera

The robot arm is attached in imaging camera. The pixels of the received image by robot arm can be changed by moving robot arm. The proposed system uses dense optical flow for anomaly detection [22].

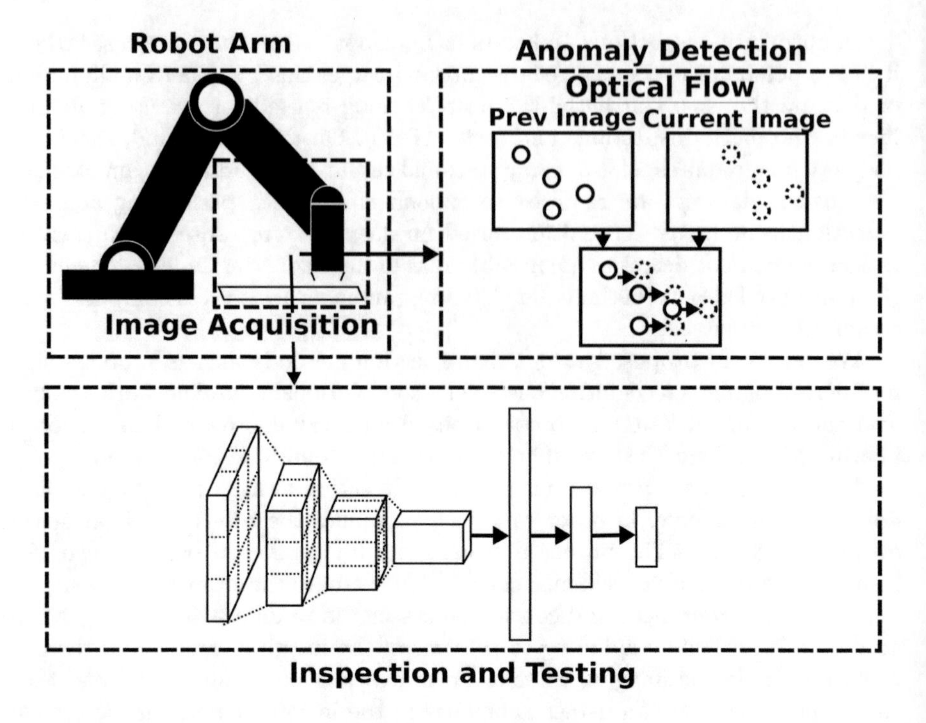

Fig. 1. Proposed system.

The anomaly detection is determined based on a vector output by dense optical flow considering previous and currently received images while moving the robot arm. In addition, the image is divided in a 9×9 area and the area is considered as anomaly if there is not any vector output for all pixels. The proposed system visualizes the output of vectors by dense optical flow as an image.

2.2 Visualization and Danger Notification for Robot Arm

The proposed system is implemented as client-server model to inform administrators. It receives the image from the robot arm attached to the imaging camera. Then, it sends the received image to the main server by TCP/IP protocol. After receiving the image, the main server investugates the anomalies of the image by dense optical flow. Then, the main server stores the calculation results and anomaly time. The proposed system visualizes the current image and the output of vectors on the web page. The anomaly area is visualized as the red area and a text written as "Anomaly".

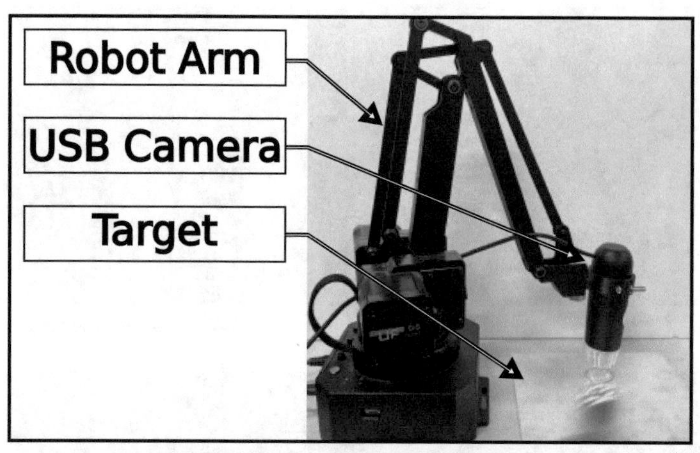

(a) Environment from vertical direction.

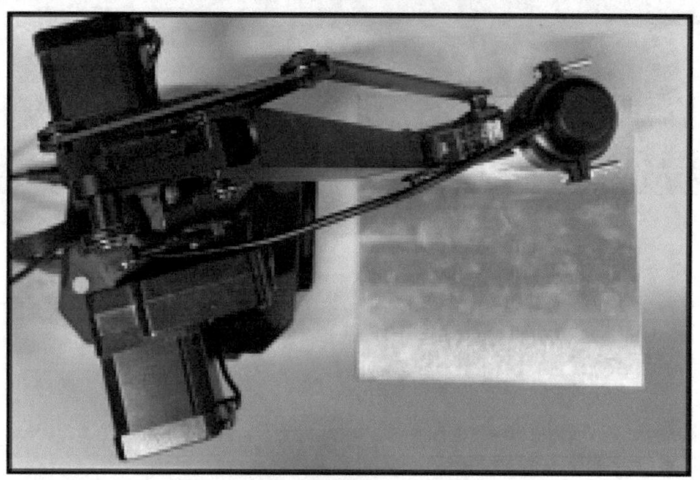

(b) Environment from horizon direction.

Fig. 2. Experimental environment.

Table 1. Components used for experiment.

Components	Models
Robot Arm	uArm Swift Pro Standard with 4°C of Freedom
Imaging Camera	5MP Digital Microscope USB 2.0
Single-board Computer	Jetson Nano 2 GB

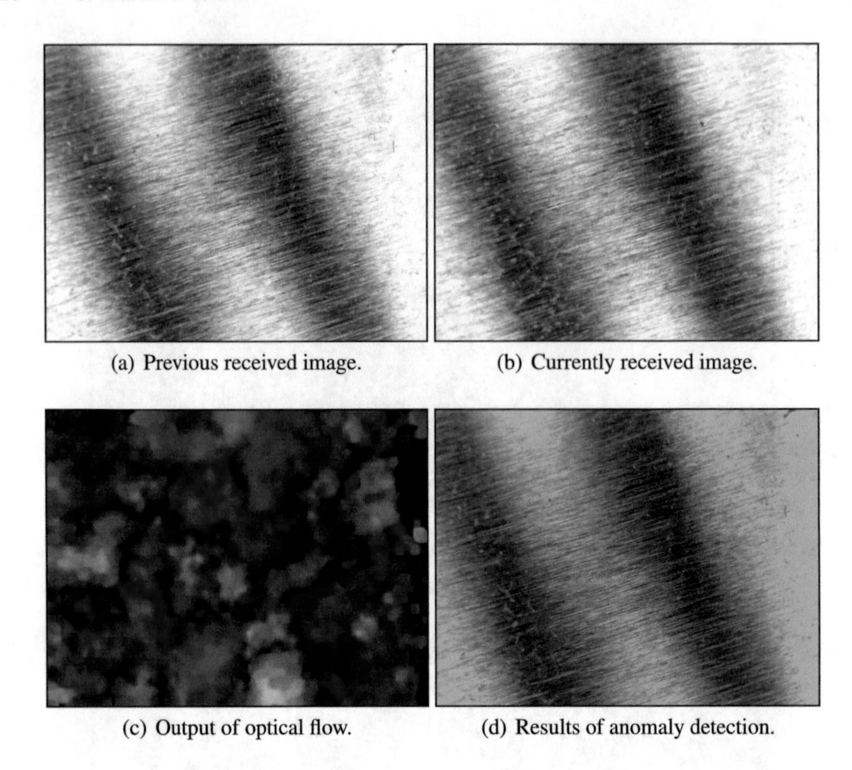

(a) Previous received image. (b) Currently received image.

(c) Output of optical flow. (d) Results of anomaly detection.

Fig. 3. Normal case.

3 Experimental Results

The experimental environment is shown in Fig. 2. While, Table 1 shows the components used for the experiment.

In this experiment, we evaluate a normal case of imaging camera, a case of black noise in imaging camera and a case of an object adhering to the imaging camera. The results for the normal case are shown in Fig. 3. In Fig. 3(a) and Fig. 3(b) are shown the images before and after the robot arm movement. From Fig. 3(c) can be seen that all pixels have vector output calculated from the optical flow based on Fig. 3(a) and Fig. 3(b). The results of anomaly detection are shown in Fig. 3(d).

The results for the case of black noise in the imaging camera are shown in Fig. 4. In Fig. 4(a) and Fig. 4(b) are shown the previous and curretly received images. From Fig. 4(c) can be seen that the black noise area has no vector output from the optical flow. In Fig. 4(d), the red are is considered as the anomaly area.

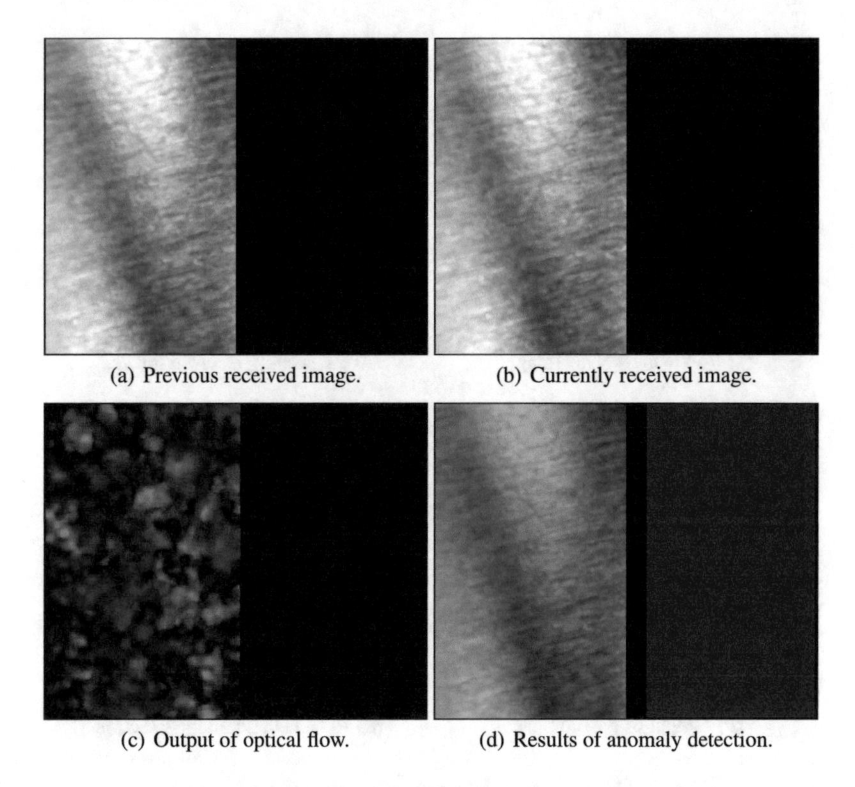

Fig. 4. Case of black noise in imaging camera.

The results for the case of an object adhering in the imaging camera are shown in Fig. 5. In Fig. 5(a) and Fig. 5(b) are shown the previous and curretly received images. It can be seen from Fig. 5(c) that there is no vector output from the optical flow.

The result of anomaly detection in the case of black noise in the imaging camera is shown in Fig. 6. It can be seen that when an anomaly has occured, the proposed system can detect it.

The results of visualization on the Web page are shown in Fig. 7. In Fig. 7(b) is shown the normal case. While in Fig. 7(b) is shown an anomaly case. We see that the proposed system is able to visualize anomaly area and text as “Anomaly”.

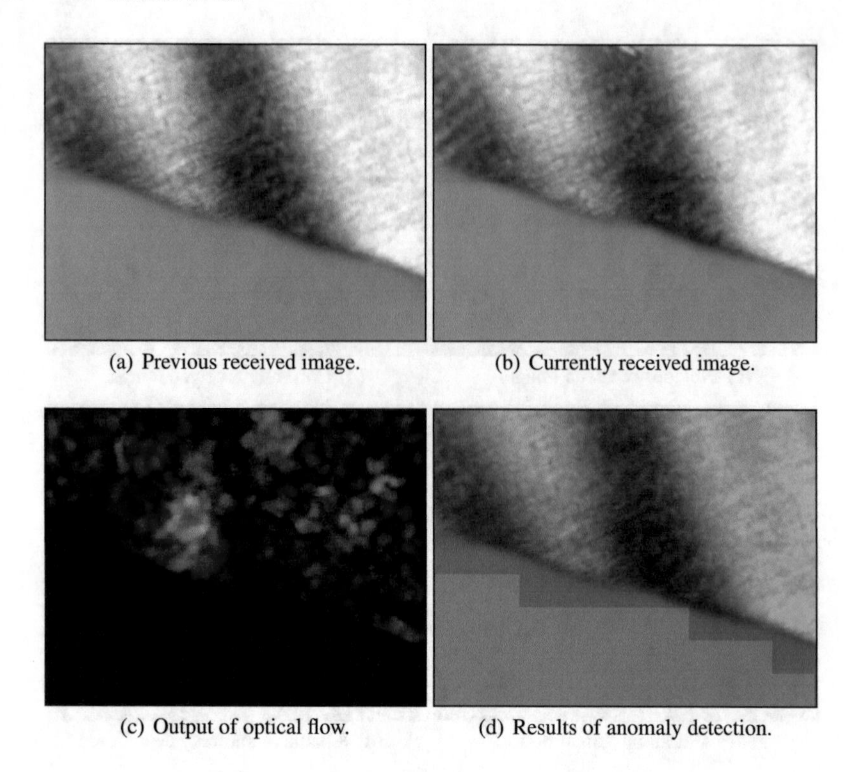

(a) Previous received image.
(b) Currently received image.
(c) Output of optical flow.
(d) Results of anomaly detection.

Fig. 5. Case of an object adhering in imaging camera.

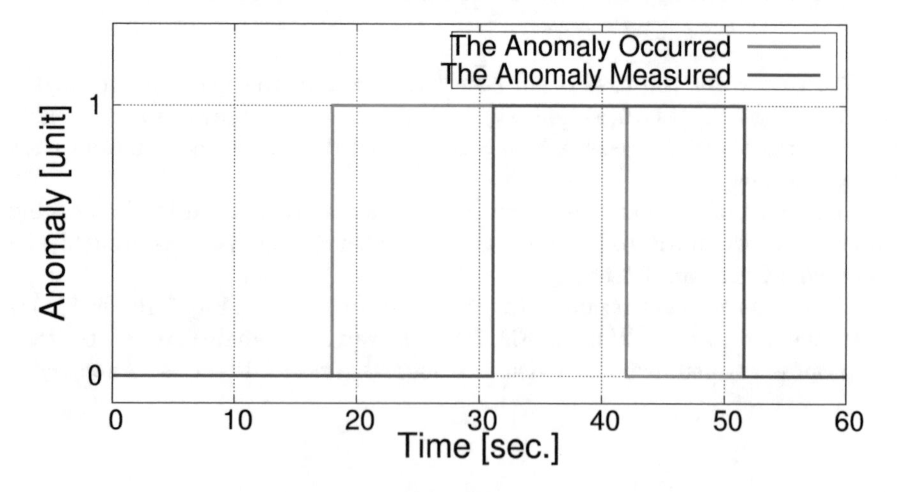

Fig. 6. Results of anomaly detection.

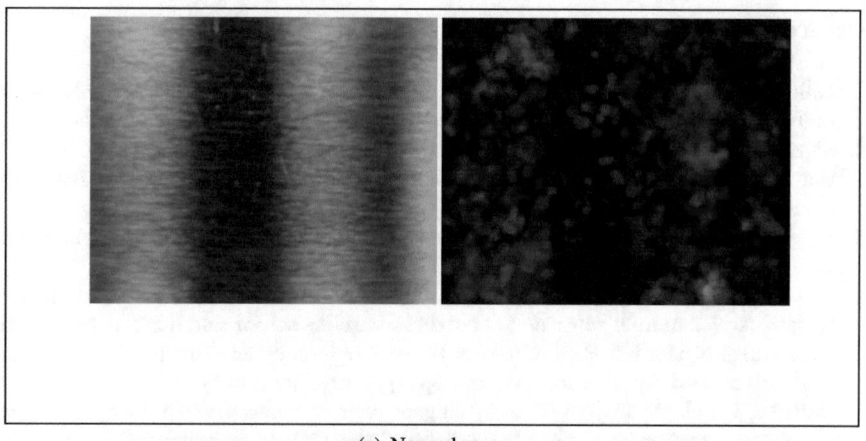

(a) Normal case.

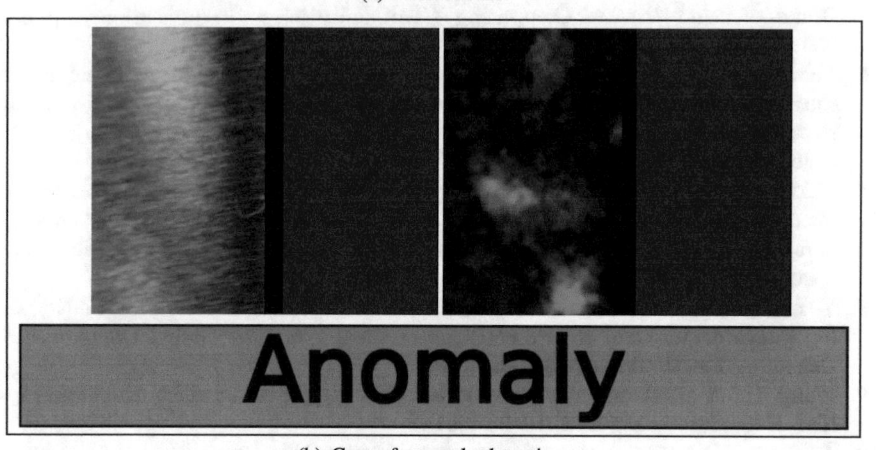

(b) Case of anomaly detection.

Fig. 7. Visualization results.

4 Conclusions

In this paper, we proposed an anomaly detection system for imaging camera by using optical flow. We evaluated the proposed system by an experiment. The experimental results show that the proposed system can detect anomalies in the imaging camera. In addition, the proposed system can visualize the results of the anomaly detection on the Web page. In the future, we would like to carry out extensive experiments in various environments.

Acknowledgement. This work was supported by JSPS KAKENHI Grant Number 20K19793.

References

1. Dalenogare, L., et al.: The expected contribution of industry 4.0 technologies for industrial performance. Inter. J. Product. Econ. (IJPE-2018) **204**, 383–394 (2018)
2. Shang, L., et al: Detection of rail surface defects based on CNN image recognition and classification. In: The IEEE 20th International Conference on Advanced Communication Technology (ICACT), pp. 45-51 (2018)
3. Li, J., et al.: Real-time detection of steel strip surface defects based on improved yolo detection network. IFAC-PapersOnLine **51**(21), 76–81 (2018)
4. T. Oda. et al.: Design and implementation of a simulation system based on deep Q-network for mobile actor node control in wireless sensor and actor networks. In: Procedings of the IEEE 31st International Conference on Advanced Information Networking and Applications Workshops, pp. 195-200 (2017)
5. Yukawa, C., et al.: Design of an intelligent robotic vision system for optimization of robot arm movement. In: Proceedings of the 17th International Conference on Broadband and Wireless Computing, Communication and Applications, pp. 353-360 (2020)
6. Yukawa, C., et al.: An intelligent robot vision system for recognizing micro-roughness on arbitrary surfaces: experimental result for different methods. In: Proceedings of the 14th International Conference on Intelligent Networking and Collaborative Systems, pp. 212-223 (2021)
7. Yukawa, C., et al.: Evaluation of a fuzzy-based robotic vision system for recognizing micro-roughness on arbitrary surfaces: a comparison study for vibration reduction of robot arm. In: The 25th International Conference on Network-Based Information Systems, pp. 230-237 (2021)
8. Yukawa, C., et al.: Design of a fuzzy inference based robot vision for CNN training image acquisition. In: Proceedings of the IEEE 10th Global Conference on Consumer Electronics, pp. 806-807 (2021)
9. Wang, H., et al.: Automatic illumination planning for robot vision inspection system. Neurocomputing **275**, 19–28 (2018)
10. Zuxiang, W., et al.: Design of safety capacitors quality inspection robot based on machine vision. In: 2017 First International Conference on Electronics Instrumentation & Information Systems (EIIS), pp. 1-4 (2017)
11. Li, J., et al.: Cognitive visual anomaly detection with constrained latent representations for industrial inspection robot. Appl. Soft Comput. **95**, 106539 (2020)
12. Ruiz-del-Solar, J., et al.: A survey on deep learning methods for robot vision, arXiv preprint arXiv: 1803.10862 (2018)
13. Yosinski, J., et al.: How transferable are features in deep neural networks?, arXiv preprint arXiv: 1411.1792 (2014)
14. Dosovitskiy, A., et al.: An Image is Worth 16x16 Words: Transformers for Image Recognition at Scale, arXiv preprint, arXiv: 2010.11929 (2020)
15. Sudharshan, D.P., et al.: Object recognition in images using convolutional neural network. In: 2018 2nd International Conference on Inventive Systems and Control (ICISC), pp. 718-722 (2018)
16. Radovic, M., et al.: Object recognition in aerial images using convolutional neural networks. J. Imaging **3**(2), 21 (2017)
17. Schlegl, T., et al.: f-AnoGAN: Fast unsupervised anomaly detection with generative adversarial networks. J. Med. Image Analysis, vol. 54, pp. 30-44 (2019)
18. Cohen, N., Hoshen, Y.: Sub-image anomaly detection with deep pyramid correspondences, vol. 130(4), pp. 947-969, arXiv preprint, arXiv: 2005.02357 (2020)

19. Bergmann, P., et al.: Beyond dents and scratches: logical constraints in unsupervised anomaly detection and localization. J. Comput. Vis. **130**(4), 947–969 (2022)
20. Yunpeng, C., et al.: Video anomaly detection with spatio-temporal dissociation. Pattern Recogn. **122**, 108213 (2022)
21. Ramachandra, B., et al.: A survey of single-scene video anomaly detection. IEEE Trans. Pattern Anal. Mach. Intell. **44**(5), 2293–2312 (2022)
22. Farnebäck, G.: Two-frame motion estimation based on polynomial expansion. In: Image Analysis: 13th Scandinavian Conference, SCIA, pp. 363–370 (2003)

A Framework for Blockchain-Based Scalable E-Voting System Using Sharding and Time-Slot Algorithm

R. Madhusudhan(✉) and Vishnu K. K.

Department of Mathematical and Computational Sciences, National Institute of Technology, Karnataka, Mangalore, India
madhu@nitk.edu.in

Abstract. People in the present era are always engaged in their jobs and other tasks. Any technology breakthrough that lowers their time wastage is greatly appreciated. Voting is an unavoidable process that takes place regularly in a democracy. Voters invest a lot of time in the current voting system for a variety of reasons, like the need for voters to physically be present at the polling booth, long queue for voting, physical procedures involved in casting a ballot, etc. Technologies which are time-efficient and highly secure are required in this era. Several researchers have suggested several blockchain-based electronic voting methods. The majority of blockchain-based electronic voting systems used now include problems of latency and non-scalability. It is a tedious task to implement a blockchain-based electronic voting system (e-voting) that scales well based on transactions per second (tps). Some researchers have suggested using sharding technology to improve scalability. But only with sharding, the system cannot handle the casting of huge amounts of votes, which is greater than the maximum transaction per second (tps) of the system, at the same time. An e-voting architecture which consists of a consortium blockchain, an Interplanetary File System (IPFS) has been proposed. The architecture has incorporated a fog layer to increase scalability by minimising the latency. Additionally, an algorithm has been proposed that enhances the system's functionality and effectiveness in a nation with a sizable voting population.

Keywords: Blockchain · E-Voting · IPFS · Latency · Scalability

1 Introduction

An election, according to the Oxford Dictionary, is the process of selecting a candidate or group of candidates via voting for a post, particularly a political one [1]. Citizens in a democratic nation have the right to vote, giving them the power to choose the government. By voting, people can truly understand the essence of citizenship. Voting is crucial to preserving a democratic country's political structure. It is how the democratic system is activated. The information available

L. Barolli (Ed.): AINA 2024, LNDECT 204, pp. 432–443, 2024.
https://doi.org/10.1007/978-3-031-57942-4_42

indicates that significant changes have been made to the voting procedure to enhance its flexibility, availability, security, speed, and cost-effectiveness, mainly regarding voter registration verification, voting, and tabulation [2].

Primarily, there are three methods of voting. They are traditional voting, traditional electronic voting (e-voting) and blockchain-based e-voting. The voting method that uses paper ballots is known as traditional voting. Ballot boxes and papers are used here to collect the votes [1]. Voters must first register with the Identification Authorities (IA) to cast a vote. To do this, the IA officials physically go to each home, verify the voters there and enlist his/her votes in the specified polling station. Voters who have registered to vote may visit the appropriate polling place on election day. On the election day, ballots are given to voters. The voter marks their choice on the ballot, which will be collected in the ballot box. The paper voting box will be tallied and tabulated manually [3]. The use of paper ballots for voting has several limitations, such as human error in manual counting, manipulation and tampering of paper ballots, difficulties with paper ballot distribution logistics, time-consuming manual counting, etc [4].

Traditional e-voting involves the use of digital electronic voting machines (EVMs) in place of ballot papers and boxes. Voters can choose from a list of candidates through voting button on an EVM. EVMs have the advantages of effective vote tallying and speedy result posting. The central authority is in charge of maintaining the EVMs. Consequently, have concerns about EVM tampering. Vulnerabilities in the EVM's hardware or software might result in hacking attempts, tampering of votes, or illegal access. Furthermore, it offers less transparency [4].

Blockchain's decentralization, immutability, and integrity make it an excellent choice for e-voting system deployment. Every blockchain node stores a copy of the ledger [5]. The decentralized design of the blockchain-based electronic voting system addresses the issues brought about by the centralized architecture of conventional electronic voting systems. Blockchain is a type of distributed ledger in which users may directly store and exchange data with each other without having to know or trust other parties. Blockchain technology has several built-in features, such as data openness, privacy, replication, and tamper-proof of the ledger [6]. These characteristics make them suitable for resolving issues with traditional e-voting systems. Traditional e-voting systems have more scalability than e-voting systems built on blockchain. This is because a consensus process must run on each peer node in a blockchain-based e-voting system before data is committed to the ledger, which will take some time. Because of this reason, the tps are far less than centralized traditional e-voting systems. When a huge number of people vote at the same instance, especially in national elections, the system will perform poorly. We consider the scalability and performance issues with blockchain-based e-voting systems in this paper. An architecture and a timeslot algorithm which improves the scalability and performance of the system especially when a lot of voters come to vote simultaneously has been proposed.

The rest of this paper is organized as follows: Sect. 2 presents preliminaries for the work. Section 3 presents the literature review. Section 4 describes the proposed methodology. Finally, Sect. 5 gives a concise conclusion of the paper.

2 Preliminaries

2.1 Blockchain Technology

Blockchain technology is an information-sharing and decentralized computing platform that lets different authority domains cooperate, interact, and take part in a fair decision-making process even when they don't trust one another. It facilitates the non-tampering storage of user data through the use of cryptographic primitives like hashing. Blocks of data are kept and linked together by hash values that are generated using a cryptographically safe hash function. A connection between successive blocks in the ledger is created by storing the hash value of the previous block in the current block. Every blockchain node keeps a local copy of the global data sheet which is up to date. Consistency between the local copies is guaranteed by the system. The hashing technique led the ledger data to be immutable [7].

Public, private, and consortium blockchains are the three primary categories of blockchain technology. Anyone can join a public blockchain network without requiring authorization or authentication. On the other hand, access to a private blockchain is restricted to certain, authenticated users. In a consortium blockchain, many public or private blockchains are joined to address business issues that arise between different organisations. Different organisations hold these private and public blockchains; if they want to work together on decision-making, they can establish consortium blockchains. The general concept of blockchain technology is distributed architecture based on cryptographic principles. The blockchain contains smart contracts, Merkle tree, digital signature and consensus algorithm. The validation of users in the blockchain is done through digital signature. Here, if the sender wants to send data, he/she needs to sign the data with his/her private key so that all the receivers can verify the sender by applying the sender's public key to that signed data. The public key of the sender is available to all. Merkle tree is a special organization of the data in the block. The internal node of the Merkle tree contains the combination of hashes of its child nodes. Each leaf node contains the hash of individual data. The Merkle trees' root will be stored in that block as metadata. The consensus algorithm is used to agree with the distributed nodes in the blockchain for various decision-making processes. The consensus algorithm running on each node ensures that the local copies are consistent and updated [7].

2.1.1 Consensus: Every node in a distributed system is free to make its own decisions, and rebel nodes can occasionally act unpredictably. Reaching a consensus in a malicious node-filled environment is not an easy task. When there are malfunctioning nodes present, consensus aids in ensuring proper operation.

The properties of the consensus in a distributed environment is termination, validity, integrity and agreement [8]. The termination property specifies that after the consensus, each valid node must make a choice. According to the validity property, if the individuals propose the same decision, then all correct individuals should follow that decision. According to the integrity property, each right person can only make one choice, and that choice needs to be put forth by a few nodes. According to the agreement property, each right person must agree on the same choice. Some of the consensus algorithms are:

- Proof of Work (PoW)
- Proof of Stake (PoS)
- Proof of Burn (PoB)
- Proof of Elapsed Time (PoET)

The public blockchain network mostly uses the consensus algorithms listed above. The reason for this is that public blockchains lack an authentication mechanism and their participant count is set arbitrarily. Therefore, a consensus method based on challenge-response will function well. The consensus algorithms shown above are challenge-response based. In e-voting technology, a private blockchain network is required because the number of participants is predetermined and limited. Because there are fewer members in private blockchain networks, a consensus process based on a message-passing mechanism is sufficient. Examples of message-passing based consensus algorithms are PBFT (Practical Byzantine Fault Tolerance), RAFT (Reliable, Replicated, Redundant, And Fault-Tolerant), etc [9].

2.2 Interplanetary File System (IPFS)

IPFS is a distributed web architecture that builds huge storage by combining tiny storage that is available to users. Each user node holds a reference to the data that is stored in other user nodes in addition to its data. This system provides high service availability while preventing data duplication. Every piece of data is referenced using its distinct cryptographic hash value. IPFS monitors each file's version history and eliminates duplications throughout the network. Each node contains some indexing information that helps in determining who is keeping what. IPFS has a routing mechanism based on name resolution [10]. IPFS is used in this work to decentralize the National Election Commission database, which has the voters' information. IPFS provides the availability of voters' data on election day without failure.

2.3 Blockchain Sharding

A blockchain is divided into several shards via sharding technology, sometimes referred to as horizontal scaling, which enables nodes to process and store transactions from a small number of shards. Through the use of sharding technology, which enables partial transaction processing and storage on a single node, the throughput of the entire blockchain may increase linearly as the number of nodes increases. This technology is crucial for the adoption of blockchains, which provide the general public with significant volumes and high-quality services inside expansive networks with an increasing number of nodes [11].

3 Literature Review

There have been various attempts to address the scalability and security issues related to voting systems. Some works that are successful in improving scalability, security, etc. for blockchain-based e-voting systems are discussed in this section:

Kohad et al. [5] proposed employing the multiobjective genetic algorithm to generate side-chains and increase the e-voting system's scalability and performance on a blockchain. They compared the suggested model with the processing and storage costs of the current models. The length of the chain is shortened by the suggested approach, which lowers the overall processing cost and latency for voting. The shorter chain length also results in a decrease in storage costs.

Russo et al. [12] suggested a thorough and adaptable architecture for creating safe online voting systems. Linkable ring signatures provide voter anonymity and authenticity, while specialised smart contracts on a blockchain provide decentralisation, transparency, determinism, and unalterability of votes. These also protect against duplicate voting when combined with appropriate smart contract limits. The article provides a detailed presentation of its concept, emphasising security assurances and design decisions that enable it to expand to a wide voter base. Additionally, they demonstrated a proof-of-concept application using the suggested framework.

Jafar et al. [13] suggested a mechanism for expanding the main chain or shard the existing blockchain to achieve security. The suggested approach functions as an off-chain that is synchronised with district-level databases and is implemented using Ethereum. Authorities should regularly update the main chain. In terms of data storage and scalability, the paradigm has done well. Limitations include the requirement that people have faith in the authorities to prevent malicious assaults.

Divya et al. [14] combined a blockchain with an electronic voting mechanism online. The method assisted in keeping the ballots of the voters safe on a blockchain. Voters could register to vote for any candidate they choose. Through the website, anybody may see all extra voter data, such as the voter's name, city, and total number of votes. Voters became more confident in the system, which allowed them to cast their ballots safely and securely. It also created a sense of security and avoided repeat voting.

Malkawi et al. [15] proposed the use of a blockchain-based voting system (BBVS) for Jordan's legislative elections. Voters in the proposed system cast ballots at two levels on a hierarchical, centralized, private blockchain. A group can access the first level, and individual group members can access the second level. The authors offered a cutting-edge, safe blockchain-based electronic voting system that assesses the efficacy, accuracy, and integrity of the electoral process. The work implemented novel algorithms to maintain a level of performance both when generating and casting votes for voters.

Kumar et al. [16] suggested an electronic voting system that would allow the Indian government to stream the entire election process online. The authors proposed the use of a decentralised application (dApp) for all elections conducted by the Election Commission of India. Candidates would submit their nominations in accordance with their preferences. The request may have been accepted or rejected by the organisers. Voters found it simpler to cast their ballots from anywhere because of the accessibility of the voting procedure. Additionally, it made voting possible for voters from other nations, raising the percentage of votes cast overall. Because the system's creators employed blockchain technology, it guaranteed usability, transparency, security, portability, dependability, and trustworthiness.

Huang et al. [17] provided a thorough analysis of blockchain-based voting systems, categorising them according to the sorts of blockchains utilised, the number of participants, and the consensus techniques employed. After conducting a thorough comparison of the various voting methods, the authors noted significant gaps in the literature and areas for further research. The survey provided a thorough analysis of the use of blockchain technology in voting systems and made recommendations for future research directions.

Mehboob et al. [18] investigated the settings for transaction vulnerability attacks in the blockchain-based e-voting system. To aid in the creation of suitable defences, the authors attempted to draw attention to the circumstances that lead to a system assault. The presentation focused on the accomplishment of a transaction malleability attack on a blockchain that accommodated an electronic voting application. The trials revealed the importance of block production rate and network latency for effectively executing transaction malleability attacks and indicated the paths for further study.

Dimitriou [19] suggested a blockchain-based voting system that was safe, scalable, and useful and that accomplished the characteristics of a large-scale election without requiring a lot from the voters. The ballot, which serves as a black box for the user, was created using a randomizer token to guarantee receipt-freeness and resistance to coercion. Because of blockchain, the append-only structure is also guaranteed.

Praveen et al. [20] created a private blockchain network configuration that allows data to be shared across peer blocks in the network. The authors created an application for democratic voting that processed, stored, and executed transactions using blockchain technology. Smart contracts were also implemented. Additionally, consistency, time, and scalability factors were taken into account while analysing the performance of the Ethereum clients Parity and Geth. For a very long time, electoral fraud has been a serious problem for the voting process. Many countries are currently using e-governance and e-voting.

Baudier et al. [21] emphasised the significant contributions made by blockchain to a reliable and safe voting mechanism that promotes world peace.

Regrettably, the intricate and laborious technology led to discord among many participants in the electoral process. Therefore, in order to determine the technology's advantages and disadvantages, the writers conducted a qualitative investigation through interviews with blockchain specialists and election watchers. The outcomes demonstrated the significance of human elements and voter confidence.

Abuidris et al. [22] proposed a system PSC-B chain, a hybrid consensus model that comprised Proof of Credibility and Proof of Stake. The proposed model addressed the scalability through sharding, efficiency, and latency problems of an e-voting system. To guarantee the precision and security of the voting procedures, smart contracts were used to offer a reliable and secure computing environment.

Shadab et al. [23] suggested an electronic voting mechanism that verifies user identification using an Aadhar card and blockchain. The concept is more decentralised and adheres to the fundamentals of electronic voting. The security issues with conventional e-voting systems can be resolved by the proposed blockchain-based e-voting system with Aadhaar card identification, improving the effectiveness, transparency, and security of elections in India.

Khedkar et al. [24] presented a system that optimises the time for vote mining, authentication, and verification using a consensus process known as Proof of Authority (PoA). The Proof of Authority (PoA) consensus method is used in the proposed system in place of the Proof of Work (PoW) consensus mechanism in the current system. It turned out to be almost nine seconds more efficient than the current setup. This study so demonstrates how the current blockchain-based electronic voting method is lagging and how the suggested approach reduces the amount of time needed and speeds up the procedure.

Zhu et al. [25] suggested a blockchain-based approach for voter registration that uses the system's authentication to verify each voter. Second, they created a two-tier blockchain architecture, with the votes of electors recorded in the upper layer and voters in each district recorded in the lower layer. Subsequently, they outline the voting procedure and assess the availability and security of the suggested plan. The experimental findings demonstrate that the suggested plan is capable of meeting the requirements of multi-district elections.

Many number of work have been done to increase the scalability of the blockchain-based e-voting system. However, they have limitations to implement directly in an election where a huge number of voters are voting, like in India. The limit in scalability (tps) is mainly because of the time-consuming consensus algorithm present in the blockchain network, but these consensus algorithms are inevitable. So many researchers have proposed a less complex consensus algorithm, where nodes can come to a consensus faster. They can increase the transaction per second of the system. However, these systems still have a limit on scalability in terms of transactions per second, which is less than what traditional centralized e-voting systems have. So a methodology is needed which maintains the performance of the system even when a huge number of voters come to vote at the same time.

4 Proposed Algorithm and Architecture

Improving scalability is an essential thing for the success of blockchain e-voting in the future. Sharding is an essential technology which proves the improvement in scalability. Here an architecture which consists of sharding blockchain, IPFS and a fog layer has been proposed. The proposed Framework for an IPFS-based blockchain e-voting system is shown in Fig. 1.

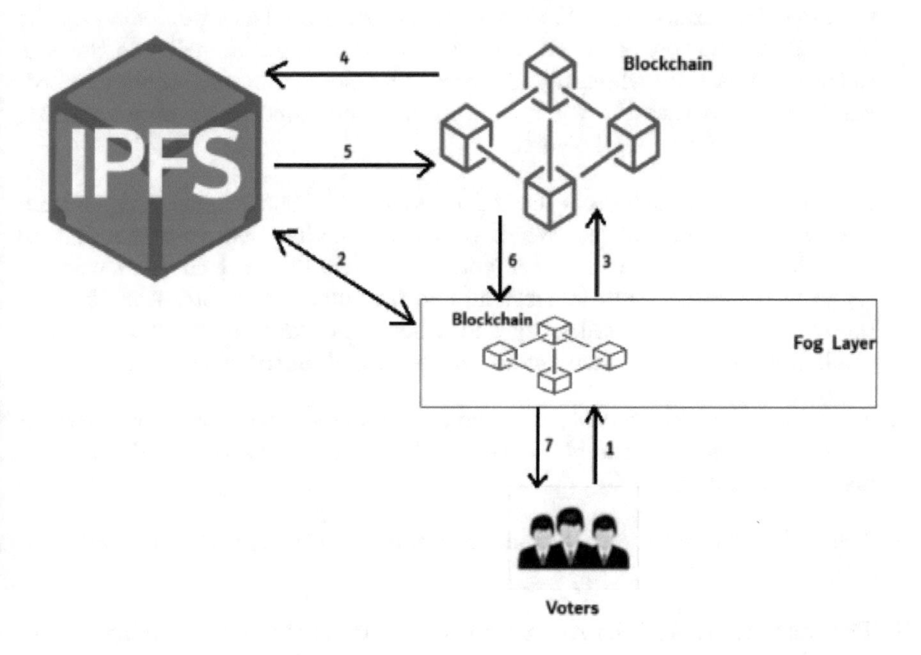

Fig. 1. Proposed Framework for IPFS-based blockchain E-Voting system

The blockchain used in the framework is consortium blockchain which contains private blockchains equal to the number of state election commissions present in the country. The private blockchain is present in the fog layer which is controlled by the state election commission. The introduction of a fog layer helps to reduce the time latency between the blockchain and voters because the fog layer is implemented in the states of the nation. The distance between the voter and the corresponding private blockchain is less. Each private blockchain is a sharding blockchain, which provides high scalability. All the information about the registered voter is kept in the IPFS by the election commission, which again reduces the overhead of the blockchain network. Normally, the voters' information is maintained by the election commission in a centralized database, which may cause a single point of failure at the time of voting. To ensure the high availability of the voters' details at the time of voting, IPFS is introduced. The flow of data in the above architecture is described below:

1. Voter casts his/her vote to state-level blockchain by providing voter identification details and the candidate to whom he/she voting.

2. The blockchain checks whether the voter already voted in the blockchain by querying the blockchain with the information provided by the voter. If voted, then reject the voter request for voting. If not voted, the blockchain finds the hash value of voters' identification details. Using this hash value, it queries the IPFS to find the voters' data maintained by the election commission. If the same data exists in IPFS, the voter is authenticated as a valid voter by the blockchain. The vote of the voter is legal and the blockchain will add the vote to the candidate for whom he/she votes. The blockchain also stores a record, which indicates that the voter did vote and the name of the corresponding candidate to whom he/she voted.

3. A consortium blockchain is formed by state-level blockchains and national election commission nodes. Which performs functions like vote counting, to find whether the same voter has voted from two different constituencies. If they did, the consortium blockchain can invalidate that vote. Also, it finds the winner of each constituency and other data like the vote share of each candidate and party. These details will be stored on the blockchain.

4. The consortium blockchain can validate the data of the voters by querying the IPFS. This helps the blockchain to find whether the voter only voted in his/her constituency.

5. The IPFS returns the data to the blockchain and the blockchain can validate.

6. The consortium blockchain can send the results to the corresponding private blockchains for storing there also.

7. If the voter asks for information like the voting result, the blockchain will forward the data to the voter which he/she is authorized to view.

The below given Time-Slot algorithm helps to further increase the performance of the e-voting system

Algorithm 1. smart contract for giving Time-Slot for e-voting

1: Calculate the total number of registered voters in the e-voting system.
2: Input the total time duration of the voting on the election day.
3: Divide that duration into slots and allocate the slot to voters such that the number of voters in a time slot should not exceed the maximum transaction per second (tps) of the blockchain.
4: If a voter tries to vote in a time slot for which he/she is not allocated, then the voter's attempt will be cancelled by the system.
5: The voters who follow the time slot can vote and their vote will be valid and stored in the system

The performance of the blockchain is under threat in a situation when the number of transactions which are coming simultaneously to the blockchain exceeds the limit that it can handle. This limit can be increased by using various methodologies like sharding. But with only sharding, the scalability of the blockchain improves but the situation encountered above persists. This is because there is always a chance that the number of simultaneous transactions coming to the blockchain exceeds the limit. Therefore, it is necessary to have a methodology which allows the system to maintain its performance when such situations occur. The proposed idea is to limit the number of transactions (votes) coming into the blockchain. Time slots can be given to voters in such a way that all voters cannot vote simultaneously. To do this, it is required to include a fixed number of voters in each time slot such that the number of voters in the time slot should be less than the maximum transaction per second (tps) of the blockchain.

5 Conclusion

A necessary component of contemporary democracy is voting. Voting may prove unpleasant for some due to the lengthy physical processes. Blockchain-based e-voting systems have been proposed by several researchers as a solution to these issues. These systems' shortcomings, meanwhile, include latency and non-scalability issues. The goal of the proposed design is to create a system that outperforms the current one in terms of performance and scalability. The implementation of the time-slot algorithm in a national election with a larger voter turnout is very promising. The proposed time-slot algorithm boosts the blockchains' efficiency and provides assurances for deploying the system in a big country like India.

References

1. Pramulia, D., Anggorojati, B.: Implementation and evaluation of blockchain based e-voting system with Ethereum and Metamask. In: 2020 International Conference on Informatics, Multimedia, Cyber and Information System (ICIMCIS), Jakarta, Indonesia, pp. 18–23 (2020). https://doi.org/10.1109/ICIMCIS51567.2020.9354310

2. Cheema, M.A., Ashraf, N., Aftab, A., Qureshi, H.K., Kazim, M., Azar, A.T.: Machine learning with blockchain for secure e-voting system. In: 2020 First International Conference of Smart Systems and Emerging Technologies (SMARTTECH), Riyadh, Saudi Arabia, pp. 177–182 (2020). https://doi.org/10.1109/SMART-TECH49988.2020.00050
3. Farooq, M.S., Iftikhar, U., Khelifi, A.: A framework to make voting system transparent using blockchain technology. IEEE Access **10**, 59959–59969 (2022). https://doi.org/10.1109/ACCESS.2022.3180168
4. Bin Kaha, P.R.F., Rahayu, S.B., Wardhana, A.A., Lee, M.-G., Lokman, A.L.A., Azmi, N.N.: Designing an e-voting framework using blockchain: a secure and transparent attendance approach. In: 2023 IEEE 8th International Conference on Software Engineering and Computer Systems (ICSECS), Penang, Malaysia, pp. 371–376 (2023). https://doi.org/10.1109/ICSECS58457.2023.10256270
5. Kohad, H., Kumar, S., Ambhaikar, A.: Scalability of blockchain based e-voting system using multiobjective genetic algorithm with sharding. In: 2022 IEEE Delhi Section Conference (DELCON), New Delhi, India, pp. 1–4 (2022). https://doi.org/10.1109/DELCON54057.2022.9753019
6. Benabdallah, A., Audras, A., Coudert, L., El Madhoun, N., Badra, M.: Analysis of blockchain solutions for e-voting: a systematic literature review. IEEE Access **10**, 70746–70759 (2022). https://doi.org/10.1109/ACCESS.2022.3187688
7. Bhutta, M.N.M., et al.: A survey on blockchain technology: evolution, architecture and security. IEEE Access **9**, 61048–61073 (2021). https://doi.org/10.1109/ACCESS.2021.3072849
8. Ali Syed, T., Alzahrani, A., Jan, S., Siddiqui, M.S., Nadeem, A., Alghamdi, T.: A comparative analysis of blockchain architecture and its applications: problems and recommendations. IEEE Access **7**, 176838–176869 (2019). https://doi.org/10.1109/ACCESS.2019.2957660
9. Gan, B., Wu, Q., Li, X., Zhou, Y.: Classification of blockchain consensus mechanisms based on PBFT algorithm. In: 2021 International Conference on Computer Engineering and Application (ICCEA), Kunming, China, pp. 26–29 (2021). https://doi.org/10.1109/ICCEA53728.2021.00012.
10. Psaras, Y., Dias, D.: The interplanetary file system and the filecoin network. In: 2020 50th Annual IEEE-IFIP International Conference on Dependable Systems and Networks-Supplemental Volume (DSN-S), Valencia, Spain, p. 80 (2020). https://doi.org/10.1109/DSN-S50200.2020.00043.
11. Yu, G., Wang, X., Yu, K., Ni, W., Zhang, J.A., Liu, R.P.: Survey: sharding in blockchains. IEEE Access **8**, 14155–14181 (2020). https://doi.org/10.1109/ACCESS.2020.296514
12. Russo, A., Anta, A.F., Vasco, M.I.G., Romano, S.P.: Chirotonia: a scalable and secure e-voting framework based on blockchains and linkable ring signatures. In: 2021 IEEE International Conference on Blockchain (Blockchain), Melbourne, Australia, pp. 417–424 (2021). https://doi.org/10.1109/Blockchain53845.2021.00065.
13. Jafar, U., Aziz, M.J.A., Shukur, Z., Hussain, H.A.: A cost-efficient and scalable framework for e-voting system based on ethereum blockchain. In: 2022 International Conference on Cyber Resilience (ICCR), Dubai, United Arab Emirates, pp. 1–6 (2022). https://doi.org/10.1109/ICCR56254.2022.9996026.
14. Divya, K., Usha, K.: Blockvoting: an online voting system using block chain. In: 2022 International Conference on Innovative Trends in Information Technology (ICITIIT), Kottayam, India, pp. 1–7 (2022). https://doi.org/10.1109/ICITIIT54346.2022.9744132.

15. Malkawi, M., Bani, M., Bataineh, A.: Blockchain-based voting system for Jordan parliament elections. Int. J. Electr. Comput. Eng. **4325–4335**, 11 (2021)
16. Kumar, A., Goyal, M.: Sustainable E-Infrastructure for Blockchain-Based Voting System, pp. 221–251. Wiley, Hoboken (2021)
17. Huang, J., He, D., Obaidat, M.S., Vijayakumar, P., Luo, M., Choo, K.K.R.: The application of the blockchain technology in voting systems: a review. ACM Comput. Surv. **1–28**, 54 (2021)
18. Mehboob, K., Arshad, J., Mubashir, M.: Empirical analysis of transaction malleability within blockchain-based e-voting. Comput. Secur. **100**, 102081 (2021)
19. Dimitriou, T.: Efficient, coercion-free and universally verifiable blockchain-based voting. Comput. Netw. **174**, 107234 (2020). https://doi.org/10.1016/j.comnet.2020.107234 (https://www.sciencedirect.com/science/article/pii/S1389128619317414). ISSN 1389-1286
20. Dhulavvagol, P.M., Bhajantri, V.H., Totad, S.G.: Blockchain ethereum clients performance analysis considering e-voting application. Procedia Comput. Sci. **167**, 2506–2515 (2020). https://doi.org/10.1016/j.procs.2020.03.303 (https://www.sciencedirect.com/science/article/pii/S1877050920307699). ISSN 1877-0509
21. Baudier, P., Kondrateva, G., Ammi, C., Seulliet, E.: Peace engineering: the contribution of blockchain systems to the e-voting process. Technol. Forecast. Soc. Change **162**, 120397 (2021)
22. Abuidris, Y., Kumar, R., Ting, Y., Joseph, O.: Secure large-scale E-voting system based on blockchain contract using a hybrid consensus model combined with sharding. ETRI J. **357–370**, 43 (2021)
23. Shadab, M., Kumar, P., Kumar, S.: A blockchain-based e-voting system for India: addressing security challenges with Aadhaar card authentication. In: 2023 3rd International Conference on Pervasive Computing and Social Networking (ICPCSN), Salem, India, pp. 1226–1231 (2023). https://doi.org/10.1109/ICPCSN58827.2023.00207.
24. Khedkar, S., Mahajan, K., Shirole, M.: Optimization of blockchain based e-voting. In: 2023 8th International Conference on Business and Industrial Research (ICBIR), Bangkok, Thailand, pp. 700–705 (2023). https://doi.org/10.1109/ICBIR57571.2023.10147663.
25. Zhu, H., Feng, L., Luo, J., Sun, Y., Yu, B., Yao, S.: BCvoteMDE: a blockchain-based e-voting scheme for multi-district elections. In: 2022 IEEE 25th International Conference on Computer Supported Cooperative Work in Design (CSCWD), Hangzhou, China, pp. 950–955 (2022). https://doi.org/10.1109/CSCWD54268.2022.9776193.

Hop-Constrained Oblivious Routing Using Prim's-Sollin's Algorithm

Mehak[(✉)] and Dharmendra Prasad Mahato

Department of Computer Science and Engineering, National Institute of Technology, Hamirpur, Hamirpur, Himachal Pradesh 177 005, India
{22mcs114,dpm}@nith.ac.in

Abstract. Our objective is to optimize routing efficiency and scalability within extensive networks through the development of streamlined oblivious routing methods, ensuring optimal load balancing and the utilization of compact routing tables. This initiative aims to enhance overall routing efficiency, reduce source demands, and elevate network reliability and availability. Addressing the challenge of implementing hop-constrained oblivious routing within near-linear time represents a pivotal step in advancing network routing capabilities, mitigating congestion, and minimizing path lengths.

The open problem of constructing hop-constrained oblivious routing in $O(m^{1+O(1)})$ time provides an opportunity for innovative solutions [12]. This challenge is effectively tackled through the application of Prim's-Sollin's algorithm, resulting in a time complexity approaching linearity while maintaining minimal congestion and dilation. The algorithm, rooted in the principles of finding the smallest spanning tree in a graph, adheres to specified hop limitations and preserves the shortest distance between hops. Hop- constrained oblivious routing, when implemented with Prim's-Sollin's algorithm, demonstrates a commendable time complexity of $O(mloglogn)$, underscoring its remarkable efficiency and effectiveness in large-scale network environments.

Keywords: Hop-constrained oblivious routing · minimum spanning tree · Prim's algorithm · Prim's-Sollin's algorithm · competitive ratio

1 Introduction

The oblivious routing method serves as a deterministic algorithm integral to congestion reduction within network systems. This technique entails the independent and random selection of packet routing, employing the Competition ratio as a performance metric, often yielding optimal results in both undirected and directed graph scenarios [2]. Oblivious routing is pivotal in enabling efficient routing in expansive distributed systems, eliminating the need for centralized control and frequent re-configurations. Current strategies have enhanced load distribution and algorithmic efficiency but suffer from unwidely routing tables, hindering scalability. Recent research focuses on streamlined oblivious routing schemes that maintain optimal load balancing while requiring small routing tables. These impact schemes are

L. Barolli (Ed.): AINA 2024, LNDECT 204, pp. 444–452, 2024.
https://doi.org/10.1007/978-3-031-57942-4_43

well-suited for emerging large-scale networks, offering efficient routing with minimal source demands [27]. The question of whether it is feasible to construct hop-constrained oblivious routing in nearly linear time, denoted as $O(m^{1+O(1)})$, remains an open problem in the field [12].

The term hop constrained denotes that this routing method systematically selects a path with a predetermined length. Hop constraints represent restrictions imposed on the number of links connecting two nodes in a network. These constraints play a vital role in enhancing the quality of service and bolstering network availability and reliability [15].

Hop-constrained oblivious routing represents a network routing strategy where the routing of individual packets is conducted independently, devoid of any interdependencies. The objective of this routing scheme encompasses the mitigation of congestion, defined as a maximal number of paths that traverse a specific network edge, and the minimization of dilation, which measures the maximal path length in terms of hop count [12].

1.1 Our Contributions

The use of the Prim's-Sollin algorithm is highly successful in optimizing hop-constrained static routing for oblivious routing schemes. These algorithms play an immense role in reducing both congestion and dilation by selecting independent paths that are optimal for individual packets regardless of other packets present in the network.

This makes use Prim's-Sollin algorithm to minimize the competitive ratio better and more effectively; the competitive ratio shows how much the highest congestion exceeds the optimal one. The application of Prim's-Sollin's algorithm presents an efficient technique for generating a minimum spanning tree on a graph; Sollin's primary approach towards this method requires incremental addition of shortest edges to identify a minimal spanning tree among all possible ones.

On the contrary, Prim starts with an initial vertex and repeatedly adds the shortest edges between any vertex within the expanding tree and one outside it. In this way, all vertices will be included in the tree. Thus, to guarantee that maximum possible traffic flows are routed through less congested channels, it is necessary to construct minimal spanning trees using these methods within specific hop limits.

2 Related Work

Oblivious routing is ideal for networks that do not require centralized control or frequent modifications. It is a solution to the problem of routing traffic without knowing how much traffic is expected in advance.

Routing a packet from sources s to destinations t such that completion time is minimized is the problem of routing a packet and to route a packet we have to do two things. The first thing is route determination, i.e., the path $(p_{s,t})$ for the transmission of the packet from s to t and the second thing is scheduling of the packet traversal timing along the path $(p_{s,t})$, i.e., exact timing along which the packet traverses along each edge (e) of the path ($p_{s,t}$). This study extends the results of Räcke [24,25], which were $log(n)$-competitive in one dimension and unbounded in the other, to further the field of oblivious routing.

The authors establish the existence of an oblivious routing scheme that achieves poly-logarithmic competitiveness concerning some relevant metrics, effectively a previously unresolved query within the domain of oblivious routing. It remains an open problem whether a hop-constrained oblivious routing can be constructed in almost-linear, $O(m^{1+o(1)})$ time [4,12].

We consider a set of predetermined sources and destinations, expressed as $D = \{(s_i, t_i)\}_i$, that will be collectively referred to as demand. Assuming full knowledge on $\{s_i\}_i$ and $\{t_i\}_i$ in advance, and also assuming known demand in advance, the solution for the route-selection problem can be optimal and polynomial. This provokes, according to an idea introduced by Srinivasan and Teo [29], a set of routes characterized by congestion and dilatation within a constant factor from the ideal solution. It results in a continuous approximation strategy for the completion time in packet routing, but it also requires introducing centralized network control which combines route selections of the various requests. Most of requests for packet delivery are regular, and from many places in the network, so it is a rather ordinary and spread out situation.

A unique scheduling method is presented in the paper "Packet Routing and Job-Shop Scheduling in $O(Congestion + Dilation)$ Steps" by F. Leighton, Bruce Maggs, and Satish Rao for routing packets along edge-simple channels in networks. In $O(c + d)$ steps, this routing is carried out, where c denotes path congestion and d is the longest path length [19]. With low congestion and a compact stretch, our oblivious routing algorithm produces pathways with little $c + d$. The oblivious approach proposed by Maggs et al. [20] also targets d-dimensional meshes (grids) and achieves a congestion level of $O(d * c * logn)$. Several expansions have been investigated in the setting of generic networks in the wake of the study described in [5,6,16,20,24,25]. These efforts have produced more sophisticated oblivious algorithms that have achieved congestion levels that are close to ideal.

Both adaptive and oblivious routing models have investigated the routing problem in great detail. An depth analysis of Oblivious Routing, a static traffic management technique without knowledge of user requests, has been done in academic literature. Randomized routing systems on hypercubes were first developed by Valiant and Brebner in the 1980s, attaining a competitive ratio of $O(logn)$. According to Racke's study, the greatest congestion in the context of generic undirected graphs is within an $O(log^3 n)$ factor of the smallest feasible congestion. This competitive ratio was eventually adjusted to $O(logn)$ [3,16,22,24,25,31] and $O(log^2 nloglogn)$. This encompasses expanders, Cayley graphs, fat trees, meshes, and more, as detailed in references [7–9,23,28,30]. For an extensive exploration of related studies, consult Scheideler's thesis [28] and refer to Räcke's comprehensive survey [26].

A lot of studies have been done on unaware routing in large-scale network situations. Räcke's seminal 2002 study [24] demonstrated poly-logarithmic scaling with network size and confirmed the existence of competing oblivious routing systems.

Aiming to increase competitiveness [16], further research introduced effective algorithms to create oblivious routing schemes within these competitive boundaries [5,6,16]. A considerable improvement was made by Räcke in 2008, who presented an oblivious routing method with an $O(logn)$ competitive ratio. This accomplishment was made possible by a simple, quick technique built on multiplicative weights and the FRT's random approximation of generic metric spaces using tree metrics [10].

By broadening the scope of the research, Gupta, Hajiaghayi, and Räcke [13] revealed a poly-logarithmic ($polylog(n)$) competitive ratio applicable to unaware routing protocols independent of both traffic patterns and related edge cost functions. Even though these contributions have been excellent at minimizing congestion in typical network contexts, they have not yet explicitly addressed the complexities of throughput, latency, or the joint design of network infrastructure and routing frameworks.

Hajiaghayi, Kleinberg, Leighton, and Räcke [14] established a minimum competitive ratio of $\Omega(log\ n.loglog\ n)$ for controlling throughput in oblivious routing strategies, primarily for expansive network topologies. It's important to note their distinct interpretation of throughput, focusing on the cumulative flow rate servicing all sender-receiver pairs. Regarding latency, another study [17] thoroughly explored the competitive ratio, especially in terms of average latency in extensive network settings. Their latency model differs, involving resistance values assigned to individual edges. They also exclusively present an oblivious routing strategy achieving a competitive ratio of $O(log(n))$ for routing to a single destination.

3 Methodology

Our approach aims to improve the speed of the algorithm by using a slimmed-down version of Prim's-Sollin's. By combining the two algorithms, we can gradually increase the performance of the minimum spanning tree. This in turn will lead to an optimal solution to the problem that significantly reduces the computational time and improves the overall performance of the algorithm.

In paper [21], using the algorithms of Prim's and Sollin's, the Minimum Spanning Tree (MST) primary power distribution network was computed and simulated. Then we found that when we apply Prim's-Sollin's algorithm together, our competitive ratio is minimized.

3.1 Proposed Algorithm

We have modified the existing algorithm proposed in [12] as below. In Algorithm 3.1 proposed by Ghaffari et al., the authors have used Prim's algorithm for a minimum spanning tree. In my paper, we have modified the algorithm by using Prim's-Sollin's algorithm (See Algorithm 3.1) in place of Prim's algorithm.

3.2 How the Proposed Algorithm Work

Algorithm 3.1 is modified from the existing algorithm proposed by [12]. In this algorithm, our contribution is that we have used Prim's-Sollin's, a hybrid algorithm for the minimum spanning tree in line 5 and lines $8-11$ (see Algorithm 3.1).

Algorithm 1. Sample a path $p \sim R_{s,t}$ given a graph $G, s, t \in V(G)$, and a hop constrained $h \geq 1$ [12]

1: Create a "completion" $H = (V, \binom{V}{2}, c_H)$ of $G = (V, E, c_G)$.
2: $c_H(e) := c_G(e)\ if\ e \in E$,
3: $c_H(e) := n^{-O(1)}.min_{e \in e}.c_G(e)$ otherwise.
4: Let $\mathcal{T}_1$ be a $\mathcal{D}^{(1)}$-router on $\mathcal{H}$ with exclusion probability $\epsilon_1 = \frac{1}{4h}$.
5: Sample $r := O(log\ n)$ trees $T_1, T_2, \ldots, T_r \sim \mathcal{T}$ conditioned $s, t \in V(T_i)$.
6: Assign $q_1 := (T_1)^G_{s,t}, q_2 := (T_1)^G_{s,t}, \ldots, q_r := (T_1)^G_{s,t}$.
7: Let $\mathcal{T}_\in$ be a $\mathcal{D}^{(1)}$-router on $\mathcal{H}$ with exclusion probability $\epsilon_2 = \frac{1}{O(hlog^4 n)}$. 8: Simple a tree $(F, F^G) \sim \mathcal{T}_\in$ and let $p := F^G_{s,t}$.
9: Simplify p by eliminating all cycles.
10: Repeat the Sampling of F and p if $\bigcup_{i=1}^r V(q_i) \nsubseteq V(T)$
11: Repeat the Sampling of F and p if $E(p) \nsubseteq E(G)$.
12: Return p.

Algorithm 2. MST by Prim's - Sollin's

Step 1: Start with Sollin's algorithm

(a) Create a list of trees, each comprising a single vertex.
(b) Create a weight of edges, each comprising a single vertex.
(c) While the list contains more than one tree, find the smallest edges connecting trees to Graph and add them to the minimum spanning tree.
(d) Merge pairs of trees in the list, reducing the number of trees by half.
(e) Repeat the above steps for $O(log\ log\ n)$ passes.

Step 2: Switch to Prim's algorithm

(a) Initialize a heap to store values for connecting the large tree with the trees in the list.
(b) Build one large tree by connecting it with the smaller trees in the list while maintaining the minimum-weight edge in the heap for each tree in the list.
(c) Ensure the number of remove-min operations in the heap equals the number of trees left in the list after Sollin's algorithm $(O(n/logn)$

Step 3: The final result is a minimum spanning tree constructed with the combined approach.

3.3 Analysis of This Algorithm

In Step 1 –(a), Make a list of trees with a single vertex in each where n is the number of vertices in the graph, and this step takes $O(n)$ time. This is so because all that has to be done is compile a list of n single-vertex trees.
Step (b), m is the number of edges in the graph, and the time required for this step is $O(m)$. This is so because all that has to be done is compile a list of every edge in the graph.

Step (c), Time for this step is $O(mlogn)$. This is because it requires $O(m)$ time to locate the shortest edge linking any two trees in the list, and $O(logn)$time to merge those two trees. Until there is just one tree remaining in the list, this process is repeated.

Step (d), This step takes $O(n)$ time. This is because it simply involves merging pairs of trees in the list, which takes $O(logn)$ time for each pair, and there are $n/2$ pairs of trees.

Step (e), This step takes $O(loglogn) \cdot (O(mlogn) + O(n))$, and it involves repeating steps c and d.

In Step 2 –(a), This step takes *O(n log n)* time. This is because it requires building a heap in which the minimum-weight edge joining each list tree to the main tree is stored.

Step (b), This Step requires *O(n log n)* time. This is because it continuously eliminates the minimum-weight edge from the heap and joins the two trees it connects. The minimum-weight edge that joins each tree to the main tree is kept at the top of the heap through maintenance.

Step (c), The because is *O(n/ log n)*. This is because it involves removing the minimum-weight edge from the heap until there are no more trees left in the list.

In Step 3, Time for this step is *O(1)*. This is due to the fact that it only requires giving back the big tree that was built in step 2b.

The Total Running time is:
$O(n) + O(m) + O(mlogn) + O(n) + O(loglogn(mlogn + n) + O(nlogn) + O(1) = O(mloglogn)$

Overall, the total time complexity of the hybrid algorithm is $O(m\ loglog\ n)$ [1,11,18].

3.4 How Efficient Prim's-Sollin's Algorithm for Minimum Spanning Tree

The Sollin's part is known for its ability to find a minimum spread tree by iteratively connecting elements of a graph, effectively reducing the number of elements per iteration. The Prim's part is known for its ability to able to grow a minimum spanning tree from an arbitrary starting node by successively adding the shortest edge that connects this new vertex and already established one.

In Sect. 3.3 analysis of Prim's-Sollin algorithm for finding a minimum spanning tree, it is shown that the competitive ratio has been reduced.

Prim's-Sollin algorithm is a combination of primitives and simple algorithms that address the problem of finding a minimum spanning tree in a graph. In this algorithm, the congestion and dilation are minimized where each packet establishes its route independently of other packets.

3.5 Comparative Study

In this section, we have compared the complexities of existing algorithms with our proposed algorithm. Table 1 shows the comparative study of complexities or competitive rato of different existing algorithms for oblivious routing.

Table 1. Comparative Study of Complexities of Competitive Ratio of Different Existing Algorithms for Oblivious Routing

Sl. No	Authors, algorithms proposed, and year	Complexities
1	Yossi Azar, Edith Cohen, Amos fiat, Harald Racke [5]	$Poly(logn)$
2	Mohammad Taghi Hajiaghayi, Robert D.kleinberg, Harald Racke and Tom Leighton [6]	$O(\sqrt{k}logn)$
3	Prahladh Harsha, Thomas P. Hayes, Hariharan Narayanan, Harald Räcke, Jaikumar Radhakrishnan [17]	$O(logn)$
4	Anupam Gupta, Mohammad T. Hajiaghayi, Harald Racke [13]	$Olog^2n$
5	Bernhard Haeupler, Harald Räcke, Mohsen Ghaffari [13]	$sub-polynomial$
6	Mohsen Ghaffari, Bernhard Haeupler, Goran Zuzic [12]	O(logn)
7	Our proposed Algorithm	$O(mloglogn)$

4 Conclusions and Future Directions

Our conclusion to this study includes proposing oblivious routing schemes for hop-constrained static routing. Our competitive ratio has been minimized using Sollin-Prim algorithms and we have also successfully reduced congestion and dilation. For instance, these algorithms are capable of determining individual packet paths without being influenced by other packets. Moreover, Prim's-Sollin's algorithm has been virtually used in our research to demonstrate its effectiveness in determining the smallest spanning tree in a graph. This is significant because it minimizes the number of hops while respecting the specified hop restriction. The implementation of Algorithm 3.1 as well as Algorithm 3.1 may not lead to an almost linear time complexity; however, it guarantees such routing construction in $O(mloglogn)$. The future possibility of achieving exact linear time complexity with Prim's-Sollin's algorithm on hop-constrained oblivious routing remains unknown. Although there might be some progress due to developments in algorithms and computation that will result in achieving $O(m^{1+O(1)})$ complexity, further investigations should be carried out to establish its feasibility.

References

1. ICS 161. Design and analysis of algorithms lecture notes for 6 February 1996 (Year Published/ Last Updated). https://ics.uci.edu/~eppstein/161/960206.html
2. Amir, D., Wilson, T., Shrivastav, V., Weatherspoon, H., Kleinberg, R., Agarwal, R.: Optimal oblivious reconfigurable networks. In: Proceedings of the 54th Annual ACM SIGACT Symposium on Theory of Computing, pp. 1339–1352 (2022)
3. Applegate, D., Cohen, E.: Making intra-domain routing robust to changing and uncertain traffic demands: understanding fundamental tradeoffs. In: Proceedings of the 2003 Conference on Applications, Technologies, Architectures, and Protocols for Computer Communications, pp. 313–324 (2003)
4. Aspnes, J., et al.: Eight open problems in distributed computing. Bull. EATCS **90**, 109–126 (2006)
5. Azar, Y., Cohen, E., Fiat, A., Kaplan, H., Racke, H.: Optimal oblivious routing in polynomial time. In: Proceedings of the Thirty-Fifth Annual ACM Symposium on Theory of Computing, pp. 383–388 (2003)
6. Bienkowski, M., Korzeniowski, M., Räcke, H.: A practical algorithm for constructing oblivious routing schemes. In: Proceedings of the Fifteenth Annual ACM Symposium on Parallel Algorithms and Architectures, pp. 24–33 (2003)
7. Busch, C., Magdon-Ismail, M.: Optimal oblivious routing in hole-free networks. In: Zhang, X., Qiao, D. (eds.) Quality, Reliability, Security and Robustness in Heterogeneous Networks, pp. 421–437. Springer, Heidelberg (2012). https://doi.org/10.1007/978-3-642-29222-4_30
8. Busch, C., Magdon-Ismail, M., Xi, J.: Oblivious routing on geometric networks. In: Proceedings of the Seventeenth Annual ACM Symposium on Parallelism in Algorithms and Architectures, pp. 316–324 (2005)
9. Busch, C., Magdon-Ismail, M., Xi, J.: Optimal oblivious path selection on the mesh. IEEE Trans. Comput. **57**(5), 660–671 (2008)
10. Fakcharoenphol, J., Rao, S., Talwar, K.: A tight bound on approximating arbitrary metrics by tree metrics. In: Proceedings of the Thirty-Fifth Annual ACM Symposium on Theory of Computing, pp. 448–455 (2003)
11. Gabow, H.N., Galil, Z., Spencer, T., Tarjan, R.E.: Efficient algorithms for finding minimum spanning trees in undirected and directed graphs. Combinatorica **6**(2), 109–122 (1986)
12. Ghaffari, M., Haeupler, B., Zuzic, G.: Hop-constrained oblivious routing. In: Proceedings of the 53rd Annual ACM SIGACT Symposium on Theory of Computing, pp. 1208–1220 (2021)
13. Gupta, A., Hajiaghayi, M.T., Räcke, H.: Oblivious network design. In: Proceedings of the Seventeenth Annual ACM-SIAM Symposium on Discrete Algorithm, pp. 970–979 (2006)

14. Hajiaghayi, M.T., Kleinberg, R.D., Leighton, T., Räcke, H.: New lower bounds for oblivious routing in undirected graphs. In: Proceedings of the Seventeenth Annual ACM-SIAM Symposium on Discrete Algorithm, pp. 918–927. Citeseer (2006)
15. Hao, F., Kodialam, M., Lakshman, T.: Hop constrained maximum flow with segment routing, US Patent 10,374,939 (2019)
16. Harrelson, C., Hildrum, K., Rao, S.: A polynomial-time tree decomposition to minimize congestion. In: Proceedings of the Fifteenth Annual ACM Symposium on Parallel Algorithms and Architectures, pp. 34–43 (2003)
17. Harsha, P., Hayes, T.P., Narayanan, H., Räcke, H., Radhakrishnan, J.: Minimizing average latency in oblivious routing. In: Proceedings of the Nineteenth Annual ACM-SIAM Symposium on Discrete Algorithms, pp. 200–207 (2008)
18. Karger, D.R., Klein, P.N., Tarjan, R.E.: A randomized linear-time algorithm to find minimum spanning trees. J. ACM **42**(2), 321–328 (1995)
19. Leighton, F.T., Maggs, B.M., Rao, S.B.: Packet routing and job-shop scheduling in o (congestion+ dilation) steps. Combinatorica **14**(2), 167–186 (1994)
20. Maggs, B.M., auf der Heide, F.M., Vocking, B., Westermann, M.: Exploiting locality for data management in systems of limited bandwidth. In: Proceedings 38th Annual Symposium on Foundations of Computer Science, pp. 284–293. IEEE (1997)
21. Marpaung, F., et al.: Comparative of prim's and Boruvka's algorithm to solve minimum spanning tree problems. J. Phys.: Conf. Ser. **1462**, 012043 (2020). IOP Publishing
22. Németh, G.: On the competitiveness of oblivious routing: a statistical view. Appl. Sci. **11**(20), 9408 (2021)
23. Rabin, M.O.: Efficient dispersal of information for security, load balancing, and fault tolerance. J. ACM **36**(2), 335–348 (1989)
24. Racke, H.: Minimizing congestion in general networks. In: Proceedings of the 43rd Annual IEEE Symposium on Foundations of Computer Science, 2002, pp. 43–52. IEEE (2002)
25. Räcke, H.: Optimal hierarchical decompositions for congestion minimization in networks. In: Proceedings of the Fortieth Annual ACM Symposium on Theory of Computing, pp. 255–264 (2008)
26. Räcke, H.: Survey on oblivious routing srtategies. In: Ambos-Spies, K., Löwe, B., Merkle, W. (eds.) Mathematical Theory and Computational Practice, pp. 419–429. Springer, Heidelberg (2009). https://doi.org/10.1007/978-3-642-03073-4_43
27. Räcke, H., Schmid, S.: Compact oblivious routing. arXiv preprint arXiv:1812.09887 (2018)
28. Scheideler, C.: Universal Routing Strategies for Interconnection Networks. Springer, Heidelberg (1998)
29. Srinivasan, A., Teo, C.P.: A constant-factor approximation algorithm for packet routing, and balancing local vs. global criteria. In: Proceedings of the Twenty-Ninth Annual ACM Symposium on Theory of Computing, pp. 636–643 (1997)
30. Upfal, E.: Efficient schemes for parallel communication. J. ACM **31**(3), 507–517 (1984)
31. Valiant, L.G., Brebner, G.J.: Universal schemes for parallel communication. In: Proceedings of the Thirteenth Annual ACM Symposium on Theory of Computing, pp. 263–277 (1981)

Harnessing the Advanced Capabilities of LLM for Adaptive Intrusion Detection Systems

Oscar G. Lira, Alberto Marroquin, and Marco Antonio To(✉)

Research Laboratory in Information and Communication Technologies, Universidad Galileo, 7a. Av. Final, Calle Dr. Suger, Zona 10, Guatemala
{rlict,oscargonzalez,amarroquin,marcoto}@galileo.edu
http://rlict.galileo.edu/

Abstract. The integration of Machine Learning (ML) and Deep Learning (DL) techniques has significantly advanced Intrusion Detection Systems (IDSs) across diverse domains such as networking, cybersecurity and industrial control systems. These approaches have played a key role in the development of Artificial Intelligence (AI) within IDSs. Moreover, the emergence of Large Language Models (LLMs) has gained prominence, encompassing both ML and AI capabilities. These models are trained on extensive datasets, enabling them to generate human-like text generation and autonomous decisions taking. This paper is intended to evaluate the capacity of LLMs in the context of IDSs for networking. LLMs exhibit the ability to process and comprehend large volumes of network log data, autonomously learn, adapt to evolving network behavior, and effectively differentiate between regular activities and potential threats. Emphasizing the substantial role of ML and AI by enhancing the adaptability and performance of IDSs technologies. The present work also underscores the potential of LLMs and their fine-tuning to reinforce IDSs capabilities, while addressing the associated challenges.

Keywords: Cybersecurity · Network Security · Intrusion Detection Systems · Large Language Models · Artificial Intelligence

1 Introduction

The evolving landscape of cybersecurity demands adaptive solutions to effectively address emerging vulnerabilities. In response to the increasing cybersecurity threats, there is a growing demand for robust IDSs. The integration of ML and DL enhances IDSs capabilities across various domains, addressing limitations of traditional methods [20,21]. LLMs with advanced machine learning and natural language processing provide a promising solution, revolutionizing IDS by effectively detecting, classifying, and proactively mitigating cyber threats, including unknown attack patterns [1]. The contributions of the work of Y. Charalambous, et al. [2] and X. Zhao, et al. [3] provide evidence of how LLMs

L. Barolli (Ed.): AINA 2024, LNDECT 204, pp. 453–464, 2024.
https://doi.org/10.1007/978-3-031-57942-4_44

can effectively be employed in the network security field, particularly by classifying attacks [17]. Also, K. Saurabh, et al. [4] and X. Huang [5] show how Long Short-Term Memory (LSTM) can improve drastically the accuracy of detection in the IDS. LSTM models have the ability to handle long-term dependencies in sequential data. Furthermore, these works emphasize how the role of LLMs have potential to be implemented in this area. As traditional methodologies often struggle to cope with the rapid evolution of sophisticated attack vectors, the integration of ML and AI [6], LLMs become imperative to enhance the adaptability of IDSs. This technique, exhibits a unique capability to directly process logs generated from monitoring tools, enabling real-time analysis and swift response to emerging threats.

This paper introduces a Bidirectional Encoder Representations form Transformers (BERT) [7] based model that improves attack detection on a well known dataset compared to other ML methodologies. The model defined as BERTIDS was fine-tuned to be capable of identifying and classifying network attacks within text logs. Leveraging the BERT-based architecture, BERTIDS is designed to serve as a foundation component within the security framework. Its role is extended to be the initial line of defense in the identification of potential security threats. Harnessing the formidable capabilities of LLMs, BERTIDS provides a dynamic and responsive approach to intrusion detection.

Our approach, is an innovative LLM-based IDS model specifically crafted for network security applications. The experimental findings presented in this paper provide valuable insights through a comparative analysis of BERTIDS against other IDSs models, encompassing both ML-based and traditional approaches. The contribution of the present work can be summarized as:

- BERT-based model capable of identifying normal and anomalous traffic from different types of attacks.
- BERT-based model capable of classifying 23 different types of attacks.
- IDS model capable of identifying and classifying attacks with better accuracy and low False Positive rate than other researched models.

The remainder of this paper is organized as follows. In Sect. 2, we present background information related to the application of LLMs models in the network security context. We present background information related to LLM fine-tuned solutions for intrusion detection domain. In Sect. 3, we describe our proposed architecture including the domain-specific dataset, the hyperparameters selection and fine-tuning process. Then we present the results of the model and a brief discussion. Finally, in Sect. 4 we provide the conclusions of our findings by comparing BERTIDS results with other ML-based IDSs models and discuss possible future work.

2 Related Work

The integration of LLMs into the domain of Network Intrusion Detection Systems (NIDSs) is a promising technology to be included in the area, addressing problems in anomaly detection and attack identification domain. In this section, we review relevant literature that lays the foundation for our work and provides context within the broader landscape of LLMs applied to NIDSs.

The intersection of ML and network security has witnessed notable advancements for the past years [18–21]. HuntGPT proposed a specialized intrusion detection dashboard that integrates LLMs for proactive threat hunting. Their work emphasizes the significance of ML in reducing detection and response times, and the incorporation of Explainable AI (XAI) frameworks for enhancing the interpretability of alerts. This approach is critical for effective decision-making in the context of NIDSs. Their work focused on the use of the LLM in the process of generating human-interpretable explanations for a Random Forest classifier. HuntGPT encounters challenges during its training process, maintenance, and the management of frequent false positives [8]. Wang et al. explored the application of LLMs in code vulnerability detection with DefectHunter, leveraging the Conformer mechanism. But their work faces limitations in capturing high-dimensional relationships and acquiring sufficient feature information [9]. While their focus is on code vulnerabilities, they prove how effective LLMs can be in the task of classify anomalies. Zaboli et al. presented a case for using LLMs, including ChatGPT, in the cybersecurity of smart grid applications. They studied particularly IEC 61850-based digital substation communications [10]. Their solution has better results among a range of six different industrial control datasets and two of them being very complex, outperforming traditional identification models. Their insights into multicast messages and the development of an LLM-based cybersecurity framework with Human-In-The-Loop (HITL) allows them to perform comparative analysis of detected anomalies in smart grid applications using different LLMs.

Jüttner et al. proposed ChatIDS, an approach leveraging LLMs to make IDS alerts understandable to non-experts [11]. This approach is based on the need for explainability as a paramount for network security solutions, as different stakeholders would require to understand how the model is taking its decisions. ChatIDS is used as a model for improving user understanding and response to alerts generated by a NIDS. Although, it confronts potential issues associated with trust, privacy, and ethics that necessitate resolution before practical implementation. Ameri et al. introduced CyBERT, a cybersecurity contextualized model that can process textual threat, attack and vulnerability information for the cybersecurity community. CyBert, encounters constraints tied to the quality and quantity of labeled sequences derived from industrial control systems (ICS) device documentation used for training [12]. This model is a BERT-based fine-tuned model. CyBERT robustness and accuracy can proof how well LLMs can be integrated in network security architecture.

Yin et al. presented ExBERT, a framework for predicting the exploitability of vulnerabilities [13]. While their emphasis is on vulnerabilities, the concept of transfer learning and predicting exploitability can inform advancements in understanding network-based threats. Their results provide insights in the high capacity LLMs have and how they can outperform the state-of-the-art model for identifying possible threats. Nonetheless, the model is delimited by the correctness standards applicable in diverse jurisdictions and the quality of the expert opinions it relies upon.

One model that has gained popularity due to the effectiveness in the domain of cybersecurity is SecureBERT, developed by Aghaei et al. this model is a domain-specific language model tailored for cybersecurity tasks [14]. By training on a large corpus of cybersecurity data, SecureBERT exhibits proficiency in solving critical Natural Language Processing (NLP) tasks in cybersecurity, its main problem depends on the quality of the extensive corpus of cybersecurity text on which it was trained. As many other works have been exploring the opportunities that LLM can integrate in the domain of cybersecurity, there is not any specific work leveraging LLMs as an attack classification tool. There have been some recent works that effectively have used BERT for classification purposes. Lin et al. which introduces ET-BERT, a novel approach for encrypted traffic classification. ET-BERT leverages the power of transformer models to pre-train a deep contextualized datagram-level representation from large-scale unlabeled data. This work establishes the potential of BERT model in the domain of network traffic analysis [15].

These identified limitations underscore the motivation to continue researching and developing applications based on LLMs like BERT in network IDS environment, particularly exploring its capacity to process raw logs from monitoring tools for the purpose of detecting and classifying attacks. From our review, we found a lack of language models specifically fine-tuned as NIDS. While these type of solutions are promising, some works mentioned the use of BERT as and IDS classifier but no information of how the model was trained or what the results are is explained. In this study, we developed a network log LLM IDS specifically for network security domain, BERTIDS. We utilized a highly accepted dataset to generate synthetic network logs to train the model. Finally some tests are made to evaluate the model capabilities in identifying and classifying attacks.

3 Bidirectional Encoder Representations from Transformers Intrusion Detection System

3.1 Methodology

BERTIDS uses the capabilities of BERT model to classify text in the task of analysing network logs containing relevant information of the network status. In this case we use the contained values in a well known dataset for intrusion detection purposes. The BERT model which stands for Bidirectional Encoder Representations from Transformers, is a ML model for natural language processing. It was designed by researchers at Google AI Language in 2018 [7].

Since then multiple application related to language processing have been done with this specific model. BERT is a transformer-based architecture that uses the Transformer encoder to process each token of input text in the full context of all tokens before and after. This bidirectional processing of input text allows BERT to understand the context and structure of the language better. A

Transformer model is a type of deep learning model that was introduced in 2017 [1]. It is a neural network architecture based on the multi-head attention mechanism. Unlike previous recurrent neural architectures the Transformer model does not include any recurrent units, which reduces the training time. When using BERT for IDS purposes the idea remains in the model analyzing network logs, separating this logs into tokens and finally for each token, that refers to a unit of text that has been extracted or split from the original log, can be considered as a piece of information or a feature in the network log. These features could be IP addresses, timestamps, protocol types, or any other relevant information being shared in the log.

BERT does not isolate these tokens; rather, it considers the full context of all tokens before and after it. This is essential because the significance of a particular feature in a network log often depends on what other features are present. For example, a certain combination of features might indicate a potential network attack, while the same features in a different context might be perfectly normal. This become crucial in the task of classifying attacks with low data available for training the model as it can better capture the patterns of each attack.

The Transformer model follows an encoder-decoder structure. The encoder maps an input sequence to a sequence of continuous representations, which is then fed into a decoder. The decoder generates an output sequence, consuming the previously generated symbols as additional input when generating the next sequence. This architecture is now used not only in natural language processing but also in computer vision, audio, and multi-modal processing. It has further led to the development of pre-trained systems, such as generative pre-trained transformers (GPTs) and BERT.

These models are usually pre-trained on a large corpus of text and then fine-tuned for specific tasks. This pre-training and fine-tuning process enables BERT to accurately predict sentiments in sentences it encounters. The strength of BERT in text classification comes from its ability to understand the context of a word based on its surroundings, i.e., words at the left and right in a sentence. The self-attention mechanism in BERT permits to weigh the importance of each feature in the context of the others. It learn how to pay more attention to certain features that are more indicative of a network attack and less attention to irrelevant features. This ability to understand the context and assign appropriate weights to different features makes BERT a powerful tool for detecting network attacks.

The architecture, with layers like feed-forward and attention layers, evolves with an increasing number of parameters, enhancing the model ability to capture intricate patterns. In classification tasks, LLMs prove versatile in zero-shot learning and classification logic learning, highlighting adaptability. Applied to network security, LLMs show potential as efficient tools for IDS, leveraging generated logs for enhanced performance [16,17].

3.2 BERTIDS Mechanism

Algorithm 1 shows the process of the log creation taking as input the whole dataset and then making the concatenation of each label with is actual value, to generate the synthetic log for just the selected features.

Algorithm 1. Log Creation

```
procedure LogCreation(Data)
    Input: NSL-KDD dataset
    Output: List of synthetic logs
    synthetic_logs ← []
    for i in range(len(Data)) do
        log ← ''
        for selected_feature in Data.keys() do
            log ← log + selected_feature + ': ' + str(Data[selected_feature][i]) + ' '
        end for
        synthetic_logs.append(log)
    end for
    return synthetic_logs
end procedure
```

Algorithm 2 is a brief description of the initialization phase and values set for the process of fine-tuning and all the required setups to correctly train the model. As a batch size a value of 32 was used and a total epoch of 1.

Algorithm 2. Fine-Tuning BERT for Intrusion Detection: Initialization Phase

```
procedure FineTuneBERTIDModelInit(logs, labels)
    Input: Synthetic Logs, Attack Labels
    Output: Tokenizer, Model, Optimizer, Number of Epochs
    tokenizer ← InitializeTokenizer()                              ▷ Initialize tokenizer
    model ← InitializeModel()                                      ▷ Initialize BERT model
    optimizer ← InitializeOptimizer(model.parameters())            ▷ Initialize optimizer
    num_epochs ← SetNumberOfEpochs()
    return tokenizer, model, optimizer, num_epochs
end procedure
```

Algorithm 3 is a description of the training or fine-tuning process. It starts by converting the input text into the format used by BERT using the tokenizer function, then the model is initialized specifying the labels which are going to be used to train the model. This labels correspond in the first case as normal or abnormal traffic in a binary classification, and in the multi class classification there are 23 different labels corresponding to normal traffic and 22 different attack types. The optimizer, in this algorithm, is utilized to update the model parameters trying to minimize the loss function.

Algorithm 3. Fine-Tuning BERT for Intrusion Detection: Training Phase

1: **procedure** FineTuneBERTIDModelTrain(*tokenizer, model, optimizer, num_epochs, logs, labels*)
2: **Input:** Tokenizer, Model, Optimizer, Number of Epochs, Synthetic Logs, Attack Labels
3: **Output:** Fine-tuned BERT IDS Model
4: *tokenized_logs* ← TokenizeLogs(*logs, tokenizer*) ▷ Tokenize logs
5: *Split data into training and testing sets*
6: *Convert data to tensors*
7: *train_data* ← CreateDataset(*train_inputs, train_labels*) ▷ Create dataset
8: *train_dataloader* ← CreateDataLoader(*train_data, batch_size=32, shuffle=True*)
9: **for** *epoch* **in** range(*num_epochs*) **do**
10: *model.train()*
11: *total_loss* ← 0
12: **for** *batch* **in** tqdm(*train_dataloader, desc=*'Epoch {*epoch*}') **do**
13: *inputs, labels* ← *batch*
14: *optimizer.zero_grad()*
15: *outputs* ← ForwardPass(*model, inputs, labels*) ▷ Perform forward pass
16: *loss* ← *outputs.loss*
17: *total_loss* + = *loss.item()*
18: *loss.backward()*
19: *optimizer.step()*
20: **end for**
21: print(f'Epoch {*epoch*}, Average Loss: {*total_loss / len(train_dataloader)*}')
22: **end for**
23: *model.eval()*
24: **return** *model*
25: **end procedure**

3.3 Implementation

To evaluate the capabilities of LLMs in the context of IDSs, we utilized the NSL-KDD dataset [22]. This dataset is used extensively as a benchmark for assessing various IDS techniques, encompasses diverse network traffic scenarios, including both normal and malicious activities. As base model we employed the *bert-base-uncased*, a variant of the BERT model available on Hugging Face. This model is recognized for its effectiveness in natural language processing tasks and its pre-trained weights and deep transformer architecture make it well-suited for complex language understanding tasks. This includes classification task, as shown in related work, this specific model has been highly used in the context of networking security [23].

Prior to integrating the LLM, a comprehensive feature exploration was carried out to identify key variables contributing significantly to the detection of attacks. Using a correlation matrix and statistical analysis, a set of relevant features were identified taken in account their potential to provide valuable insights into network behavior and potential security threats.

The fine-tuning process involved training the LLM on the selected features extracted from the NSL-KDD dataset. This process was conducted in a Google Colab notebook with a T4 GPU with 12 GB available, 15 GB of RAM and a total of 70GB of free space to simulate a limited computational resource environment [24]. Iterative training involved feeding the LLM with synthetic log data generated by the dataset values and corresponding labels, enabling the model to learn and adapt to different network patterns and characteristics. The log generation function included an additional feature to classify normal and abnormal traffic for the fine-tuning process, but this feature was not present in the logs used during the testing phase for the binary classification task. This process of using the additional feature show better results than just sending the log and its corresponding binary label for the training phase. At the multi class classification just the selected features were used.

To assess the performance of the LLM-based IDS, standard evaluation metrics, including accuracy, precision, recall, and F1-score, were employed [18]. Additionally, the false-positive rate was monitored to understand the model ability to classify normal and abnormal network traffic, as well as the ability to classify attacks. The evaluation was conducted on a reserved portion of the dataset not utilized during the fine-tuning phase, ensuring an unbiased assessment of the LLM performance. One relevant aspect of this specific dataset is that traditional ML-based IDS struggles to capture the pattern of some attacks as the dataset is unbalanced, this constraint can be avoided in the LLM as they manage to learn patterns with less data available. In the subsequent sections, we delve deeper into the results of our LLM-based IDS evaluation. We compare the performance of our model with various other IDS techniques highlighted in the existing literature. For binary classification we focused on deep learning methods that were used for this task. We aim to illustrate the superior capabilities and potential of LLMs in addressing the limitations of traditional IDS methods and paving the way for more robust and adaptive security solutions.

3.4 Results and Comparison

The results of our LLM-based IDS were compared within an analysis made for binary classification using Neural Networks (NNs) following the work done by E. Magdy, et al. [28]. Our LLM-based IDS demonstrated to have a relative good performance in identifying anomalies, surpassing several other NNs architectures but utilizing the whole text log and with less features selected, not needing the prepossessing step as in traditional ML techniques. The results are shown in the Table 1.

The results of our LLM-based IDS were also compared in multi class classification with various existing approaches as outlined in the article [29]. Our LLM-based IDS demonstrated a significantly higher accuracy in identifying anomalies, surpassing several existing methodologies, including Deep Neural Network (DNN), Convolutional Neural Network (CNN), and Autoencoder (AE) as shown in Table 2. Results not showing in Table 2 occur because the metric is not being measured in the model.

Table 1. Comparison Models with Binary classification in NSL-KDD

NN Architecture	No of features	Epoch	Accuracy
34	41	60	79
14	41	469	77
17	41	541	74
27	41	181	78
23	**41**	**178**	**80**
25	29	2466	79
17	29	1212	79
17	**29**	**265**	**80**
32	29	90	78
LLM-IDS	**12**	**0.20**	**80**

Table 2. Comparison of Intrusion Detection Systems

Approach	Accuracy	Precision	Recall	F1-score	FP rate
DNN	75.75	83	76	75	–
NDAE	89.22	92.97	89.22	90.76	10.78
CNN	79.48	23.4	68.66	–	27.90
HCPTC-IDS	89.75	–	86.71	–	6.23
STL	84.96	96.23	76.57	85.28	–
GRU-RNN	89	91	90	90	–
FEU and FFDNN	87.74	–	–	–	–
DNN, SHIA	80.1	69.2	96.9	80.7	–
ICVAE-DNN	85.97	97.39	77.43	86.27	2.74
AE	84.24	87	80.37	81.98	**0.4**
DLS-IDS	83.57	96.46	78.12	86.32	3.57
DSSTE	82.84	83.94	82.78	81.66	–
5 Layer AE	90.61	86.83	**98.43**	92.26	–
LGBM K-Means	90.41	84.78	86.9	90.96	–
DT-PCADNN	88.64	–	84.56	–	–
Enhanced random forest	78.47	–	78	–	–
GMM-WGAN	86.59	88.55	86.59	86.88	–
RT-IDS	81.87	96.45	70.71	81.59	–
BERTIDS	**98.01**	**98.31**	98.01	**98.09**	1.48

Furthermore, the LLM-based IDS exhibited a notable precision and recall balance, emphasizing its capacity to reduce false positives while effectively identifying potential security threats. The integration of LLMs as an IDS solution contributed to a more adaptive and robust system, capable of taking care of the challenge of correctly classifying various attacks using this dataset.

3.5 Model Predictions

The results suggest that the model performs better than many traditional IDS solutions. The testing subset comprises only 12 out of the total 23 training classes, and the model accurately classifies 11 of them. The single unclassified instance involves a unique entry correctly identified as anomalous but misclassified in terms of the specific attack type. In the subsequent sections, we offer a comprehensive analysis of the specific results obtained, highlighting the strengths and areas of improvement for the LLM-based IDS approach. We also discuss the implications of our findings in the context of enhancing the overall security posture of network infrastructures and mitigating potential cybersecurity risks.

4 Conclusion

The integration of LLMs in IDS offer several practical advantages that significantly enhance its capabilities. First, LLMs are highly adaptive and can continuously learn and update their knowledge over time as new threats emerge. This attribute ensures that the IDS remains effective in detecting novel and sophisticated attacks that could evade traditional rule-based detection mechanisms. Second, LLMs offer streamlined automation, enabling the implementation of complex security policies with minimal human intervention. This automation reduces errors associated with manual configuration, ultimately reducing misconfigurations that can potentially result in security breaches. Third, LLMs provide a comprehensive understanding of network traffic behavior, making it possible for IDS to detect anomalies that may not be immediately apparent. This characteristic equips the IDS to identify behaviors that could indicate potential security threats or faulty configurations in network devices, leading to proactive and preventive measures taking place before an attack occurs.

However, incorporating LLMs into IDS is not without its limitations and challenges. As stated before, the complexity of the model makes them difficult to understand how they take a particular decision. This limitation can lead to trust issues among cybersecurity professionals who rely on transparent explanations to validate the effectiveness of their security measures. Another limitation is that fine-tuning LLMs for specific networks requires significant computational resources, making it an expensive and time-consuming process if not sized correctly. Additionally, the high cost of model development can hinder organizations from implementing LLM-based IDS solutions, particularly in resource-constrained environments. Finally, overfitting is a major challenge associated with using LLMs in IDS implementation, leading to poor generalization and reduced performance when exposed to unseen data. This issue arises because LLMs are highly specialized for specific use cases and may not be effective in identifying novel threats or network configurations that deviate from the training data used during model development.

In future research, we seek to enhance the IDS by training our models to handle different types of logs and to study the capabilities of integrating multiple specialized attack detection models. This could be an important step in making networks more secure without needing continuous human involvement.

References

1. Vaswani, A., et al.: Attention is all you need, *arXiv preprint*, arXiv:1706.03762, https://arxiv.org/abs/1706.03762 (2023)
2. Ferrag, M.A., Ndhlovu, M., Tihanyi, N., Cordeiro, L.C., Debbah, M., Lestable, T.: Revolutionizing cyber threat detection with large language models, *arXiv preprint*, arXiv:2306.14263, https://arxiv.org/abs/2306.14263 (2023)
3. Li, F., Shen, H., Mai, J., et al.: Pre-trained language model-enhanced conditional generative adversarial networks for intrusion detection. Peer-to-Peer Netw. Appl. (2023). https://doi.org/10.1007/s12083-023-01595-6
4. Saurabh, K.: LBDMIDS: LSTM based deep learning model for intrusion detection systems for IoT networks. IEEE World AI IoT Congress (AIIoT). **2022**, 753–75 (2022). https://doi.org/10.1109/AIIoT54504.2022.9817245
5. Huang, X., Zhang, Y.: Network intrusion detection based on an improved long-short-term memory model in combination with multiple patiotemporal structures. Wirel. Commun. Mobile Comput. **2021**, 6623554 (2022).https://doi.org/10.1155/2021/6623554
6. Salih, A., Zeebaree, S.T., Ameen, S., Alkhyyat, A., Shukur, H.M.: A survey on the role of artificial intelligence, machine learning and deep learning for cybersecurity attack detection. In,: 7th International Engineering Conference "Research & Innovation amid Global Pandemic" (IEC). Erbil, Iraq vol. 2021, pp. 61–66 (2021). https://doi.org/10.1109/IEC52205.2021.9476132
7. Devlin, J., Chang, M.-W., Lee, K., Toutanova, K.: BERT: pre-training of deep bidirectional transformers for language understanding, *arXiv preprint*, arXiv:1810.04805, https://arxiv.org/abs/1810.04805 (2019)
8. Ali, T., Kostakos, P.: HuntGPT: integrating machine learning-based anomaly detection and explainable AI with large language models (LLMs). arXiv preprint arXiv:2309.16021, https://arxiv.org/abs/2309.16021 (2023)
9. Wang, J., Huang, Z., Liu, H., Yang, N., Xiao, Y.: DefectHunter: a novel LLM-driven boosted-conformer-based code vulnerability detection mechanism." arXiv preprint arXiv:2309.15324, https://api.semanticscholar.org/CorpusID:263152779 (2023)
10. Zaboli, A., Choi, S.L., Song, T.-J., Hong, J.: ChatGPT and other large language models for cybersecurity of smart grid applications. arXiv preprint arXiv:2311.05462, https://arxiv.org/abs/2311.05462 (2023)
11. Jüttner, V., Grimmer, M., Buchmann, E.: ChatIDS: explainable cybersecurity using generative AI. arXiv preprint arXiv:2306.14504, https://arxiv.org/abs/2306.14504 (2023)
12. Ameri, K., Hempel, M., Sharif, H., Lopez Jr J., Perumalla, K.: CyBERT: cybersecurity claim classification by fine-tuning the BERT language model. J. Cybersecur. Priv. **1**(4), 615-637 (2021). https://doi.org/10.3390/jcp1040031
13. Yin, J., Tang, M., Cao, J., Wang, H.: Apply transfer learning to cybersecurity: Predicting exploitability of vulnerabilities by description (2020). https://doi.org/10.1016/j.knosys.2020.106529
14. Aghaei, E., Niu, X., Shadid, W., Al-Shaer, E.: SecureBERT: a domain-specific language model for cybersecurity, arXiv preprint arXiv:2204.02685, https://arxiv.org/abs/2204.02685 (2022)

15. Lin, X., Xiong, G., Gou, G., Li, Z., Shi, J., Yu, J.: ET-BERT: a contextualized datagram representation with pre-training transformers for encrypted traffic classification (2022). https://doi.org/10.1145/3485447.3512217
16. Naveed, H., et al.: A comprehensive overview of large language models. arXiv preprint arXiv:2307.06435, https://arxiv.org/abs/2307.06435 (2023)
17. Manocchio, L.D., Layeghy, S., Lo, W.W., Kulatilleke, G.K., Sarhan, M., Portmann, M.: FlowTransformer: a transformer framework for flow-based network intrusion detection systems. arXiv preprint arXiv:2304.14746, https://arxiv.org/abs/2304.14746 (2023)
18. Shaukat, K., Luo, S., Varadharajan, V., Hameed, I.A., Xu, M.: A survey on machine learning techniques for cyber security in the last decade. IEEE Access **8**, 222310–222354 (2022). https://doi.org/10.1109/ACCESS.2020.3041951
19. Liu, H.; Lang, B.: Machine learning and deep learning methods for intrusion detection systems: a survey. Appl. Sci. **9**(20), 4396 (2019). https://www.mdpi.com/2076-3417/9/20/4396
20. Kocher, G., Kumar, G.: Machine learning and deep learning methods for intrusion detection systems: recent developments and challenges. Soft Comput. **25**, 9731–9763 (2021). Accepted 17 May (2021). 24 June 2021. Issue Date: August 2022. https://doi.org/10.1007/s00500-021-05893-0
21. Ridwan, M.A., Radzi, N.A.M., Abdullah, F., Jalil, Y.E.: Applications of machine learning in networking: a survey of current issues and future challenges. IEEE Access **9**, 52523–52556 (2022). https://doi.org/10.1109/ACCESS.2021.3069210
22. Wibowo, R.N., Sukarno, P., Jadied, E.M.: NSL-KDD Dataset. In: Proceedings (2019). https://api.semanticscholar.org/CorpusID:198166203
23. Britto, R., Murphy, T., Iovene, M., Jonsson, L., Erol-Kantarci, M., Kovács, B.: Telecom AI native systems in the age of generative AI – an engineering perspective *arXiv preprint*arXiv:2310.11770, https://aps.arxiv.org/abs/2310.11770 (2023)
24. Lv, K., Yang, Y., Liu, T., Gao, Q., Guo, Q., Qiu, X.: Full parameter fine-tuning for large language models with limited resources. arXiv preprint arXiv:2306.09782, https://doi.org/10.48550/arXiv.2306.09782 (2023)
25. Yang, J., et al.: Harnessing the power of LLMs in practice: a survey on ChatGPT and beyond. arXiv preprint arXiv:2304.13712, https://doi.org/10.48550/arXiv.2304.13712 (2023)
26. Kaddour, J., Harris, J., Mozes, M., Bradley, H., Raileanu, R., McHardy, R.: Challenges and applications of large language models. arXiv preprint arXiv:2307.10169, https://arxiv.org/abs/2307.10169 (2023)
27. Ye, W., et al.: Assessing Hidden risks of LLMs: an empirical study on robustness, consistency, and credibility. arXiv preprint arXiv:2305.10235, https://doi.org/10.48550/arXiv.2305.10235 (2023)
28. Ingre, B., Yadav, A.: Performance analysis of NSL-KDD dataset using ANN. In: 2015 International Conference on Signal Processing and Communication Engineering Systems, Guntur, India, pp. 92–96 (2015). https://doi.org/10.1109/SPACES.2015.7058223
29. Author, E., et al.: A Comparative study of intrusion detection systems applied to NSL-KDD Dataset. Egyptian Int. J. Eng. Sci. Technol. **43**(2), 88-98 (2023). https://doi.org/10.21608/eijest.2022.137441.1156

Real-Time Rogue Base Stations Detection System in Cellular Networks

Ying-Tso Wen(✉), Sz-Hsien Wu, Chia-Yi Chen, Yen-Yuan Hsu, Po-Hsueh Wang, Shun-Min Hsu, Chen-Yuan Chang, and Bo-Chen Kung

Wireless Communications Laboratory, Chunghwa Telecom Laboratories, Taipei, Taiwan
{enzowen,nealxgs,chiayi,mario,stt12706,shunmin, harrychang,bckung}@cht.com.tw

Abstract. Recently, rogue base station (RBS) attack is growing common. A RBS attack occurs when an attacker uses a fake base station (FBS) to mimic a legitimate base station, luring phone users to connect and facilitating activities like stealing personal information and sending spam messages. This type of attack poses a significant threat to users privacy and security and has a serious impact on the operation and management of mobile networks. To address RBS attacks, it is essential to create a RBS/FBS detection system. In this paper, we proposed three different approaches to detect RBS/FBS, including the user equipment (UE) approach, core network (CN) approach, and Radio Access Network (RAN) approach. Finally, we integrate the three approaches mentioned above into a monitoring user interface (UI). This UI can instantly display the locations where FBS are detected and track the moving trajectories of FBS through heatmaps over a period of time. Our proposed RBS/FBS detection system can detect RBS in real time, block attacks, and ensure service quality.

Keywords: Fake Base Station · Rogue Base Station · Cellular Network Security · Autonomous Network

1 Introduction

Globally, scam activities are an everyday occurrence, with phishing [1] constituting a notable and widespread technique. Often, these scam groups pretend to be trusted institutions such as government agencies, financial institutions, or e-commerce platforms, and disseminate emails or text messages as a means of deceit. Users may operate under the assumption that they are accessing websites affiliated with legitimate organizations, thereby inputting confidential data such as bank details, passwords, credit card details, or other personal information. Moreover, it furnishes scam groups with the opportunity to exploit this sensitive information further in their illegal activities. According to a Gartner report [2], 98% of texts are read by users, compared to only 20% of emails. Additionally,

L. Barolli (Ed.): AINA 2024, LNDECT 204, pp. 465–474, 2024.
https://doi.org/10.1007/978-3-031-57942-4_45

texts have a 45% response rate, while emails only have a 6% response rate. This implies that text messages are a more effective medium for scam groups than emails.

When scam groups opt to utilize text messages as a medium, they may encounter two obstacles. Firstly, users can identify whether the sender is legitimate through the text message sending number. Secondly, law enforcement officers can trace the source of the text message and arrest the criminals. To address these issues, scam groups have begun establishing FBS. When they send messages through FBS, they can impersonate any legitimate organization, making it impossible for users to verify the authenticity through the sending number. Scam groups can set up their own FBS, making it impossible for law enforcement officers to trace the source of the text message senders. In Taiwan, the first case involving the use of FBS to commit crimes was identified in April 2023 [3]. In response, we immediately began developing a real-time RBS detection system to monitor network services and quickly identify any similar crimes that might be occurring.

FBS exploits the weaknesses of Global System for Mobile Communications (GSM) [4], the second-generation mobile communication standard, whose first version was completed in 1990. Notably, GSM initially featured a one-way authentication design. That is, only the network will authenticate the identity of the UE, while the UE does not authenticate the network. In such a structure, the UE cannot verify the authenticity of the base station (BS). Once the UE successfully connects to an FBS, the scam group can send text messages. Later, the third-generation mobile communication standard, Universal Mobile Telecommunications System (UMTS), incorporated mutual authentication to prevent such problems.

GSM is already an outdated standard. In the current era dominated by Long Term Evolution (LTE) and New Radio (NR), GSM FBS systems need to attract UEs to connect. Signal jammers can be set up to block the signals from telecom operators' BSs. In this way, UEs are forced to connect only to GSM FBS. Besides establishing a GSM FBS, the scam group can also create an LTE FBS. If this LTE FBS offers a better signal quality than the legitimate LTE BS, it can exploit the handover mechanism's vulnerability to initiate a denial-of-service (DoS) attack [5,6]. This method guides the UE to the GSM FBS, enabling text scams.

In a mobile network, there are three main parts: UE, CN, and RAN. Figure 1 shows how these components interact. Our system uses three main elements to monitor and potentially identify FBS from different aspects. Regarding the UE aspect, since we have completely phased out GSM, we have developed an application (app) to detect GSM signals. Once a GSM signal is detected, it can almost determine FBS. Details will be introduced in Sect. 3.1. Regarding the CN aspect, we analyze the signaling to check the legality of the location area code (LAC). We also hypothesize that when the FBS is shut down, many people will re-register with our network simultaneously, therefore we analyze if there is an abnormal increase in registration behavior. Details will be presented in Sect. 3.2. In terms of RAN, we examine various performance counters at the BS to detect any abrupt fluctuations in a short period. Details will be discussed in Sect. 3.3. Finally, We created a web application that displays real-time UE, CN, and RAN monitoring data directly on a map. In this UI, we can track the location and movement paths of FBS, aiding law enforcement units in catching scam groups. Details will be discussed in Sect. 4.

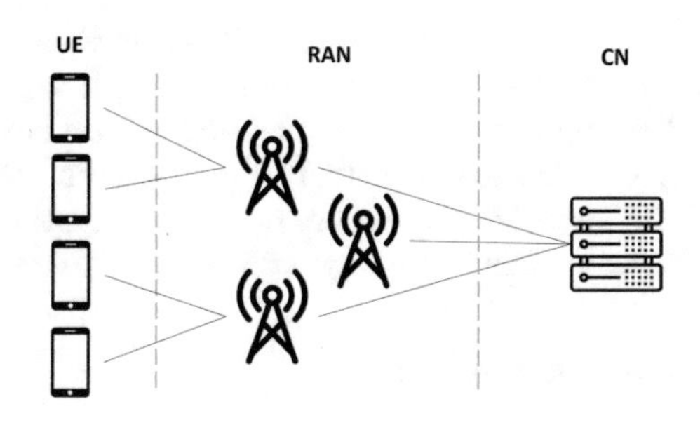

Fig. 1. Main network components in cellular network.

2 Related Work

In [7], Hua et al. use the UE to detect FBS. When the UE is affected by FBS, there will be an abnormal enhancement in signal strength. This method involves using the measured signal strength and the distance from the legitimate BS to calculate the anomaly ratio of signal strength to distance. If the value exceeds the normal range, it is identified as being affected by FBS. To apply this method, the UE needs to identify the nearest legitimate BS, and must first acquire the transmission power as well as the longitude and latitude of the BS to carry out the calculations. In [8], FBS-Radar be proposed. This system performs several types of analysis upon receiving a text message, including evaluating the signal strength at the time, BS ID, BS location, and sender number, and incorporates FBS-Radar into the widely-used Baidu PhoneGuard app. In [9], Ali et al. introduced a radio frequency (RF) fingerprinting technique that, unlike common methods, uses phase noise and clock synchronous carrier frequency offset to directly detect FBS at the physical layer instead of the application layer. All the mentioned methods are terminal-based and require deployment in the UE.

In [10], Steig et al. described a technique for identifying FBS through analyzing unexpected neighboring cells in UE-to-BS transmission reports. In [11], Shin et al. performed FBS detection in the Automatic Neighbor Relation (ANR) procedure, which is a key feature in Self-Organizing Networks (SON). In [12], the issue of using location area code (LAC) to detect FBS was proposed. While the LAC is transmitted when the UE reconnects to the legitimate network, false positives may arise if the user boards a flight. The methods mentioned above are all network-based detection methods. Telecom operators can analyze the available data to detect FBS. The real-time RBS detection system we developed combines both network-based and terminal-based detection methods, and utilizing different aspects of FBS detection will yield more reliable results.

3 Detection Approach

The real-time RBS detection system we developed carries out FBS detection from two aspects. One of the aspects is terminal-based; we develop an app in the UE to detect GSM signals. Another approach we use is network-based; we evaluate the legality of LAC and registration patterns in CN, analyze performance metrics in RAN, and utilize telecom operators' existing data to pinpoint potential FBS locations. In this section, we will describe these methods in detail.

3.1 User Equipment Approach

In this subsection, we explore the techniques for detecting GSM (2G) FBS using a mobile app. To validate the proposed FBS detection technology, we emulated a 2G FBS using the open-source software YateBTS and the BladeRF 2.0 SDR development board.

Fig. 2. A 2G RBS emulator approach on YateBTS software and bladeRF SDR.

For testing, we set up the system by connecting the Nuand BladeRF 2.0 Micro xA9 SDR development board to a computer equipped with Ubuntu 22.04 x64, using a USB 3.0 interface. The simulated 2G FBS shows in Fig. 2. In this system, we integrate the open source software YateBTS to realize complete GSM or General Packet Radio Service (GPRS) BS and CN functions, achieving the purpose of serving as a 2G FBS. When the victim's mobile phone is blocked by a jammer from connecting to the public network, it will start searching for available mobile communication networks. Alternatively, it might be lured by malicious handovers, eventually connecting successfully to the 2G FBS. At this time, the FBS sends a scam text message to the victim UE, inducing the victim to be deceived.

The design of the operating parameters of the 2G FBS will affect its attack capability. We set the corresponding public land mobile network (PLMN) ID

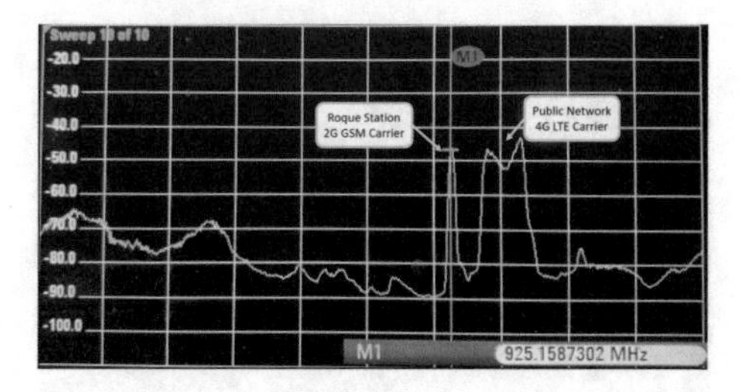

Fig. 3. Frequency response diagram of 2G RBS and LTE mobile network operating carrier.

on the 2G FBS based on the Subscriber Identity Module (SIM) card installed in the victim mobile phone. Thus, when the victim mobile phone searches for mobile communication networks, it will try first Access the 2G FBS broadcasting the same PLMN ID. In additional, it is also necessary to consider whether the 2G FBS will be overshadowed by the existing mobile network bands. In Fig. 3, we eventually used 925.2 MHz as the downlink signal frequency at GSM 900, staggering it with the commercial 4G LTE mobile network band, to ensure that the 2G FBS can transmit and receive normally.

The Principle of Client App Detecting 2G FBS. At present, mobile platforms like Android and iOS offer software development kits (SDKs) that enable the identification of the network type being used by the user's device, such as 5G, 4G, 3G, or 2G. On Android and iOS platforms, you can use the **TelephonyManager** and **CTTelephonyNetworkInfo** components, respectively, to detect the current network type through event-triggered methods. The app we developed uses the mentioned components to instantly identify the network type utilized by the user's device in real-time. Under the premise that the network type used by the terminal can be obtained instantly, coupled with the fact that the commercial 2G network service in Taiwan was completely shut down in 2017, if a connection to a 2G network is detected within the geographical boundaries of Taiwan, it can be determined that the connected 2G network service is provided by a 2G FBS. In Fig. 4, our app will promptly alert users before the FBS sends a scam text message to the victim. At the same time, our app will also record the parameters of the current scam text message received by the user equipment, including time, location and mobile network information, and automatically send it back to the back-end system to generate further analysis results, which can be used by law enforcement agencies to combat scam groups.

3.2 Core Network Approach

Starting from the perspective of the mobile core network, it is possible to detect FBSs and their nearby areas by observing user signaling information, such as signaling types, area codes, and so forth, and analyzing signaling behaviors, such as registration frequency.

Fig. 4. When the app detects a 2G FBS, it will provide an alarm.

Most known RBS attacks use interference to force user devices to connect to the FBS. During this process, the location area might change, affecting the tracking area identity (TAI) and the location area identity (LAI), or the user might switch to a different network, like the GSM network. Once the user leaves the interference area, their device can return to the legitimate telecommunications provider's network. According to the 3GPP Spec 24.301 specification, when the user's location or RAN changes, the registration process should be re-executed and the location information of the last successful registration should be added to the registration signaling. The FBS detection steps in CN are listed below:

1. Signal interference from FBS attacker
2. Victim UE disconnected with operator
3. The victim UE identified the FBS and tried to register; thus, the victim UE was affected by FBS attacks.
4. Victim UE went back to operator BS and re-initiated registration procedure with old LAC/TAC information
5. Detected the old LAC/TAC and the ratio of user re-register attempts if FBS existed.

The core FBS detection steps corresponding diagram is shown in Fig. 5.

Therefore, the core network can detect FBS through this feature of signaling using the following two mechanisms.

- Mechanism 1: Determine through the trend of base station user registration ratios
  After the user device is interfered with and loses contact with the legitimate base station, it will search again for available base stations and attempt to register. Regardless of whether this registration process is successful or not,

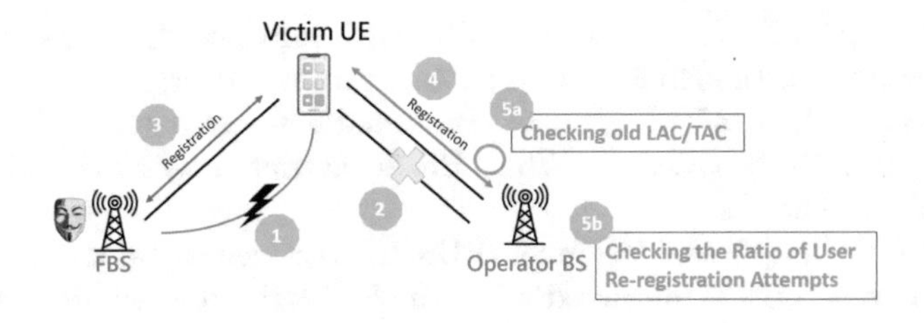

Fig. 5. The FBS detection steps in core network.

the terminal will definitely re-register after it leaves the interference range and returns to the telecommunications provider's network. The more users affected by FBS, the higher the proportion of users who re-register. Therefore, by monitoring whether the proportion of base station user registrations increases abnormally, we can filter out base stations that may have FBS in nearby areas.

- Mechanism 2: Determine through the legality of the previous successfully registered location
  In common FBS attack methods, the FBS exploits vulnerabilities in the one-way authentication between the user device and the network, inducing user devices to receive scam text messages sent by FBS. During this, the user device logs the FBS's location code, and adds this information to the old LAI section when it reconnects to the official network. Then, the core network checks if this old LAI information is on the authorized list of the telecommunications provider. After excluding reasonable situations, such as the user registering with a foreign operator network, our mechanism will send a warning or notification for more analysis. We can pick out phone numbers likely affected by FBS and pinpoint the area where the FBS are based on these numbers' locations.

Combining the analysis results of the above two mechanisms, it is possible to detect common appearance areas of FBSs from the proportion and content of registration signals sent by UE.

3.3 Radio Access Network Approach

Due to FBS wanting to attract UE from the legitimate BS to the FBS, it may interfere with the signals of the legitimate BS to facilitate UE connection to the FBS. Therefore, we have selected five performance indicators related to signal interference on the BS to determine whether the BS has been maliciously interfered with. Furthermore, FBS might exploit the vulnerabilities in the handover process [5,6] to execute DoS attacks. Consequently, we have included an additional performance indicator concerning handover. The six performance indicators are list as follow:

- Random-access channel (RACH) Success Rate: When BS is interfered with, the RACH success rate will decrease.

- E-UTRAN Radio Access Bearer (ERAB) Setup Success Rate: When BS is interfered with, the ERAB setup success rate will decrease.
- Media access control (MAC) hybrid automatic repeat request (HARQ) uplink Block Error Rate (BLER): When BS is interfered with, the MAC HARQ BLER will increase.
- Physical Uplink Shared CHannel (PUSCH) received signal strength indicator (RSSI): When BS is interfered with, the PUSCH RSSI will decrease.
- Physical Uplink Shared CHannel (PUSCH) signal-to-interference-plus-noise ratio (SINR): When BS is interfered with, the PUSCH SINR will decrease.
- Handover (HO) preparation phase fail times: When the FBS uses a stronger signal to attract the UE for handover, the number of handover preparation failures increases due to unsuccessful attempts.

Our system will monitor these six performance indicators. If any indicator experiences a sudden increase or decrease over a short period of time, the system will suspect that the BS may be experiencing interference from an FBS and hypothesize that there might be an FBS in the area. In addition, during this clustering process, we also determine whether the cluster contains results detected by multiple methods. Clustering areas detected by multiple detection methods also indicate a greater possibility of FBS.

4 Integrated Monitoring User Interface

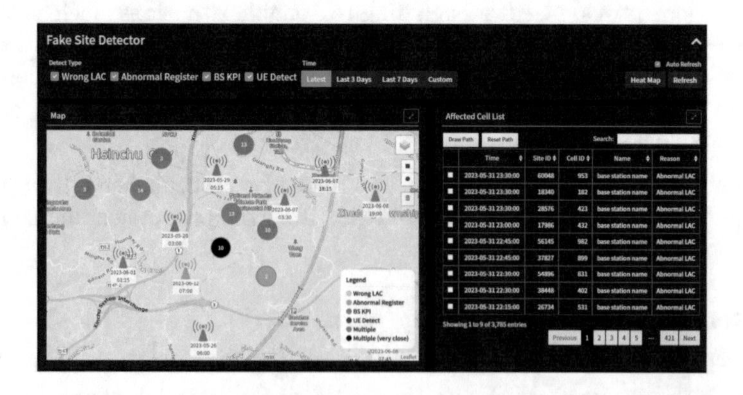

Fig. 6. The main page of the real-time RBS detection system.

We have developed a real-time RBS detection system to display FBS detection results. The main page of the system is shown in Fig. 6. In this monitoring system, users can choose the detection methods themselves, including incorrect LAC, abnormal registration behavior, abnormal BS performance indicators, and UE detecting GSM signals. Users have the flexibility to mix and match these four detection methods as they prefer, with the results subsequently being displayed on the map. On this system, in addition to presenting the latest data (within 24 h), we also provide historical data for query, so that these detection results can be further analyzed afterwards.

In Fig. 6, the detection results are displayed on the map of the real-time RBS detection system. A marker will represent any BS or UE affected by FBS, as detected by one method at one time. In this case, the map will have many markers because multiple UEs and BSs can be affected by FBS, and a single BS affected by FBS may be detected by various methods. Moreover, our system will display historical data within 24 h on the map, and the markers on the map will be very messy. In order to solve this problem, we used Leaflet.markercluster [13] to perform clustering. The more markers contained in a cluster, the greater the possibility of FBS in the area. Clustering areas detected by multiple detection methods also indicate a greater possibility of FBS. If clustering methods show affected BS or UE affected by FBS are close, they're marked black; if not, they're marked gray.

The final goal of detecting FBS is to capture scam groups. An FBS will not stay in one place, waiting for the law enforcement agencies to come and arrest it; it will keep moving. To capture the movement routes of the scam groups, we tried to plot their trajectories. However, backtracking the path of the FBS is challenging because the current detection system can only approximate the FBS's location based on the affected BSs and UEs. We use Leaflet.heat [14] to represent the detection results at a certain timestamp using a heat map. Heat maps with different timestamps are shown in Fig. 7. Through these heat maps, we can observe the range of FBS's impact and how this affected area changes. This can be used to roughly estimate the movement path of the FBS.

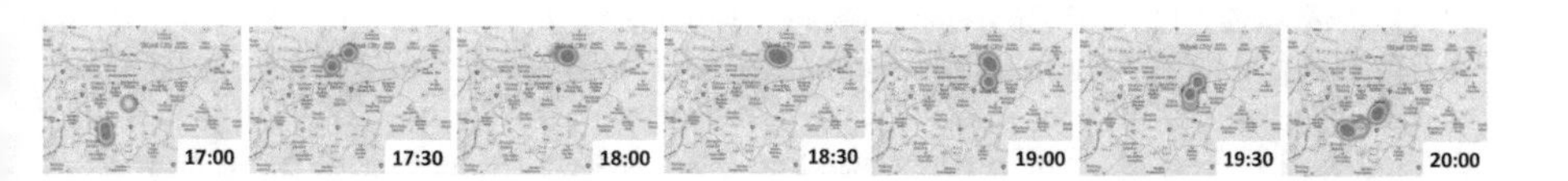

Fig. 7. The FBS monitoring system presents the detection results from 17:00 to 20:00 using a heat map.

5 Conslusion

In response to scam groups using FBS to send scam text messages and causing financial losses to users, we have developed a real-time RBS detection system. This system conducts FBS detection based on the three major components in the mobile network, namely UE, RAN, and CN. On the UE, we detect GSM signals to identify FBS. On the RAN, we use the BS's performance indicators to analyze whether the BS is being interfered with by FBS. On the CN, we use the LAC and registration behavior in the signaling to determine whether users have disconnected from the FBS and re-registered with the legitimate network. Moreover, we have developed a real-time monitoring user interface that presents the integrated detection results from the three major components on a map. The map incorporates clustering to show the impact of FBS in specific areas. We also introduced heat map analysis to estimate the range and movement path of FBS. These detection results can also serve as references for law enforcement agencies, aiding in the combat against illegal FBS setups.

References

1. Ollmann, G.: The Phishing Guide-Understanding and Preventing Phishing Attacks, September 2004
2. Gartner. Tap Into The Marketing Power of SMS (2016). https://www.gartner.com/en/marketing/insights/articles/tap-into-the-marketing-power-of-sms
3. Huang, T.: Taiwan busts fraudster using fake base station to scam victims (2023). https://www.taiwannews.com.tw/en/news/4857528
4. Mouly, M., Pautet, M.: The GSM System for Mobile Communications. Telecom publishing, USA (1992)
5. Bitsikas, E., Pöpper, C.: Don't hand it over: vulnerabilities in the handover procedure of cellular telecommunications. In: Annual Computer Security Applications Conference, pp. 900–915 (2021)
6. Mawaldi, I., Anugraha, T., Ginting, I., Karna, N., et al.: Experimental security analysis for fake eNodeB attack on LTE network. In: 2020 3rd International Seminar On Research of Information Technology and Intelligent Systems (ISRITI), pp. 140–145 (2020)
7. Hua, Y., Zhang, Y., Liu, Q., Liu, Y., Song, Y.: The Fake BTS detection and location on smart phones. In: Proceedings Of The 2017 International Conference On Industrial Design Engineering, pp. 116–120 (2017)
8. Li, Z., et al.: FBS-Radar: uncovering fake base stations at scale in the wild. NDSS (2017)
9. Ali, A., Fischer, G.: The phase noise and clock synchronous carrier frequency offset based RF fingerprinting for the fake base station detection. In: 2019 IEEE 20th Wireless and Microwave Technology Conference (WAMICON), pp. 1–6 (2019)
10. Steig, S., Aarnes, A., Van Do, T., Nguyen, H.: A network based IMSI catcher detection. In: 2016 6th International Conference On IT Convergence and Security (ICITCS), pp. 1–6 (2016)
11. Shin, J., Shin, Y., Park, J.: Network detection of fake base station using automatic neighbour relation in self-organizing networks. In: 2022 13th International Conference On Information and Communication Technology Convergence (ICTC), pp. 968–970 (2022)
12. Dabrowski, A., Petzl, G., Weippl, E.: The messenger shoots back: network operator based IMSI catcher detection. In: Research In Attacks, Intrusions, and Defenses: 19th International Symposium, RAID 2016, Paris, France, 19–21 September 2016, Proceedings 19, pp. 279–302 (2016)
13. Leaver, D.: Leaflet.markercluster. GitHub Repository (2012). https://github.com/Leaflet/Leaflet.markercluster
14. Agafonkin, V.: Leaflet.heat. GitHub Repository (2015). https://github.com/Leaflet/Leaflet.heat

A Brief Review of Machine Learning-Based Approaches for Advanced Interference Management in 6G In-X Sub-networks

Nessrine Trabelsi(✉) and Lamia Chaari Fourati

Laboratory of Signals, systeMs, aRtificial intelligence, and neTworkS (SM@RTS), Digital Research Center of Sfax (CRNS), Sfax 3021, Tunisia
nessrinetrabelsi@gmail.com

Abstract. 6G is envisioned to take the form of a 'network of networks', where wireless networks of different types and operational ranges are to be seamlessly integrated. In-X sub-networks are short-range low power radio cells to be installed at the very edge of such "network of networks" inside of different entities like a production module or a vehicle, and supporting highly localized yet different services with extreme requirements. Though benefiting from decision-making capabilities of the umbrella 6G network when available, these sub-networks are by default independent mobile cells sharing the same spectrum and thus leading to dense scenarios with severe and time-varying interference that should be carefully handled. However, managing such interference is an intractable challenge to traditional approaches. Therefore, driven by the benefits and efficiency of Machine-Learning (ML), this article investigates ML-based approaches, including supervised, unsupervised, and reinforcement learning approaches for advanced interference management in 6G in-X sub-networks. To this end, we overview the proposed schemes in the literature and discuss their methodologies, strengths and weaknesses. Furthermore, we highlight related open issues and future research directions, and we propose some useful guidelines for an ML-based solution.

1 Introduction

1.1 General Context and Motivations

Compared to 5G, 6G networks are expected to support revolutionary services with much more demanding communication requirements [11]. Introduced in [3, 8], 6G in-X sub-networks have been recognized as a particularly promising radio technology for providing seamless wireless coverage to meet extreme connectivity requirements such as 100 Gbps data rates, sub millisecond latencies, and seven to nine nines reliability. In-X sub-networks are presented as dedicated, low-power, short-range cells, comprising of a controller acting as the Access Point (AP) for multiple devices. As shown in Fig. 1, they are to be installed inside entities

L. Barolli (Ed.): AINA 2024, LNDECT 204, pp. 475–487, 2024.
https://doi.org/10.1007/978-3-031-57942-4_46

such as robots, production modules, vehicles or human bodies [5], and can be integrated into a larger 6G network or run independently in case of critical applications, like intra-body heart rate control or intra-vehicle break control [9].

6G In-X sub-networks may however, result in high density scenarios, such as in-vehicle sub-networks at a road intersection [9,11]. Sometimes, sub-networks might be located in the same physical area, e.g. an in-body subnetwork inside a person traveling on an airplane with its own in-airplane sub-network [9]. Such dense deployments will likely result in high interference levels that may significantly impact the expected performance [13]. Moreover, the dynamic and sometimes uncontrolled mobility of in-X sub-networks, would generate highly dynamic interference behavior in comparison to typical cellular configurations with static base stations (BS). Interference management is therefore extremely important in 6G in-X sub-networks to provide particularly ultra-reliable low-latency applications [9,11].

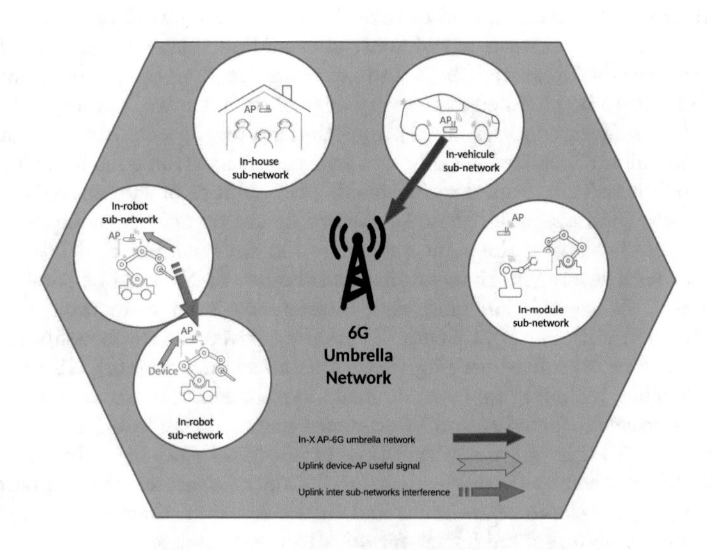

Fig. 1. Examples of 6G In-X Sub-networks.

Recently, there have been a number of published works presenting interference management schemes in 6G in-X sub-networks. For example, in [10], assuming that the sub-networks transmit at a fixed power, the authors proposed a hybrid framework for radio channel selection, where decisions are distributed between a global agent at the umbrella 6G network, and a local agent at each AP, depending on the scenario density, battery status to perform local sensing, and backhaul connection for information exchange. The study used heuristic policies, such simulated annealing and graph coloring at the global agent, and greedy and random selection locally at each sub-network. However, such rule-based heuristics are generally sub-optimal and insufficient to ensure the fulfillment of multiple conflicting and extreme requirements. This calls for novel adaptive policies exploiting local sensing capabilities and addition information provided at the umbrella 6G network to proactively manage radio resource allocation and handle interference in in-X sub-networks. In this context, data-driven ML-based approaches can be explored to intelligently learn the specific operational environment dynamics and make proactive adequate decisions to manage the radio resources and induced interference.

1.2 Machine Learning Overview

Radio resource management in wireless networks typically involves non-convex objective functions, and is known to be NP-hard combinatorial problem. Over the last few years, to perform such optimization and provide near-optimal solutions, a large number of studies started to apply ML-based methods including Supervised Learning (SL), Unsupervised Learning (UL), and Reinforcement Learning (RL).

Using pre-labelled datasets, SL is generally applied to classify data or predict certain outcomes. UL models are built to discover hidden patterns in unlabelled data without the need for human intervention and are generally applied for clustering, association and dimension reduction tasks. Self-supervised learning is presented in some literature as one approach to UL where labels can be automatically obtained. Reinforcement Learning, considered as the third ML basic paradigm, is generally modeled as a Markov Decision Process (MDP) or one of its variants. In RL, one or multiple agents interact with an environment, and based on some state information, choose an action that results in a new environment state, and a reward feedback signal. Through this trial-and-error process, each agent would learn the optimal policy, i.e., a state-action mapping, maximizing its expected cumulative discounted reward. Depending on the specific problem formulation, there have been multiple algorithms in the literature for Single-Agent RL and Multi-Agent RL (MARL).

Using two separate phases for training and execution, these ML-based methods achieve competitive performance with reduced complexity in comparison to heuristic schemes. In the training stage that is generally performed offline, in a centralized, distributed or federated fashion, the models' parameters are optimized to minimize some loss function. Then, the trained models are executed either at a centralized controller or at distributed entities in the inference stage.

Recently, Deep Learning (DL), a branch of ML, has proved its learning capabilities from massive data to build complex parameterized models using Deep Neural Networks (DNNs). These DNNs are basically multiple layers of interconnected artificial neurons that model complex non-linear relationships, and can be trained like any other ML algorithm. During the last decade, different types of neural networks have been proposed in the literature. Among them, Graph Neural Networks (GNNs) have emerged as an effective way to handle graph-structured data such as wireless mobile networks. Based on message passing paradigm, the nodes iteratively exchange information with their neighbors and update their embeddings. Deep neural networks may be used within SL, UL, and RL approaches to help performing complex tasks relying on massive input data with hidden patterns connections.

1.3 Related Surveys and Main Contributions

Prior to this work, many survey papers dealt with interference management in wireless networks, such as [13] which provided a comprehensive review of the different interference types and appropriate management schemes in 5G and beyond networks, and presented some insights on interference challenges in 6G networks. However, to the best of the authors' knowledge, a very few surveys focused on interference management in 6G in-X sub-networks.

[5] presented an initial study on interference avoidance, coordination and mitigation methods in 6G in-X sub-networks, and evaluated the potential of several techniques including blind-repetition with pseudo-random frequency hopping and interference-aware dynamic channel allocation.

In [9], Berardinelli et al. presented their vision for 6G in-X sub-networks using a top-down approach. They first introduced the motivations, possible use cases and deployments of the short-range low power cells. Then, they focused on the lower layer aspects of the design, with a particular emphasis on the interference issues. They suggested that interference management should be proactive while operating on frequency domain. The authors also discussed coordination schemes and concluded that they should rely on hybrid approaches to seamlessly switch between centralized schemes when using licensed spectrum in the coverage of a 6G network, and distributed schemes with implicit coordination when using unlicensed spectrum or being out of coverage.

While these two previous studies provided an overview of 6G in-X sub-networks and related interference issues, they either focused on a general framework concept, or presented non-ML, sub-optimal heuristics and traditional mechanisms for interference management.

Driven by the benefits and efficiency of Machine-Learning, in this review paper, we investigate ML-based approaches, including supervised, unsupervised, and reinforcement learning approaches for advanced interference management in 6G in-X sub-networks. To this end, we classify and overview the proposed schemes in the literature and discuss their methodologies, strengths and weaknesses. Furthermore, we highlight related open issues and future research directions, and we propose some useful guidelines for an ML-based interference management solution in 6G in-X sub-networks.

2 Supervised Learning-Based Schemes

In [4], the authors proposed a supervised DNN-based classification method for distributed dynamic channel group allocation with limited sensing in TDD asynchronous in-X sub-networks. While the intra-sub-network interference is avoided due to orthogonal resource allocation within a sub-network, the asynchronous time division duplex between the different cells gives rise to cross-link interference between DownLink and UpLink transmissions in addition to intra-mode interference as in time-aligned networks. Therefore, the suggested solution aimed to mitigate such interference while maintaining the outage probability below a specified target. The solution involved a Centralized Graph Coloring (CGC) to generate offline training data, and a DNN which was trained using the generated dataset and then deployed at each sub-network for distributed channel group selection. Specifically, using interference measurements between sub-network pairs, a conflict graph, connecting each sub-network with its K-1 strongest interferers (K being the number of channel groups), was created. Then, K-coloring of the conflict graph permitted the mapping of aggregate interference power measurements at each sub-network to a class of frequency channel groups. These training examples, where each one corresponds to a pair of interference power vector on all channels at one sub-network and its corresponding channel group assigned by CGC, were collected in a training data matrix and label matrices. A DNN was then designed and trained using the generated dataset to approximate the relationships between aggregate interference power levels in each sub-network and the color assigned by the CGC scheme. Applying an appropriate gradient descent algorithm, the DNN parameters were optimized until achievement of 80% accuracy. Once trained, the DNN was deployed at the controller of each sub-network. While the centralized training required the conflict graph of all sub-networks, in the execution phase, each sub-network only needed inputs of its own measured aggregated interference power on all groups, to dynamically select a suitable channel group using its trained DNN model. The proposed solution performance was compared to centralized CGC and heuristics algorithms. Results showed that the DNN-based solution achieved similar performance as the centralized baseline up to a probability of loop failure of 6×10^{-5}, while only requiring local interference power measurements. The robustness of the solution was also confirmed when the propagation conditions in the test phase were different from training. It is worth-noting, however, that the DNN-based solution heavily relies on the accurate pre-modeling of the propagation environment and the generation of a large representative dataset. Although this may be possible in an indoor factory environment, it would be challenging in outdoor scenarios. Furthermore, the proposed method showed no improvement over other heuristic approaches, as it only relied on static interference information and did not consider the dynamic nature of wireless propagation networks. Thus, learning to proactively perform channel allocation in dense deployments of 6G in-X sub-networks may significantly be improved by learning temporal dependencies in sensing measurements and channel selection decision.

3 Unsupervised Learning-Based Schemes

Based on self-supervised learning as a category of UL, [1,2] explored novel solutions for power control to alleviate interference in industrial wireless sub-networks.

The first study [1] presented a centralized power control solution based on message passing graph neural networks with limited Channel State Information (CSI). In particular, the authors considered a Central Resource Management (CRM) that is responsible for selecting the transmit power of all sub-networks with the objective to maximize the network's sum Spectral Efficiency (SE). The CRM updated a graph representation of the sub-networks deployment consisting of the desired link channel gains as the nodes' features and the interfering link distances as the edges' features. Using the message passing paradigm, the power control GNN (PCGNN)-based scheme provided a node embedding for each sub-network and computed the transmit power for all nodes. During the training, complete CSI was required to compute the network performance and the loss function. However, in the execution phase, since mutual distance information between sub-networks was available in advance, only the channel gain of the desired links was needed for inference. The authors used 10,000 samples for training and 50,000 samples for testing, while randomly varying the sub-networks deployment and channel characteristics. Simulation results showed that the proposed scheme reduced average power while enhancing the spectral efficiency, resulting in increased energy efficiency.

As an extension to the previous study, [2] proposed four graph attribution approaches that required different levels of channel information. Specifically, in the first method referred to as channel gain attribution (hH), the node feature was represented by the gain of the desired link channel, while the gain of the interfering link channel corresponded to the edge feature. In the second method, called distance-based attribution (dD), the geometric distances of the interfering and useful links were used to assign the interference graph. The third approach consisted of an hybrid graph attribution (hD) similar to the one proposed in [1]. In the last method referred to as sum channel gain attribution ($h\Sigma H$), the node feature was represented by the combination of the gain of the desired link channel and the sum of the gains of the interfering links channels, while the edges were not assigned. Given the interference graph, the objective was to determine the transmission power that optimizes the sum of SE with respect to minimum required SE per sub-network. To solve such constrained optimization problem, the authors applied the Lagrangian method, then trained in a centralized fashion the PCGNN based on message passing and following a self-supervised penalty approach. Compared to the GNN model in [1] which only considered the sum SE maximization with no minimum SE constraint, the PCGNN model in [2] was far more complex. After the training, the performance of the different PCGNN variants and their complexity trade-offs in terms of sensing and signaling were evaluated, from both the centralized and decentralized perspectives. Simulation results demonstrated that the second and third variants didn't meet the optimization objective, and that having the desired links channel gain information along with the channel gain of the strongest interfering links or the sum of interference was required. Furthermore, it was shown that both PCGNN-hH with only the strongest interference link channel gain as an edge feature, and PCGNN-$h\Sigma H$ are scalable and robust to multiple industrial channel models, while incurring lower sensing, signaling and computational costs in comparison to benchmark schemes.

4 Reinforcement Learning-Based Schemes

In [6], the authors proposed a fully distributed multi-agent table-based Q-learning (MAQL) solution to joint channel and power selection while adopting a two-level quantization of the sensing information. They further developed an alternative heuristic method that randomly selects the resources with similar constrained information exchanges. The joint optimization problem was first cast as multi-objective optimization tasks involving the simultaneous maximization of the lowest achieved capacity at each sub-network. These multiple conflicting non-convex objective functions, coupled with the restriction of using a single bit per channel for sensing information exchanges, made the problem even harder to solve. The problem was then formulated as a decentralized partially observable MDP, where each agent, i.e., sub-network AP, based solely on its local state defined as two-level quantized SINR on each channel, chooses one channel and one discrete power level, with the goal to maximize its reward defined as the capacity of the worst link. Based on centralized training with distributed execution (CTDE) paradigm, a single Q-table to evaluate all state-action pairs was then trained for all sub-networks in the system. While sharing the same Q-table between all agents permits to accelerate the training, the optimization is not proved to converge to a global optimum, as from the perspective of individual agents, the environment is continually changing due to other unknown agents' actions. To minimize the environment instability and ping-pong effects, the authors introduced switching delays and varying sensing intervals. Once the training is completed, the Q-table is copied to all in-X sub-networks for a fully distributed execution. Simulations in industrial factory settings proved that depending on appropriate selection of the SINR quantization threshold, MAQL improved the lowest link capacity compared to heuristics baseline.

While the results of the last study underlined the great potential of Q-learning for resource allocation, the tabular-based RL approach still lacks scalability to high-dimensional problems, and suffers from the impact of state quantization on the overall performance. Therefore, in [7], the authors proposed a multi-agent double deep Q-network (MADDQN) scheme for frequency channel selection using the aggregate interference power sensed at each sub-network, to maximize data rate for each sub-network, while maintaining a specified target data rate per device. While the execution was fully distributed, the authors studied both centralized training, which consisted of training a single shared network and then duplicating its parameters on all agents, and decentralized training where multiple DDQN networks were trained simultaneously. To model the state representation of each sub-network, the authors proposed two variants. The first one assumed no communication between sub-networks, and that each agent only observes its local sensing information defined as the measured aggregate interference power on each channel. In the second variant, resource selection was performed with limited cooperation between neighboring sub-networks, as each agent observe its own sensing measurements and the ones performed by other sub-networks from its neighboring set. At the training, when an agent observes its state and selects one frequency channel, it receives an individual reward signal defined as: (i) the sum rate achieved by its associated devices if the target rate constraint per device is satisfied, and (ii) a weighted linear function of the sum rate and a penalty measuring the difference between achieved and minimum rate, otherwise. While such reward definition permits to guide each agent towards achieving its own goal, it does not favor inter-agents cooperation as a common reward would, and it may lead to greedy agents and sub-optimal overall system performance. At the training, ϵ-greedy exploration, experience replays and random switching delays were utilized to balance the exploration-exploitation and improve the learning stability and convergence. Simulation results demonstrated that an agent achieves a marginal improvement in averaged reward at convergence while requiring longer training to learn the optimal policy using sensing measurements from multiple sub-networks than using only local measurements. In terms of per-device or averaged rate, the proposed MADDQN scheme outperformed random resource allocation and achieved similar or marginally worse compared to other centralized benchmarks while requiring much lower signaling overhead. The authors also provided a robustness and complexity analysis. It was concluded that all studied schemes were similarly impacted by variations in the number of sub-networks and the shadowing standard deviation. Furthermore, MADDQN reduced time complexity by a factor of 2,000 compared with the iterative surrogate optimizer and was considered as more appropriate than centralized coloring for deployments with large number of sub-networks. Presuming a fixed time cost for the exchange of sensing measurements, and a limited number of available channels compared to the number of sub-networks, then the signaling complexity for MADDQN was upper bounded by $\mathcal{O}(N)$, with N equals to the number of sub-networks.

While [7] applied multi-agent deep RL for frequency allocation, the authors in [12] proposed a novel deep multi-agent RL scheme with attention mechanisms and GNN framework for dynamic frequency and power allocation in 6G in-X sub-networks, using Received Signal Strength Indicator (RSSI) at each spectrum band, without requiring any prior knowledge about source output power and channel gains. The problem was first formulated as a partially observable Markov Game with the goal of maximizing the sub-network payload delivery probability within a certain time constraint. Each sub-network (agent) observes the RSSI at each channel, i.e., the known traffic signal and the total interference, the remaining traffic payload and time budget, and the previous action, then selects one channel and a discrete transmission power value. The reward was defined as the sum of transmission rate of channels divided by the target payload if any payloads remain to be transmitted, or a positive constant value, otherwise. To limit the non-stationarity of multi-agent setting, CTDE paradigm was adopted, where the agents were allowed to observe other agents' states and actions in the training phase to update their own distributed policies. Furthermore, the centralized training was shifted to the BS to acquire additional historical information of all sub-networks. To suit the multi-agent setting, the proposed scheme extended the well-known Soft Actor Critic (SAC) algorithm, where each sub-network trains critic and actor networks. To effectively decouple and reason the inter- sub-networks interference relationships based on current and historical observations of all agents, the authors proposed a novel Graph-Attention Network (GA-Net) which combines a GNN with enhanced attention layers. The proposed GA-Net permitted the obtaining of an information-condensed joint state encoding by selectively paying attention to the attributes of other sub-network agents, which was then used as input to each agent's critic. The critic and actor networks were centrally trained using SAC algorithm, then the trained policy models taking as input the local observations were downloaded from the BS to be distributively executed at each sub-network. The proposed MARL-based resource management solution was tested and compared with traditional schemes and Deep RL-based approaches such as QMIX and MADDPG. The authors proved that their approach converges the fastest to a greater stable reward than other approaches, and provides the lowest outage rate, in particular for a large number of sub-networks. Moreover, the policies learned using the novel method have been shown to enhance channel utilization owing to the coordinated selection of channel and power level. The centralized training using an attention-based GNN permitted to learn the topology information of the environment and to derive critical interference features from RSSI, which helped to learn better polices.

5 Discussion and Lessons Learned

Table 1 compares all the previously listed studies presenting ML-based schemes for advanced interference management in 6G in-X sub-networks. The majority of existing literature considered industrial use cases where the sub-networks are located inside robots or production modules. Only inter- sub-networks interference was considered, as orthogonal resources were allocated to the devices within the same sub-network and thus avoiding intra- sub-network interference. Generally, all studies dealt with device-to-AP interference in time-aligned cells. Cross-link (UpLink-to-DownLink and vice versa) interference was also considered in addition to intra-mode interference as in [4]. To efficiently manage such interference, the surveyed papers proposed to optimize the allocation of frequency, power, or both resource types. Different objectives were formulated, including the minimization of outage probability, or the maximization of link reliability, network spectral efficiency or lowest achieved capacity per sub-network. Some studies also specified some constraints such as minimum required SE per sub-network or target rate per device.

Supervised-learning based schemes rely on the offline generation of labeled datasets, which may limit their usage especially if no large, labelled, and representative dataset is available in public. While unsupervised-learning based schemes do not require labeled data, one has still to provide extensive training dataset for training. Online model-free RL-based schemes permit the agent

Table 1. Comparison of Proposed Machine Learning-based Solutions

ML Type	Ref	Algo	Resource Domain	Training – Inputs	Execution – Inputs	Objective	Strengths	Weaknesses
SL	[4]	DNN-based classification	Frequency	Centralized – Conflict graph of all sub-networks	Distributed – Local aggregated interference power	maintain the outage probability below a specified target	Robustness + minimal required inputs at execution stage	Similar performance to other heuristics + relies on accurate pre-modeling and static interference information
UL	[1]	PCGNN	Power	Centralized – CSI of desired and interfering links	Centralized – CSI of desired links	maximize the network SE	Simplicity and minimal inputs requirements at inference	Centralized solution with signaling costs
	[2]	PCGNN	Power	Centralized – CSI of desired and interfering links	Centralized/Distributed – Different degrees of channel information	maximize the network SE with minimum required SE constraint	Different graph-attribution variants and possible distributed execution	Minimum SE not guaranteed in certain cases
RL	[6]	MAQL	Frequency + Power	Centralized with shared Q-table – A two-level quantized SINR on each channel	Distributed – Same as in training	maximize the lowest achieved capacity at each sub-network	Simplicity + improved performance compared to heuristics baseline	Non-scalability of tabular-based RL + dependency on SINR quantization threshold
	[7]	MADDQN	Frequency	Centralized/Distributed – (1) Local sensing information, (2) Own and neighbors sensing information	Distributed – Same as in training	maximize the per-sub-network rate subject to a target rate constraint for each device	Different variants and good performance with reduced complexity compared to benchmarks	Lack of common reward favoring cooperation between sub-networks
	[12]	MA-SAC with GA-Net	Frequency + Power	Centralized – Remaining traffic payload and time budget, current and historical RSSI of all sub-networks and their actions	Distributed – Current local RSSI and remaining traffic payload and time budget	maximize the sub-network payload delivery probability within a time constraint	Fast and stable convergence + improved performance	Solution complexity

to interact with the environment and learn via trial-and-error process an optimal policy maximizing its long term reward. Generally, this requires however accurate simulations tools to emulate the environment, especially when it can be unsafe and highly expensive to directly deploy such RL mechanisms in real world. To model the interaction of multiple sub-networks with the wireless environment, variants of MDP and Markov Game have been used in multiple studies, such as [6,7,12]. Then, the authors applied well known RL algorithms that have been extended to the multi-agent settings. While the proposed schemes have outperformed other benchmarks, some design limitations, including the observation representation and multi-objective reward definition, can be further enhanced. Moreover, some known MARL challenges need to be thoroughly studied to improve the learning stability and convergence, and to accurately account for the per-agent credit in the global system reward.

The integration of DL proved its efficiency especially for large problem dimensions, compared to standard ML schemes suffering from non-scalability. However, traditional DNN-based schemes, such as in [4], fail to exploit the underlying dynamic topology of wireless networks, and their performance decrease significantly for large networks. Due to their ability to handle complex graph-structured data, mine the deep information and non-linear relationships between nodes, and generalize to unseen environments, emerging GNN have been used in recent studies such as [2,12] achieving near-optimal performance with much fewer training samples than traditional DNN. Furthermore, GNN can be executed in both centralized and distributed manner, making them more suitable for resource management in in-X sub-networks depending on their 6G coverage and allowed exchange overhead. These GNN-based models can either be applied in self-supervised learning schemes to readout the optimized resource allocation such as in [2], or coupled with RL-based approaches to model the state representation and account for the non-linear dynamic interference relationships between sub-networks such as in [12].

6 Open Challenges and Future Scope

In this section, we highlight the open challenges encountered in interference management in 6G in-X sub-networks and ML related issues. Furthermore, we present insights on the possible research directions to address the identified challenges, and provide general guidelines for an ML-based interference management framework.

In the most existing literature, the system model and interference management problem formulation in 6G in-X sub-networks were limited to some specified scenarios, especially industrial use cases, considering only inter-sub-networks interference, and only optimizing either frequency or/and power resources. Thus, we need a more generalized system model and problem formulation that consider:

- different use case scenarios, with a careful propagation modeling, including the coexistence of hybrid and possibly overlapping in-X sub-networks;
- different sub-networks mobility patterns, Quality of Service (QoS) requirements and optimization objectives depending on the use case;
- the use of multiple APs per sub-network, distributed MIMO and different possible access schemes per cell that may lead to intra-sub-network interference;
- non-cellular interference types such as impulsive noise, and jamming;
- the use of higher frequency bands and unlicensed spectrum;
- PHY-MAC cross-layer optimization including beamforming weights, frequency-time-power radio resources, and different MAC and PHY parameters.

While ML-based schemes have demonstrated their efficiency in dynamically handling large optimization problems, they still face some challenges that need to be addressed. In the context of in-X sub-networks resource management, one of the fundamental elements in SL, UL and offline RL, is to obtain massive, high quality training data. Furthermore, online RL based approaches highly rely on accurate pre-modeling of the environment and on compliant simulators to efficiently emulate the real channel propagation gains and generate sufficient training samples.

As previously mentioned, GNNs have gained huge interest as a DL method permitting the expansion to non-Euclidean space and thus efficiently modeling the complex relationships inter the different nodes in wireless networks. However, when based on static graph structures that don't model the temporal dependencies in the input data, their performance is rather limited. Spatio-temporal GNN, accounting for the time factor, may be a better solution for dynamic in-X sub-networks.

Another challenge in ML-based schemes and more particularly in MARL, is the training paradigm, as centralized training may rise some privacy concerns, whereas decentralized one may lead to learning instability and convergence issues. Some possible solutions include the use of abstracted and compressed information to preserve privacy in centralized training, and the use of fingerprints tracking the policies of other agents in the observation space to enhance the stability in distributed training. Federated learning can also be applied to tackle the privacy challenges for mitigating data leakage, and to permit a faster and more efficient learning.

As 6G in-X sub-networks may support life-critical services with extreme requirements, such as intra-body heart control or intra-vehicle break control, ML-based approaches have to be risk-free and ensure safe decisions. Therefore, safe ML-based interference management methods should be further studied where the risks of output actions are accurately measured and where unsafe actions are excluded.

As multiple heterogeneous mobile sub-networks may be sharing the same radio resources while being on coverage or out-of-coverage of an umbrella 6G network, we envision a hierarchical solution based on deep MARL and spatio-temporal GNN, that exhibit the following characteristics: (i) Different types of agents (one central agent at the 6G network, and distributed agents at each AP) would cooperate at different time-scales with different state/action spaces to achieve common and individual goals; (ii) The local state information observed by each agent would be represented using spatio-temporal GNN to efficiently learn the spatial and temporal dependencies between the agents, and permit the generalization to other unseen environments. It must also consider the fact that feedback may be partial and obsolete; (iii) The action space of the agents should contain hybrid discrete-continuous actions that consist in the selection of different resources domains including frequency, power, time, and space. The central agent may be allocating some envelope resources for each AP which performs per device scheduling at each time step. Action masking should also be applied to exclude risky actions that would deteriorate the system performance; (iv) A multi-objective reward function should consider the different metrics of reliability, latency, Age of Information (AoI), and throughput, and accurately model the contribution of each agent to the common reward; (v) The ML-based framework should deliver explainable, Pareto-optimal, and safe distributed policies.

7 Conclusion

While wireless in-X sub-networks are considered as a key 6G technological enabler to provide capillary coverage and support demanding services with extreme QoS requirements, they are facing multiple challenges towards their deployment, particularly due to their uncontrolled mobility resulting in highly dense scenarios with severe dynamic inter-sub-network interference. Driven by the benefits of ML in radio resource management, we investigated the application of supervised, unsupervised and reinforcement learning schemes for advanced interference management in 6G in-X sub-networks. We further identified the open challenges regarding system model, problem formulation and specific ML-related issues for efficient radio resource management in in-X sub-networks. We finally provided useful guidelines for an ML-based framework using MARL and GNN to manage such interference.

References

1. Abode, D., Adeogun, R., Berardinelli, G.: Power control for 6G industrial wireless subnetworks: a graph neural network approach. In: 2023 IEEE Wireless Communications and Networking Conference (WCNC), pp. 1–6 (2023)
2. Abode, D., Berardinelli, G., Adeogun, R.: Power control for 6G in-factory subnetworks with partial channel information using graph neural networks. IEEE Trans. Mach. Learn. Commun. Network. (2023)
3. Adeogun, R., Berardinelli, G., Mogensen, P., Rodriguez, I., Razzaghpour, M.: Towards 6G in-X subnetworks with sub-millisecond communication cycles and extreme reliability. IEEE Access **8**, 110172–110188 (2020)
4. Adeogun, R., Berardinelli, G., Mogensen, P.: Learning to dynamically allocate radio resources in mobile 6G in-X subnetworks. In: 2021 IEEE 32nd Annual International Symposium on Personal, Indoor and Mobile Radio Communications (PIMRC), pp. 959–965 (2021)
5. Adeogun, R., Berardinelli, G., Mogensen, P.: Enhanced interference management for 6G in-X subnetworks. IEEE Access **10**, 45784–45798 (2022)
6. Adeogun, R., Berardinelli, G.: Multi-agent dynamic resource allocation in 6G in-X subnetworks with limited sensing information. Sensors (Basel, Switzerland). **22**(13), 5062 (2022)
7. Adeogun, R., Berardinelli, G.: Distributed channel allocation for mobile 6G subnetworks via multi-agent deep Q-learning. In: 2023 IEEE Wireless Communications and Networking Conference (WCNC), pp. 1–6 (2023)
8. Berardinelli, G., Mogensen, P., Adeogun, R.: 6G subnetworks for life-critical communication. In: 2020 2nd 6G Wireless Summit (6G SUMMIT), pp. 1–5 (2020)
9. Berardinelli, G., et al.: Extreme communication in 6G: vision and challenges for 'in-X' subnetworks. IEEE Open J. Commun. Soc. **2**, 2516–2535 (2021)
10. Berardinelli, G., Adeogun, R.: Hybrid radio resource management for 6G subnetwork crowds. IEEE Commun. Mag. **61**, 148–154 (2023)
11. Bernardos, C., et al.: European vision for the 6G network ecosystem (2021)
12. Du, X., et al.: Multi-agent reinforcement learning for dynamic resource management in 6G in-X subnetworks. IEEE Trans. Wirel. Commun. **22**(3), 1900–1914 (2022)
13. Trabelsi, N., Fourati, L.C., Chen, C.: Interference management in 5G and beyond networks: a comprehensive survey. Comput. Networks **239**, 110159 (2024). https://www.sciencedirect.com/science/article/pii/S1389128623006047

A Comparative Study of Adaptive and Switched Beamforming for 5G Mobile Network Applications

Meryem Hamdi[1,2(✉)], Mohamed Hayouni[1,2], and Emna Jebabli[2]

[1] Institut Supérieur d'Informatique du Kef, Université de Jendouba, Kef, Tunisia
{hamdi,mohamed.hayouni}@isikef.u-jendouba.tn
[2] Innov'COM Research Laboratory, École Supérieure des communications de Tunis, Université de Carthage, Ariana, Tunisia
emna.jebabli@supcom.tn

Abstract. Beamforming antennas are considerd as good candidates for 5G mobile network applications in mm-wave band. They provide a spectrum efficiency, a high data rate and an important coverage. The beamforming can be analog, digital or hybrid. A comparative study of analog and digital beamforming is discussed. Indeed, we have used a 4×4 Butler matrix (BM) to steer the beam. However, a digital beamforming antennas has been achieved by using adaptive algorithms. As a result, the 4×4 BM gives four predifined beam steering in the range of $\pm 29^\circ$ whereas the adaptive algorithms show a good flexibility in which they can adapt the radiation pattern in real time but their weakness is a high power consumption.

Keywords: Adaptive algorithms · Adaptive Beamforming · Analog Beamforming · Beamforming · Butler matrix · Digital Beamforming · Switched Beamforming

1 Introduction

A beamofming is a an important technology for 5G applications. It uses a narrow beam toward the intended user or direction whereas nulls in the other direction [1].

In analog beamforming technology (ABF), the radiation pattern steered toward the desired direction by using a beam splitters such as Butler matrix, rotman lenses, etc. [2]. The main drawback of ABF is in a given instant only one radiation pattern is possible [3]. Thus, in the digital beamforming (DBF) the phases are adjustable digitally by using complex algorithm [4]. But in the most cases it not feasible due to the high power consumption [5]. However, the hybrid beamforming is a combination of analog and digital beamforming [6,7], which presents as a possible solution to avoid a significant power consumption [8].

Switched beamforming [9], is a subset of analog beamforming in which the system uses predefined beamforming patterns and switched between them depending on the

L. Barolli (Ed.): AINA 2024, LNDECT 204, pp. 488–497, 2024.
https://doi.org/10.1007/978-3-031-57942-4_47

location of the target. Hence, the Butler Matrix is the most popular used technique in switched beamforming. However, the adaptive beamforming [10], DBF, is a technology that allows the antenna array to dynamically modify its beamforming parameters in response to reduce the interference. It reduces noise and interference and maximizes the target signal's transmission or reception.

In this paper, we have studied and compared the benefits and drawbacks of analog beamforming made possible by Butler Matrices with adaptive digital beamforming using adaptive algorithm such as LMS, NLMS, RLS, etc. It organized as follows. In Sect. 2, Switched beamforming will be discussed. In Sect. 3, we will present the adaptive beamforming. Section 4, a comparison between switched and adaptive beamforming. Finally, Sect. 5 it as going to be a conclusion.

2 Switched Beamforming

The Butler Matrix (BM) is a passive beamforming circuit [11]. Its radiation patterns can be created by generating output signals at different phases [11].

In [12] a proposed 4×4 BM has been designed and simulated. Figure 1 show its representation in which it composed by four input port(1,2,3,4), four output port (5,6,7,8), four hybrid couplers of 3 dB, two $-45°$ phase shifters and two crossovers [11]. As given by Fig. 2, its simulation results show that a predefined four radiation patterns oriented according to the angles $\pm 4°$ and $\pm 29°$ with main lobes magnitude varies from 13 dB to 14.5 dB at 28 GHz.

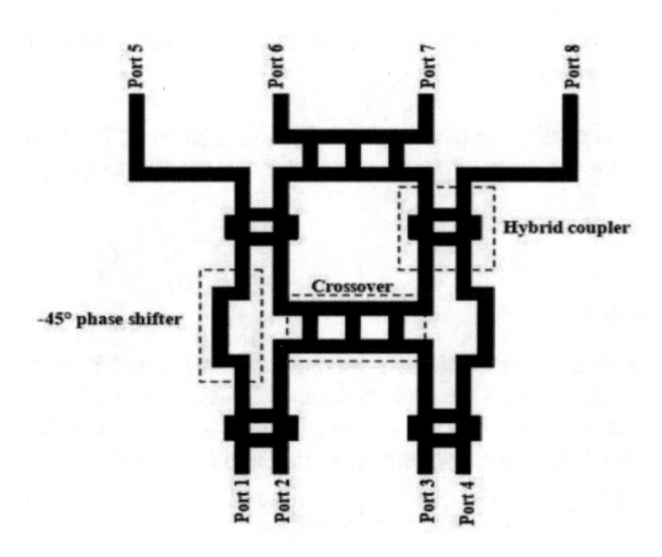

Fig. 1. 4×4 BM [11].

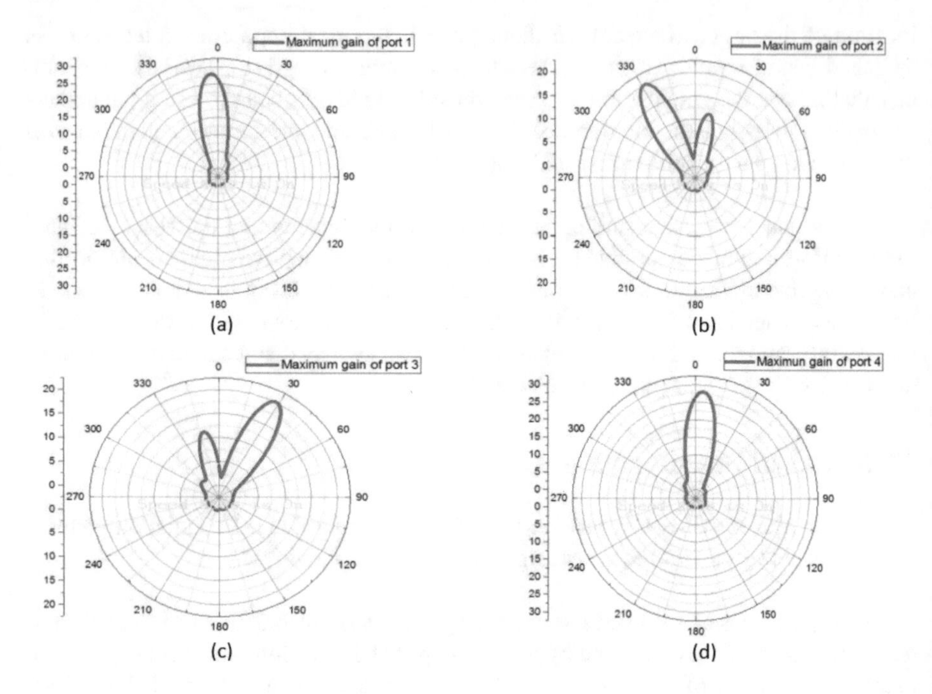

Fig. 2. Simulation results of BM beamforming.

3 Adaptive Beamforming

Adaptive beamforming is consider as a digital beamforming [13] in which it employs adaptive algorithms to provide beamforming that effectively tracks the desired signal while constantly lowering interference.

For all simulation results, we have performed the simulation on MATLAB2023Ra for 1000 iteration, for a number of antenna elements equal to 8 with a step parameter value of 0.01.

3.1 Algorithm LMS

The LMS implements a stochastic gradient descent technique in which the signal samples received from each antenna are linearly combined after being scaled by the optimum weighted and linearly combined to generate the wanted direction and to delete the interference direction [14]. The output y(n) at an instant n is given by Eq. 1:

$$y(n) = w^H(n)x(n) \tag{1}$$

where: W(n) as the weight vector and x(n) as the data samples. The desired signal at time n is represent by d(n). The error signal e(n)is given by Eq. 2.

$$e(n) = d(n) - y(n) \tag{2}$$

The weight updating at time n+1 given by Eq. 3:

$$w(n+1) = w^{(n)} + \mu e(n)x(n) \tag{3}$$

where μ is the step parameter that settles the convergence speed.

Slow convergence to the optimal value is indicated by small values of μ, while for many values of μ the speed of convergence is fast.However, it is less stable and not limited.

We have made an application of the LMS algorithm with desired user signal arrived at angles +30° at 28 GHz as shown in Fig. 3. It's clear that this algorithm respond to the desired direction with a good accuracy.

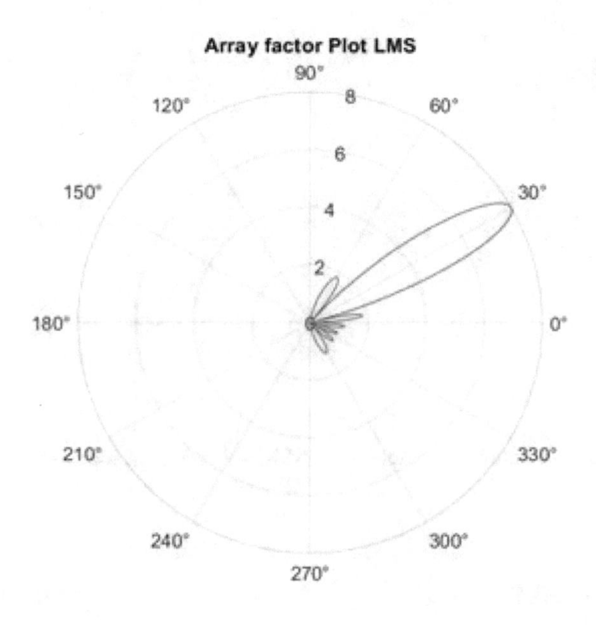

Fig. 3. Simulation result of the LMS algorithm at angle +30°.

3.2 Algorithm NLMS

The NLMS (normalized-least-mean-square algorithm) is the evolution of the LMS algorithm used to obtain good stability and faster convergence than the LMS algorithm. The

NLMS algorithm is like the LMS algorithm by modifying the correction term used in the weight update equation given by Eq. 4:

$$w(n+1) = w(n) + \frac{\beta x(n)}{\alpha + x(n)x^H(n)} e^*(n) \tag{4}$$

Example of attended the desired direction +45° by using the NLMS algorithm. As shown in Fig. 4, this algorithm respond to the It's clear that this algorithm respond to the desired direction at the desired angle.

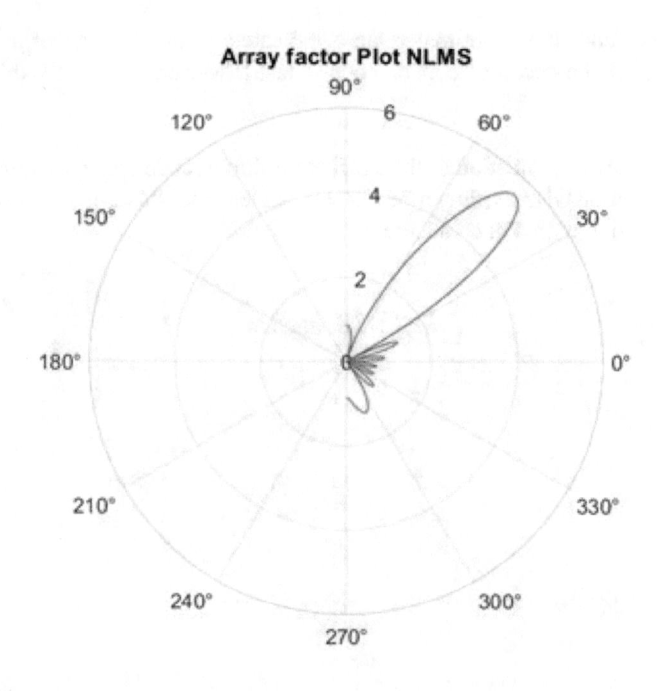

Fig. 4. Simulation result of the NLMS algorithm at angle +45°.

3.3 Algorithme RLS

It is a recursive implementation of the least-squares (LS) solution [14]. RLS initializes the inverse auto correlation matrix P(n) which is generally taken as an identity matrix of size N (number of antenna elements) and will calculate the Kalman gain K(n) show in Eq. 5:

$$k(n) = \frac{P(n-1)x(n)}{\lambda + x^H P(n-1)x(n)} \tag{5}$$

After that the weights of the antenna array are updated according to the calculated Kalman gain and the error e(n) Eq. 6:

$$w(n) = w(n-1) + k(n)e(n) \tag{6}$$

By using this algorithm we have oriented the radiation pattern of the antenna at the desired user signal arrived at angles +60° at 28 GHz. As shown in Fig. 5, this algorithm respond to the desired direction.

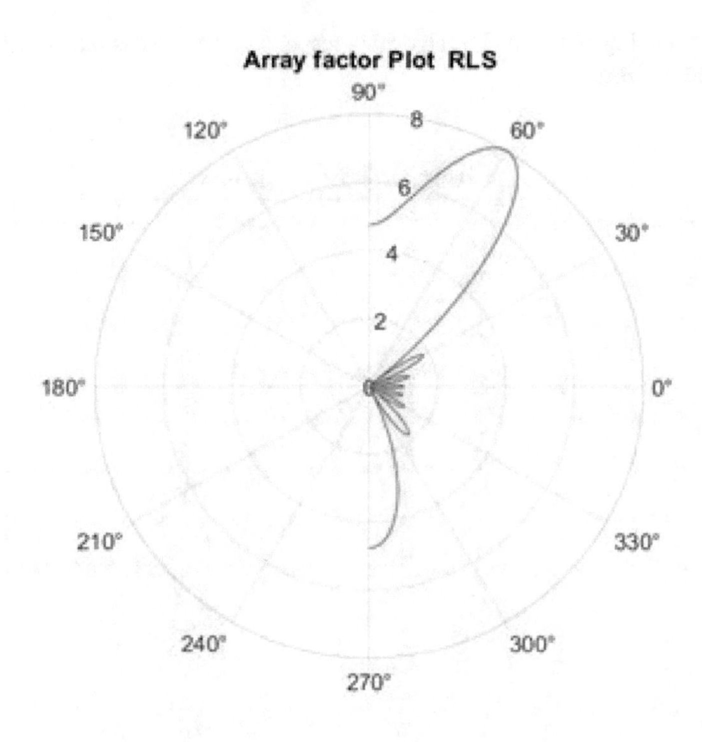

Fig. 5. Simulation result of the RLS algorithm at angle +60°.

3.4 The MUSIC Algorithm

The Subspace-based Multiple Signal Classification (MUSIC) algorithm is a widely used technique with excellent computing efficiency, high precision, and adaptability. To provide precise direction of arrival (DOA) estimations for uncorrelated signal sources, it depends on prior information of the system response, the quantity of signal sources, and background noise. MUSIC first configures an adaptive network with predetermined antenna configurations and captures signals from several sources. It divides the signal and noise subspaces using correlation matrix decomposition into Eigen values and

Eigen vectors, and then computes the MUSIC spectrum using the eigen vectors linked to the biggest eigen values [15]. The MUSIC spectrum defined by Eq. 7 [15]:

$$P_{\text{MUSIC}}(\theta) = \frac{1}{A^H V_n V_n^H A} \tag{7}$$

With V_n is the eigen vectors of the noise subspace, A is is the network direction matrix (N x K), where N is the number of antennas and K is the number of sources. Thus, finding peaks in this spectrum provides estimates of the DOA.

As shown in Fig. 6, the MVDR algorithm respond to the two desired direction −40° and +40° at 28 GHz.

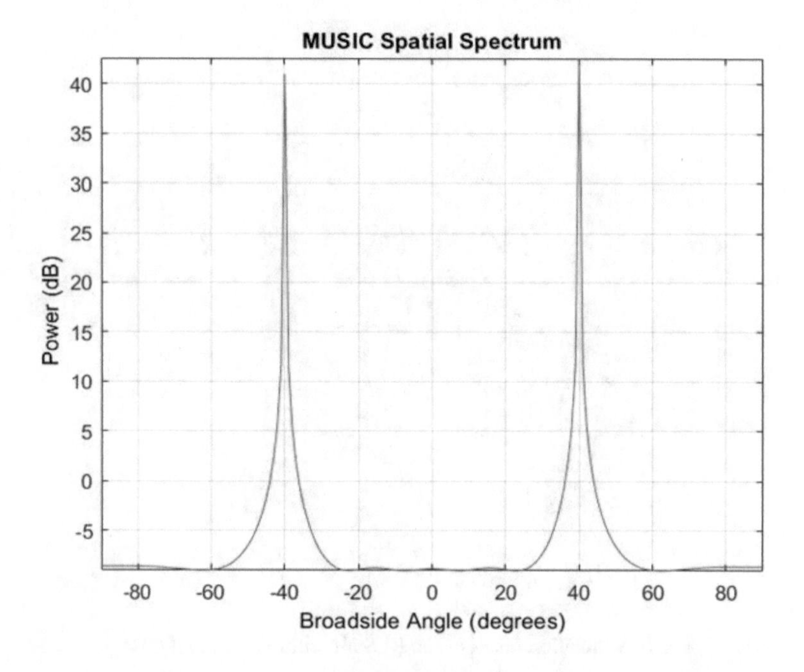

Fig. 6. Simulation result of the MUSIC algorithm at angle −40° and +40°.

3.4.1 The MVDR Algorithm The objective of the MVDR method is to reduce noise and interference while preserving a constant gain in the direction of observation. By weighting the network response vector for each direction and using correlation matrix calculations and subspace separation, it computes the MVDR spectrum [15]. The MVDR angular spectrum is defined by Eq. 8 [15]:

$$P_{\text{MVDR}} = \frac{1}{A^H R_{xx}^{-1} A} \tag{8}$$

With R_{xx} is the signal correlation matrix (K×K).

Here our desired direction are $-20°$ and $+20°$. As shown in Fig. 7, the Music algorithm repond to the two desired direction at 28 GHz.

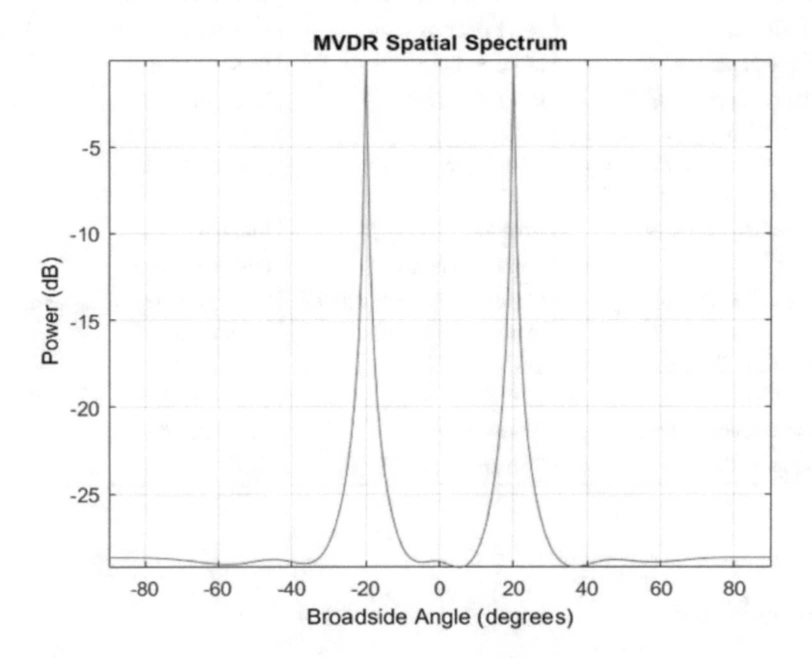

Fig. 7. Simulation result of the MVDIR algorithm at angle $-20°$ and $+20°$.

4 Comparative Between Switched and Adaptive Beamforming

Based on the simulation results that we have carried out, it's obviously that the switched beamforming has a fixed radiation pattern and has a very simple structure whereas the adaptive beamforming has no static beam, it respond to a defined user in real time at any desired angle of direction, but this type of beamforming is very complex and very costly.

Table 1, gives a comparative between Switched and adaptive beamforming in terms of different performance.

Table 1. Comparative between Switched and adaptive beamforming

Characteristics	Adaptive beamforming	Switched beamforming
Computation Time	Slower	Faster
Latency	Higher latency	Generally lower
Beamforming Method	Adjusts antenna weights	Uses fixed antennas
Flexibility	More flexible	Less flexible
Implementation	More complex	Less complex
Noise Suppression Capability	More effective	Less effective
Performance in Changing Environments	More effective	Less effective
Coverage Capacity	Broader	Limited
Cost	More expensive	Less expensive
Algorithm Complexity	More complex and difficult to implement	Less complex and easier to implement
Requires Channel Information	Yes	No
Interference Reduction	More effective	Less effective
Adaptability	Dynamic	Static

5 Conclusion

The switched beamforming providing a fixed radiation pattern, and switched between them to respond for a desired user, while the adaptive beamforming offer a dynamically beam that responds to the desired user in real time. Hence, to create a comparison between, them we have given an example of application of 4×4 Butler Matrix and adaptive beamforming such as LMS, NLMS and RLS, etc. The 4 × 4 BM give four static beam steering pointing at different four angle $\pm 4°$ and $\pm 29°$. Thus, the adaptive beamforming it respond to any desired direction angle in real time. Finally, we have made a comparative study between the two types of beamforming and defined the advantages and limitations of each theory. Indeed, the switched beamforming has a simple configuration whereas the adaptive beamforming has a complex structure and very costly.

References

1. Rao, L., Pant, M., Malviya, L., Parmar, A., Charhate, S.V.: 5G beamforming techniques for the coverage of intended directions in modern wireless (2021)
2. Nilsen, C.-I.C., Hafizovic, I.: Digital Beamforming Using a GPU, p 612 (2009). https://doi.org/10.1109/ICASSP.2009.4959657
3. Venkateswaran, V., Van Der Veen, A.-J.: Analog beamforming in MIMO communications with phase shift networks and online channel estimation. IEEE Trans. Signal Process. **58**(8), 4131–4143 (2010). https://doi.org/10.1109/TSP.2010.2048321

4. Elhefnawy, M.: Design and simulation of an analog beamforming phased array antenna. IJECE **10**(2), 1398 (2020). https://doi.org/10.11591/ijece.v10i2.pp1398-1405
5. Nickel, U.: Properties of Digital Beamforming with Subarrays, p. 5 (2006). https://doi.org/10.1109/ICR.2006.343380
6. Molisch, A.F., et al.: Hybrid Beamforming for Massive MIMO - a Survey (2017). Accessed 17 Dec 2023. http://arxiv.org/abs/1609.05078
7. Molu, M.M., Xiao, P., Khalily, M., Cumanan, K., Zhang, L., Tafazolli, R.: Low-complexity and robust hybrid beamforming design for multi-antenna communication systems. IEEE Trans. Wireless Commun. **17**(3), 1445–1459 (2018). https://doi.org/10.1109/TWC.2017.2778258
8. Roth, K., Pirzadeh, H., Swindlehurst, A.L., Nossek, J.A.: A comparison of hybrid beamforming and digital beamforming with low-resolution ADCs for multiple users and imperfect CSI (2017). https://doi.org/10.1109/JSTSP.2018.2813973.
9. Imtiaj, S., Misra, I.S., Bhattacharya, S.: Revisiting smart antenna array design with multiple interferers using basic adaptive beamforming algorithms: comparative performance study with testbed results. Eng. Rep. **3**(1), e12295 (2021). https://doi.org/10.1002/eng2.12295
10. Joshi, R., Dhande, A.: Adaptive beamforming using LMS algorithm (2013)
11. Jebabli, E., Hayouni, M., Choubani, F.: Phased millimeter-wave antenna array for 5G handled devices. In: Zbitou, J., Hefnawi, M., Aytouna, F., El Oualkadi, A., Eds., Advances in Mechatronics and Mechanical Engineering, IGI Global, pp. 97-118 (2023). https://doi.org/10.4018/978-1-6684-5955-3.ch005
12. Jebabli, E., Hayouni, M., Choubani, F.: A low profile beamforming microstrip antenna array for 5G applications using butler matrix. In: International Wireless Communications and Mobile Computing (IWCMC). Dubrovnik, Croatia 2022, pp. 116–120 (2022). https://doi.org/10.1109/IWCMC55113.2022.9824417
13. Hossain, S., Islam, M.T., Serikawa, S.: Adaptive beamforming algorithms for smart antenna systems. In: 2008 International Conference on Control, Automation and Systems, Seoul, South Korea, pp. 412–416. IEEE (2008). https://doi.org/10.1109/ICCAS.2008.4694679
14. Thomas, J.A., Mini, P.R., Kumar, M.N.A.: DOA estimation and adaptive beamforming using MATLAB and GUI. In: 2017 International Conference on Energy, Communication, Data Analytics and Soft Computing (ICECDS), Chennai, pp. 1890-1896. IEEE (2017). https://doi.org/10.1109/ICECDS.2017.8389778
15. Akbari, F., Shirvani Moghaddam, S., Tabataba Vakili, V.: MUSIC and MVDR DOA estimation algorithms with higher resolution and accuracy. In: 2010 5th International Symposium on Telecommunications, Tehran, Iran, pp. 76–81. IEEE (2010). https://doi.org/10.1109/ISTEL.2010.5734002

Author Index

L. Barolli (Ed.): AINA 2024, LNDECT 204, pp. 499–501, 2024.
https://doi.org/10.1007/978-3-031-57942-4

Printed in the United States
by Baker & Taylor Publisher Services